Progress in Colloid & Polymer Science · Vol. 80

PROGRESS IN COLLOID & POLYMER SCIENCE
Editors: H.-G. Kilian (Ulm) and G. Lagaly (Kiel)

Volume 80 (1989)

Relaxation in Polymers

Guest Editors: M. Pietralla and W. Pechhold (Ulm)

Springer-Verlag Berlin Heidelberg GmbH

ISBN 978-3-662-15097-9 ISBN 978-3-7985-1689-2 (eBook)
DOI 10.1007/978-3-7985-1689-2
ISSN 0340-255-X

Originally published by Dr. Dietrich Steinkopff Verlag GmbH & Co. KG, Darmstadt in 1989

Softcover reprint of the hardcover 1st edition 1989

Chemistry editor: Dr. Maria Magdalena Nabbe; Copy editor: James Willis; Production: Holger Frey.

Type-Setting: K+V Fotosatz GmbH, D-6124 Beerfelden

Preface

The Sixth International Seminar on Polymer Physics focused on relaxation phenomena in polymers. This theme is central for understanding the behavior and properties of polymers; it embraces a vast field of theoretical and experimental approaches and methods. This progress volume presents the major part of invited lectures from the seminar.

It was a pleasure to take part in this meeting which attracted about 70 participants from eastern and western Europe, and which was held for the first time in the Federal Republic of Germany. The small village of Gomadingen provided a pleasant setting for lively discussions and comraderie throughout the meeting.

The seminar was held in memory of F. H. Müller, one of the pioneers in the study of relaxation phenomena who was also the initiator of the famous Marburger discussions that set the standard for such meetings.

Appropriately, the seminar was begun by Prof. W. Holzmüller, who spoke on the history of polymer physics. He enthusiastically took us back to the early days, and paid honor to Müller as a friend and worthy competitor.

M. Pietralla
W. Pechhold

Contents

Progress in Colloid & Polymer Science Progr Colloid & Polymer Sci 80:1–7 (1989)

Shear induced crystallization, a relaxation phenomenon in polymer melts*

G. Eder, H. Janeschitz-Kriegl*) and G. Krobath

Institute of Chemistry, Linz University, Linz, Austria

Abstract: A survey is given of the experimental evidence of shear induced lamellar crystallization of polymers. This survey forms the basis for a theory by two of the authors. In this theory elements of polymer melt rheology and polymer crystallization kinetics are combined. It turns out that the capability of the melt of crystallizing in the typical lamellar form, is only gradually developed during flow. Similarly, relaxation occurs after cessation of flow. The obtained expressions are introduced into Avrami's formalism in order to define an induction time. In principle the model can be applied to any flow and temperature history. The case of isothermal flow at constant shear rate and at a certain level of supercooling is worked out in more detail. The obtained result favorably compares with experiments by Lagasse and Maxwell. Implications for the application in industrial processes are discussed.

Key words: Polymers, shear flow, crystallization, induction time, relaxation, injection moulding

1. Introduction

As has been shown many years ago [1], the wall of an injection moulded article is structured. If a microtome cut of such a wall is investigated under a polarizing microscope, one clearly notices a kind of sandwich structure comprising at least three layers, of which the outer ones can be proved to be induced by the shear flow during the process of mould filling. Usually, a more or less fine grained spherulitic structure is found in the core. The said surface layers are undesirable from the point of view of the end user of the article: The impact strength of the article is influenced in a negative sense by these rather cleavable layers [2]. Because of their high degree of molecular orientation they can also be responsible for undesirable warping [3]. These effects are minimized by the use of resins tailored for special application in injection moulding: Their average molar mass should not be too large and their molar mass distribution should not be too broad. The high molar mass tail of the molar mass distribution was made responsible for the formation of oriented surface layers [4].

*) Dedicated to Prof. J. Meissner on the occasion of his sixtieth birthday

Obviously, these qualitative experiences deserve to be underpinned by a model consideration, hopefully leading to a more quantitative description of the phenomenon of shear-induced crystallization. In the past it had been assumed that the oriented layer was partly caused by the so-called fountain flow occurring at the flow front during the filling operation [5]. However, in the experimental part of this lecture it will be shown that these layers can be produced by faultless shear flow. We wish to restrict ourselves to this very simple flow situation.

Before doing so, however, we wish to point to the fact that a so-called isothermal model, suitable for the case of a fixed degree of supercooling, cannot be the ultimate goal of such an investigation: It can only be a first step in the right direction. From the easier situation of the crystallization in a quiescent melt we recently could learn how difficult it is to treat heat conduction and crystallization simultaneously. In fact, the pertinent differential equations (equation of heat conduction and rate equations for nucleation and growth) are coupled by the release of the latent heat of crystallization [6–8]. In this respect one has to discern between the limiting cases of the propagation of a supercooled crystallization front and the formation of a dispersed crystallization zone [9–11]. In principle, with shear-induced crystallization additional mathematical problems will arise in those cases where convective transport of heat and nucleated fluid are of importance. As it seems that the time scale of flow induced nucleation differs by orders of magnitude from the one of nucleation in quiescent melts [12, 13], one can hope to find a simplified treatment of the convection problems, at least for very high rates of nucleation.

After this degression into possible future tasks some final remarks of a more practical nature should not be omitted. Stretch induction seems to be related to shear induction of crystallization and occurs in melt spinning [14]. The fibrillar structure of the obtained crystallites is not necessarily of a negative influence on the fibre quality. Recently, self re-enforcement of plastics is claimed for the induction of fibrillar structures in plastics articles by the introduction of proper flow-paths into the process [15].

2. Model experiments

Apparently, experimental investigations into the induction of crystallization by deformation and flow have encountered great difficulties. This become obvious from the fact that pertinent papers appeared only irregularly and with large intervals [16–18]. The most important progress was made by Lagasse and Maxwell [19]. According to these authors the onset of shear induced crystallization is, irrespective of the applied rate of shear, the consequence of the total shear accumulated. This critical shear is a function of the degree of supercooling, the type of polymer and its molar mass (distribution). As will be seen below, this behavior is characteristic for linear polyethylene and forms a limiting case of a more general behavior.

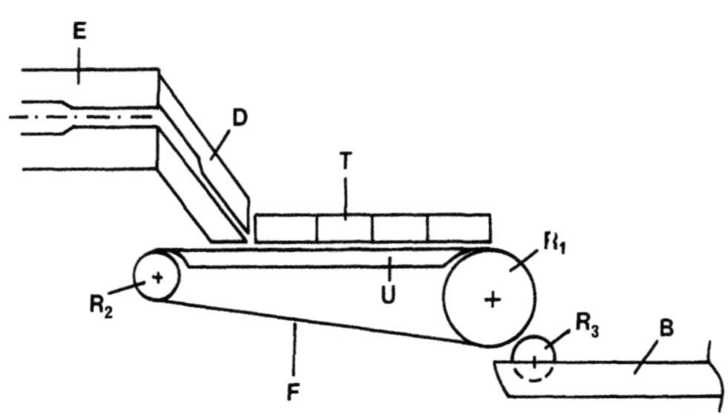

Fig. 1. Apparatus for the relaxation experiments: E = extruder, D = slit die (exchangeable), T = tunnel (in segments), U = heatable pad, F = conveyor-belt, R_1 = chill-rôle, R_2, R_3 = pulleys, B = bath (from [21], courtesy of Alfred-Hüthig-Verlag, Basle)

From experiments, as carried out by Krobath and Liedauer [20, 21] in the laboratory of Linz University, it becomes obvious that the said induction process must be a kind of elastico-viscous relaxation process. In Fig. 1 the apparatus, as developed by the authors, is shown. The polymer melt is plasticized in a screw extruder and extruded through a duct with rectangular cross-section of a large aspect ratio (10:1 in millimetres). The length of this duct can be varied between 2 and 111 mm. The shear treatment of the melt in the duct can be varied with the aid of the extrusion speed and of the residence time distribution over the slit cross-section. The latter can be uncoupled from the extrusion speed by the choice of different duct lengths. To minimize the remembrance of the action of the screw, the polymer melt is allowed to relax before entering the duct.

In these experiments, the shear treatment is carried out at a temperature above the melting point. The shear induced crystallization occurs afterwards on account of the remembrance of the melt by the application of a quench. For the purpose, the extrudate is first deposited on a conveyor-belt made of a stainless steel foil of a thickness of only 0.025 mm. This conveyor-belt slides over a heated pad U. The quench occurs when the belt reaches the chill-rôle R_1. The whole draw-down unit can be moved in a horizontal direction. In this way the exit of the duct can also directly be placed above the chill-rôle. On the pad a segmented tunnel can be placed. This arrangement permits the carful administration of a temper to the previously sheared melt. The use of various tunnel lengths enables the uncoupling of the residence time of the relaxing melt in the tunnel from the extrusion speed (to which the speed of the belt must always be adjusted).

The investigations were carried out with an industrial polypropylene of Petrochemie Danubia (Daplen KS 10). In Fig. 2 the results are shown which were obtained when the extrudate was immediately quenched on the chill-rôle. The temperature of the coolant was mainly 10 °C, the temperature of extrusion was 200 °C. As the adjustment of the speed of the draw-down to the extrusion speed (= average speed of fluid in the duct) is never perfect, the extrudate is always stretched to some extent. In order to eliminate the influence of this varying (elastic) stretch on the thickness of the shear induced surface layer, the latter is given in percentages of the final thickness of the extruded tape (of about 0.7 mm). From Fig. 2 one notices that the thickness of the surface layer increases considerably with the extrusion speed. Below a minimum speed there is no surface layer at all. Very sur-

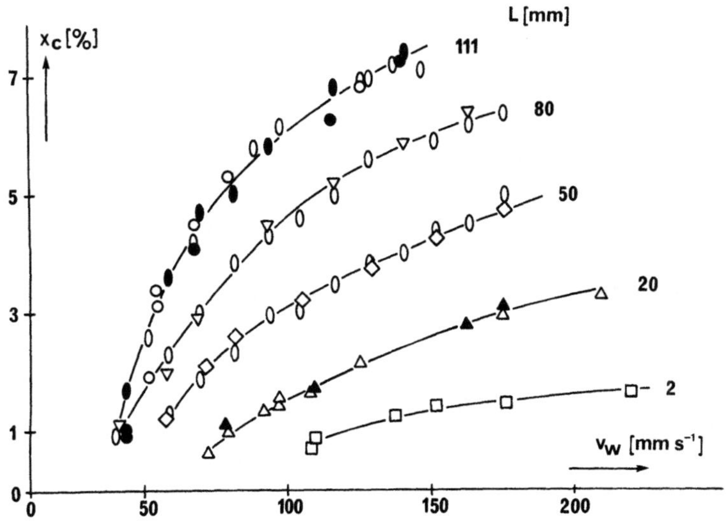

Fig. 2. Thickness of boundary layer as a function of the extrusion speed (= speed of conveyor-belt) for various die lengths L, extrusion temperature 200 °C, chill-rôle temperatures: ○ = 100 °C, ● = 50 °C, other symbols = 10 °C. (Polypropylene Daplen KS 10; (from [21] courtesy of Alfred-Hüthig-Verlag, Basle)

prising is the influence of the duct length. If, say at V = 150 mm/s a cross-section is made through the graph, one does not find a saturation of the thickness of the layer with further increasing duct length. This reflects a continuous change of the internal structure of the melt with increasing shearing times.

Apparently, this structural change of the melt can be observed only indirectly by the increasing capability of the melt to the lamellar type of crystallization so characteristic for the shear induced crystallization. This structural change is not reflected by the rheology of the melt or the degree of orientation of the macromolecules, as caused by shear flow. The latter fact is shown in the following way. To the end of the duct an adapter is mounted, in which glass windows are placed in the opposite larger side walls of the (unchanged) cross-section. In this way the flow birefringence of the fluid could be measured in the 1,3-plane [22]. The pertinent results are shown in Fig. 3. These measurements were extended to other than only 200 °C extrusion temperature, as used for the preparation of Fig. 2. From Fig. 3 it becomes obvious that the variation of the duct length between 7 and 118 mm had no influence on the measured birefringence. The same holds for the degree of orientation which is reflected by the birefringence effect. This result is in accordance with the experience that normal entrance effects die out in a duct in general within about 5 duct heights [22]. The preliminary conclusion is that chain orientation is certainly no sufficient condition for

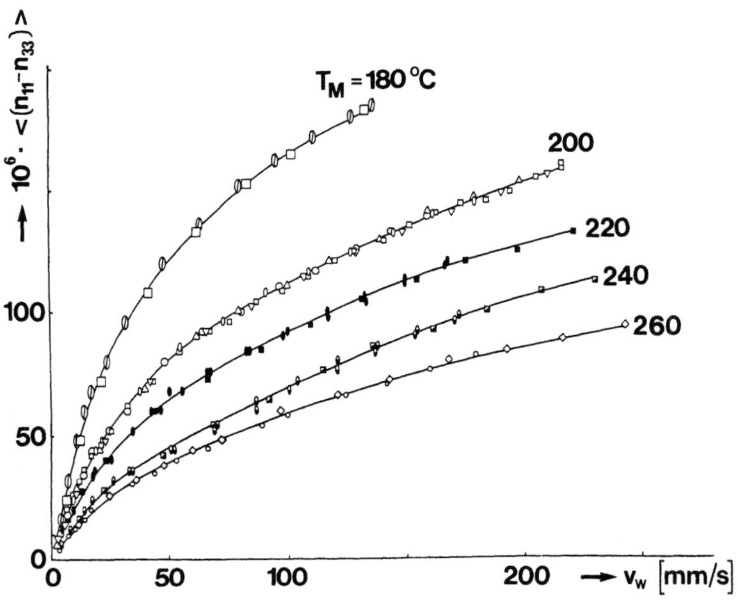

Fig. 3. Flow birefringence (average over die cross-section) vs extrusion speed, as measured at a distance of 5 mm from the die exit for polypropylene Daplen KS 10 at the temperatures T_M indicated near the curves. Aspect ratio of die cross-section is 10 : 1 (in millimeters). Die lengths in millimeters are: ○, ○, ●, ◒ = 121, □, ■, □, ◇ = 12, ○ = 90, △ = 60, ▽ = 30

shear induced crystallization. The structural change leading to lamellar crystallization must be of a more subtle nature.

Finally, the results of the relaxation in the tunnel (of varying length) will cast more light on the nature of the process leading to the capability of lamellar crystallization. These relaxation experiments are carried out exclusively on extrudates from the longest duct of 111 mm at an extrusion temperature of 200 °C. For various extrusion speeds, as indicated near the curves of Fig. 4, the thickness of the obtained surface layer is plotted against the residence time in the tunnel. For the lower part of the figure the tunnel temperature was equal to the extrusion temperature of 200 °C. In this way the (partial) relaxation occurred at the same temperature at which the capability for this crystallization was built up. Only for two of the three uppermost curves, which are obtained for an extrusion speed of 150 mm/s, lower tunnel temperatures were chosen. One observes that at a tunnel temperature of 200 °C the capability for a lamellar crystallization relaxes completely within about 4 s. At lower tunnel temperatures this relaxasion time considerably increases. At 130 °C tunnel temperature no signs of relaxation are found within the residence times available in our apparatus. This explains why the chill-rôle temperature had no influence in the previous experiments where the extrudate was directly quenched on the chill-rôle. For practical reasons – one had to avoid that the extruded tape stuck to the rôle – the chill-rôle temperature could never be chosen higher than 100 °C. From the relaxation experiments, however, we could learn that even at 130 °C, relaxation was so slow that no observable changes could be found within 4 s, which is a longer time than the residence time on the chill-rôle. So the results of Fig. 2 are finally determined by the temperature of the bath which was kept at zero degrees by the addition of ice to the cooling water.

An extraordinary feature of some of these "relaxation curves" is their horizontal initial tangent. (See the curves for the highest extrusion speeds of 150, 125, and 100 mm/s in Fig. 4.) As will be shown below, this feature is borne out by the model to be described in the next section.

3. Theoretical considerations

From theoretical considerations, which were published by the authors [23], it follows that one needs at least three parameters in order to predict the described phenomena qualitatively. The authors started with the following differential equation:

$$\tau\dot{\phi} = (\dot{\gamma}/\dot{\gamma}_a)^2(1-\phi)-\phi \; . \tag{1}$$

In this equation τ is a relaxation time and $\dot{\gamma}_a$ is a critical shear rate of activation, whereas ϕ is a dimensionless variable describing the internal state of the melt. For a quiescent melt in equilibrium one has $\phi = 0$, for $(\dot{\gamma}/\dot{\gamma}_a) = \infty$ one obtains $\phi = 1$. Function ϕ is a kind of probability for lamellar crystallization being nucleated. A quadratic dependence on $\dot{\gamma}$ is chosen as the internal state of the melt cannot depend on the sign of $\dot{\gamma}$. For constant shear rate $\dot{\gamma}$, Eq. (1) can easily be integrated. One obtains a function $\phi(t)$ of time t starting for $t = 0$ at $\phi(0) = 0$ and levelling off at

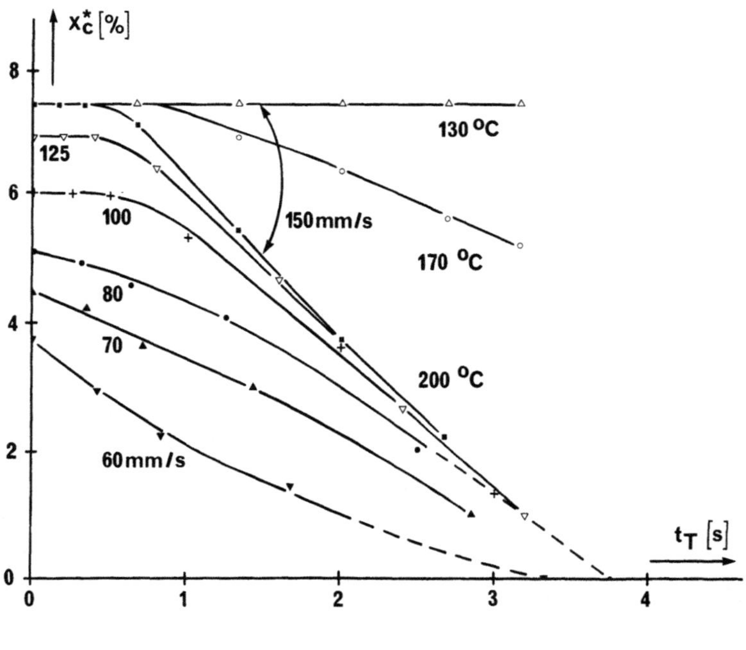

Fig. 4. Thickness of boundary layer as a function of the residence time t_T in the tunnel: extrusion from a slit-die of $L = 111$ mm, extrusion temperature of 200 °C and chill-rôle temperature of 10 °C. Parameters at the curves: extrusion speeds (mm s^{-1}) and tunnel temperatures. (Polypropylene Daplen KS 10; from [21], courtesy of Alfred-Hüthig-Verlag, Basle)

$$\phi(\infty) = (\dot{\gamma}/\dot{\gamma}_a)^2/[1+(\dot{\gamma}/\dot{\gamma}_a)^2] \tag{2}$$

when time progresses. The rate at which this function levels off increases with $(\dot{\gamma}/\dot{\gamma}_a)$ (exponential dependence).

As a next step, Avrami's ideas are introduced [24]. The speed of nucleation and the growth speed of crystallites are assumed to be proportional to ϕ. The induction time t_i is defined as the time of half conversion, as Lagasse and Maxwell [19] suggested previously. Going through this procedure one finally arrives at the following simple equation:

$$\int_0^{t_i} \phi(t')dt' = t_\infty \ , \tag{3}$$

where t_∞ is the induction time under the most favorable conditions ($\phi \cong 1$) for $\dot{\gamma} \gg \dot{\gamma}_a$). This t_∞ is the third parameter of the theory after τ and $\dot{\gamma}_a$. For all cases where ϕ did not reach its limiting value one obviously has $t_i > t_\infty$. Using the solution of Eq. (1) for $\dot{\gamma}$ = constant one can rather easily invert Eq. (3) analytically and obtains t_i/t_∞ as a function of $\dot{\gamma}/\dot{\gamma}_a$ with $\hat{C} = t_\infty/\tau$ as a parameter. This functional dependence is shown in Fig. 5.

One notices that there exists a wider range of shear rates where $t_i \propto (1/\dot{\gamma})$ only if $\hat{C} \ll 1$. It is easy to recognize that this means $t_i\dot{\gamma} = \gamma_i$ = constant. This is the special case where crystallization is induced after a constant total shear γ_i as verified by Lagasse and Maxwell for linear polyethylenes. Looking at our condition $\hat{C} = t_\infty/\tau \ll 1$ we have no mental difficulties knowing polyethylene as a very fast crystallizer amongst the polymers.

Another case, which can be treated rather easily, is the following. Because of the fact that ϕ relaxes at a finite rate when flow is stopped, one can expect that the shearing time t_s (at constant $\dot{\gamma}$) can be kept shorter than induction time t_i, as found for the chosen value of $\dot{\gamma}$. In other words, it should not be necessary to shear up to the very moment t_i where crystallization occurs: One should still obtain shear-induced crystallization when flow is stopped somewhat earlier (at t_s). This can be put into the following equation:

$$\int_0^{t_s} \phi(t')dt' + \int_{t_s}^{t_{ip}} \phi_r(t_s, t')\, dt' = t_\infty \ , \tag{4}$$

where $\phi_r(t_s, t)$ is the relaxing function ϕ for $t > t_s$ and t_{ip} is the prolonged induction time. (It seems quite clear that t_{ip} increases above t_i if flow is stopped

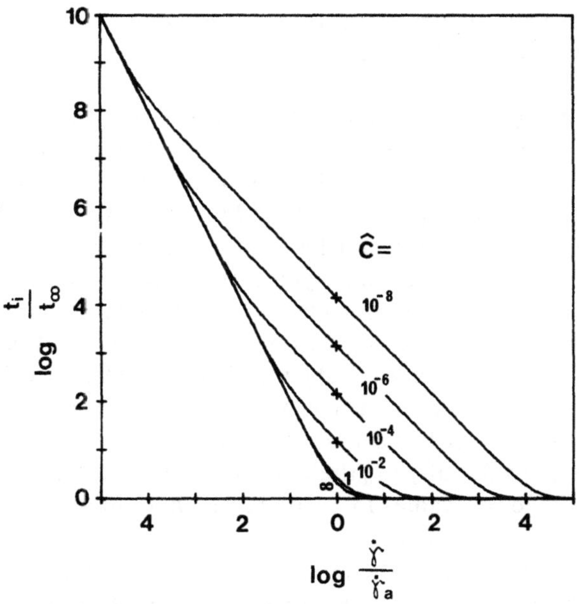

Fig. 5. Dimensionless induction time (t_i/t_∞) after continuous shearing as a function of reduced shear rate $(\dot{\gamma}/\dot{\gamma}_a)$ for various values of the parameter $\hat{C} = t_\infty/\tau$: t_∞ = induction time for $\dot{\gamma} \gg \dot{\gamma}_a$; $\dot{\gamma}_a$ = critical shear rate of activation; τ = relaxation time of probability function ϕ

earlier.) Also, Eq. (4) can be solved for t_{ip}, if the previous shear rate $\dot{\gamma}$ is constant. In Fig. 6 the dependence of t_{ip} on t_s is shown for six values of the ratio $\dot{\gamma}/\dot{\gamma}_a$ and $\hat{C} = 10^{-8}$. One notices that with increasing shear rate the minimum value of t_s at which t_{ip} goes to infinity decreases more and more, which means that at

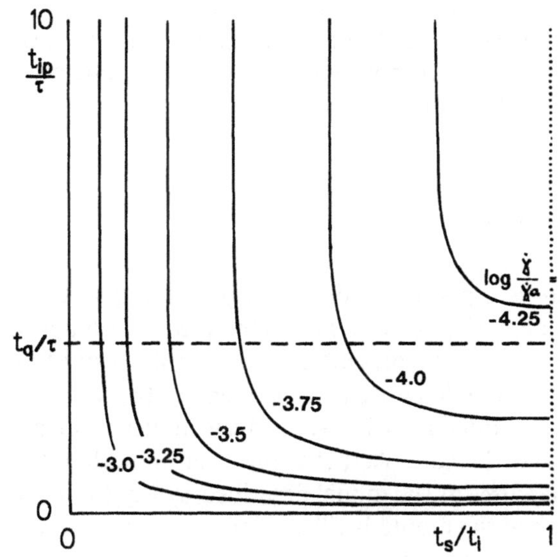

Fig. 6. Prolonged induction time t_{ip} vs shearing time $t_s \leq t_i$ for various values of shear rate $\dot{\gamma}$ for $\hat{C} = 10^{-8}$. Dashed horizontal line = fictitious induction time for (spherulitic) crystallization in quiescent melt t_q. The way in which the indicated parameters are scaled can be read from the graph. (For symbols see legend of Fig. 5)

extremely high rates of shear a very short period of shearing would suffice for the induction of lamellar crystallization.

In Fig. 6 a dashed horizontal line is shown; the position of this line represents a fictitious value for the induction time t_q in a quiescent melt (for spherulitic crystallization). As Lagasse and Maxwell have pointed out, spherulitic and lamellar crystallization do not occur simultaneously: one type excludes the other. But this means that, in principle, waiting for shear-induced crystallization until t_{ip} becomes infinite, means that before then the normal spherulitic crystallization will occur. From Fig. 6 one can see from the intersections of the dashed horizontal line over the full lines for several values of $\dot{\gamma}/\dot{\gamma}_a$ that, dependent on the value of t_q, the lowest values for t_s are larger than those where t_{ip} goes to infinity.

In Fig. 7 the pertinent results are plotted in a more lucid way. This figure seems of great importance for the design of future experiments even if our model considerations are oversimplified. The great pity is that the dimensionless variables of this graph are so well separated. The induction time t_∞ for the most favorable conditions of shear induced crystallization scales the ordinate axis, whereas the critical shear rate $\dot{\gamma}_a$ of activation scales the abscissa. If the range of available values of $\dot{\gamma}$ is insufficient there is little chance to separate these parameters. Measurable quantities are: t_i, t_s, t_q and $\dot{\gamma}$. If $\hat{C} = t_\infty/\tau$ were known, one could guess $\dot{\gamma}_a$ from the ratio t_i/t_s. However, t_∞ is not accessible by direct measurement, whereas τ can certainly be estimated. Future studies will have to show whether such a complete separation of parameters is necessary for practical purposes.

4. Conclusion

There is a good chance that a puzzling qualitative experience of injection moulders can be explained with the aid of the presented theory. In the wall of an injection-moulded article on frequently observes rather sharp boundaries between shear-induced surface layers and the core. Actually, injection moulding has many features of the model described by Eq. (4): Flow stops when the mould is filled. There is no reason why lamellar crystallization should not proceed after this moment. If t_q is very large, as with polypropylene at medium levels of supercooling, one can expect a sharp boundary since t_{ip} will go almost to infinity. At this minimum value of t_s the progress of lamellar crystallization loses the match against relaxation.

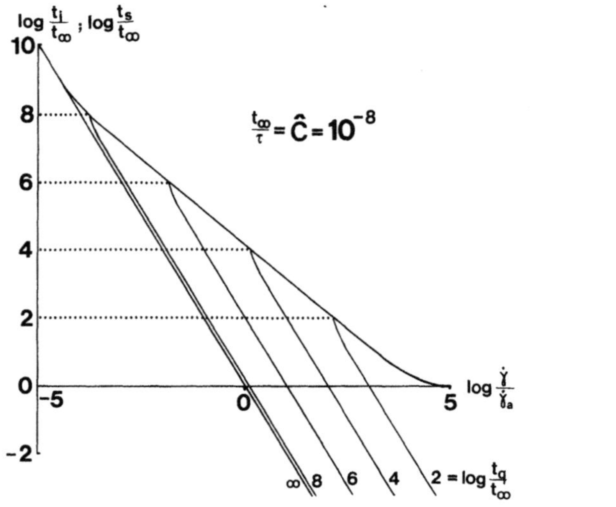

Fig. 7. Shortest required shearing time t_s for shear induced crystallization vs shear rate $\dot{\gamma}$ for various values of induction time t_q for the quiescent melt, if $\hat{C} = 10^{-8}$. Above the points of departure of the steeper lines from the line for t_i vs $\dot{\gamma}$, as taken from Fig. 5 for $\hat{C} = 10^{-8}$, we have spherulitic crystallization. (For scaling parameters see legend of Fig. 5)

Another hope for the usefulness of the presented approach is based on the observation that some of the "relaxation curves" of Fig. 4 show zero initial slopes. In fact, these curves do not depict a pure relaxation. They seem to show that relaxation is counterbalanced initially by the progress of crystallization, if the shear rates are sufficiently high. This would be in accordance with the remark made in the last section with respect to the effectiveness of very high shear rates in reducing the minimum shearing time.

At the moment the most serious obstacle for the application of the presented theory in practical numerical problems is the lack of data. The parameters of the theory will certainly be strong functions of temperature. Unfortunately, the available range of temperatures in which the processes proceed at reasonable speeds, is small for all polymers of practical interest. Moreover, one can expect that parameters like τ and $\dot{\gamma}_a$ are strongly dependent, not only on the nature of the polymer, but also on their molar mass (distribution). Extensive experimental work is in progress at the Institute of Chemistry of Linz University. We hope a report on this work can be published in the near future.

Acknowledgements

The authors are very much indebted to the Austrian "Fonds zur Förderung der wissenschaftlichen Forschung" for sponsoring this work under project no. S 33 02 in the course of a national research program S 33 on injection-moulded articles.

References

1. Kantz MR, Newman HD, Stigale FH (1972) J Appl Polym Sci 16:1249
2. Narbeshuber F, Kügler F (1984) Österr Kunststoff-Ztschr 15:90
3. Altendorfer F, Seitl E (1984) Plastverarbeiter 35:10
4. Altendorfer F, Seitl E (1983) Plastverarbeiter 34:511
5. Tadmor Z (1974) J Appl Polym Sci 18:1753
6. Malkin AYa, Beghishev VP, Keapin IA, Bolgov SA (1984) Polym Engng Sci 24:1396 and Malkin AYa, Beghishev VP, Keapin IA, Andrianova ZS (1984) Polym Engng Sci 24:1402
7. Schneider W, Köppl A, Berger J (1988) Intern Polym Proc 2:151
8. Berger J (1988) Erstarrung von Kunststoffen unter Einfluß von Wärmeleitung und Kristallisationskinetik. Doctoral thesis, Vienna University of Technology
9. Eder G, Janeschitz-Kriegl H (1984) Polym Bull 11:93
10. Berger J, Schneider W (1986) Plastics and Rubber Processing and Applications 6:127
11. Janeschitz-Kriegl H (1989) One Hundred Years of Chemical Engineering. Peppas NA, ed. Kluwer/Reidel, Dordrecht
12. Haas TW, Maxwell B (1969) Polym Engng Sci 9:225
13. White JL, Cakmak M (1986) Adv Polym Technol 6:295
14. Ziabicki A (1976) Fundamentals of Fiber Formation. Wiley, New York
15. Ehrenstein GW (1987) Technologie der Eigenverstärkung von Thermoplasten. Kassel
16. Katayama K, Murakami Sh, Kobayashi K (1976) Bull Inst Chem Res Kyoto Univ 54:81
17. Amano T, Kajita S, Katayama K (1975) Progr Colloid and Polym Sci 58:108
18. Wereta A, Gogos CG (1971) Polym Engng Sci 11:19
19. Lagasse RR, Maxwell B (1976) Polym Engng Sci 16:189
20. Janeschitz-Kriegl H, Eder G, Krobath G, Liedauer S (1987) J Non-Newtonian Fluid Mech 23:107
21. Janeschitz-Kriegl H, Wimberger-Friedl R, Krobath G, Liedauer S (1987) Kautschuk+Gummi, Kunststoffe 40:301
22. Janeschitz-Kriegl H (1983) Polymer Melt Rheology and Flow Birefringence, p. 85. Springer, Berlin Heidelberg New York
23. Eder G, Janeschitz-Kriegl H (1988) Colloid and Polymer Sci 266:1087
24. Avrami M (1939, 1940, 1941) J Chem Phys 7:1103, 8:212, 9:117

Received January 23, 1989;
accepted February 14, 1989

Authors' address:

Prof. Dr. H. Janeschitz-Kriegl
Universität Linz
Institut für Chemie
Altenberger Straße 69
A-4040 Linz, Austria

Discussion

GODOVSKY:

In your theoretical treatment of crystallization kinetics with the Avrami equation you have suggested that the induction period of crystallization is equal to the crystallization half-time. It is a very unusual suggestion. What does it mean from the physical point of view?

JANESCHITZ-KRIEGL:

The three parameters of our theory are $\dot{\gamma}_a$, τ, and τ_∞. The latter parameter is given by $t_\infty = (\ln 2/C_r^3 C_2^3 C_1)^{1/4}$ with C_r, C_1, C_2 being crystallization parameters (see [23]). If our definition of the induction time $t_i = t_{1/2}$ is revised as follows $t_i = t_{1/10}$, one obtains ln (10/9) instead of ln 2 in the above equation. This means that such a change of definition does not influence our results, as the constants C_r, C_2, C_1 are unknown. If on the other side, t_i is defined by some other convention, our mathematical equations become less tractable. Finally, it should be said that Lagasse and Maxwell really measured a time which must be close to $t_{1/2}$, as the shear force only increases considerably if the bulk of material crystallizes. This is our physical point of view.

Contour-length fluctuation of polymers: NMR experiments and statistical physics

R. Kimmich and M. Köpf

Sektion Kernresonanzspektroskopie, Universität Ulm, Ulm, FRG

Abstract: A series of NMR experiments concerning chain fluctuations in concentrated polymer systems is briefly reviewed. There is growing evidence for the influence of longitudinal displacements and contour-length fluctuation as fundamental mechanisms. On this basis a segment orientation correlation function has been established which describes and explains a broad variety of dependences and phenomena. This concept involves a new interpretation of the Doi-Edwards tube. The crucial element is the existence of fluctuating chain folds. It is argued that these folds should thermodynamically be stable. The primitive path of a chain is considered to be quasi-static in the time scale of fold fluctuations. The dynamic and conformational degrees of freedom of the fluctuating folds therefore dominate the partition function. This function has been calculated by the aid of a simplified model merely retaining the features essential for principal statements and conclusions. In order to discuss microscopic irreversibility, i.e., globally all degrees of freedom other than longitudinal displacements, nonlinear Schrödinger equations have been used for the derivation of the dynamic part of the partition functions. The time-scale effect inherent to the present treatment should generally be relevant if dynamic degrees of freedom are partly suppressed by topological restraints.

Key words: Contour-length fluctuation, reptation, polymer melts, statistical physics, partition function, nuclear magnetic resonance, Schrödinger-Langevin equation

1. Introduction

Polymer chains entangled in a matrix of neighboring molecules suffer topological restraints which are represented by the-so-called "tube" [1]. Depending on the definition of the tube, lateral degrees of freedom are considered to be more or less strongly damped. With the "tubelike regions" introduced by Doi and Edwards [1], all topological restraints in concentrated polymer systems are projected to a loose matrix of entanglement points with the chains free to fluctuate in between. The axis of the shortest tube having the entanglement points of a given chain on its surface has been called the "primitive path" by those authors. The length of the primitive path is designated the "contour length", which of course, fluctuates in time. Thus the contour-length fluctuation (CLF) must be regarded as a mechanism reducing the effective length of a chain which is actually confined to a tube in a time scale exceeding that of the CLF (Fig. 1). We may imagine this process as a kind of breathing motion permitting the chain-end blocks to disengage from the original tube prior to the release of the central block [2]. Doi and Edwards [1] furthermore treat the entropy of a chain in a given tube by considering all conformations in an *empty* box of the dimensions of the tubelike region.

In the course of our own work we have developed a somewhat different view based on the conclusions of a series of NMR experiments with concentrated polymer systems [3,4]. The finding that NMR relaxation techniques are sensitive to *global* chain motions, though only interactions localized in the Ångström regime are relevant, proves that segment motions are anisotropic even in the time scale of the global chain motions. This anisotropy can only be understood as a restraint effect of "tight tube" as originally proposed by de Gennes [5].

The tight tube is defined by the dense packing of the chains (Fig. 2) in a polymer melt, for instance. The diameter of the tight tube is directly measurable as the mean distance between nearest neighbor chains. Clearly, the existence of fluctuations of the contour

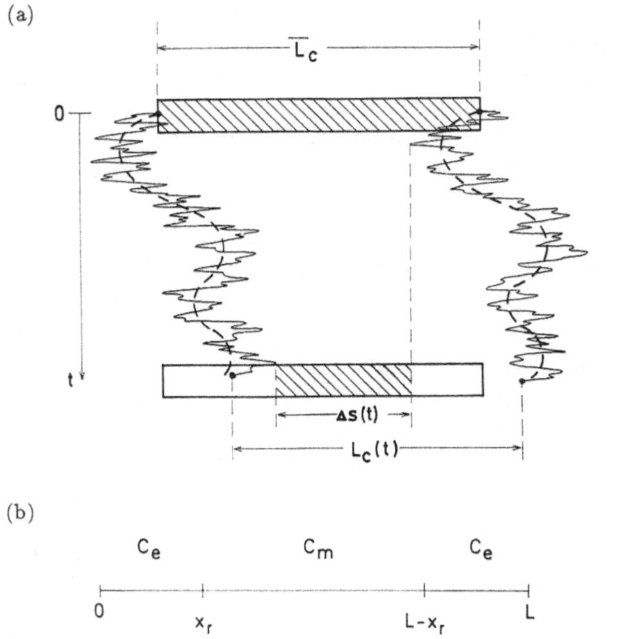

Fig. 1. a) Schematic representation of contour-length fluctuation (after [1]). The primitive path is represented by a bar with the average length L_c. Contour-length fluctuation is a kind of breathing motion of the chain and leads to disengagement of the chain-end blocks from the initial tube faster than by whole-chain reptation. The hatched part of the lower bar thus represents the primitive chain block which has not been affected by this process in the indicated time interval. b) Dynamic three-block structure of an entangled chain

length requires any kind of lateral degrees of freedom. For a tight tube this means that "folds", i.e., lateral excursions through which no neighboring chains are wound, must be present. Such folds can grow or shrink or even vanish completely and reappear at any other position of the chain. The definition of the *contour length* L_c is then the length of a modified *primitive path* given by the tight tube disregarding the length of the folds (Fig. 2). L_c fluctuates as a consequence of the fluctuating ensemble of folds of a chain.

In the following we deal with two questions. First we interrogate the experimental evidence of the processes mentioned above. We restrict ourselves virtually to NMR investigations. It is shown that the data can be interpreted by the aid of a suitable segment-orientation correlation function. Although this function has been derived for the description of NMR experiments, elements of it can be extracted for comparison with data of the other techniques (of rheology, for instance).

The second question concerns the thermodynamic stability of the fluctuating folds as a prerequisite of the applicability of this scheme. Partition functions are derived for the cases with and without friction.

Prior to the detailed treatment of these problems let us make some general remarks serving the plausibility of the starting point. There is a simple thermodynamic argument suggesting contour-length fluctuation. It takes a long time until a chain disengages by the reptation mechanism [5] from its initial tube. The time scale in which different primitive paths are probed therefore is extremely long for entangled chains. The formation and fluctuation of folds, on the other hand, does not require the complete disengagement of the chains from their initial tubes. This is rather a mechanism providing conformational degrees of freedom while the primitive path may be considered to be quasi-static in the corresponding short-time scale. *Each* conformation of the primitive path is compatible with *many* different fold arrangements.

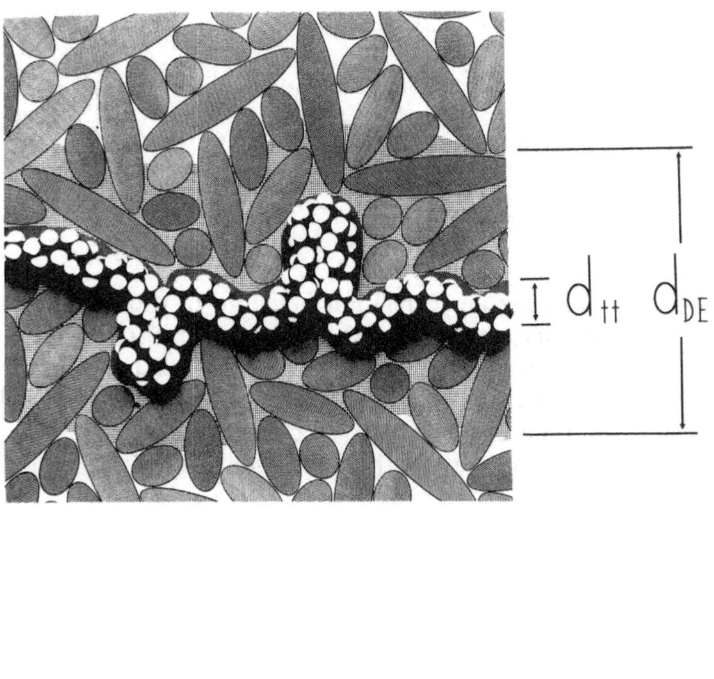

Fig. 2. Chain folds, the tight tube, and the Doi-Edwards tube (after [4]). The latter may be considered to represent the tight tube with the average extension of the chain folds

In addition, folds open translational degrees of freedom other than those of the whole chain. The chain sections between folds are displaced relative to each other as a consequence of fold fluctuations. Thus, though folds reduce the number of conformations of the primitive path which are accessible in the long-time limit, they increase, on the other hand, the number of states that can be probed in short times. They disintegrate a chain into more or less independent "subchains". The correlation time τ_r^c of contour length fluctuation is much shorter than the correlation time of tube disengagement, τ_r^r, or

$$\tau_r^c \ll \tau_r^r \ . \tag{1}$$

Ergodicity so to say is fulfilled in two steps corresponding to the different time scales in which different sections of the phase space are populated.

2. Three-component/three-block orientation correlation function

Most of the interactions relevant for NMR quantities are short-range couplings. Practically they are effective only in the segment regime. What we need for the description of NMR experiments is therefore the segment-orientation correlation function. The term "orientation" refers to the laboratory frame represented for instance by the external flux density $\vec{B}_0$.

Chain fluctuations in condensed polymer systems can be analyzed in three competitive components to which we attribute the letters A, B, C [3,4]. *Component A* refers to the reorientation of segments about the axis of the Kuhn chain. It is therefore an *anisotropic* and *local*, i.e., a molecular weight (M) independent process with a finite residual correlation in the long-time limit. Compared with the other components this process is *fast.*

Reptational displacements along the tight tube, which is considered to be static for the moment, leads to reorientations according to the coiled conformation of the Kuhn chain. This mechanism is called *component B.* Its most important feature is that it is a *one-dimensional* process leading to an exceptional *non-exponential* correlation function. The nature of component B may be characterized as semi-local in the sense that folds subdivide a chain in subchains dynamically more or less independent from each other.

The correlation function of component B as a one-dimensional diffusion process decays extremely slowly. The reason is that a segment has a good chance to return to its initial position again and again. The final correlation loss, however, is caused by the third component, C. It comprises two competitive mechanisms: *Disengagement* of the chain from its initial tube and *contour-length fluctuation.* Both mechanisms are *global,* i.e., chain-length dependent. Theoretically they can be described by relaxation modes with very *slow* contributions.

A consequence of the contour-length fluctuation is also that the chain-end blocks dynamically behave differently from the central block (Fig. 1a and b). The chain ends escape from the initial tube almost instantaneously while segments in the middle of the chain are confined until the reptational displacements exceed their curvilinear distance from the chain ends. Therefore the chain-end blocks of a certain length x_r practically do not underlie tube restraints and are assumed to obey Rouse dynamics.

The correlation function derived on the basis of all these components is given by [4]

$$G(t) = A(t)B(t)C(t) \tag{2}$$

$$A(t) = a_1 \exp(-t/\tau_s) + a_2 \tag{3}$$

$$B(t) = \exp[X(t)] \operatorname{erfc}\sqrt{X(t)} \tag{4}$$

$$X(t) = \frac{t + 2(\pi t T_d)^{1/2}}{2\tau_l} \tag{5}$$

$$C(t) = \begin{cases} C_e(t) & \text{for } 0 \le s_n \le x_r \\ & \text{and } L - x_r \le s_n \le L \\ C_m(t) & \text{for } x_r < s_n < L - x_r \end{cases} \tag{6}$$

$$C_m(t) = C_c(t) C_r(t) \tag{7}$$

$$C_k(t) = \frac{8}{\pi^2} \sum_{j,\text{odd}} \frac{1}{j^2} \exp(-tj^2/\tau_k) \tag{8}$$

$$\tau_k = \begin{cases} \tau_R^c & \text{for } k \equiv e \\ \tau_r^c & \text{for } k \equiv c \\ \tau_r^r & \text{for } k \equiv r \end{cases} \tag{9}$$

$$x_r = \tfrac{1}{2}(L\Delta_0)^{1/2} \tag{10}$$

The partial correlation functions of the three components are designated by $A(t), B(t), C(t)$. τ_s is the correlation time and $a_2 = 1 - a_1$ the residual correlation of component A in the long-time limit. The time constant characteristic for component B is the mean diffusion time over the correlation length l of the tight tube

$$\tau_l = \frac{l^2}{2D_l^s} \propto M^x , \quad (0 \le x \le 1) \tag{11}$$

while the "defect equilibration time" [4, 5]

$$T_d = \frac{N_K^2 b^2}{\pi^2 D_d} \propto M^2 \tag{12}$$

often has a negligible influence. ($D_1^s \propto M^{-x}$, curvilinear diffusion coefficient of a subchain for short-scale displacements; $N_K \propto M$, number of segments of the Kuhn chain; b, root mean square length of a Kuhn-chain segment; D_d defect-diffusion coefficient).

Component C depends on different series of modes. The block in the middle of a chain is affected by contour-length fluctuation (C_c) and tube disengagement by reptational displacements (C_r) as well. The chain-end blocks underlie Rouse dynamics with a correlation function C_e. The longest (and dominating) correlation times are

$$\tau_R^c = \frac{\zeta_K x_r^2}{3\pi^2 k_B T} \propto M \tag{13}$$

$$\tau_r^c = \frac{N_p a^2}{3\pi^2 D_1^p} \propto M^2 \tag{14}$$

$$\tau_r^r = \frac{L^2[1-(\Delta_0/L)^{1/2}]^2}{\pi^2 D_1^K} \tag{15}$$

for C_e (Rouse dynamics), C_c (contour-length fluctuation), and C_r (reptational tube disengagement), respectively. (ζ_K, friction coefficient of a Kuhn-chain segment; N_p, number of segments of the primitive path; a, root mean square length of a primitive-chain segment; $D_1^p \propto M^{-1}$, diffusion coefficient along the primitive path; $L \propto M$, mean length of the Kuhn chain; Δ_0, mean length of a subchain, i.e., of the Kuhn chain per chain fold; $D_1^K \propto M^{-1}$, curvilinear diffusion coefficient of the Kuhn chain). The molecular-weight dependence of τ_r^r effectively is somewhat stronger than $\propto M^3$ [3].

Note that we have distinguished the curvilinear diffusion coefficient of a subchain, D_1^s, from that of the whole Kuhn chain, D_1^K. The reason is that the subchains dynamically are more or less decoupled as concerns the short-scale displacements. As a result a weaker or even vanishing M dependence of the subchain diffusion coefficient is expected for short displacements.

3. Comparison with experiments

The three-component/three-block correlation function given above represents functional relationships specific for the assumed processes. In particular the peculiar partial correlation function describing component B [Eq. (4)] is characteristic for a one-dimensional diffusion process. It therefore leads to a concerted action of all three components resulting in certain limiting cases and dependences. There are limits in which the individual components dominate and thus reveal themselves in an unsuperimposed way. Our argument in favor of this concept, however, is rather based on the total set of experimental data available up to now. Table 1 summarizes the dependences which have been studied in the diverse NMR experiments, and which can be explained by the three-component/three-block concept. In particular, phenomena as striking as the characteristic molecular weights and temperatures [4] can completely be understood on this basis.

The above scheme can also be applied to non-NMR experiments. The zero-shear viscosity, for instance, is generally related to component C [3, 7]. The corresponding expression explains the molecular-weight dependence in the full range. The "critical molecular weight" M_C turns out to be the consequence of the crossover between the two limits where component C is dominated by C_e and C_m, respectively. M_C is found

Table 1. List of NMR experiments which can be explained on the basis of the three-component/three-block concept

NMR experiment	References
$T_1 = T_1(\nu_0)$	[4], [10], [11]
$T_{1\varrho}(\nu_1)$	[12], [13]
$T_1 = T_1(M_w)$	[14], [15]
$T_2 = T_2(M_w)$	[3], [4], [14], [20]
$D = D(M_w)$	[3], [16], [17]
$T_1 = T_1(\theta)$	[15], [18]
$T_{1\varrho} = T_{1\varrho}(\theta)$	[13]
$T_2 = T_2(\theta)$	[4], [18]
$\frac{T_2}{T_1} = f(c)$	[4], [19]
$T_2 = T_2(s)$	[9]

(T_1, spin-lattice relaxation time in the laboratory frame; $T_{1\varrho}$, spin-lattice relaxation time in the rotating frame; T_2, transverse relaxation time; D, self-diffusion coefficient; ν_0, Larmor frequency in the laboratory frame; ν_1, Larmor frequency in the rotating frame; M_w, weight average of the molecular weight; θ, temperature; c, polymer concentration; s, curvilinear distance from a chain end) to coincide with the corresponding NMR quantity. We therefore use the same symbol [3,4]. Contour-length fluctuation also explains [2] the fractal exponents of the molecular-weight dependence of the zero-shear viscosity [8]

$$\eta \propto M^{3.4} \ . \tag{16}$$

A direct verification of contour-length fluctuation is possible by the aid of block copolymers of the ABA type with hetero blocks A labelling the chain ends [9]. In the case of polystyrene/polyisoprene block copolymers, ^{13}C spectroscopy permits the distinction of the blocks. It turned out that the chain-end blocks do not show any crossover from "entangled" to "unentangled" behavior upon dilution. The chain-end blocks obviously do not underlie topological restraints in either cases (compare Fig. 1).

Folds are expected to partially "decouple" the dynamics of the chain sections between them [4,12,21]. This explains the tendency of D_1^s and hence of τ_l (Eq. 11) to be independent of the molecular weight.

The fluctuation of folds is expected to be connected with density fluctuations on a molecular scale. The consequence is that fluctuations of the local magnetic susceptibility and, therefore, of the magnetic-field homogeneity occur. The NMR lines are broadened correspondingly. A broadening up to two orders of magnitude has been reported [22]. The inhomogeneity can be estimated by the aid of a simple model assuming spherical voids. The maximum value of the relative inhomogeneity of the magnetic flux density is found to be [22]

$$\frac{\Delta B}{B_0} = -3\chi \frac{1+\chi}{3+2\chi} \approx -\chi \ . \tag{17}$$

For polyethylene one therefore expects $\Delta B/B_0 \approx 10^{-5}$ in good agreement with the experimental finding.

4. Formulation of the statistical problem

The question is now whether the assumption of fluctuating folds is justified from the thermodynamic point of view. We have to show that they are compatible with the general equilibrium conditions, at least in principle. The equilibrium properties of the dynamic structure of a chain are characterized by the partition function Z. Parameters of the system can be derived by searching the maximum of Z at constant temperature. In the following, we treat Z as dependent on the mean contour length L_c and the mean fold length f.

The formation of a fold of length f reduces the length of the primitive path, i.e., the contour length L_c, by an amount $2f$. The degeneration factor of the partition function of the primitive path therefore *decreases* by the factor

$$\Gamma_d = \exp\left[-\frac{\alpha}{a}(L-L_c)\right] , \tag{18}$$

where a is the step length of the primitive path as discussed in [4] and α is the multiplicity of the primitive path [1, p. 206].

On the other hand, folds represent new dynamic and conformational degrees of freedom. Therefore the state of the primitive-path chain is degenerated and the partition function is *increased* by a factor Γ_i. Folds subdivide a chain into subchains wchich can locally fluctuate back and forth. We may represent them by a "gas of particles", so that

$$\Gamma_i = \frac{1}{n!}\left[g \sum_k \exp\left(-\frac{\varepsilon_k}{k_B T}\right)\right]^n . \tag{19}$$

n is the mean number of folds per chain,

$$n = \frac{L-L_c}{2f} \ . \tag{20}$$

Equation (19) is the partition function for n independent particles [e.g., 23]. The sum is carried out over all possible states k of a subchain between two neighboring folds for a given primitive path; g is the degeneracy of the states due to different fold conformations; ε_k is the energy of a subchain in the state k; k_B is the Boltzmann constant. With the factors given above, the total partition function in principle is

$$Z = Z_0 \Gamma_d \Gamma_i \ , \tag{21}$$

where Z_0 is the partition function expected for a free chain. This function implies the total ergodicity of the system.

From the above considerations it became clear that the range of conformations which can be probed by a chain depends on the time scale considered. For $t \ll \tau_r^r$ the primitive path must be assumed to be quasi-static. It has a given conformation being virtually fixed in this time scale. Not so the chain itself. The existence of fluctuating folds guarantees that many different arrangements and dynamic states are probed in this time scale. We have a situation which is only ergodic in part. The mean fold length f_e corresponding to the thermal equilibrium with respect to this partially ergodic situation therefore must not be derived from the total partition function Eq. (21). Rather the partial partition function Γ_i, Eq. (19) has to be considered. It refers to all states which are accessible for a given primitive path. (For simplicity we will assume that the contour length L_c has a fixed rather than a fluctuating value in this calculation.)

Having the equilibrium fold length, we can then derive the equilibrium contour length L_c^e of the primitive path. Different primitive paths can only be probed in the time scale $t \gg \tau_r^r$ for which the full ergodicity is relevant. We therefore consider the total partition function Eq. (21). It depends on the equilibrium fold length f_e which, on the other hand, is derived on the basis of the partially ergodic situation being valid for each of the allowed primitive paths and for every value of the contour length.

The problem is now to find an expression for the energies ε_i. For this purpose we simplify the model further. As already mentioned, a chain is considered to be subdivided into n subchains by folds. The subchains are thought to be replaced by a one-dimensional gas of "particles" of mass

$$m = \frac{M}{n} . \tag{22}$$

The particles can be displaced within a range in the order of the length stored in a fold, i.e., $2f$. This free path may be regarded as a kind of "free volume" in analogy to treatments of molecular motions in gases or liquids. Thus the particles are assumed to be confined in one-dimensional square wells of length $2f$, and we arrive at a standard problem of physics. For simplicity we assume that f has the same value for all n folds of a chain.

Apart from the translational degrees of freedom of the subchains, many other degrees of freedom exist, of course, which are not related to chain folds. The dissipation of translational energy of the subchains into these degrees of freedom may be characterized by the introduction of a friction coefficient γ. We will first treat the problem without friction and then suggest an actual ansatz for the solution of eigenvalue problems implying friction.

5. Equilibrium fold length

In the time scale of fold fluctuation the number of *accessible* primitive paths is only 1. It takes a long time until the chain disengages from a given tube, so that it has the possibility to probe another conformation of the primitive path. For the moment, we can therefore ignore the reduction factor Γ_d and, on the contrary, consider the *increase* of the number of conformations due to fold fluctuation.

For large fold numbers n, Eq. (19) can be approximated by

$$\Gamma_i = \left[\frac{e}{n} g \sum_k \exp\left(-\frac{\varepsilon_k}{k_B T}\right)\right]^n . \tag{23}$$

The energies of a particle in a one-dimensional box of length $2f$ are

$$\varepsilon_k = \frac{k^2 h^2}{32 m f^2} \quad (k \text{integer}) . \tag{24}$$

The conformational degeneracy of a fold is

$$g = \exp\left(\frac{\alpha}{a} f\right) . \tag{25}$$

α is the multiplicity of the random walks of the folds with the mean step length a. Inserting Eqs. (24) and (25) in Eq. (23) and approaching the sum by an integral leads to

$$\Gamma_i \approx \left[\frac{e}{n} \exp\left(\frac{\alpha}{a} f\right) \frac{[2\pi M k_B T]^{1/2}}{h n^{1/2}} 2f\right]^n = \left\{\frac{e}{h} \frac{[2\pi M k_B T]^{1/2}}{h(L-L_c)^{3/2}} (2f)^{5/2} \exp\left(\frac{\alpha}{a} f\right)\right\}^{(L-L_c)/2f} . \tag{26}$$

Γ_i has a maximum for the equilibrium fold length

$$f = f_e \equiv \frac{1}{2} \left[\frac{2\pi \hbar e^3}{M k_B T} (L-L_c)^3\right]^{1/5} . \tag{27}$$

The derivation of the partition function when the subchain translations are connected with friction is presented in the appendix on the basis of nonlinear Schrödinger equations (24–26). The result is

$$\Gamma_i = \exp\left(\frac{\alpha(L-L_c)}{2a}\right) \times \exp\left\{\frac{5(L-L_c)}{4} \left[\frac{\ln(Bf)}{f} - \frac{E\gamma^2 f^2}{2}\right]\right\} \tag{28}$$

with

$$B = 2\left(\frac{e}{h}\right)^{2/5} \frac{[2\pi M k_B T]^{1/5}}{(L-L_c)^{3/5}} \tag{29}$$

and

$$E = \frac{4M}{5(L-L_c)\kappa k_B T} . \tag{30}$$

κ is a numerical constant of order 1. The condition for an extremum with respect to f leads to the equation

$$\frac{Bf}{e}\exp(E\gamma^2 f^3) = 1 . \tag{31}$$

The expansion of the exponential function gives

$$f + E\gamma^2 f^4 + \frac{E^2\gamma^4 f^7}{2} + \frac{E^3\gamma^6 f^{10}}{6} + \ldots = \frac{e}{B} . \tag{32}$$

In the lowest order, i.e., for $\gamma = 0$, the result – Eq. 27 – is reproduced. The next higher order is already quadratic in γ, so that the neglect of friction appears to be justified provided that $L-L_c$ is large enough. This is expected for $M \gg M_c$.

At the opposite limit, $\gamma \to \infty$, the equilibrium fold length tends towards zero. For finite or negligible values of the friction coefficient, f_e in any case takes a value neither vanishing nor diverging. This is a remarkable result. It means that finite fold lengths are consistent with thermal equilibrium.

6. Equilibrium contour length

Based on Γ_i alone, the contour length would tend to zero. The reason is that only a single primitive path conformation is implied. Rather we must now take into account all possible primitive path conformations. These can be probed in the time scale of global motions, i.e., for $t \gg \tau_r^r$. It then becomes clear that a strong reduction of the length of the primitive path also leads to a reduction of the conformational contribution, so that the opposite tendency appears.

Inserting the equilibrium fold length Eq. (27) in the expression for Γ_i [Eq. (26)] and this in the total partition function Eq. (21) leads to

$$Z = Z_0 \exp\left\{\frac{5}{2e}\left[\frac{2e^2\pi M k_B T}{h^2}(L-L_c)^2\right]^{1/5} - \frac{\alpha}{2a}(L-L_c)\right\} . \tag{33}$$

The condition for the maximum of Z determines the equilibrium contour length

$$L_c = L_c^e \equiv L - \frac{1}{e}\left(\frac{M k_B T}{2\pi\hbar^2}\right)^{1/3}\left(\frac{2a}{\alpha}\right)^{5/3} , \tag{34}$$

which turns out to be finite as required. Note that Eq. (34) is a crude formula based on extreme simplifications of the system to be treated. Quantitative estimations on the basis of Eq. (34) therefore must not be taken too seriously.

Nevertheless, the qualitative behavior implied in this relation is considered to reflect the essence of the concept proposed above. In particular, inserting the equilibrium contour length, Eq. (34), in the expression for the equilibrium fold length Eq. (27), gives

$$f_e = \frac{a}{\alpha} . \tag{35}$$

This is a quite reasonable result: It states that the fold length neither depends on L nor T as it must be. f_e is a property solely determined by the segment parameters and the topological restraints. It is remarkable that the equilibrium fold length Eq. (35) is of the same order of magnitude as the diameter of the Doi-Edwards tube. These authors could derive a similar expression for the tube dimension [1, p. 206] despite a completely different starting point.

7. Conclusion

The three-component/three-block orientation-correlation function successfully describes many dependences and phenomena observed in NMR and other experiments. Component B directly reflects longitudinal or reptational displacements. From the more or less decoupled dynamics of the subchains and the one-dimensional displacements along a Gaussian path one concludes that the chain dynamics well approaches the following limits:

$$D_1^s \propto M^x , \quad x \approx 0 , \tag{36}$$

$$\tau_l \gg T_d , \tag{37}$$

$$\tau_l \propto M^x , \quad x \approx 0 , \tag{38}$$

$$\langle r^2 \rangle \propto t^{1/2} , \tag{39}$$

where $\langle r^2 \rangle$ is the three-dimensional mean-square displacement as a result of the longitudinal displacements along the tight tube. Equation (39) manifests itself in the $\nu^{1/2}$ frequency dependence of T_1 observed for high molecular weights and low frequencies [3, 4, 10, 11].

Superimposed to the component B displacements, those due to fold fluctuation have to be considered. In total a short scale tendency towards a Rouselike time dependence of $\langle r^2 \rangle$ [see Eq. (39)] is expected as it has been demonstrated also with neutron scattering experiments [27].

The finite equilibrium fold length arises because the topological restraints of an entangled chain impose different time scales for the conformations of the primitive path and of the chain itself. The same is true for the dynamic contributions. We are thus dealing with a time-scale effect in the thermodynamic equilibrium of this particular class of system. It should always be relevant when dynamic degrees of freedom are partly suppressed due to structural conditions.

The quantum mechanical equations used in the appendix for the derivation of the dynamical partition function in the case of friction are still under discussion. This paper does not make any final decision concerning the universal validity of one of these equations. We have instead shown that contour-length fluctuation and folds of polymer chains are in agreement with fundamental principles represented by any of these equations.

The finding that the equilibrium fold length Eq. (35) was found to be of the same order as the tube diameter estimated by Doi and Edwards [1] sheds new light on the question of how the Doi-Edwards tube must be interpreted – in view of the above consideration it represents the space region over which fold fluctuations are spread.

8. Appendix: Partition function implying friction

We start with the quantum mechanical formulations of the Langevin equation, which proved to be useful with related problems. Kostin [24] developed the so-called Schrödinger-Langevin equation

$$i\hbar \frac{\partial \psi}{\partial t} = \left[-\left(\frac{\hbar^2}{2m}\right) \nabla^2 + U + \frac{\hbar \gamma}{2i} \ln \frac{\psi}{\psi^*} \right] \psi \,, \tag{40}$$

with the potential U and the friction constant γ. In the present problem, U does not imply any stationary potential. Rather it represents the so-called stochastic potential. The classical limit $\hbar \to 0$ of Eq. (40) leads to the Hamilton-Jacobi equation [25]. On the other hand, this equation fails to provide the right frequency for the damped oscillator. Schuch et al. [26] therefore proposed another nonlinear Schrödinger equation for the description of dissipative systems. It reads

$$i\hbar \frac{\partial \psi}{\partial t} = \left[-\frac{\hbar^2}{2m} \nabla^2 + U - i\gamma\hbar(\ln \psi - \langle \ln \psi \rangle) \right] \psi \,. \tag{41}$$

We will designate this equation as nonlinear Schrödinger equation II (NLSEII). It provides the right reduction of the oscillator frequency due to friction. It is also compatible with Heisenberg's uncertainty relation. Thus it appears to conform to basic physical principles.

The two Schrödinger-type equations differ merely in the analytical form of their dissipative terms. In order to have an argument not so intimately related to the actual form of these terms, we will present treatments for both equations. We will thus demonstrate that the analytic form of the friction terms virtually does not affect the resulting mean values as already discussed by Schuch et al. [26].

a) Classical limit

Before deriving the stationary solutions of Eqs. (40) and (41) we will investigate some properties inherent to the Schrödinger-Langevin equation as an example. In particular we are interested in the classical limit and in the convergence of the time-dependent solutions. As in the following we set $U = 0$ and $m \equiv \langle m(x,t) \rangle = \text{const}$. The substitution

$$\psi(x,t) = A \exp\left[\frac{i}{\hbar} W(x,t)\right] \tag{42}$$

transforms Eq. (40) to

$$\frac{\partial W}{\partial t} = \frac{1}{2m} \left[-(\nabla W)^2 + i\hbar \nabla^2 W \right] - \gamma W \,. \tag{43}$$

The limit $\hbar \to 0$ leads to a real equation equivalent to the classical Hamilton-Jacobi equation in a version modified for dissipative systems [25]. This shows again that a treatment on the basis of the above equations is a reasonable starting point. Substituting

$$W(x,t) = g_1(t) + g_2(x) \tag{44}$$

leads to

$$\frac{\partial g_1}{\partial t} + \gamma g_1(t) + \frac{1}{2m} (\nabla g_2)^2 + \gamma g_2 = 0 \,. \tag{45}$$

This equation is fulfilled if

$$\frac{\partial g_1}{\partial t}+\gamma g_1=-C_0 \tag{46}$$

and

$$\frac{1}{2m}(\nabla g_2)^2+\gamma g_2=C_0 , \tag{47}$$

where C_0 is a real constant. The solution for g_1 is

$$g_1(t)=-\frac{C_0}{\gamma}[1-\exp(-\gamma t)] . \tag{48}$$

In the short-time limit, $t \ll 1/\gamma$, we have

$$g_1(t)\approx -C_0 t . \tag{49}$$

This result together with the above substitutions identifies C_0 as the initial energy. From this we learn that the energy eigenvalues to be drived from the stationary solutions from the (complex) quantum mechanical equations are the real parts of a complex constant C appearing in the following treatments equivalently as above.

In the opposite limit, $t\rightarrow\infty, g_1(t)$ becomes constant:

$$g_1(t)=-\frac{C_0}{\gamma} , \tag{50}$$

so that the action $W(x,t)$ also approaches a time-independent limit. In total, $g_1(t)$ describes the irreversible time development of the system as it must be.

In the following we will not use the classical equation (47) for the derivation of the stationary solutions of the problem. It is rather more suitable to solve the quantum mechanical equations to obtain the complete set of eigenvalues needed for the calculation of Γ_i.

b) Partition function on the basis of the NLSEII equation

We are interested in the stationary solutions of

$$i\hbar\frac{\partial\psi}{\partial t}=-\left[\frac{\hbar^2}{2m}\nabla^2+i\gamma\hbar(\ln\psi-\langle\ln\psi\rangle)\right]\psi . \tag{51}$$

As a separation ansatz we use

$$\psi=F(t)G(x) \tag{52}$$

with x representing the position of the moving subchain. Inserting Eq. (52) in Eq. (51) leads to

$$i\hbar\frac{\partial}{\partial t}\ln F=-\frac{\hbar^2}{2m}\frac{\nabla^2 G}{G}-i\gamma\hbar(\ln G-\langle\ln G\rangle) . \tag{53}$$

Equating either of the two sides of Eq. (53) with the complex quantity $C=C'+iC''$ leads to new relations for F and G alone:

$$i\hbar\frac{\partial}{\partial t}\ln F=C , \tag{54}$$

$$-\frac{\hbar^2}{2m}\frac{\nabla^2 G}{G}-i\gamma\hbar(\ln G-\langle\ln G\rangle)=C . \tag{55}$$

Equation (55) can be modified by replacing

$$G=\exp\{V(x)\} \tag{56}$$

with

$$V\equiv V'+iV'' . \tag{57}$$

We then have

$$-\frac{\hbar^2}{2m}[\nabla^2 V+(\nabla V)^2]-i\gamma\hbar(V-\langle V\rangle)=C . \tag{58}$$

With the definition

$$\langle V\rangle=\langle V'\rangle+i\langle V''\rangle , \tag{59}$$

all complex quantities in Eq. (58) can be expressed by their real and imaginary parts. Adding Eq. (58) to its conjugate complex version yields

$$-\frac{\hbar^2}{2m}[\nabla^2 V'+(\nabla V')^2-(\nabla V'')^2] + \gamma\hbar(V''-\langle V''\rangle)=C' . \tag{60}$$

Similarly the subtraction provides

$$-\frac{\hbar^2}{2m}[\nabla^2 V''+2\nabla V'\nabla V'']-\gamma\hbar(V'-\langle V'\rangle)=C'' . \tag{61}$$

Equations (60) and (61) have solutions of the form

$$V'=V_0=\text{const} , \tag{62}$$

$$V'' = k_0 + k_1|x| + k_2x^2 \ . \tag{63}$$

Below we will choose the phase constant $k_0 = \pi/2$.

From Eq. (62) it follows $\langle V' \rangle = V_0$ and the last left-hand term of Eq. (61) vanishes. V_0 follows from the normalization relation

$$\int_{-f}^{f} |G|^2 \, dx = 2f \exp(2V_0) = 1 \tag{64}$$

to be

$$V_0 = -\tfrac{1}{2}\ln(2f) \ . \tag{65}$$

Equations (62) to (65) permit the derivation of $\langle V \rangle$ and, hence, of $\langle V' \rangle$ and $\langle V'' \rangle$:

$$\langle V \rangle = \frac{1}{2f}\int_{-f}^{f} V dx = V_0 + i[k_1 f + \tfrac{4}{3}k_2 f^2] \ , \tag{66}$$

i.e.,

$$\langle V' \rangle = V_0 \ , \tag{67}$$

$$\langle V'' \rangle = k_1 f + \tfrac{4}{3}k_2 f^2 \ . \tag{68}$$

Inserting Eqs. (62), (63), and (67) in (60) and (61) gives

$$C'' = k_2 \frac{\hbar^2}{m} \tag{69}$$

and

$$\frac{\hbar^2}{2m}(k_1^2 + 4k_1k_2x + 4k_2^2x^2) + \gamma\hbar(k_1x + k_2x^2 - \langle V'' \rangle) = C' \ . \tag{70}$$

The x and x^2 terms vanish for

$$k_2 = -\frac{\gamma m}{2\hbar} \ , \tag{71}$$

and we have

$$C' = \frac{\hbar^2}{2m}k_1^2 - \gamma\hbar\langle V'' \rangle \ . \tag{72}$$

We will now determine the quantities k_1 and $\langle V'' \rangle$. The boundary conditions

$$\Re\{\psi(0)\} = \Re\{\psi(2f)\} = 0 \tag{73}$$

are fulfilled for

$$V''(\pm f) = k_0 + 2k_1 f + 4k_2 f^2 \doteq j\pi \ , \quad (j \text{ integer}) \ , \tag{74}$$

or with Eq. (71)

$$k_1 = \frac{1}{2f}\left[j\pi + \frac{2\gamma m}{\hbar}f^2\right] \ . \tag{75}$$

Inserting Eqs. (71) and (75) in (68) leads to

$$\langle V'' \rangle = j\frac{\pi}{2} + \frac{\gamma m}{3\hbar}f^2 \ . \tag{76}$$

Equations (71), (75), and (76) finally determine C' and C'' expressed by Eq. (72) and (69), respectively:

$$C' = \frac{j^2\pi^2\hbar^2}{8mf^2} + \frac{\gamma^2 m}{6}f^2 \ , \tag{77}$$

$$C'' = \frac{\gamma\hbar}{2} \ . \tag{78}$$

C' and C'' have the dimension of energy. In the frictionless limit $\gamma = 0$, C' simply represents the discrete energy eigenvalues of a particle in a one-dimensional box of length $2f$, while C'' vanishes. The real part of C therefore is the set of energy eigenvalues relevant for the partition function.

Inserting Eq. (77) in Eq. (23) yields, after approaching the sum by an integral

$$\Gamma_i \approx \left\{\frac{e}{h}\frac{[2\pi M k_B T]^{1/2}}{(L-L_c)^{3/2}}(2f)^{5/2} \times \exp\left(-\frac{\gamma^2 M f^3}{3k_B T(L-L_c)} + \frac{a}{a}f\right)\right\}^{(L-L_c)/2f} \ . \tag{79}$$

c) Partition function on the basis of the Schrödinger-Langevin equation

The Schrödinger-Langevin equation is given by Eq. (40). As above we set $U = 0$ and $m \equiv \langle m(x,t) \rangle = \text{const}$. We then have

$$i\hbar\frac{\partial}{\partial t}\psi = \left[-\frac{\hbar^2}{2m}\nabla^2 + \frac{\gamma\hbar}{2i}\ln\frac{\psi}{\psi^*}\right]\psi \ . \tag{80}$$

The separation ansatz Eq. (52) leads to two equations analogous to Eqs. (54) and (55):

$$i\hbar\frac{\partial}{\partial t}\ln F + \frac{\gamma\hbar}{2i}[\ln F - \ln F^*] = C \tag{81}$$

and

$$-\frac{\hbar^2}{2m}\frac{\nabla^2 G}{G}+\frac{i\gamma\hbar}{2}\ln\frac{G^*}{G}=C \tag{82}$$

with $C=C'+iC''$. Let us first consider Eq. (81). Adding it to its conjugate complex version yields

$$\frac{\partial}{\partial t}(\ln F+\ln F^*)=\frac{C-C^*}{i\hbar}=\frac{2C''}{\hbar}\ . \tag{83}$$

Substituting

$$W=W'+iW''\doteq\ln F \tag{84}$$

and subtracting Eq. (81) from its conjugate complex version gives

$$\frac{\partial}{\partial t}W''+\gamma W''=-\frac{C'}{\hbar}\ . \tag{85}$$

Let us now turn to Eq. (82). Substituting

$$A\equiv-\frac{\hbar^2}{2m}\ ,\qquad B\equiv\frac{\gamma\hbar}{2}\ ,\qquad G=\exp(V) \tag{86}$$

leads to

$$A[\nabla^2 V+(\nabla V)^2]+iB[V^*-V]=C\ . \tag{87}$$

Inserting the solution ansatz

$$V=i(a_0+a_1x+a_2x^2)-a_3x-a_4x^2 \tag{88}$$

yields

$$\begin{aligned}&A[2ia_2-2a_4-a_1^2-4a_1a_2x-4a_2^2x^2+a_3^2+4a_3a_4x\\&\quad+4a_4^2x-2i(a_1+2a_2x)(a_3+2a_4x)]\\&\quad+2B(a_1x+a_2x^2)C\ .\end{aligned} \tag{89}$$

We add the conjugate complex equation and obtain

$$\begin{aligned}&[A(-2a_4-a_1^2+a_3^2)-C']\\&\quad+x[-4a_1a_4A+4a_3a_6A+a_2a_1]\\&\quad+x^2[-4Aa_2^2+4Aa_4^2+2Ba_2]=0\ .\end{aligned} \tag{90}$$

All three bracket terms must be zero for themselves. From the coefficients of x and x^2 it follows

$$a_4=0 \tag{91}$$

and

$$a_2=\frac{B}{2A}=-\frac{\gamma}{2\hbar}\ . \tag{92}$$

The boundary conditions

$$\Re\{G(0)\}=\Re\{G(2f)\}=0 \tag{93}$$

lead with

$$G\equiv\exp(V) \tag{94}$$

to

$$a_0+a_1f+a_2f^2=j\pi\qquad (j\ \text{integer})\ . \tag{95}$$

Normalization requires

$$\int_0^\infty|G|^2\,dx=\int_0^\infty\exp(-2a_3x)\,dx\doteq1\ , \tag{96}$$

so that

$$a_3=\frac{1}{2}\ . \tag{97}$$

Identifying the energy eigenvalues ε_j with C' gives

$$\varepsilon_j=\frac{j^2\pi^2\hbar^2}{8mf^2}+\frac{j\pi\gamma\hbar}{4}+\frac{\gamma^2f^2m}{8}\ , \tag{98}$$

where we have neglected $\hbar^2$ against γ^2f^2. The partition function Eq. (23) is then

$$\Gamma_i\approx\left[\frac{e}{n}g\exp(-b_1)\int_0^\infty\exp(-j^2b_2-jb_3)\,dj\right]^n \tag{99}$$

with

$$b_1=\frac{\gamma^2f^2m}{8k_BT}\ , \tag{100}$$

$$b_2=\frac{\pi^2\hbar^2}{8mf^2k_BT}\ , \tag{101}$$

$$b_3=\frac{\pi\gamma\hbar}{4k_BT}\ . \tag{102}$$

(103)

The integral can be approximated by

$$\int_0^\infty\exp(-j^2b_2-jb_3)\,dj\approx\frac{2f}{\lambda} \tag{104}$$

with

$$\lambda = \left(\frac{2\pi\hbar^2}{m k_B T}\right)^{1/2} . \tag{105}$$

The partition function thus turns out to be

$$\Gamma_i \approx \left\{\frac{e}{h}\frac{[2\pi M k_B T]^{1/2}}{(L-L_c)^{3/2}}(2f)^{5/2} \times \exp\left(-\frac{\gamma^2 M f^3}{4k_B T(L-L_c)}+\frac{a}{a}f\right)\right\}^{(L-L_c)/2f} \tag{106}$$

Apart from a numerical factor in the exponential function, Eq. (106) coincides with Eq. (79).

Acknowledgements.

This work has been supported by the Deutsche Forschungsgemeinschaft.

References

1. Doi M, Edwards SF (1986) The Theory of Polymer Dynamics. Clarendon Press, Oxford
2. Doi M (1981) J Polymer Sci, Polymer Letters Edn 19:265–273
3. Kimmich R, Bachus R (1982) Colloid Polymer Sci 260:911–936
4. Kimmich R, Schnur G, Köpf M (1988) Progr NMR Spectr 20:385–421
5. de Gennes PG (1971) J Chem Phys 55:572–579
6. Rouse PE (1953) J Chem Phys 21:1272–1280
7. Doi M (1974) Chem Phys Letters 26:269–272
8. Ferry ID (1980) Viscoelastic Properties of Polymers. Wiley, New York
9. Köpf M, Schnur G, Kimmich R (1988) Macromolecules 21:3340–3341
10. Preißing G, Noack F (1975) Progr Colloid Polym Sci 57:216–224
11. Kimmich R, Schmauder Kh (1977) Polymer 18:239–243
12. Callaghan PT (1988) Polymer Preprints 29:37
13. Ramik T (1988) Diploma Thesis, University of Ulm
14. Kimmich R, Rosskopf E, Schnur G, Spohn K-H (1985) Macromolecules 18:810–812
15. Hiller W, Schenk W (1986) Polymer 27:1353–1358
16. Bachus R, Kimmich R (1983) Polymer 24:964–970
17. Fleischer G (1987) Colloid Polymer Sci 265:89–95
18. Schnur G, Kimmich R (1988) Chem Phys Letters 144:333–338
19. Köpf M, Schnur G, Kimmich R (1988) J Polym Sci C: Polym Letters 26:319–323
20. Cohen-Addad JP, Guillermo A (1984) J Polym Sci, Polym Phys Ed 22:931–955
21. Kimmich R (1984) Polymer 25:187–194
22. Bachus R, Kimmich R (1983) Polymer Communications 24:317–320
23. Eyring H, Henderson D, Jones Stover B, Eyring EM (1982) Statistical Mechanics and Dynamics. Wiley, New York
24. Kostin MD (1975) J Stat Physics 12:145–151
25. Yasue K (1978) Ann Physics 114:479–496
26. Schuch D, Chung K-M, Hartmann H (1983) J Math Physics 24:1652–3092
27. Richter D, Baumgärtner A, Binder K, Ewen B, Hayter JB (1981) Phys Rev Lett 47:109–113

Received October 31, 1988; accepted February 24, 1989

Authors' address:

R. Kimmich
Sektion Kernresonanzspektroskopie
Universität Ulm
Oberer Eselsberg
D-7900 Ulm, FRG

Discussion

REINEKER:

Is it possible to derive specific parameter values for the tube model from your theory?

Comment to the tube model:

A recent molecular dynamics calculation by K KREMER et al. seems to support the tube model.

KIMMICH:

The "tight tube" is characterized by the residual correlation of the local motions (component A) in the long-time limit. This component refers to spin interactions modulated by anisotropic segment reorientations about the tube axis. The diameter of the tight tube is expected to be of the order of magnitude of the nearest neighbor chain distance. The folds of the tight tube, on the other hand, fluctuate in a much wider range. The thermodynamic treatment leads to the same order of magnitude as derived by Doi and Edwards from a different point of view. The diameter of this tube-like region corresponds to the step-length of the primitive path.

The tight tube manifests itself also by the quasi one-dimensional displacement behavior of a chain, as revealed by the specific frequency and temperature dependences of the relaxation times. The tube-like region covered in the average by the fluctuating folds is mirrored by the contour-length fluctuation with the corresponding time constant and molecular weight dependence. The consequence in particular is the appearance of the critical molecular weight M_c.

WARTEWIG:

As I learned from literature, there are now discussions against the tube model. The motions of chains should be more or less isotropic. Would you comment on that?

KIMMICH:

The fact that one observes strong molecular weight dependences in diverse NMR experiments, and the fact that the interactions determining the NMR relaxation times are restricted to a range of a few Ångströms leave no other conclusion than that the constraints which the local segment reorientations underlie persist up to the time scale of global, that is, molecular weight-dependent chain modes.

The parameter characterizing the anisotropy of component A is the residual correlation a_2. If a_2 would be zero, one would not be able to observe any molecular weight dependence by NMR relaxation (with the exception of free volume effects).

SAUTTER:

Can you give information about the number of folds per chain from fitting your approach to the experimental data?

KIMMICH:

The dynamic three-block structure is characterized by the parameter Δ_0, the mean chain length per fold. Δ_0 determines the critical molecular weight M_c which reveals itself in the molecular weight dependence of T_2, for instance. The data of polyisoprene as an example can well be described assuming $\Delta_0 = 900$ atomic mass units. A real least-square fit taking into account the whole set of the available data would be extremely time-consuming and has therefore not been carried out up to now.

Progress in Colloid & Polymer Science Progr Colloid & Polymer Sci 80:21–25 (1989)

Photoelectric response in the pTS diacetylene monomer-polymer single crystals

M. Orczyk, J. Sworakowski and M. Bertault*)

Institute of Organic and Physical Chemistry, Technical University of Wrocław, Wrocław, Poland

Abstract: The polymerization-dependent photoconductivity in single crystals of pTS diacetylene was studied. pTS has been known to polymerize topochemically in the solid state, and its polymerization kinetics has been known. Since the monomer is non-photoconducting, a study of the photoconductivity build-up during the polymerization provides information about the transport of photogenerated charge carriers, and, indirectly, about changes of the average chain lengths in function of the polymer contents. A simple model put forward by the present authors allows for a comparison with the estimates of the chain lengths published earlier.

Key words: Polydiacetylene; polymerization; photoconductivity; chain length

1. Introduction

Poly(diacetylenes) (PDA) constitute a unique class of polymers which relatively easily may be obtained in form of macroscopic-size single crystals. Since the pioneering work of Wegner (e.g., [1,2]), there have been several papers published dealing with, among others, changes of structural, spectral and electrical properties of diacetylenes (DA – $R_1-C{\equiv}C-C{\equiv}C-R_2$, R_1 and R_2 being side groups, usually bulky and flexible) upon their polymerization (see, e.g., [3,4] for recent reviews and corrections of papers).

Single crystals of PDA are usually obtained by the solid-state polymerization of single crystals of respective monomers, the latter being grown from solutions in appropriate solvents. The topochemical polymerization yields single crystals of reasonable quality consisting of extended systems of conjugated π bonds $-(-C{\equiv}C-\underset{R}{\overset{R}{C}}={C}-C{\equiv}C-)_n-$, well separated from one another [5]. This feature of the structure of PDA results in a remarkable anisotropy of physical properties of the polymer (e.g., [4]), thus making it a good candidate for studying properties of quasi-1-dimensional systems. In particular, transport of charge carriers was investigated by several groups (see, e.g., [6–8]).

On the other hand, crystals of monomers behave in many respects as typical molecular solids, exhibiting moderate anisotropies of physical properties. Thus the system offers a unique opportunity of studying a modification of crystal properties in the course of the solid state polymerization. Most research on diacetylenes was carried out on bis(p-toluenosulphonate) of 2,4-hexadiyne-1,6-diol, commonly referred to as pTS ($R_1 = R_2 = -CH_2-OSO_2-\phi-CH_3$), for which the kinetics of solid-state polymerization was studied in detail (see [9] and references therein). Monomer pTS crystallizes in the monoclinic system (space group $P2_1/c$ [10] with the molecules forming columns along the crystallographic ***b*** direction. Upon polymerization, the growing polymer chains also extend in the same direction, the resulting crystal structure remaining essentially unchanged [5]. It is interesting to mention that poly(pTS) is a good photoconductor [6–8], whereas the monomer does not exhibit any appreciable photoconductivity.

Recently, we published a paper [11] describing the evolution of the photoconductivity during the solid-state polymerization of pTS. The photoconductivity build-up was interpreted as being controlled by a percolation between growing polymer chains. In the present paper, we extend the results presented in [11].

*) Groupe de Physique Cristalline, Université de Rennes I, Campus de Beaulieu, Rennes, France

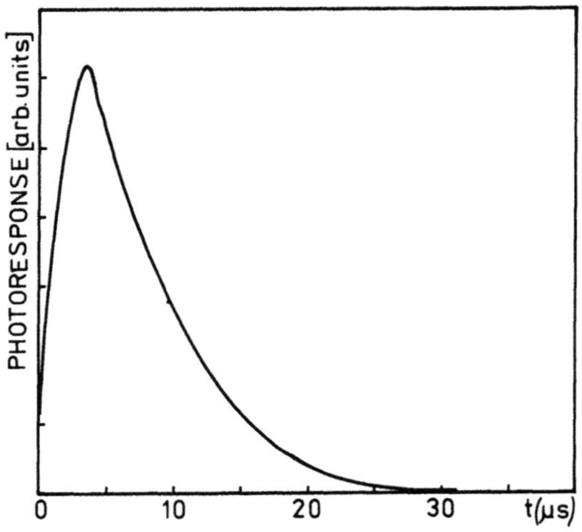

Fig. 1. A typical photocurrent transient in polymerizing pTS samples

2. Experimental

Crystals of pTS monomer were obtained from acetone solutions by a controlled evaporation of the solvent as described in [12]. The crystals obtained were typically as large as a few hundred mm^3, the crystallographic directions being easily identifiable. Samples of a desired orientation were then cut from as-grown crystals with a wire saw. Since pTS is easily thermally and photo-polymerizable, the samples were stored at low temperatures and handled in the dark or under a weak red illumination to minimize any uncontrolled polymerization. For the same reason, measurements of polymerization-dependent photoresponse of pTS samples were carried out in the sandwich geometry using a weak pulse illumination, and employing a typical time-of-flight setup [13].

The experiments were carried out on samples cut in such a way that the electric field was applied either parallel to the ***b*** crystallographic direction (i.e., parallel to the direction of growing polymer chains) or along the ***a**** or ***c*** directions (i.e., perpendicular to the chains). The measurements of the polymerization-dependent temporal evolution of the photoconductivity were performed at 333 K, i.e., at the temperature for which the kinetics of thermal polymerization had been known from independent measurements (e.g., [12, 14, 15]).

A more detailed description of the crystal growth and experimental procedure was given in [11].

3. Results and discussion

As was mentioned above, monomer pTS is non-photoconducting, thus monomer samples freshly mounted in the experimental setup exhibited no measurable photoconductivity. Appreciable photoresponses appeared in samples only after ca. 19 h of annealing at 333 K, and then their magnitude quickly increased to saturate after ca. 26 h of annealing.

A typical photocurrent transient is shown in Fig. 1. The shapes (but *not* the magnitudes) of the photocurrent transients were found independent of the polymer contents and direction of measurements. Moreover, the shapes of the dependence of the photocurrent magnitude on the polymerization time were also identical for all directions of measurements to within the experimental accuracy. We found, however, a pronounced anisotropy of the magnitudes of the currents: in fully polymerized samples, the ratio of signals measured under identical conditions amounted to $J_{a*}:J_b:J_c \cong 1:1000:2.5$.

A (photocurrent vs annealing time) curve can be converted into a (photocurrent vs polymer contents) dependence provided the polymerization kinetics are known, which is indeed the case for the thermal polymerization of pTS [12, 14, 15]. We employed the curve determined by Bertault [12]; the resulting $J(x)$ curve, determined for measurements carried out in the direction parallel to the polymer chains is shown in

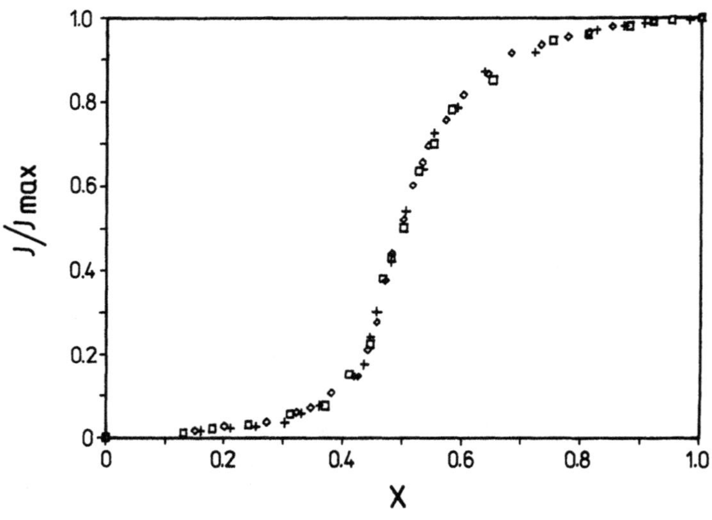

Fig. 2. The photocurrent in polymerizing pTS samples plotted in function of the polymer contents. The measurements were performed with the electric field directed parallel to the crystallographic ***b*** axis; different symbols denote results obtained on different samples. The currents were normalized to their maximal values measured on fully polymerized samples.

Fig. 2. It should be stressed once more that the shapes of the $J(x)$ dependences were identical also for measurements carried out in the directions perpendicular to the chains, the magnitudes of currents being, however, about three orders of magnitude lower.

The shapes of the (photocurrent vs polymer contents) dependences can be explained assuming a simple model put forward in [11]. According to the model, the monomer molecules constitute an "inactive" matrix, the generation and transport of charge carriers taking place on polymer chains only. Consequently, one essentially deals with a percolation-type problem. However, the $J(x)$ dependences found in our experiments could not be fitted by simple equations resulting from the percolation models. In the previous paper [11], we suggested that these discrepancies can be explained taking into account the increase of the average length of the polymer chains during polymerization. According to the model [11], the photocurrent is given by the equation

$$J(x) = \eta N_{av}(x)[1-p_d p(x)]^{-1} , \tag{1}$$

where x is the polymer fraction, N_{av} stands for an average unperturbed chain length at a given polymerization stage, p – for a probability that a randomly chosen polymer chain has at least one polymer chain in the nearest-neighbour position, p_d – for a probability of a charge transfer between nearest-neighbouring polymer chains, and η is an overall efficiency of the charge carrier photogeneration under the experimental conditions (throughout this paper, we shall assume η to be independent of the polymer contents). To a good approximation, one may assume p_d to be independent of the monomer-to-polymer conversion, whereas p is a function of the conversion approximately fulfilling the asymptotic relations

$$\begin{aligned} p(x) &\sim x \quad \text{for} \quad x \approx 0 , \\ p(x) &\approx 1 \quad \text{for} \quad x \to 1 . \end{aligned} \tag{2}$$

Equation (1) is subject to experimental verification once the chain length is known in function of the polymer contents. Unfortunately, the existing experimental results for $N_{av}(x)$ [16] cover a narrow range of the polymer contents around $x \approx 0$, and the only results available are those obtained from model calculations carried out by Baughman in 1978 [15], and then Baessler [8], and Kollmar and Sixl [17]. We employed the results given in [8, 15, 17] to verify our model.

According to Eq. (1)

$$\frac{N_{av}(x)}{J(x)} \sim 1 - p_d p(x) \tag{1a}$$

The ratio N_{av}/J should therefore be a linear function of x at $x \approx 0$ and should saturate for $x \to 1$. Thus, taking values of $N_{av}(x)$ from the literature one can test the above prediction and hence the applicability of our model. The calculated N_{av}/J quotients, normalized to unity at $x = 0$, are shown in Fig. 3. It should be noted that the calculated curves follow the predicted dependence for $x < 0.8$.

Fitting the curves calculated from Eq. (1a) to those published earlier in the literature allows us to estimate the inter-chain jump probability p_d (see Fig. 3); however, the chain lengths cannot be determined *directly* from our measurements as the curves must be "calibrated", thus we can only make estimates of the course of the $N_{av}(x)$ dependences.

It should be noted that our results reasonably fit the literature curves only over a limited range of polymer contents: for low x, we cannot make any comparison

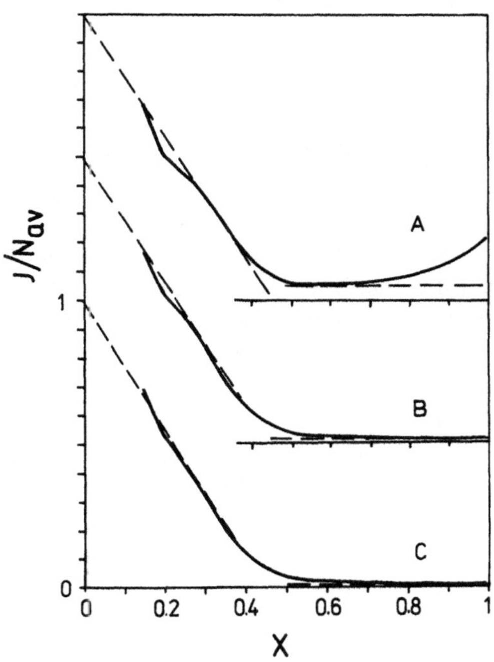

Fig. 3. The quotient J/N_{av} plotted in function of the polymer contents. The values $N_{av}(x)$ were taken from Baessler [8] (curve A), Baughman [15] (curve B), and Kollmar and Sixl [17] (curve C). The quotients were normalized to unity at $x = 0$ and vertically shifted for the sake of clarity. The asymptotic fits are indicated with broken lines; the values of the inter-chain jump probability estimated from the fits amount to 0.95, 0.99 and 0.99 for curves A to C, respectively

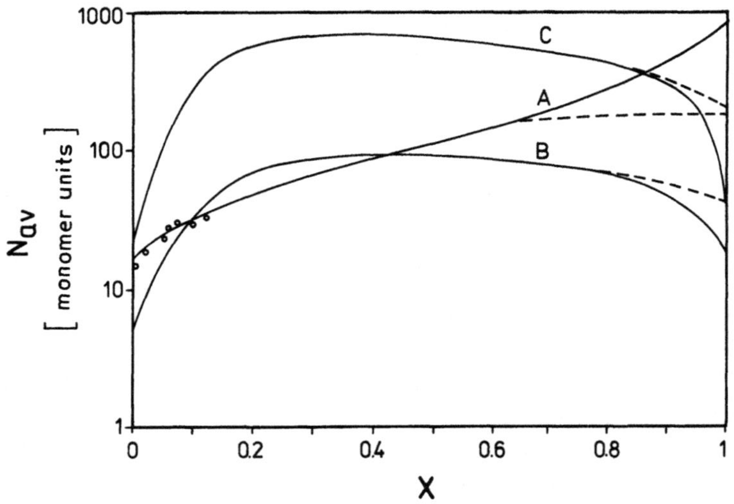

Fig. 4. Average chain lengths in polymerizing pTS crystals as estimated by Baessler [8] (curve *A*), Baughman [15] (curve *B* – note that this curve has been given in arbitrary units only) and Kollmar and Sixl [17] (curve *C*). The dashed lines indicate the course of "corrected" curves, modified in such a way so as to fit our $j(x)$ dependence. The points give average chain lengths as determined form an x-ray experiment by Albouy et al. [16]

for lack of data, whereas for x approaching unity our results deviate from the literature curves. The latter discrepancy can be explained assuming that the models of Baughman [15], Baessler [8], and Kollmar and Sixl [17] break down for high conversions. Our data may now serve to modify the course of the input $N_{av}(x)$ dependences so that they fit our $J(x)$ curve and the postulated $p(x)$ dependence. Such a "correction" was made, and the resulting curves are shown in Fig. 4. One can notice that the "corrected" curves saturate at high conversions. One can offer the following explanation for the saturation: Our estimates are based on measurements of charge carrier transport, and thus, once the concentrations of polymer is high enough, the average length of "easy pathways" is monitored, irrespective of the concentration of short chains. Therefore, one should expect the chain length as estimated from our measurements to saturate at high conversions, as indeed results from the fit.

Acknowledgement.

The research described in this paper was supported in part by the Polish Academy of Sciences within the Programme CPBP 01.14.

References

1. Wegner G (1971) Makromol Chem 145:85–94
2. Wegner G (1972) Makromol Chem 154:35–48
3. Cantow HJ (ed) (1984) Polydiacetylenes. Springer, Berlin
4. Bloor D, Chance RR (eds) (1985) Polydiacetylenes: synthesis, structure and electronic properties. Nijhoff, The Hague
5. Kobelt D, Paulus EF (1974) Acta Cryst. B30:232–234; Enkelmann V (1977) Acta Cryst. B33:2842–2846
6. Wegner G (1979) In: Hatfield WE (ed) Molecular Metals. Plenum Press, New York, pp 297–314
7. Bloor D (1979) Phil Trans Roy Soc London A314:51–68
8. Baessler H (1984) In: Cantow HJ (ed) Polydiacetylenes. Springer, Berlin, pp 1–48
9. Enkelmann V (1985) In Cantow HJ (ed) Polydiacetylenes. Springer, Berlin, pp 91–136
10. Aimé JP, Lefebvre J, Bertault M, Schott M, Williams JO (1982) J Physique (Paris) 43:307–322
11. Orczyk M, Sworakowski J, Schott M, Bertault M (1988) Chem Phys 121:245–254
12. Bertault M (1983) Thèse. Université Paris VII, France; Bertault M, Schott M, Brienne MJ, Collet A (1984) Chem Phys 85:481–490
13. Kepler RG (1960) Phys Rev 119:1226–1229
14. Bloor D, Koski L, Stevens GC, Preston FH, Lando DJ (1975) J Mat Sci 10:1678–1688
15. Baughman RH (1978) J Chem Phys 68:3110–3121
16. Albouy PA, Keller P, Pouget P (1982) J Am Chem Soc 104:6556–6561
17. Kollmar C, Sixl H (1984) In: Sixl H (ed) Proc VW-Symposium "Photoreaktive Festkörper", Kleinwalsertal, Wahl, Karlsruhe, pp 327–356

Received February 2, 1989;
accepted March 1, 1989

Authors' address:
Prof. J. Sworakowski
Politechnika Wrocławska, I-4
50-370 Wrocław, Poland

Discussion

PECHHOLD:

Can you imagine that the levelling off of $N(x)$, for $x > 0.6$, could be brought about by the formation of tight folds during polarization?

SWORAKOWSKI:

In fact, the value of $N_{av}(x)$ found from our experiments should be considered as *an average length of unperturbed sequence of conjugated bonds.* Perturbations of the conjugation result in a perturbation of the carrier transport, and – since we are unable to identify various defects – we cannot decide whether the carriers are localized on, say, chain ends, or on any other type of defects. On the other hand, please note that the polymerization of diacetylenes, as carried out in our experiments, is a topochemical process occurring in single crystals, where the arrangement of molecules makes creation of such a type of defects hardly possible.

KILIAN:

Could you not characterize the distortions within the crystals by applying x-ray line width or line profile analysis methods in the wide-angle range? Prof. Wilke in Ulm has investigated pTS crystals. It came out that defects or strain-induced broadening is measurable.

SWORAKOWSKI:

I am not aware of any measurements of this type carried out on poly(diacetylenes). It seems therefore that the results mentioned by you might be worth publishing.

ROTH:

I have a question concerning the polymerization of the diacetylene in the crystalline state.

i) Is the polymerization a radical one?

ii) Is it possible to initiate polymerization by laser radiation and to do it in a way that you localized form polymers inside the diacetylene monomer crystal without changing the monomer crystal structure in the surroundings?

SWORAKOWSKI:

i) The thermal solid-state polymerization of diacetylenes is indeed a radical reaction. Details of its kinetics were investigated by several groups; reference to the most important work can be found, for example, in [2].

ii) Yes, it is possible. Such experiments were already performed in Prof. Schwoerer's laboratory in Bayreuth (e.g., Kohler BE, Bauer HD, Kohler BE, Güttler W, Schwoerer M (1986) Chem Phys Letters 125:251–256)

Characterization of network structure by branching theory and correlation with viscoelastic properties

K. Dušek and M. Ilavský

Department of Polymer Physics, Institute of Macromolecular Chemistry, Czechoslovak Academy of Sciences, Prague, Czechoslovakia

Abstract: The theoretical methods of characterization of detailed network structure based on the theory of branching processes and correlation with viscoelastic behavior in main transition and rubbery regions for polyurethane and flexible epoxy-amine networks are reviewed. The theory was amended to calculate the averages of the molecular mass of dangling chains. The comparison with viscoelastic data reveals that the length of both the elastically active network chains (EANC) and dangling chains contribute to the width of the retardation spectrum in the main transition and rubbery zone. When the effect of EANCs is factored out, the increase of the width correlates with the molecular mass of dangling chains.

Key words: Theory of branching processes, elastically active network chains, dangling chains, viscoelasticity, polyurethane networks, epoxy networks

Introduction

In the last few years, it has been firmly established that the viscoelastic properties of polymer networks in the main transition and rubbery region depend on their detailed structure and not only on the number or length of elastically active network chains (EANC) (cf., e.g., [1–6]). Recent development of the branching theories has made it possible to characterize in relatively great detail the network structure in terms of the number, size, and structure of EANC's, dangling chains, junctions, or number of contacts between segments of EANC's [5, 6].

Examination of quantitative correlations between structural parameters of networks and parameters that characterize viscoelastic behavior is the main current issue in this field of investigation. Thus, Vallés and Macosko [7] have found that the viscosity of the pregel polydimethylsiloxane system correlates with the average longest path through the branched molecules, a quantity which is closely related to the mass-average molecular mass of the system. Krakovský et al. [6] found that the width of the main transition region of polyurethane networks correlates well with the average molecular mass of dangling chains. A more detailed investigation of the viscoelastic behavior of these networks revealed that the transition to equilibrium can be separated into two relaxation processes [8]: the faster one is associated with segmental mobility responsible for the glass transition and the other slower one is determined by the mobility changes due to longer range correlations in the structure associated with dangling chains. The finding by Winter and Chambon [9] that, at least for some systems undergoing crosslinking, the storage and loss moduli cross when gelation sets in also points to the sensitivity of viscoelastic behavior to long-range correlations. However, a molecular explanation of this phenomenon is not yet available.

In this article, the method of calculation of the structural parameters of networks is reviewed and expanded and correlations are made with viscoelastic behavior of polyurethane and flexible epoxy-amine networks.

Calculation of structural parameters of networks

Within the framework of the theory of branching processes, branched and crosslinked structures are built-up from structural (sometimes monomer) units differing in the number and type of bonds that bind them to neighboring units. The type of bond is usually specified by the type of units they are connected with. For example, for A- and B-type units, the bonds may be of AA, BB, and AB type. Beyond the gel point one can distinguish between bonds having *infinite* and *finite continuations.* An infinite continuation means that one can proceed via a given bond looking out of the unit to the surface of the sample which means statistically, to infinity. A finite continuation means that the structure looking out of the unit is only finite and is terminated by unreacted functional groups. The gel, which exists beyond the gel point, is considered an infinite structure. Any bond leading to the gel structure has infinite continuation. Within the theory of branching processes, the distributions of building units with respect to the number and type of issuing bonds are described by *probability generating functions* (pgf) (cf. e.g., [5, 10]). For example, for a single component system with one type of bond, this distribution is described by the pgf $F_0(z)$

$$F_0(z) = \sum a_i z^i \tag{1}$$

where z is the auxiliary variable of the pgf and the coefficient in front of z^i, a_i, is the probability of finding a unit with i groups engaged in bonds to neighboring units. For classification of units in the gel, one has to distinguish whether they have a finite or infinite continuation. This is governed by *extinction probability*, which by definition is a probability that, given the existence of a bond, has only finite continuation. The extinction probability, v, is obtained by solving the following equation

$$v = F(z = v) = F(v) \tag{2}$$

where $F(z)$ is the pgf for the number of bonds issuing from a unit in generation $g>0$ to units in generation $g+1$ [5, 10]. The pgf $f(z)$ is obtained from $F_0(z)$ by differentiation.

Knowing v, one can obtain the distribution of units with respect to the number of bonds with finite and infinite continuation, given by the pgf P_0, making the following substitution

$$P_0(z_v, z_{1-v}) \equiv \sum_{j,k} p_{jk}\, z_v^j\, z_{1-v}^k$$

$$= F_0(z = v z_v + (1-v) z_{1-v}) \tag{3}$$

which gives

$$P_0(z_v, z_{1-v}) = \sum_i^f \sum_j^i a_i \binom{i}{j} v^j (1-v)^{i-j} z_v^j\, z_{1-v}^{i-j} \tag{4}$$

$$i - j = k\ ,$$

so that

$$p_{jk} = a_i \frac{i!}{j!\,k!} v^j (1-v)^k\ ,\quad j+k = i \tag{5}$$

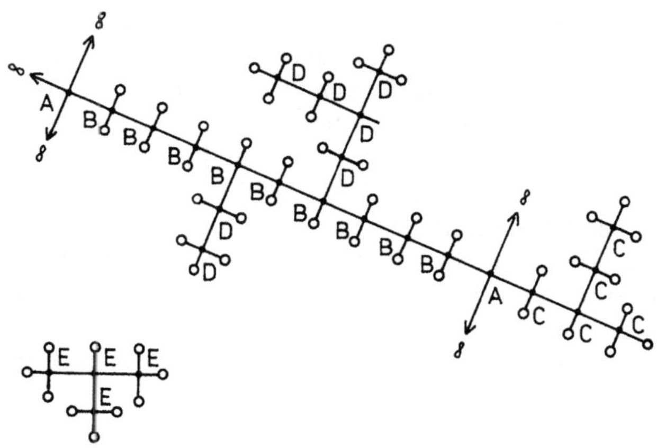

Fig. 1. Schematic representation of a part of the structure of a single-component network. Units: A = elastically active junctions, B = backbone of elastically active chains, C, D = dangling chains, E = sol

where f is the functionality of the monomer. Various structural formation can be classified as follows (cf., also Fig. 1):

Fractions of units	*Structural formation*
p_{j0}	sol
p_{j1}	dangling chains
p_{j2}	backbone of elastically active network chains
$p_{jk}(k \geq 3)$	elastically active crosslinks

Figure 1 shows various structural formations in a part of a network formed from a tetrafunctional monomer. Units A with three and more bonds with infinite continuation represent elastically active junctions, units B form the backbone of elastically active network chains, D and C represent the units in dangling chains and units E are contained in the sol. Arrows show continuation to infinity (connection to the gel structure).

Thus, knowing the pgf F_0, i.e., the set of probabilities (fractions of units), $\{a_i\}$, one can calculate the extinction probability v and the probabilities (fractions) p_{jk} which characterize the network structure.

If the intensity of pairwise interchain contacts is to be calculated, one takes the products of the fractions of the respective units. For example, the interaction between units of EANC's backbone is proportional to $(\sum p_{j2})^2$, interaction between EANC's backbone units and dangling chain units to $(\sum p_{j2})(\sum p_{j1})$, etc.

This approach has been applied by Krakovsky et al. [6] to polyurethane systems based on poly(oxypropylene) triols (POPT) and 4,4'-diisocyanatodiphenylmethane (MDI). The treatment has been fully analogous to the single-component system respecting the fact that only bonds between the OH groups of POPT and NCO groups of MDI can be formed. Also, one prefers to express the amount of material in the respective structural formations in the weight fractions rather than number fractions because of a large difference between the molecular masses of the constituent units.

The pgf for each of the two components equivalent to F_0 [Eq. (1)] read [6]

$$F_{0A}(z_B) = (1-\alpha_A+\alpha_A z_B)^3 \tag{6}$$

$$F_{0B}(z_A) = (1-\alpha_B+\alpha_B z_A)^2 \tag{7}$$

where A and B mean the POPT and MDI components and α_A and α_B are conversions of OH and NCO groups. The subscript at z mean the directions of a bond looking out of the unit under consideration (the bond goes either from A to B or from B to A). When the continuation of the bonds is also considered, the pgf F_{0A} and of F_{0B} go over to

$$\begin{aligned} P_A(z_{Bv}, z_{B(1-v)}) &= \{1-\alpha_A+\alpha_A \\ &\quad \times [v_B z_{Bv}+(1-v_B)z_{B(1-v)}]\}^3 \\ &= \sum p_{jk} z^j_{Bv} z^k_{B(1-v)} \end{aligned} \tag{8}$$

$$\begin{aligned} P_B(z_{Av}, z_{A(1-v)}) &= \{1-\alpha_B+\alpha_B \\ &\quad \times [v_A z_{Av}+(1-v_A)z_{A(1-v)}]\}^2 \\ &= \sum q_{jk} z^j_{Av} z^k_{A(1-v)} \; . \end{aligned} \tag{9}$$

The possible states of units characterized by the fractions p_{jk} and q_{jk} are given in Fig. 2 and a part of the network structure in Fig. 3. The extinction probabilities are again given by relations

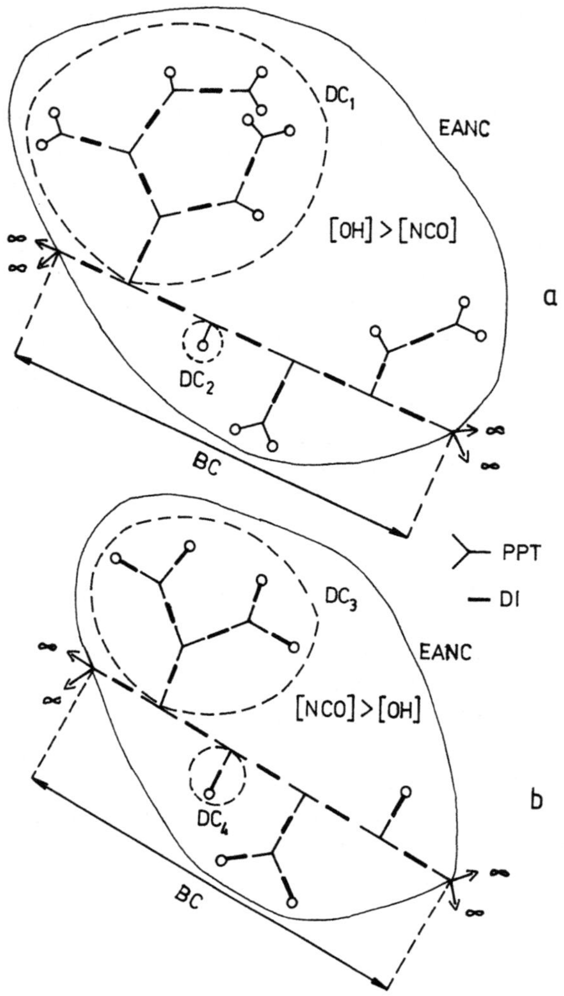

Fig. 3. Structure of a part of a POPT-MDI network with completely reacted minority groups. BC = backbone of an elastically active network chain (EANC), DC = dangling chain

$$v_A = F_A(v_B) \; ; \quad v_B = F_B(v_A) \tag{10}$$

where the pgf F_A and F_B are derived from the pgf F_{0A}

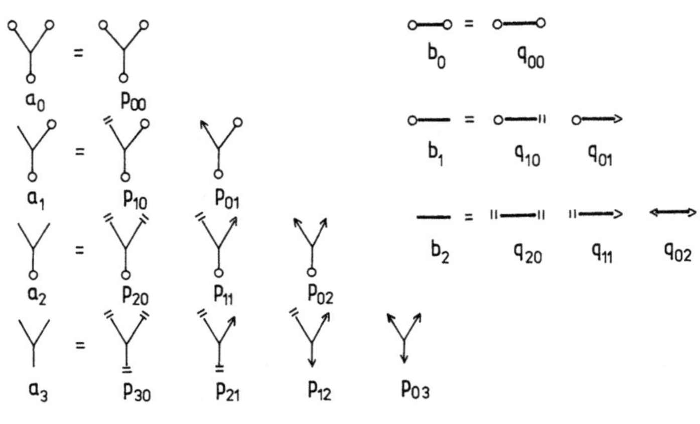

Fig. 2. Fractions of POPT and MDI units with different number of reacted functional groups; ○ is unreacted functionality, ——— is bond, ———|| is bond with finite continuation, → is bond with infinite continuation

and F_{0B} by differentiation with respect to z_B and z_A [6]. Using the above relations, it is easy to derive the weight fractions of the respective structural formations. For calculating the molecular masses of the respective formations, one has to take into account the star shape of the POPT (for details see [6]). The calculated variation of mass fraction of various structural formations is illustrated by Fig. 4.

It may be of interest to characterize the polydispersity of the EANC's or dangling chains. If the backbone EANC's are considered, their degree-of-polymerization distribution is of the most probable type since the EANC backbone is a linear sequence and first-order Markovian statistics are inherent in cascade substitution [5]. The dangling chains are, however, branched and the distribution can be much wider. In the case of POPT and MDI, the dangling chains are composed of units characterized by the following fractions:

$$p_{01}, p_{11}, p_{12}\ ;\quad q_{01}, q_{11}\ .$$

There is only one direction with infinite continuation issuing from each unit and the structure grows in this direction until it hits a unit in the EANC or EAC but an EAC must have all three bonds with infinite continuation (cf. Fig. 5). The distribution of units forming dangling chains with respect to the number and type of issuing bonds can be represented by the pgf D_{0A} and D_{0B}

$$D_{0A}(z) = (p_{01} + p_{11} z_{Bv} + p_{21} z_{Bv}^2) \times z_{B(1-v)}/(p_{01} + p_{11} + p_{21}) \tag{11}$$

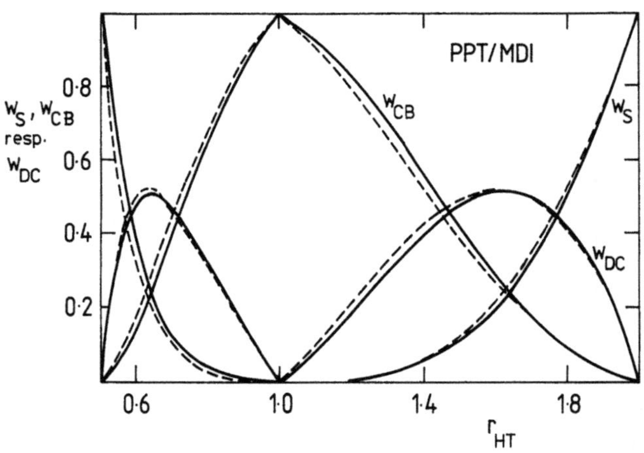

Fig. 4. Dependence of the mass fraction of the sol, w_s, mass fraction of units in EANCs backbone, w_{CB}, and mass fraction of dangling chains, w_{DC}, on the molar ratio of reactive groups, $r_{HT} = [OH]_0/[NCO]_0$, for POPT-MDI networks with completely reacted minority groups

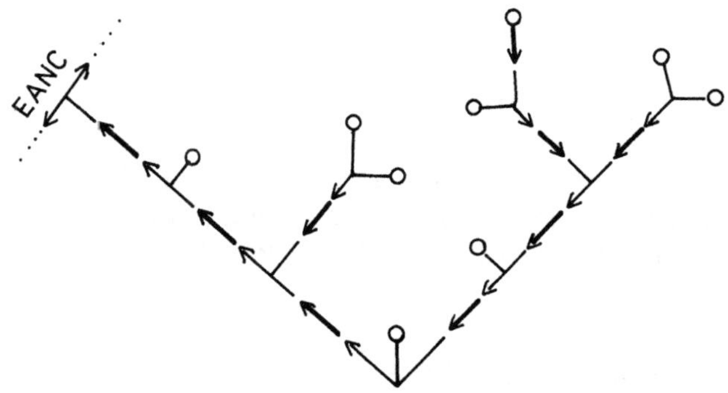

Fig. 5. Way of generation of a dangling chain. → is direction to the gel (infinity); chain growth stops when a unit of EANC is hit

$$D_{0B}(z) = (q_{01} + q_{11} z_{Av})[\Psi_A + (1 - \Psi_A) \times z_{A(1-v)}]/(q_{01} + q_{11})\ . \tag{12}$$

The factor ψ_A in Eq. (12) represents the fact that the dangling chain stops growing in the $1 - v$ direction if it hits a unit in EANC backbone; ψ_A is the probability of such an event and is given by the following relation

$$\Psi_A = p_{12}/(p_{01} + p_{11} + p_{21} + p_{12})\ . \tag{13}$$

For generation of the molecular mass distribution, one also needs the pgf, D, for the number and type of bonds issuing from units in generations $g > 0$. Here one has to distinguish between bonds with finite or infinite continuation by which the unit is rooted (looking out of the unit to generation $g - 1$)

$$D_{Av}(z) = (p_{11} + 2p_{21} z_{Bv}) z_{B(1-v)}/(p_{11} + 2p_{21}) \tag{14}$$

$$D_{Bv}(z) = z_{A(1-v)} \tag{15}$$

$$D_{A(1-v)} = (p_{01} + p_{11} z_{Bv} + p_{21} z_{B(1-v)})/(p_{01} + p_{11} + p_{21}) \tag{16}$$

$$D_{B(1-v)} = (q_{01} + q_{11} z_{Av})/(q_{01} + q_{11})\ . \tag{17}$$

The weight-fraction generating function, which is suitable for obtaining molecular weight averages, $W(z)$, is as follows

$$W(z) = m_A W_A(z) + m_B W_B(z) \tag{18}$$

where m_A and m_B are mass fractions and the components of W read

$$W_A(z) = z_A^{M_A} D_{0A}(u_{Bv}, u_{B(1-v)}) \tag{19}$$

$$u_{Bv} = z_{Bv}^{M_B} D_{Bv}(u_{A(1-v)}) \quad (20)$$

$$u_{B(1-v)} = z_{B(1-v)}^{M_B} D_{B(1-v)}(u_{Av}) \quad (21)$$

and similar expressions are valid for W_B. The mass-average molecular mass of dangling chains, $(M_{DC})_w$, is obtained by differentiation of W with respect to the variables z_k

$$(M_{DC})_w = \sum_k [\partial W(z)/\partial z_k]_{z=1} \; . \quad (22)$$

The results of calculation of the mass average and polydispersity are not displayed here, but it can be shown that the polydispersity of dangling chains expressed through $(M_{DC})_w/(M_{DC})_n$ diverges at the gel point while both the mass and number averages diverge, too. However, with respect to the relaxation behavior the EANCs, dangling chains are not expected to differ much since in the vicinity of gel point they both are very long.

Experimental part

Preparation of samples

Polyurethane networks were prepared from two samples of poly(oxypropylene) triols (POPT) (Union Carbide Niax LHT-240, $M_n = 710$, number-average functionality $f_n = 2.89$, and Niax LG-56, $M_n = 2630$, $f_n = 2.78$) and 4,4'-diisocyanatodiphenylmethane (MDI). The samples were prepared in bulk using different initial molar ratios of reactive groups, $r_{HT} = [OH]_{POPT}/[NCO]_{MDI} = 0.6-1.7$. When NCO groups were in excess, $r_{HT} > 1$, monofunctional cyclohexanol was added in an amount to keep the molar ratio [OH]/[NCO] constant and equal to 1. This was done because of additional crosslinking due to side reactions occurring when NCO groups were in excess. The reaction was catalyzed by dibutyltin dilaurate and was carried out to the highest possible conversion of the minority groups [11,12].

Epoxy networks were prepared from poly(oxypropylene)diamine or triamine: Jeffamine D-400: amine equivalent $\gamma_H = 106$ g/mol H of NH_2, $M_n = 397$, $f_n = 3.75$ and Jeffamine T-403: $\gamma_H = 78$ g/mol, $M_n = 427$, $f_n = 5.46$. The molecular mass of diglycidyl ether of Bisphenol *A* (DGEBA) was 350 and $f_n = 2$. The composition of the networks varied in molar ratio $r_A = 2[NH_2]_0/[E]_0$ between 1 and 3.5. The reaction took place at 90 °C for 16 h [13].

Measurements of viscoelastic properties

The time dependence of creep compliance, $J(t)$, was measured at several temperatures in the main transition region [2]. After performing the time-temperature superposition, the retardation spectra, $L(\tau)$, were calculated. From the maxima of the spectra, the half-width was determined at the half height of the maximum on the spectrum, L_m, i.e., $h = \log \tau_1 - \log \tau_2$, where τ_1 and τ_2 are retardation time for which it holds that $L(\tau_1) = L(\tau_2) = L_m/2$.

The measurement of the complex dynamic compliance $J^*(\omega) = J' - iJ''$ (J' is the storage and J'' the loss compliance in the main transition region) was performed using instrumentation described earlier [8]. Using the superposition frequency ω – temperature, the dependences of compliance components J' and J'' were superimposed (cf., Fig. 6). In all cases, it was possible to separate the frequency dependences of compliance into two relaxation processes of the functional form [8]

$$J_i^* = \Delta J_i/[1 + (i\omega\tau_i)^{a_i}] + J_\infty \; , \quad (i = 1,2) \quad (23)$$

where Δj_i is the relaxation strength, $\tau_i = 1/\omega_m^i$ is the characteristic relaxation time, a_i is a factor characterizing the width of the relaxation process and J_∞ is the equilibrium contribution to compliance.

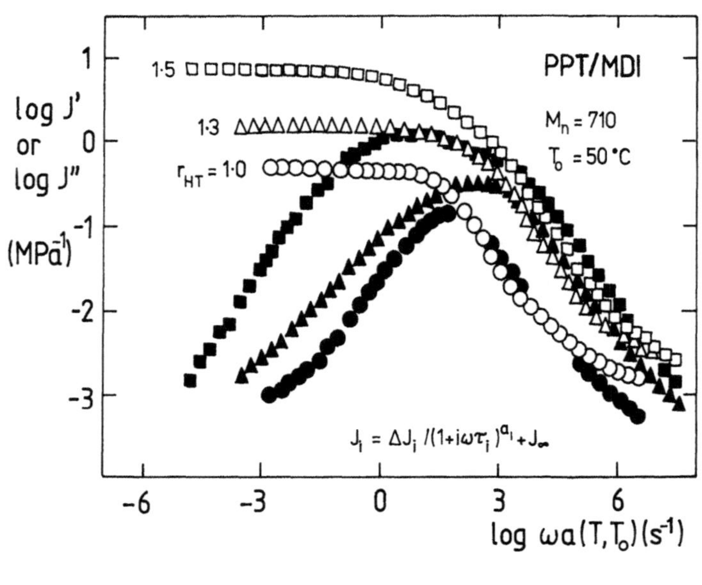

Fig. 6. Example of the dependence of superimposed components of dynamic compliance J^* on reduced frequency $\omega a(T, T_0)$

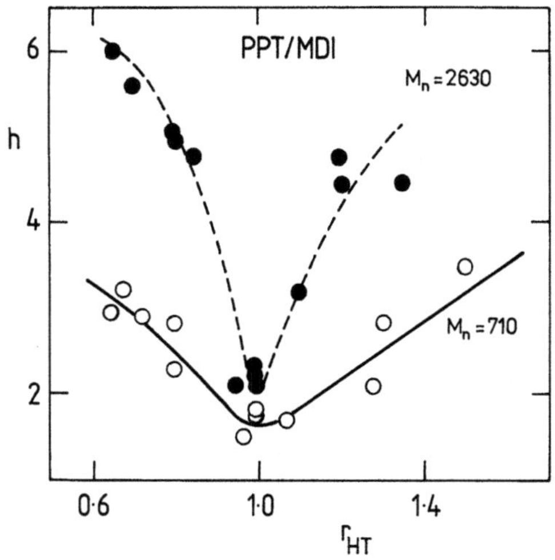

Fig. 7. Dependence of the retardation spectrum half-width, h, on the molar ratio r_{HT}

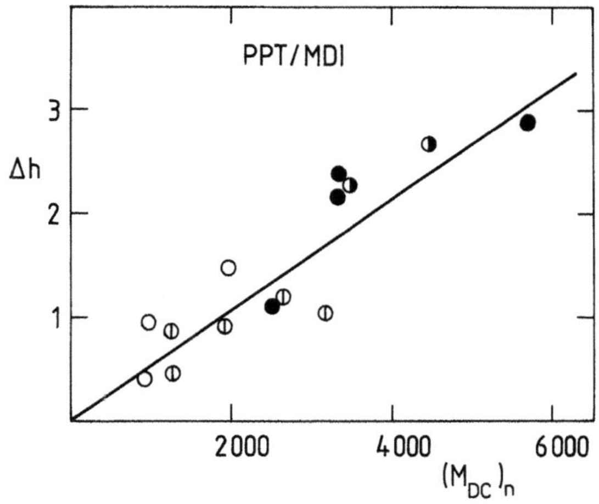

Fig. 9. Dependence of Δh on the number-average molecular mass of dangling chains $(M_{DC})_n$. ◑, ◐ $r_{HT}<1$, ○, ● $r_{HT} \geq 1$

Results and discussion

The effect of the varying structure of the network on relaxation in the main transition region can be characterized by the half-width of the retardation spectrum, h. From Fig. 7, it is evident that the smallest value of h was found for polyurethane networks prepared at stoichiometric ratio of functional groups and having the shortest elastically active network chains (EANC) which corresponds to $r_{HT} = 1$. Networks prepared from the higher molecular mass POPT exhibit higher values of h than those prepared from the smaller POPT. With increasing imperfectness of the network caused by deviation of r_{HT} from 1, the spectra width increases. This increase can be due to the growth of the length of EANCs and increasing number and length of dangling chains as well.

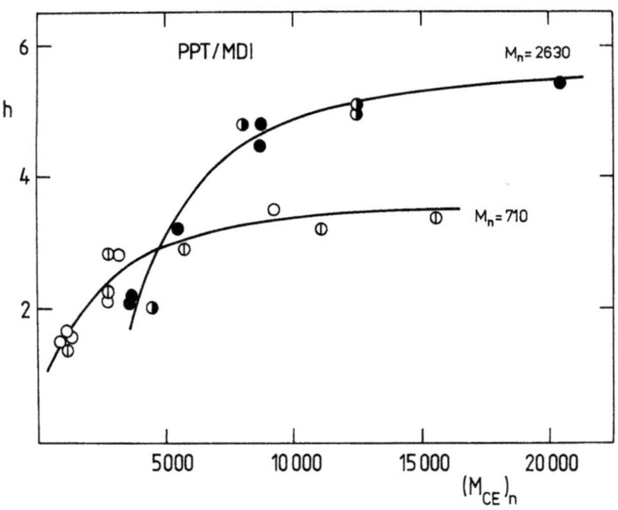

Fig. 8. Dependence of the retardation spectrum half-width, h, on the number-average molecular mass of EANCs, $(M_{CE})_n$. ◑, ◐ $r_{HT}<1$; ○, ● $r_{HT} \geq 1$

To distinguish between the effects of these two varying structural factors, h was plotted against molecular mass of EANCs (Fig. 8). If the molecular mass of an EANC including the dangling side chains, $(M_{CE})_n$ is used as a scaling factor, the dependences for the triol are different but independent of whether the imperfections were introduced by addition of a monofunctional component or excess of OH groups. A similar conclusion was obtained if h was plotted against the molecular mass of the backbone EANC, $(M_{BE})_n$. However, the difference $\Delta h = h - h_x$ (h_x is the value of h for $r_{HT} = 1$) can be correlated with the number-average molecular mass of dangling chains, $(M_{DC})_n$ (Fig. 9).

By analysis of the retardation spectra it has been found [8] that the relaxation to equilibrium can be separated into two processes [Eq. (23)]. Figure 10 shows that the relaxation strength of the faster process (glassy process), ΔJ_g, is very little affected by the value of r_{HT}. Figure 11 demonstrates that ΔJ_g is mainly determined by the content of MDI in the sample. On the other hand, the dependence of the relaxation strength of the slower relaxation process, ΔJ_d, on r_{HT} is similar to that of h on r_{HT} (cf., Figs. 10 and 7). Therefore, the existence of the slower relaxation process is to be assigned to the dangling chains. The magnitude of the increase in the relaxation strength $(\Delta J_d - \Delta J_d^{r_{HT}=1})$ correlates well with the increase in the molecular mass of the dangling chains $(M_{DC})_n$ (Fig. 12).

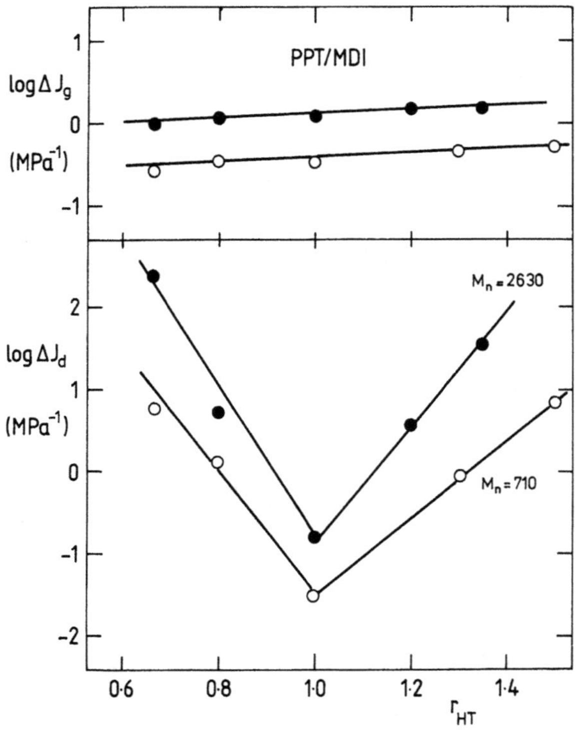

Fig. 10. Dependence of the relaxation strength of the faster ΔJ_g and slower ΔJ_d process on r_{HT}

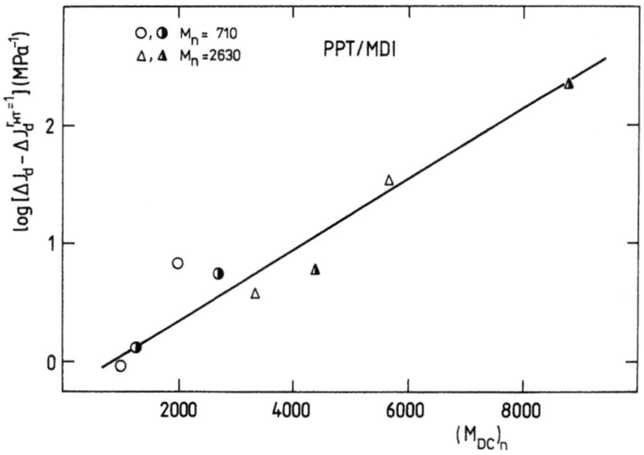

Fig. 12. Dependence of the difference of the relaxation strength $\Delta J_d - \Delta J_d^{r_{HT}=1}$ on the number-average molecular mass of dangling chains $(M_{DC})_n$. ◑, ▲ $r_{HT}<1$, ○, △ $r_{HT} \geq 1$

Correlations of the same type between the viscoelastic behavior and detailed network structure are valid also for epoxy networks prepared from poly(oxypropylene) polyamines and DGEBA (Figs. 13 and 14). Also in these systems, the half-width of the retardation spectra h increases with increasing deviation from stoichiometry and the increase in h with molecular mass of the dangling chains $(M_{DC})_n$ is the same for networks prepared from both Jeffamine D-400 and T-403 (Fig. 14). Therefore, the same conclusion can be made: the increase in the width of the retardation spectra is determined by the increase in the molecular mass of dangling chains.

It can be concluded that the viscoelastic behavior of polymer networks is determined by the detailed network structure and that the speed of approach to equilibrium is mainly determined by the length and amount of dangling chains. Additional ways of changing network structure, such as dilution during network formation, or extraction after network formation, etc., can say more about factors determining the viscoelasticity of networks.

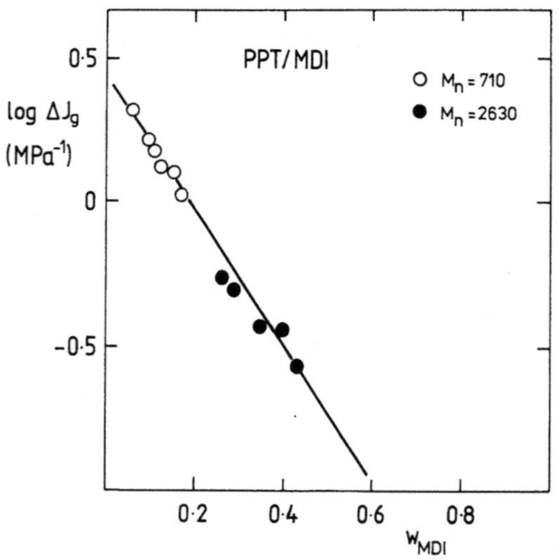

Fig. 11. Dependence of the relaxation strength of the ΔJ_g (glassy) process on the mass fraction of MDI in POPT-MDI networks

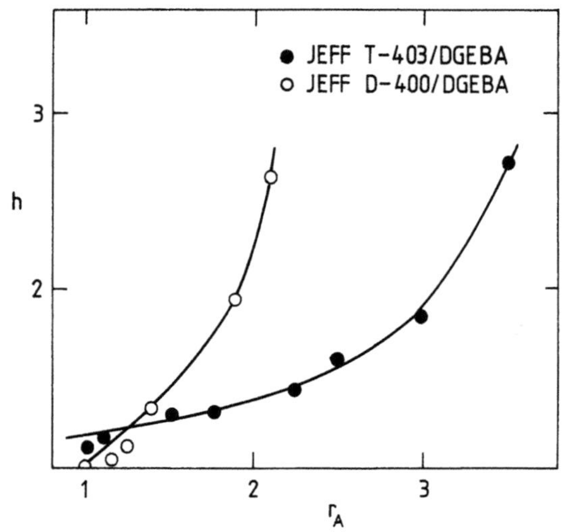

Fig. 13. Dependence of the half-width, h, of the retardation spectrum on the molar ratio $r_A = 2[NH_2]_0/[E]_0$ for DGEBA-Jeffamine networks

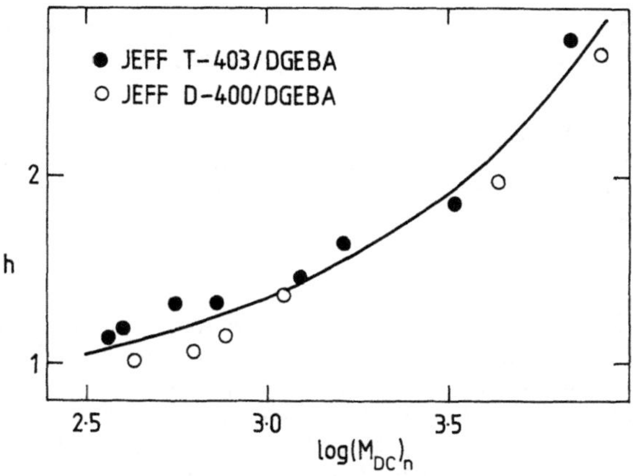

Fig. 14. Dependence of the half-width of the retardation spectrum on the number-average molecular mass of dangling chains $(M_{RC})_n$ for DGEBA-Jeffamine networks

References

1. Ferry JD (1980) Viscoelastic Properties of Polymers, 3rd ed. Wiley, New York
2. Havránek J, Nedbal J, Berčík Č, Ilavský M, Dušek K (1980) Polym Bull 3:497
3. Bibbó MA, Vallés EM (1984) Macromolecules 12:521
4. Curro JG, Pearson DS, Helfand E (1985) Macromolecules 18:1985
5. Dušek K (1984) Macromolecules 17:716
6. Krakovský I, Havránek A, Ilavský M, Dušek K (1988) Coll Polym Sci 324:266
7. Vallés EM, Macosko CW (1979) Macromolecules 17:360
8. Havránek A, Ilavský M, Nedbal J, Böhm M, von Soden W, Stoll B (1987) Coll Polym Sci 265:8
9. Winter HH, Chambon F (1986) J Rheol 30:367
10. Dušek K (1986) Adv Polym Sci 78:1
11. Ilavský M, Dušek K (1983) Polymer 24:981
12. Ilavský M, Dušek K (1982) Polym Bull 8:359
13. Dušek K, Ilavský M, Štokrová S, Matějka L, Luňák S (1987) In: Sedláček B, Kahovec J (eds): Cross-Linked Epoxies. Walter de Gruyter, Hamburg, p 269

Received February 2, 1989;
accepted February 27, 1989

Authors' address:

Cor. Member of Czechosl. Acad. Sci.
K. Dušek
Institute of Macromolecular Chemistry
Czechoslovak Academy of Sciences
16206 Prague 6, Czechoslovakia

Discussion

KILIAN:

i) Could you explain the definition of trapped entanglements? How do they behave at large strains?

ii) What is the effect of clusters on the equilibrium modulus?

DUŠEK:

i) A trapped entanglement is an entanglement between elastically active network chains (EANC) (a topological constraint) that cannot become disentangled (released) upon any kind of deformation. These trapped entanglements contribute to the equilibrium elastic response of the network, but their contribution differs from that of chemical crosslinks. With increasing deformation, these entanglements may slide along EANCs depending on the strain geometry. The stress-strain dependence for networks with trapped entanglements may be described, for example, by the slip-link model.

ii) The clusters that were mentioned in the lecture were so-called chemical clusters. These are represented by several chemically bonded units differing by their structure and properties from the surrounding structure. Hard segments in polyurethanes are one of the examples. If these hard segments contain units of functionality higher than two, they are three-dimensional and are called hard chemical clusters (cf. K. Dušek (1982) Rubber Chem Technol 55:1). The effect of hard chemical clusters on equilibrium modulus can be two-fold: 1) In networks with homogeneously distributed branch points (crosslinks), the number of elastically active network chains (EANC) is calculated from the number of elastically active crosslinks (EAC). If the EACs are buried inside the cluster, they do not contribute individually to the stress, but the whole cluster acts as a polyfunctional crosslink. As a result, the equilibrium response of the network with clusters is less than that of a homogeneous network with the same number of EACs. 2) If the clusters are large, they may exhibit a filler effect on modulus characteristic for particulate composites.

WARTEWIG:

I wonder why the slower process is determined by dangling chains?

DUŠEK:

After deformation, the mean positions of elastically active crosslinks (EAC) deform affinely with macroscopic strain. Also the instantaneous displacement of the segments is affine. The elastically active network chains (EANC) relax faster than dangling chains because they are tied to EACs at both ends. The dangling chains are only singly tied to a segment of EANC. The conformation of a dangling chain in space thus depends on the position of the segment of EANC it is tied to. Experiments show that networks with dangling chains relax more slowly than networks without dangling chains and of the same concentration of EANCs.

GOTLIB:

i) Do you have a possibility to compare the mechanical response prescribed to dangling chains with mechanical response of a model system containing nonlinked systems with the same molecular weight as that of the dangling chains?

ii) What is your opinion about the possible role of physical junctions – or long living entanglements – in the mechanical response, especially for very long dangling chains, because their dimensions may be greater than those between physical junctions?

DUŠEK:

i) Unfortunately, no molecular theory of relaxation of imperfect network systems is available at present, neither are clear-cut experiments concerning the effect of the sol fraction. The sol fraction usually makes the relaxation faster because the size of the sol molecules is usually small. At the gel point, however, the sizes of EANCs, dangling chains and sol molecules diverge.

ii) Relaxation is affected by physical interactions (entanglements) between different elements of the network structure – EANCs, dangling chains, and sol. Therefore, dangling chains may become entangled with themselves, if they are sufficiently long, or with EANCs, or sol. However these entanglements cannot become entrapped.

HOLZMÜLLER:

Is it possible to gain network structures by irradiation with radioactive materials during the time of free radical polymerization in an effective way?

DUŠEK:

Ionizing radiation is frequently used for preparation of crosslinked materials. Radiation either induces free-radical polymerization of unsaturated components or causes crosslinking of existing polymer chains (radiation vulcanization). Electron beam or gamma ray crosslinking is most frequently used, the former also on an industrial scale. The resulting networks are usually not perfect because of simultaneously occurring chain scission.

Dispersive and non-dispersive transport processes in polymers

H. Bässler

Fachbereich Physikalische Chemie, Philipps-Universität, Marburg, FRG

Abstract: Based on Monte Carlo simulation techniques and analytic theory the influence of positional and energetic disorder as well as of multiple trapping on charge carrier motion in a random solid is discussed. While under realistic conditions positional disorder will not give rise to dispersive behavior on the time-scale of a carrier transit, both E-hopping and multiple trapping do so because of energetic relaxation of the carriers within the density of states (DOS). For a Gaussian DOS of width σ dynamic equilibrium can be attained under realistic experimental conditions resulting in a non-Arrhenius-type temperature dependence of the mobility. Representative recent literature results on dispersive and non-dispersive transport in polymeric systems are discussed in relation to the various disorder cases. Finally, the temperature dependence of the viscosity of supercooled melts is shown to be tractable in terms of the concept developed for the diffusion and relaxation of excitations in a Gaussian DOS.

Key words: dispersive transport, photoconductivity, polymers, random media, viscosity

1. Introduction

Depending on the type of the substituent or dopant pendant group, polymers or molecularly doped polymers are reasonably good photoconductors [1,2]. However, the velocity a charge carrier acquires in an electric field is usually orders of magnitude lower than it is in molecular crystals. Moreover, time of flight studies indicate that its drift velocity and, concomitantly, its transit time across a sample is often not a single-valued quantity but subject to a distribution that evolves with time. Transport is thus a stochastic process reflecting the disorder of the sample. The purpose of this paper is to give an overview on the various types of disorder-controlled transport, both time dependent (dispersive) and time independent. In doing so, I draw heavily on Monte Carlo computer simulations of well-characterized model systems, as well as on analytic theories. Representative experimental results will be discussed. Finally it will briefly be shown that the transport model developed for the case of intrinsic energetic disorder is not restricted to the motion of electronic excitations but can be applied to treat also mass motion involved, for instance, in viscous flow in supercooled melts.

2. Model consideration

In a crystalline, yet incoherent transport system the probability density that a charge carrier leaves a site after time t is an exponential function of time. This is no longer true in a random medium because disorder converts the transition rate among the hopping sites into a statistical quantity accounting for the fact that there are sites that are depopulated faster or slower than average. A useful function for quantifying this property is the distribution of dwell times carriers spend on the hopping sites (waiting time distribution). Scher and Montroll (SM) [3] proposed a power law

$$\psi(t) \propto t^{-(1+\alpha)} , \tag{1}$$

$0<\alpha<1$ being the so-called dispersion parameter. Employing the continuous random walk formalism, SM demonstrated that it translates into a power law decay of a photocarrier transit pulse recorded in course of a time of flight (TOF) experiment

$$j(t) \propto \begin{cases} t^{-(1-\alpha)} & t<\tau_{tr} \\ t^{-(1+\alpha)} & t>\tau_{tr} \end{cases} . \tag{2}$$

The τ_{tr} is the transit time of the fastest carriers through the sample. It varies with electric field E and sample thickness L as

$$\tau_{tr} \propto L^{1/\alpha} , \quad \tau_{tr} \propto E^{-1/\alpha} . \tag{3}$$

Although originally developed for the case of positional (geometric) disorder of the hopping sites, the SM treatment is a quite general one yet its consistency with various disorder concepts remained to be shown.

In the following sections a different route will be followed. Employing Monte Carlo (MC) simulation techniques as well as analytic theory it will be examined how a specific type of disorder – positional or energetic disorder of the hopping sites or a distribution of trapping levels – manifests itself in the electronic properties of a sample of macroscopic dimension. MC simulation can be considered as an idealized experiment, done on a clean system with controllable properties and adjustable degree of sophistication.

The type of disorder always present in a random system is the frozen-in positional fluctuation of the hopping sites reflecting the statistics in the molecular environment (R-disorder or geometrical disorder) [5]. Because the site energy contains a contribution from the interaction of the molecule with the surrounding molecules – the polarization energy of a charge carrier or the van der Waals interaction energy of an electronically excited molecule (the solvent shift energy) [4] – positional disorder causes fluctuation of the inter-site transition rates ("off-diagonal disorder") and inevitably translates into disorder of the site energies of the bulk molecules as well (E-disorder). Depending on the magnitude of the energy fluctuation it causes the valence and conduction bands as well as the exciton band of an organic crystal to split into a distribution of localized states (DOS) when passing to a random (glassy) structure [6]. Since the site energies depend on a large number of internal coordinates, each varying randomly by small amounts, a Gaussian is the distribution function of choice for the DOS at least in weakly bound molecular solids. Direct experimental evidence comes from the Gaussian shape of inhomogeneously broadened absorption profiles of evaporated organic layers [7] and of chromophores embedded in glasses [8]. In addition to these intrinsic type of disorder, real world systems often contain traps that, by virtue of the above argument, are also distributed in energy giving rise to a distribution of trapping and detrapping times sampled by a charge carrier in course of its way across the sample (multiple trapping (MT) case).

To model disorder in a MC computer experiment, the Marburg group used a cubic lattice of $50 \times 50 \times 70$ sites (lattice constant a) with periodic boundary conditions, the site energies being chosen from a Gaussian distribution of width σ (fwhm $= 2\sqrt{2 \ln 2}\sigma$). To mimic pure positional disorder, σ was set at zero and the distribution of jump distances was generated by replacing up to 90% of all sites by sites not participating in the conduction process [9]. A charge carrier was generated at an arbitrary site and allowed to execute a random walk under the action of an applied electric potential requiring that the jump rate between two sites i and j be

$$\nu_{ij} = \nu_0 \exp -(2\gamma \Delta R_{ij}) \cdot \exp\left(-\frac{\Delta\varepsilon_{ij}}{kT}\right) \tag{4}$$

where γ is the inverse wavefunction localization radius and $\Delta\varepsilon_{ij}$ is the difference in site energies including the electric field energy. For jumps down in energy $\Delta\varepsilon_{ij}$ was set at zero. For details the reader is referred to the original work [10]. Simulated quantities include the ensemble averaged displacement of a carrier as a function of time, its mean square displacement as a measure of its diffusivity, its mean energy as a function of time and the number of sites it has visited after a time t.

An analytic approach able to cover both R-hopping and the more complicated case of E-hopping, where exchange rates among sites are no longer symmetric, has been developed by the Movaghar group. It is based on a generalized kinetic master equation in conjunction with the effective medium concept (EMA) [11–14].

3. Behavior of the model systems

3.1 The R-hopping case

Figure 1 shows a family of photocurrent transients simulated for an array of iso-energetic hopping sites as a function of the fractional concentration c of the hopping sites. c decreases as the profiles acquire dispersive features. Importantly, however, transition to non-dispersive (Gaussian) transport occurs after a time not exceeding a few percent of the carrier transit time. This concurs with the result of Marshall's [15] MC simulation performed on an undiluted geometrically disordered network of hopping sites. Both the demarcation time separation the two transport regimes and the dependence of the dispersion parameter on $\gamma\langle R\rangle$, $\langle R\rangle$ being the mean intersite distance ($\langle R\rangle = ac^{-1/3}$) are in excellent agreement with the

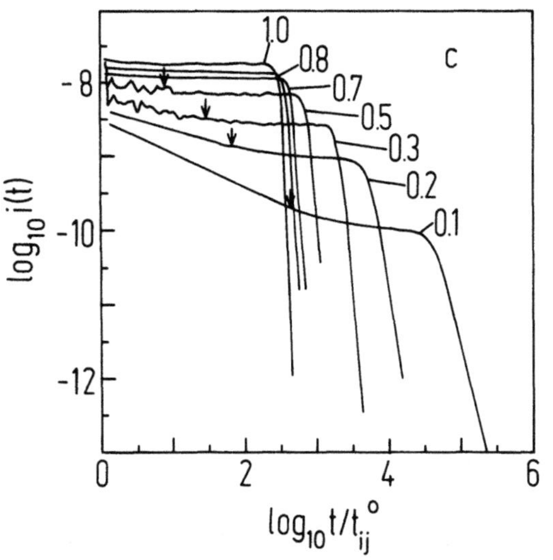

Fig. 1. Time of flight signals simulated on a spatially random system, the randomness being generated by sample dilution. c is the concentration of hopping sites, t_{ij}^0 is the dwell time of a carrier on a site in the undiluted lattice. Simulation parameters were $a = 7$ Å, $2\gamma a = 5$, $E = 10^6$ V cm^{-1}, $T = 400$ K. To exemplify the statistical error computer printouts are shown for $c = 0.5$ and $c = 0.3$ (from [9])

prediction of analytic theory developed for an amorphous network of sites [14]. Not only does this demonstrate the consistency of both approaches, in particular the approximations implied by the analytic treatment, it also proves that under realistic conditions dispersion has vanished after the carrier packet has moved across a few lattice planes only. The important message is that in a sample, characterized by hopping sites subject to geometrical disorder on a molecular level, dispersive transport will not be detectable on a macroscopic time scale.

The situation is different for fractal systems. Recently, Blumen et al. [16] applied the continuous time random walk (CTRW) concept to fractals presuming that the temporal disorder can be accounted for by a waiting time distribution as given by Eq. (1). This approach precludes recovering the transition from dispersive to non-dispersive transport. One can, however, surmise that self-similar sample topology leads to an extension of the dispersive transport regime as compared to systems that are disordered on a molecular level only. This is a field of promising future work.

3.2 The intrinsic E-hopping case

Consider a charge carrier started at an arbitrary energy within a Gaussian DOS,

$$g(\varepsilon) = (2\pi\sigma^2)^{-1/2} \exp\left[-(\varepsilon - \varepsilon_0)^2/2\sigma^2\right] , \tag{5a}$$

or an exponential DOS,

$$g(\varepsilon) = (kT_0)^{-1} \exp(-\varepsilon/kT_0) . \tag{5b}$$

While executing a random walk it will relax energetically within the DOS, presumed to remain static on the time scale of the motion which must slow down with time because fewer acceptor sites become available for a hopping carrier. Straightforward arithmetic indicates that within a Gaussian DOS, dynamic equilibrium will be attained after a relaxation time t_{rel} when the carrier has relaxed to a mean energy

$$\langle \varepsilon_\infty \rangle = -\sigma^2/kT \tag{6}$$

relative to the center of the DOS [12]. In case of an exponential DOS, equilibrium will never be attained [18] unless the DOS exhibits a cut-off.

MC, as well as analytic work [17], yields the energetic behavior portrayed in Fig. 2 for a Gaussian DOS. At short times, $\varepsilon(t)$ decays approximately according to $\Delta\varepsilon \propto \ln(t/t_0)$ law, t_0 being the jump time of a carrier within an array of isoenergetic hopping sites, to finally settle at $\langle \varepsilon_\infty \rangle$ given by Eq. (6). The

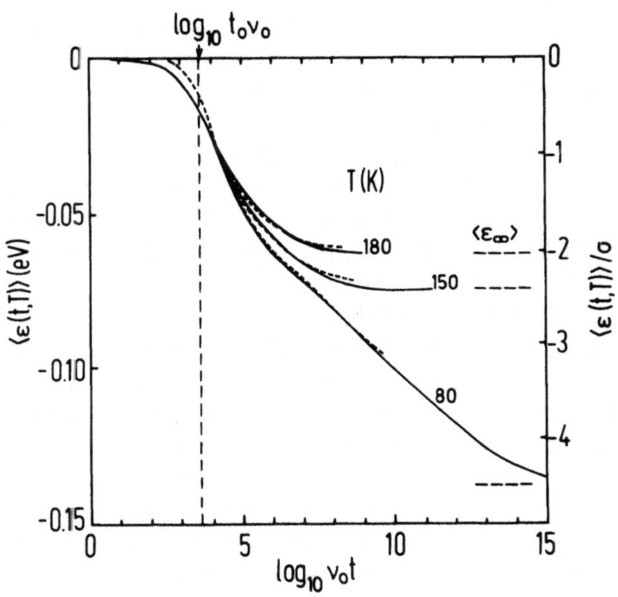

Fig. 2. Time dependence of the mean energy of an ensemble of excitations generated randomly within a Gaussian DOS of width $\sigma = 31$ meV, equivalent to 250 cm^{-1}. Parameter is the temperature. Solid curves and dashed curves represent analytic and simulation results, respectively. The theoretical $\langle \varepsilon_\infty \rangle$ (see Eq. (6) of the text) is indicated. $t_0 = \nu_0^{-1}$ is the dwell time of an excitation in an energetically ordered system (from [17])

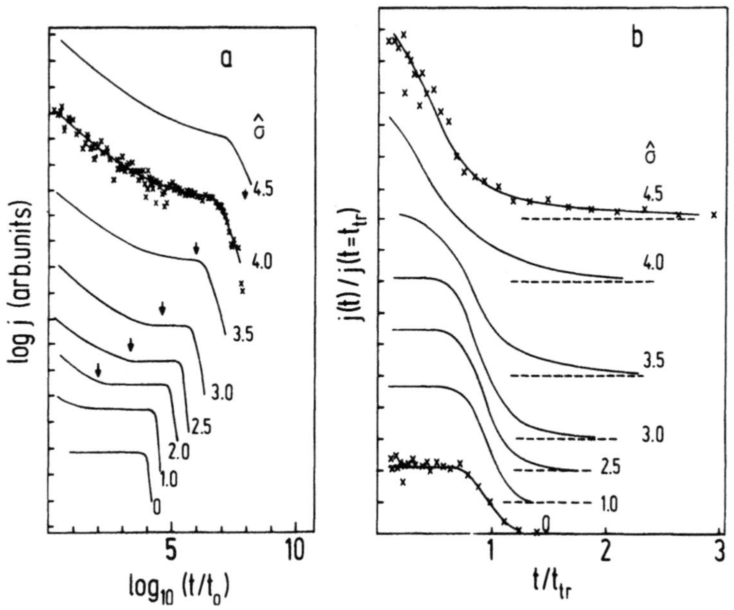

Fig. 3. Time of flight signals simulated for carrier hopping in a Gaussian DOS for variable disorder parameter $\hat{\sigma} = \sigma/kT$. The sample length was 300 sites, $E = 10^5$ V cm^{-1}

concomitant decrease of the carrier diffusivity $D(t)$ as a function of both time and disorder parameter σ/kT, characterizing the degree of intrinsic energetic disorder, is reflected in the decay of photocurrent transients shown in Fig. 3. Obviously, carrier transport is dispersive within a considerable time domain. Remarkably, the transition from dispersive to non-dispersive transport seen in experiment [19] can be accounted for in terms of variable σ/kT only. $j(t)$ and, concomitantly, $D(t)$ cannot, however, be expressed in terms of a simple power law [Eq. (2)] unless a time-dependent dispersion parameter is introduced, indicating that the SM ansatz for the waiting time distribution [Eq. (1)] represents a first order approximation to the E hopping case only.

Unfortunately, neither the evolution of its mean energy nor of the diffusivity of a packet of charge carriers is amenable to direct experimental probing because i) both intrinsic and extrinsic charge carrier generation is a two step process involving a random walk within a coulombic potential that masks the initial relaxation process within the DOS and ii) there is no optical technique to monitor the absolute energy of charge carriers moving within a manifold of states. However, a suitable substitutional probe for the dynamic behavior of an excitation migrating within a random potential is a triplet excitation. Recall that triplets move via exchange interaction as charge carriers do, suggesting that charge carrier and triplet excitation transport is subject to a scaling behavior. Following this reasoning the Marburg group probed the energetic relaxation of triplet excitations migrating in a molecular benzophenone glass as well as its time dependent diffusivity over four decades in time, employing time-resolved phosphorescence techniques [20–22]. Excellent agreement with theory has been demonstrated. The only adjustable parameter required for data fitting was the width of the DOS which came out in good agreement with the widths of inhomogeneously broadened phosphorescence band profiles.

Owing to their limited lifetime, triplet excitations are unsuitable for probing the asymptotic behavior of the random walk of a particle diffusing in a random potential. This is, however, of relevance for carrier transport across samples of macroscopic dimensions as illustrated in Fig. 4. The relaxation time t_{rel} follows an exponential dependence on $(\sigma/kT)^2$, as does the number of new sites ($\hat{S}_{rel}$) a carrier has to visit in order to attain dynamic equilibrium. The $\hat{S}_{rel}$ can be compared to the number of new sites ($\hat{S}_t$) a carrier samples on its way across a sample of length L. Values for $\hat{S}_t$ calculated for an electric field of 10^5 V cm^{-1} and assuming validity of the Einstein relation have been included in Fig. 4. Because of the deviations from the Einstein relation (see below), $\hat{S}_t$ values represent lower bounds only. We learn from Fig. 4 that for $\sigma/kT = 4$, which is a representative value for polymeric photoconductors, the critical sample length for observing a transition from dispersive to non-dispersive transport at 300 K is about 2 μm. Considering that transit time measurements on polymeric systems

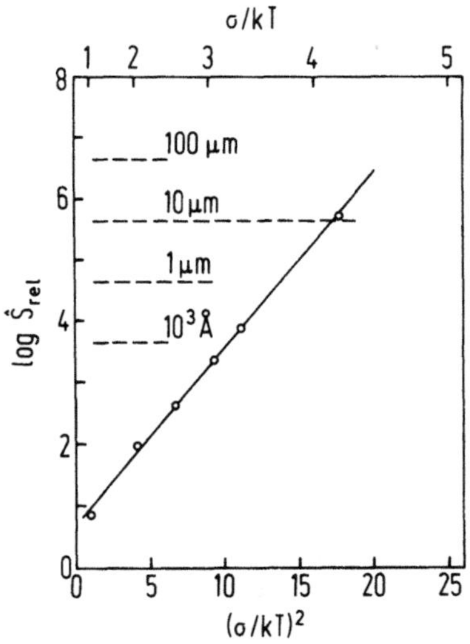

Fig. 4. Monte Carlo data for the number of new sites $\hat{S}_{rel}$ a carrier, started at random within a Gaussian DOS of variable σ/kT, has to visit in order to relax to dynamic equilibrium. To allow estimating the critical value σ/kT for which TOF signals, recorded with a sample of a given length L, change from dispersive to non-dispersive, the number of new sites ($\hat{S}_t$) a carrier visits on its way across the sample is indicated by dashed lines parametric in the sample length L

are usually performed with samples about 20-µm-thick, the case of non-dispersive transport is clearly of relevance for systems devoid of deep traps.

In the non-dispersive transport regime, carrier motion should be thermally activated. If transport were to proceed via sites located at the center of the Gaussian DOS, the expected activation energy were $\langle \varepsilon_\infty \rangle = -\sigma^2/kT$, i.e., increasing with decreasing temperature and yielding a $\mu \propto \exp[-(\sigma/kT)^2]$ behavior of the carrier mobility μ. Except for a scaling factor of about 2/3 in the exponent that arises from the statistics of the activation energies both MC simulation and EMA treatment confirm the validity of this super-Arrhenius type of temperature dependence (see Fig. 5), which, by the way, has also been derived for spin glass relaxation [23, 24]. The essential message is that in the non-dispersive transport regime the charge carrier mobility should obey a $\mu \propto \mu_0 \exp[-(T_0/T)^2]$ law, kT_0 being the equivalent to an activation energy in conventional activated transport and related to the width of the DOS via $kT_0 \simeq (2/3)\sigma$. Without quantitative analysis it should be mentioned that μ must be field dependent at large electric fields, because the field reduces the effective width of the DOS [25].

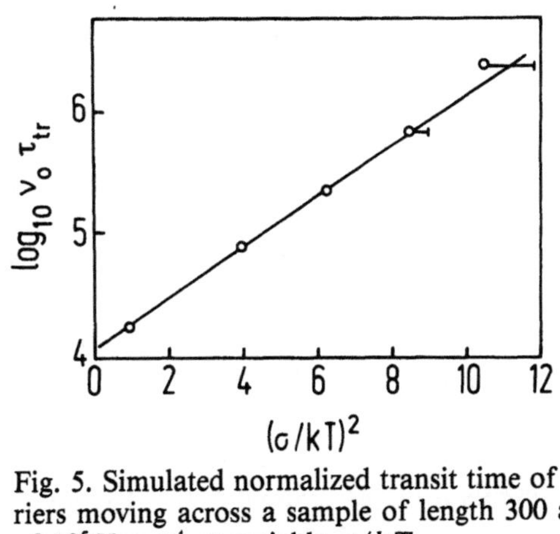

Fig. 5. Simulated normalized transit time of a sheet of carriers moving across a sample of length 300 a under a field of 10^5 V cm^{-1} at variable σ/kT

To conclude this section simulated transit pulse shapes for the case of an exponential DOS are presented in Fig. 6 [26]. Owing to the absence of dynamic equilibrium no steady transport regime is attained and the photocurrent follows the prediction of SM theory, expressed via Eqs. (2) and (3), the dispersion parameter being proportional to T/T_0.

3.3 *The multiple trapping (MT) case*

Being relevant for transport in amorphous inorganic semiconductors where the density of states tails off exponentially below a demarcation energy separating extended states from localized states, the MT trapping case has received considerable attention in the literature on disprsive transport [27–33]. It can be solved analytically and leads to SM behavior expressed via Eqs. (1) to (3), the dispersion parameter being $\alpha = T/T_0$, except at very low temperatures where the freezing out of thermally activated jumps leads to stronger dispersion [32]. We note, however, (and this illustrates the ambiguities involved in the analysis of dispersive transport) that phenomenologically, MT- and E-hopping in an exponential DOS are indistinguishable [26]. Experimentally, the MT case is well documented in the literature on inorganic systems like As_2Se_3 [32].

4. Comparison with experiment

Perhaps the cleanest example for pure dispersive transport in an organic pendant group polymer where transport proceeds via hopping among the substituents is the recent work of Schnörer et al. [34] on polysiloxane with pendant carbazole groups as transport sites. Genuine SM behavior [see Eqs. (2), (3)] is found and the value for the dispersion parameter inferred from $j(t)$, $\tau_{tr}(d)$, as well as $\tau_{tr}(E)$ are mutually consistent. Although the authors do not report on the temperature dependence, the straightforward assignment is to either the E-hopping or the MT case, assum-

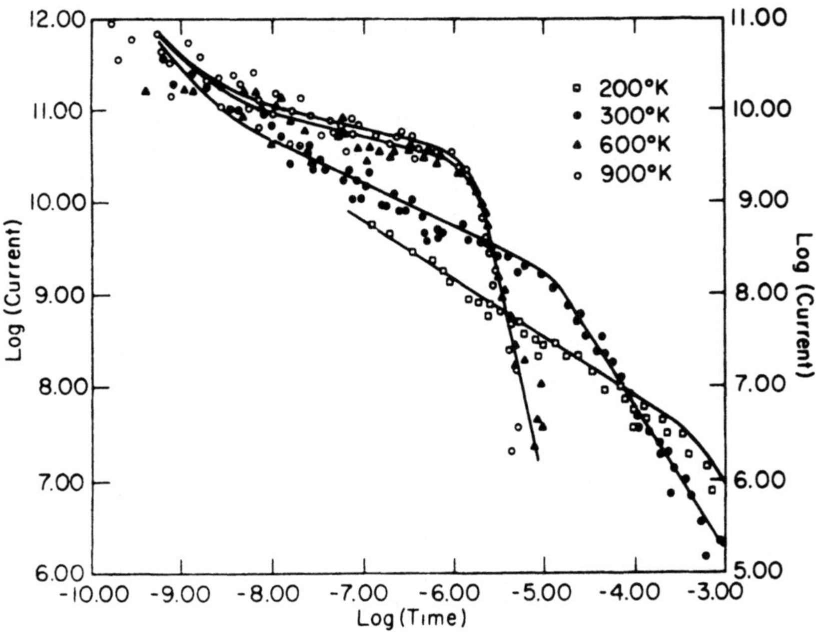

Fig. 6. Simulated TOF profiles for hopping within an exponential DOS characterized by $T_0 = 504$ K. The sample length was 70 a (a = 10 Å), the electric field was 10^4 V cm^{-1}. The scale at the righthand side applies to the 200 K curve (from [26])

ing an exponential distribution of intrinsic or trapping sites associated with the carbazole moieties and characterized by the distribution parameter $kT_0 \simeq 0.05$ eV. At first glance it seems surprising to invoke an exponential DOS for an organic photoconductor where all evidence is in favor of a Gaussian DOS. However, within the energy range probed by a relaxing charge carrier an exponential DOS may well be approximated by a set of displaced Gaussians representing different categories of conformations forming the localized states.

The situation is much more complex in the case of PVK. While Tahmasbi and Hirsch [35] report non-dispersive transport behavior, albeit at large electric fields, both Müller-Horsche et al. [36] as well as Bos and Burland [37] do see dispersive features, although of different signature. In [36] a transition from dispersive to non-dispersive transport has been observed and analyzed in terms of the MT model applying the Laplace transformation technique to recover the spectrum of trap release rates as well as the trap distribution. Remarkably, the average trap depth (~ 0.5 eV) is close to the value Stolzenburg et al. [38] inferred from measurements of delayed luminescence resulting from geminate pair recombination and comparable to the value Nespurek and Cimrova [39] report for the binding energy of a defect electron at a dynamically formed pair of carbazole groups. This would add support to the model of small polaron hopping recently invoked by Schein and Mack [40]. Polaronic transport is unlikely to occur in crystalline counterpart structures (e.g., crystalline *N*-isopropylcarbazole) where structural constraints prevent molecular reorganization in the vicinity of a temporarily charged molecule.

On the other hand, Bos and Burland [37] claim to have seen temperature-independent dispersive photocurrent transients if plotted on a normalized time scale, although the tails might carry some residual *T*-dependence. It seems conceivable that the manner in which the samples were prepared favored the formation of cluster structures establishing, as the authors assume, a geometrically disordered, fractal-like network on a supermolecular level that overrides the energetic disorder. It would certainly be useful, albeit tiresome, to conduct a more rigorous study on the correlation between charge carrier transport and sample morphology in PVK.

There is abundant evidence that the case of time-independent transport can under suitable conditions be realized in molecularly doped polymers [41–43] and silicon backbone polymers [44, 45]. The conditions are i) that there be no or only little trapping by shallow traps ensuring that the equilibration time within the DOS on intrinsic happing sites determines photocurrent relaxation, and ii) that $\tau_{tr} > t_{rel}$ (see above). A textbook example of this behavior is found in the work of Schein et al. [42] on hole transport in *p*-diethylaminobenzaldehyde-diphenylhydrazone (DEH) doped into a polycarbonate matrix at a concentration of 40 weight % (see inset in Fig. 7). Remarkably, plotting $\log \mu$ vs T^{-2} reveals a linear behavior over the entire temperature range studied (210 K to 370 K) (see Fig. 7)

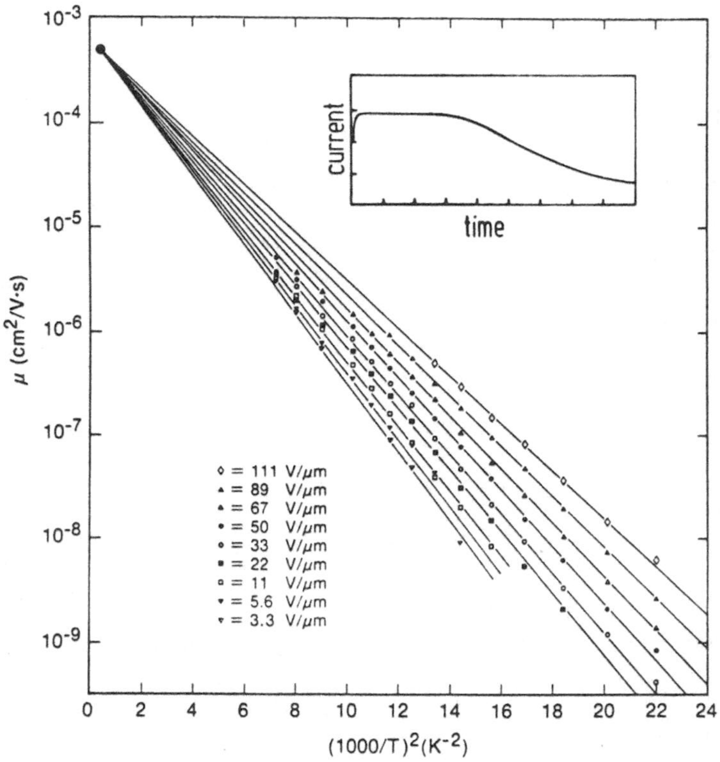

Fig. 7. Temperature dependence of the hole mobility in a molecularly doped polymer (DEH in Lexan) at various electric fields plotted on a $\log \mu$ vs T^{-2} scale. The inset shows a typical photocurrent transient (from [42])

while an Arrhenius plot shows significant deviations. The family of straight lines measured at variable electric field intersects very close to $T \to \infty$ yielding a realistic value of the activation-free mobility, contrary to the result that would follow from an Arrhenius plot. From the $E \to 0$ plot a width of the DOS of $\sigma \simeq 0.1$ eV is derived, i.e., $\sigma/kT \simeq 4$ at room temperature. Other systems yield similar values. Note that owing to the larger polarization energy of a charge carrier as compared to an optical excitation, σ is expected to be about a factor of 3 larger than widths of the DOS for transport of singlet or triplet excitations in an organic glass, known to be about 0.03 eV (see section 3.2).

A still unresolved question is the field dependence of the carrier mobility. While to first order approximation the model of stochastic random walk in a Gaussian DOS predicts an exponential variation of μ with E, recent careful experiments consistently yield $\ln \mu \propto E^n$ with n being close to 0.5 [42, 47, 48]. Since various chemically quite different systems share this pattern of behavior, interpretation in terms of a Poole-Frenkel mechanism can be dismissed. It would be highly advantageous if differently manifactured polymers contained the same amount of charged defects acting as carrier traps. Instead, it appears likely that the stochastic transport model, which has no material-specific ingredients, requires some refinement.

To conclude this section the problem of the broad tails exhibited by photocurrent transients of the "non-dispersive" type shown in the inset of Fig. 7 shall

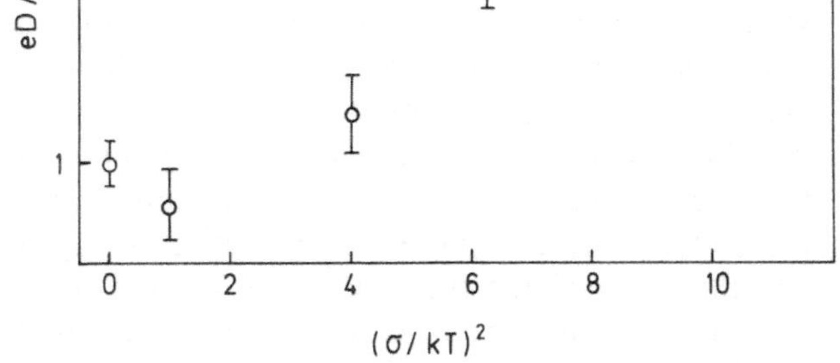

Fig. 8. The ratio $eD/\mu kT$ as a function of $(\sigma/kT)^2$ at $E = 10^5$ V cm^{-1}. Inset: Simulation result for the field dependence of the charge carrier mobility and diffusivity in the non-dispersive transport regime for a sample with Gaussian DOS of the intrinsic hopping sites ($\sigma/kT = 3$)

briefly be addressed. A sheet of carriers travelling across a sample under the action of bias voltage V spreads by diffusion. Using the Einstein relation connecting μ and diffusivity D via a $D = \mu kT$ a dispersion of transit times $\Delta\tau_{tr}/\tau_r = (2kT/eV)^{1/2}$ follows. For $V = 500$ V and $T = 310$ K, $\Delta\tau_r/\tau_{tr} = 10^{-2}$ is obtained, clearly at variance with experiment. Recent MC simulations carried out to resolve this puzzle [49] indicated significant deviations from the Einstein relation in the case of stochastic transport within a Gaussian DOS under a bias field of order 10^5 V cm^{-1} (Fig. 8). Because of the reduced local dimensionality of the microscopic motion, D is strongly field dependent while μ, probing motion on a macroscopic scale, exhibits linear response. This effect can explain at least part of the anomalous transit time dispersion seen in non-dispersive time of flight signals [44].

4. Application of the E-hopping model to viscous flow

If a non-crystallizing liquid is cooled sufficiently below the melting point the glass-forming elements will be trapped in the local minima of an extremely complex potential energy surface established during the freezing process. We adopt the notion that the elementary step involved in viscous flow is a displacive motion of a glass-forming element and make the following assumptions: i) The density of states distribution is a Gaussian; ii) the temperature of the system is low enough to ensure that the structure of the systems can be regarded static on the time scale of an elementary act, and iii) above the glass transition temperature T_g occupation of the local minima of the potential energy surface is in thermodynamic equilibrium, ensuring that the mean energy of the elements of the glass decreases with T according to $\langle\varepsilon_\infty\rangle = -\sigma^2/kT$. Viscous flow can then be viewed as complementary to motion of an electronic excitation within the energy distribution of intrinsic hopping sites except that now the moving elements are the glass constituents themselves or subunits thereof. The model puts Goldstein's earlier ideas on a quantitative basis representing that viscous flow is the consequence of thermally activated jumps, the activation energy being subject to a distribution. A crucial test for the applicability of this concept is its success in explaining the temperature dependence of the viscosity of supercooled melts [51] often explained in terms of the phenomenological Vogel-Fulcher or WLF law [52]. Figure 9 shows viscosity data representative for a broad class of systems, both organic and inorganic,

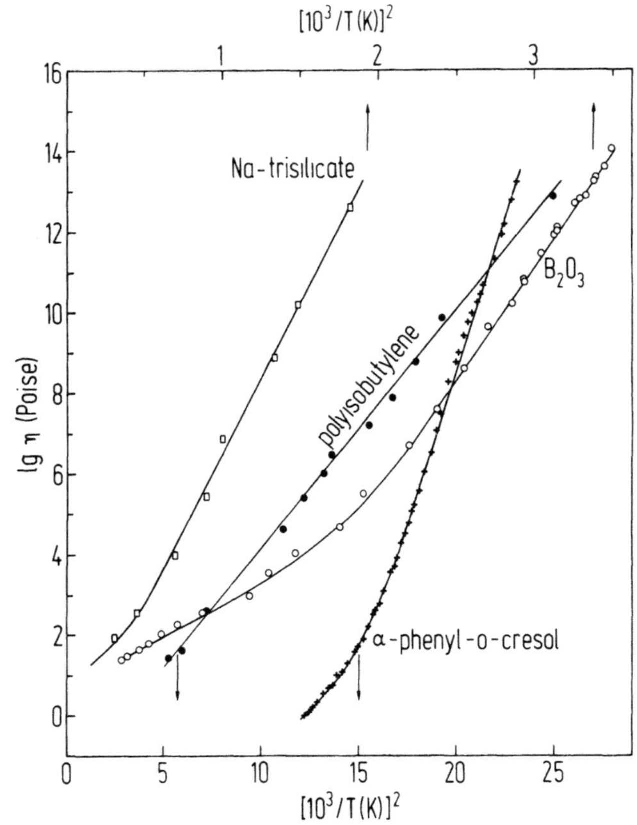

Fig. 9. Temperature dependence of the viscosity of various supercooled melts plotted on a $\log\eta$ vs T^{-2} scale (1 Poise = 0.1 Pa·s)

plotted on a $\log\eta$ vs T^{-2} scale. The fit is good over 10 orders of magnitude. Deviations occur at $T > 290$ K where the glass begins to soften, thus violating the assumption of a static DOS. At higher temperatures collective feedback effects between viscosity and local density fluctuations become important, tractable in terms of the mode coupling theory of Bengtzelius, Götze, and Sjölander [53]. It is worth noting that the relaxation behavior of polymers and supercooled molecular liquids near T_g involved, for instance, in effects like physical ageing or enthalpy relaxation can also be rationalized in terms of the concept developed for diffusion and relaxation of externally generated excitation in a random potential. A more elaborate version of these ideas will be published elsewhere [54].

Acknowledgement

I am greatly indebted to Dr. R. Richert for many clarifying discussions. Financial support by the Deutsche Forschungsgemeinschaft and the Fonds der Chemischen Industrie is acknowledged.

References

1. Stolka M, Pai DM (1978) Adv Polym Sci 29:1
2. Mort J, Pfister G (1982) In: Mort J, Pfister G (eds) Electronic Properties of Polymers. Wiley, New York
3. Scher H, Montroll EW (1975) Phys Rev B 12:2455
4. Bässler H (1981) phys stat sol (b) 107:9
5. Eiermann R, Parkinson GM, Bässler H, Thomas JM (1983) J Phys Chem 87:544
6. Mott NF, Davies EA (1971) Electronic Processes in Non-Crystalline Materials. Clarendon Press, Oxford
7. Jankowiak R, Rockwitz KD, Bässler H (1983) J Phys Chem 87:552
8. Elschner A, Bässler H (1987) Chem Phys 112:285
9. Ries B, Bässler H (1987) Phys Rev B 35:2295
10. Schönherr G, Bässler H, Silver M (1981) Phil Mag B 44:47
11. Movaghar B (1980) J Phys C 13:4915
12. Movaghar B, Schirmacher W (1981) J Phys C 14:589
13. Grünewald M, Pohlmann B, Movaghar B, Würtz D (1984) Phil Mag B (1984) 49:341
14. Movaghar B, Grünewald M, Pohlmann B, Würtz D, Schirmacher W (1987) J Stat Phys 30:315
15. Marshall JM (1978) Phil Mag B 38:335; (1981) Phil Mag B 43:401; (1983) Phil Mag B 47:329
16. Blumen A, Klafter J, White BS, Zumofen G (1984) Phys Rev Lett 53:1301
17. Movaghar B, Grünewald M, Ries B, Bässler H, Würtz D (1986) Phys Rev B 33:5532
18. Tiedje T, Rose A (1980) Solid State Comm 37:49
19. Pfister G, Griffith CH (1978) Phys Rev Lett 40:659
20. Richert R, Bässler H (1986) J Chem Phys 84:3567
21. Lange J, Ries B, Bässler H (1988) Chem Phys 128:47
22. Richert R, Bässler H, Ries B, Grünewald M, Movaghar B (1989) Phil Mag Lett 59:95
23. Young AP (1983) Phys Rev Lett 50:917
24. Larsen U (1985) Phys Rev B 32:1772
25. Bässler H, Schönherr G, Abkowitz M, Pai DM (1982) Phys Rev B 26:3105
26. Silver M, Schönherr G, Bässler H (1982) Phys Rev Lett 48:352
27. Silver M, Cohen L (1977) Phys Rev B 15:3276
28. Noolandi J (1977) Phys Rev B 16:4466
29. Schmidlin FW (1977) Phys Rev B 16:2362
30. Arkhipov VI, Iovu MS, Rudenko AI, Shuton SD (1979) phys stat sol (a) 54:67
31. Orenstein J, Kastner M (1981) Phys Rev Lett 46:1421
32. Monroe D (1985) Phys Rev Lett 54:146
33. Seynhaeve G, Adriaenssens GJ, Michiel H, Overhoff H (1988) Phil Mag B 58:421
34. Schnörer H, Domes H, Blumen A, Haarer D (1988) Phil Mag Lett 58:101
35. Tahmasbi AR, Hirsch J (1980) Solid State Comm 34:75
36. Müller-Horsche E, Haarer D, Scher H (1987) Phys Rev B (1987) 35:1273
37. Bos CF, Burland DM (1987) Phys Rev Lett 58:152
38. Stolzenburg F, Ries B, Bässler H (1987) Ber Bunsenges Phys Chem (1987) 91:853
39. Nespurek S, Cimarovà V (1988) 6th International Seminar on Polymer Physics: Relaxation in Polymers, Gomadingen
40. Schein LB, Mack JX (1988) Chem Phys Lett 149:109
41. Stolka M, Yanus JF, Pai DM (1984) J Phys Chem 88:4707
42. Schein LB, Rosenberg A, Rice SL (1986) J Appl Phys 40:4287
43. Abkowitz MA, Pai DM (1986) Phil Mag B 53:193
44. Yuh H-J, Stolka M (1988) Phil Mag B 58:539
45. Abkowitz MA, Knier FE, Yuh H-J, Weagley RJ, Stolka M (1987) Solid State Comm 62:547
46. Abkowitz MA, Stolka M, Weagley RJ, McGrane K, Knier FE, In: Advances in Silicon Based Polymer Science, ACS Book, to be published
47. Santos Lemus SJ, Hirsch J (1986) Phil Mag B 53:25
48. Abkowitz MA, Stolka M, Phil Mag Lett, in press
49. Richert R, Pautmeier L, Bässler H (1989) Phys Rev Letts, in press
50. Goldstein M (1969) J Chem Phys 51:3728
51. Bässler H (1987) Phys Rev Lett 58:767
52. Fulcher GS (1925) J Am Ceram Soc 8:339
53. Bengtzelius U, Götze W, Sjölander A (1984) J Phys C 17:5915
54. Richert R, Bässler H, submitted for publication

Received January 31, 1989;
accepted February 22, 1989

Author's address:

Prof. H. Bässler
Fachbereich Physikalische Chemie
Philipps-Universität
Hans-Meerwein-Straße
3550 Marburg

Discussion

SCHRADER:
Can you comment on the role of supertraps, which should be of great importance for charge transport in real polymeric systems in such model calculations?

BÄSSLER:
Most systems do, of course, contain traps. A time of flight signal is then the convolution of the effects of intrinsic disorder and of trapping. In case of a discrete trap the situation is simple: The trap-free mobility has then to be multiplied by a trap emptying factor in accord with the Hoesterey Letson model. The detrapping energy is the energy difference between the trap level and the energy of the carrier packet in the manifold of intrinsic hopping states in the steady-state limit. If trap energies are distributed the detrapping rate will be a statistical quantity in accord with what I said for the multiple-trapping case and in accord with what Müller-Horsche measured for PVC. Incorporating both a distribution of intrinsic hopping states and of trap levels into the simulation fails because of the increase in computer time necessary to obtain reasonable statistics.

ROTH:

I have questions concerning the generation of charge carriers by photons. If you are producing the charge carriers by light – for example from a nitrogen laser – you generate simultaneously two charge carriers with opposite signs. Why can you explain all your results by means of only one type of charge carrier? I expect that the geminate pairs of charge carriers have electron spins and are in the moment of generation in a singlet or triplet state. For that reason the recombination of geminate pairs, and with it the observable number of charge carriers, should be dependent on magnetic fields. Have you found a magnetic-field dependence in your measurements or would you expect one?

BÄSSLER:

In my treatment, I leave out the history of the charge carriers before escaping their mutual coulombic field because only those carriers that escaped geminate pair recombination can contribute to a time-of-flight signal. Anticipating low intensity conditions, I also ignored bimolecular carrier recombination which depends on the square of the charge carrier concentration. The magnetic field effects you mention affect the geminate pair recombination kinetics and modulate the concentration of free carriers. This effect is irrelevant for the motion of an individual carrier.

GOTLIB:

In the first part of Monte-Carlo simulations you have used the preposition that jumping frequency (or probability) from energy state ε_j to ε_i is proportional to $\exp(-(\varepsilon_j-\varepsilon_i)/(kT))$, for example, it depends on the initial and final state. But in reality the hopping frequency depends on the difference between activation energy in the intermediate (activated) state and the initial state $(-(\varepsilon_j^*-\varepsilon_i)/(kT))$, and that may not coincide (or be proportional) to $\varepsilon_j-\varepsilon_i$?

BÄSSLER:

This assumption is a plausible one for jumps along close pairs of molecules and supported by detrapping studies in doped molecular crystals. On the other hand, the simulation results are invariant against replacing the distribution of site energies by an equivalent distribution of activation barriers.

CHUDACEK:

We have measured the time- and temperature-dependence of the drift mobility of solution-cast and plasma polymerized PVK thin layers. In terms of your approach, only plasma polymerized samples exhibit dispersive transport as far as the time and temperature behavior is concerned, but both of them have practically the same value of μ_D.

BÄSSLER:

What you said is in line with the discrepancy seen when comparing the results of the Bayreuth and IBM groups on transient photoconductivity in PVK, which I mentioned. It supports the notion that transport is morphology-dependent. In the *E*-hopping model, changes of the morphology would affect the distribution of hopping states.

KLIMOVIC:

Consider a low temperature energy distribution of traps caused by structural disorder in a material. Now, if on heating, the depths of the traps do not change, only the trapping time of charge carriers will be altered. However, in reality, the structural disorder will be modified by the superimposed thermal disorder and the distribution will change. Will this lead to its effective broadening or narrowing?

BÄSSLER:

What you envisage is the turn-on of molecular motion, for instance, near the glass transition temperature. This should reduce the *effective* width of distribution of transport states and enhance the mobility.

Hypersonic properties of a smectic side chain polymer liquid crystal with a PA main chain

J. K. Krüger*), C. Grammes*) and J. H. Wendorff+)

*) Fachbereich Physik, Universität des Saarlandes, Saarbrücken, FRG, +) Deutsches-Kunststoff-Institut, Darmstadt, FRG

Abstract: We have investigated the elastic and the optical properties of a side chain polymer liquid crystal around the smectic-nematic and nematic-isotropic transitions at hypersonic frequencies. In contrast to classical liquid crystals the polymer materials show at both transitions a significant coupling between the static order parameter and the square of the elastic strain ε_3. The inverse shear stiffness constants do not diverge. All elastic constants behave continuously at both transitions. From certain optoacoustic properties and the refractive index functions it follows that order parameter fluctuations play no role at the nematic-isotropic transition temperature T_{ni}, but that the dynamic glass transition dominates the elastic properties at hypersonic frequencies.

Key words: Brillouin spectroscopy, Hypersonic properties, polymer side chain liquid crystals, phase transitions

1. Introduction

While classical liquid crystals (CLCs) show significant optical anisotropies [1a–c], rather small elastic anisotropies are observed within the monodomain state [2]. This holds true even at hypersonic frequencies [3,4], where the shear stiffness constants may become significant [5]. This implies that in these materials only a weak coupling takes place between the static order parameter [1b–c] and the classical elastic strain tensor components ε_j (Voigt notation).

In nematic CLCs the anisotropy of the sound velocity of quasilongitudinal polarized phonons is typically of the order 10^{-2} with a maximum along the optical axis [2]. No acoustic attenuation anomalies are observed at the nematic-isotropic transition (T_{ni}) at hypersonic frequencies. Even for the smectic A phase only small elastic anisotropies of some percent have been reported for the longitudinal moduli (e.g., [2]). It turns out that the maximum of the longitudinal stiffness appears along the preferential direction of the mesogenic molecules.

Side chain polymer liquid crystals (PLCs) contain in principle two orienting molecular elements which can affect the elastic properties:

i) The mesogenic groups, which may contribute to the elastic anisotropy as they do in the case of CLCs and
ii) the monomer units of the polymer backbone. An increase of the orientation of the polymer backbone on a macroscopic scale produces an increase of the longitudinal elastic stiffness coefficient along the molecular preferential direction (e.g., [6]).

Neutron scattering experiments [7, 8] show that in side chain polymer liquid crystals the polymer main chains have a preference to be oriented orthogonal to the optical axis. The ratio of the mean dimensions of the chain molecules perpendicular and parallel to the director was found to be about 1.1 in the nematic phase and about 4 in the smectic A one. The conclusion was that the chain backbone is predominantly confined to the smectic layers [8,9]. From these experiments and from the arguments i) and ii) above, a complicated elastic behavior was expected.

Only few results exist on the elastic properties of side chain polymer liquid crystals. First ultrasonic investigations of the nematic and smectic A phases of a siloxane PLC have been reported by Benguigui et al. [10]. Brillouin spectroscopic investigations of a PLC are available for the smectic and nematic states of one polydomain sample [11]. The elastic anisotropies of a monodomain smectic and a nematic PLC have been measured at room temperature. For the nematic material elastic data exist also for the temperature regime around the nematic-isotropic transition [12]. In agreement with the results of Benguigui et al. a maximum of the longitudinal sound velocity was observed along the preferential direction of the mesogenic groups. However, the elastic anisotropy is much more pronounced in the Brillouin data than in the ultrasonic data. It is not yet clear, whether this difference is due to different frequencies involved or just due to the fact that different materials are compared.

In this paper we present Brillouin scattering results on the velocity and polarization of sound waves in monodomain samples of the nematic and of the smetic A phases of the PLC given in Table 1. For short notation the material is denoted as PA6BOCH3. The polydispersity of the polymer molecules of both samples was $\bar{M}_w/\bar{M}_n \approx (2 \pm 0.1)$. Nearly the complete elastic stiffness tensor $\boldsymbol{c} = \{c_{ij}\}$ could be determined around the nematic–isotropic and the smectic A-nematic phase transitions in the frame of pure elasticity theory. Because of the reduced optical quality of the sample the stiffness coefficient c_{66} could not be measured yet. By comparing the refractive index functions $n_i(T)$ (i = 1, 3) with the optoacoustic D_i^{90R}-functions (see Sect. 3 and [13a, b]) the influence of relaxation processes in the vicinity of the mesomorphic phase transitions will be discussed. It will be shown, that these relaxation processes are not connected to the order parameter fluctuations at T_{ni} but can be attributed to the dynamic glass transition.

2. Experimental

The transition temperatures of our PA6BOCH3 material were determined by DSC-measurements (heating rate: 20 K/min). The quasistatic glass transition occurs at $T_g \approx 290$ K, the transitions from the smectic A phase to the nematic phase and from the nematic to the isotropic one appear at $T_{s_A n} \approx 357$ K and $T_{ni} \approx 378$ K respectively (Fig. 1). For our Brillouin investigations a thin film-like PA6BOCH3 sample (see Table 1) of thickness $d \sim 25$ µm was prepared using glass slides as support for the film. To prepare a thin homogeneously aligned monodomain sample the surfaces of the glass slides were coated with polyimide (Merck, Liquicoat PI) and rubbed along an arbitrary direction within

Table 1. Chemical formula and transition temperatures (from DSC) of PA6BOCH3

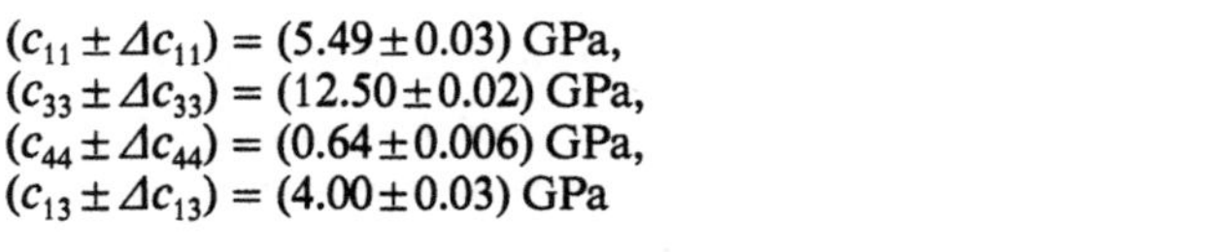

$T_g = 290$ K , $T_{s_A n} = 357$ K , $T_{ni} = 378$ K

$(c_{11} \pm \Delta c_{11}) = (5.49 \pm 0.03)$ GPa,
$(c_{33} \pm \Delta c_{33}) = (12.50 \pm 0.02)$ GPa,
$(c_{44} \pm \Delta c_{44}) = (0.64 \pm 0.006)$ GPa,
$(c_{13} \pm \Delta c_{13}) = (4.00 \pm 0.03)$ GPa

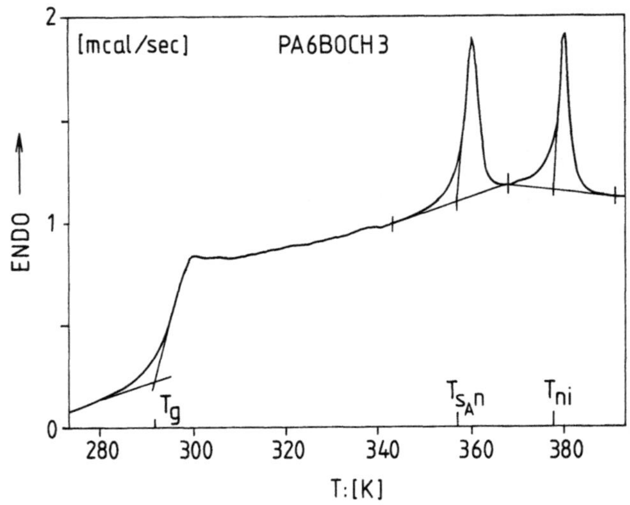

Fig. 1. DSC heating power of PA6BOCH3 at a heating rate of 20K/min

the film plane which defines the preferential direction of the mesogenic groups.

For the interpretation of the elastic stiffness data we used an orthogonal reference frame with the x_3-axis parallel to the director and the x_2-axis orthogonal to the film plane. The monodomain samples had fiber symmetry. This was confirmed by measurements of the birefringence and is consistent with the elastic behavior (see Sect. 3). The calculation of the stiffness coefficients c_{ij} (Voigt – notation) was done using similar densities as obtained for acrylate side chain polymers with an identical chemical structure of the side chain [14]: $\varrho(300\,\text{K}) = 1100$ kg/m^3. Since the volume expansion coefficient of the sample is not yet known, we neglected its temperature dependence in our calculations of

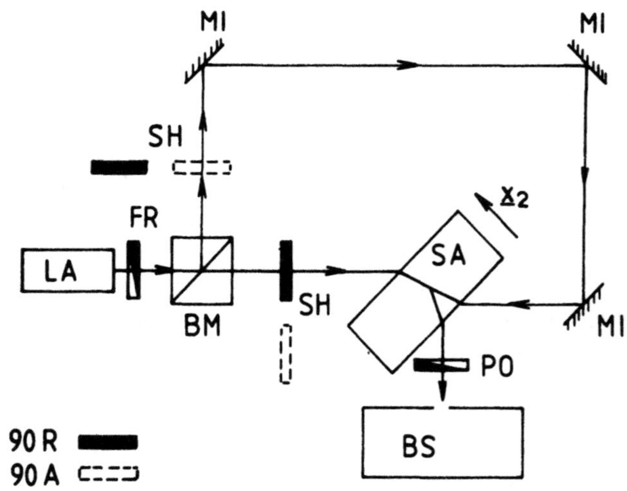

Fig. 2. Brilloin setup for a simultaneous measurement of different phonon modes. LA: argon ion laser; BM: beam splitter; FR: Faraday rotator; SH: shutters; SA: sample; MI: mirrors; PO: polarizer; BS: Brillouin spectrometer; x_2: unit wave vector along the x_2-axis of the sample coordinate system. The 90A- and 90R-methods are explained in the text and in Fig. 3

the elastic stiffness coefficients. This is justified because the influence of the volume expansion coefficients on the elastic constants is expected to be small compared to the observed elastic anomalies ($\alpha_{s_A} \approx \alpha_n \approx \alpha_i \approx 5 \times 10^{-4}\,\mathrm{K}^{-1}$, [14, 15]). The Brillouin measurements were made with a high performance Brillouin spectrometer using an argon-ion-laser at the vacuum wavelength $\lambda_0 = 514.5$ nm. The characteristics of this spectrometer were discussed elsewhere [16]. Figure 2 shows schematically the spectrometer setup which was used to obtain different realizations of the $90A$- and $90R$-scattering geometries [13b, 17, 18] described below. Using the shutters SH indicated in Fig. 2 we could discriminate between these two types of scattering geometries. Furthermore, the plate-like sample SA (Fig. 2) could be rotated in a defined way around $\boldsymbol{x}_2$. The optical polarization conditions could be chosen using a Faraday rotator FR and a polarizer PO.

The principles of the $90A$- and $90R$-scattering geometries have been discussed elsewhere [e.g., 13b]. The main features of the scattering geometries are shown in Figs. 3a and 3b, respectively. The $90A$-method requires some comments: i) Sound velocity polar plots can easily be obtained for sound wave propagation in the (x_1, x_3)-plane of the sample by simply rotating it around $\boldsymbol{x}_2$ (Fig. 2a); ii) Using proper optical polarization conditions [13b] the involved acoustic wavelength $\Lambda^{90A} = \lambda_0/\sqrt{2}$ does not depend on the optical properties of the sample. Because of the fiber symmetry of the LC-films the $90A$- and the $90R$-scattering geometries could be used to determine the optoacoustic D_i^{90R}-functions described in Sect. 3. These functions provide a sensitive measure for the presence of hypersonic dispersion or deliver directly, in the absence of dispersion, the main refractive indices n_i ($i = 1, 3$).

To avoid heating or burning of the sample, the effective laser power within the scattering volume was kept below 10 mW. The smallness of the scattering volume, together with this small illumination power, led to a rather low scattering intensity. Because of the resulting limited quality of the Brillouin spectra we abstain from a discussion of the hypersonic attenuation behavior. Excessive hypersonic attenuation can, however, be excluded even in the vicinity of the nematic-isotropic and the smectic A-nematic transition temperatures. Figure 4 shows a Brillouin spectrum for the phonon wave vector $\boldsymbol{q}$ oriented within the (x_1, x_3)-plane with an angle $\Phi \approx 45°$ between $\boldsymbol{q}$ and the x_3-axis. The solid line in the spectrum is only a guide line. The intensity of the Rayleigh line R was reduced by a factor of 10^3 by means of a Pockels-cell as light modulator [17]. Because of the small sample thickness and its reduced optical quality the relative accuracy of the sound frequencies is only about 0.4%. Concerning the determination of the elastic stiffness constants c_{ij} the accuracy is furthermore reduced by an uncertainty of the director orientation within the scattering volume. We estimate therefore an overall relative accuracy of $(\Delta c_{ij}/c_{ij}) \approx 1.5\%$. However, by evaluating the elastic stiffness data from least squares fits applied to sound velocity polar plots (see Table 1), the accuracy is increased in general.

The temperature dependent main refractive indices $n_1 = n_2$ and n_3 (Fig. 8) were determined with a prism method.

3. Results and discussion

The $90A$-scattering geometry was used to determine sound velocity polar plots at different temperatures by rotating the phonon wave vector $\boldsymbol{q}$ at fixed temperatures within the (x_1, x_3)-plane. Within the limits of error the influence of birefringence on the sound velocity results can be neglected [13b]. Figure 5 shows the data points measured at room temperature for the quasilongitudinal (I) and quasitransverse (II) sound waves.

Provided that the classical theory of elasticity can be used to describe the measured data, the fundamental relation for the determination of the elastic stiffness coefficients is the Christoffel equation. Assuming fiber symmetry for both mesomorphic phases of the

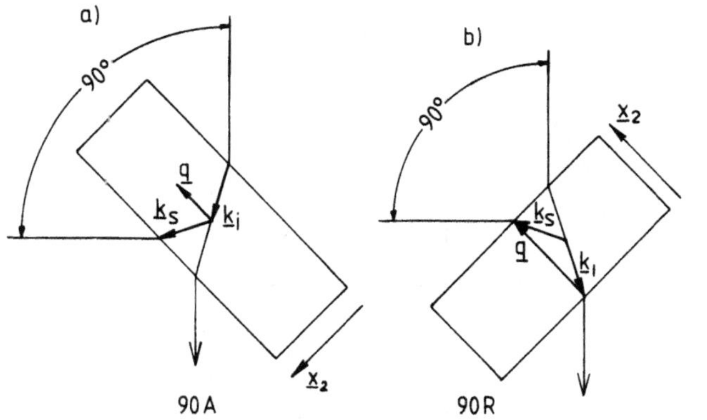

Fig. 3. Characteristics of the $90A$- and $90R$-scattering geometries (a and b, respectively). Outer scattering angle is 90°; $\boldsymbol{k}_i$: wave vector of the incident laser beam; $\boldsymbol{k}_s$: wave vector of the scattered light; $\boldsymbol{q}$: phonon wave vector; $\boldsymbol{x}_2$: unit wave vector along the x_2-axis of the sample coordinate system

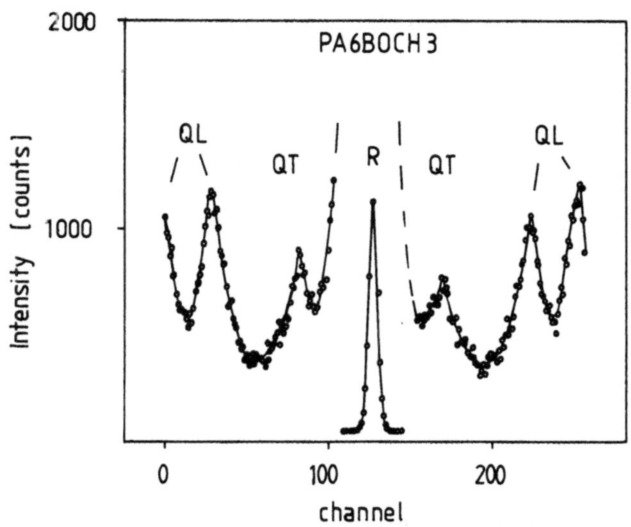

Fig. 4. Brillouin spectrum of PA6BOCH3. Phonon wave vector $\boldsymbol{q}$ oriented in the (x_1, x_3)-plane $\Phi = \sphericalangle(\boldsymbol{q}, \boldsymbol{x}_3) \approx 45°$. QL(QT): Brillouin lines of quasilongitudinal (quasitransverse) phonons; R: Rayleigh line (reduced by a factor 10^3)

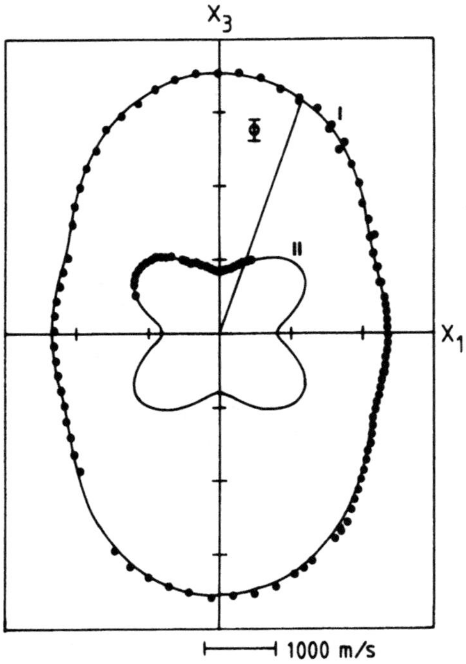

Fig. 5. Sound velocity polar plot of PA6BOCH3 for q oriented in the (x_1, x_3)-plane measured at room temperature (295 K). I: sound velocities of QL-sound waves; II: sound velocities of QT-sound waves. $\Phi = \sphericalangle(q, x_3)$

monodomain sample, this leads to the following cubic eigenvalue equation:

$$\det\,[I c I^{\mathrm{T}} - E c_{\hat{q}}] = 0 \tag{1}$$

with

$$I = \begin{pmatrix} l_1 & 0 & 0 & 0 & l_3 & l_2 \\ 0 & l_2 & 0 & l_3 & 0 & l_1 \\ 0 & 0 & l_3 & l_2 & l_1 & 0 \end{pmatrix} \tag{2}$$

$$c = c^T = \begin{pmatrix} c_{11} & c_{12} & c_{13} & & & \\ c_{12} & c_{11} & c_{13} & & & \\ c_{13} & c_{13} & c_{33} & & & \\ & & & c_{44} & & \\ & & & & c_{44} & \\ & & & & & c_{66} \end{pmatrix} \tag{3}$$

$$c_{\hat{q}} = \varrho\,\omega^2/q^2 = \varrho\,v^2$$

and

$$\hat{q} = (l_1, l_2, l_3) = l_1 x_1 + l_2 x_2 + l_3 x_3 \tag{4}$$

x_i $(i = 1,2,3)$: unit vectors in the coordinate directions of the sample coordinate system, $q = q\hat{q}$: acoustic wavevector, l_i $(i = 1,2,3)$: direction cosines of q, c_{ij} $(i,j = 1,2, \ldots, 6)$: elastic stiffness coefficients (Voigt notation), $f = \omega/2\pi$: sound frequency, v: sound velocity, ϱ: mass density, $c_{\hat{q}}$: eigenvalues which have been measured by Brillouin spectroscopy for different directions $\hat{q}$ of phonon propagation. An appropriate data analysis procedure using Eqs. (1–4) and Marquardt's nonlinear least squares fit combined with the formula of Cardano [13b] yields at each temperature all stiffness coefficients except c_{66}.

The coefficient c_{33} is within the smectic phase at room temperature more than twice as large as c_{11}, indicating a pronounced increase of the longitudinal elastic stiffness along the preferential orientation of the mesogenic side chains. The influence of the preferential orientation of the backbone molecules (orthogonal to the director) on the elastic properties are overcompensated by the ordering of the mesogenic side chains. The remarkable size of the shear stiffness c_{44} is a consequence of the frequency dependent imaginary part of the complex shear viscosity, which leads to a real elastic shear constant at high frequencies ($\omega\tau \gg 1, \tau$: relaxation time of the corresponding shear viscosity). It is obvious that $[c_{44}]^{-1}$, the coefficient describing the shear compliance between the smectic layers, does not diverge.

To get information on the temperature dependence of the hypersonic properties, sound velocity polar plots have been determined with increasing temperature up to the isotropic state. Using the procedure described above we could calculate the stiffness coefficients c_{11}, c_{33}, c_{44}, and c_{13} as functions of temperature (Fig. 6). On slow cooling through T_{ni} and $T_{s_A n}$ the same monodomain state could be reproduced.

The strongest influence of the order parameter is reflected for both mesomorphic phases by the temperature behavior of the coefficient c_{33}. The other coefficients are only slightly affected. The coefficient c_{11} in the nematic and smectic A phases behaves almost as a continuation of the longitudinal modulus c_{11}^{iso} of the isotropic phase, if it is extrapolated to the nematic- and smectic A state.

From the DSC- and the birefringence measurements it is clear that the smectic A-nematic and the nematic-isotropic transitions are of first order (Svetkov order parameter at T_{ni} $S \approx 0.25$). On the other hand the elastic stiffness coefficients (Fig. 6) behave continuously at $T_{s_A n}$ and T_{ni} (within the error limits). This discrepancy is not yet understood. From the elastic properties it follows therefore that the transition may be described approximately as a second order transition (Fig. 6) with a coupling between the static order parameter and the elastic strain component ε_3

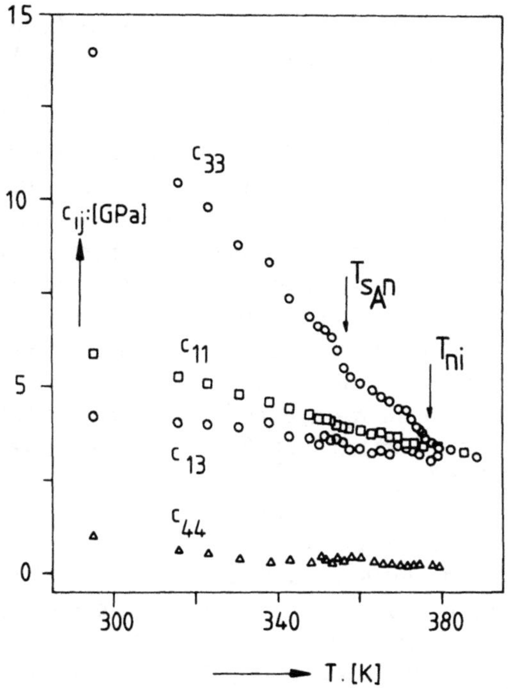

Fig. 6. Elastic stiffness coefficients c_{33}, c_{11}, c_{44}, and c_{13} of PA6BOCH3 as functions of temperature T. $T_{s_A n}$: transition temperature at the smectic A-nematic transition, T_{ni}: transition temperature at the nematic-isotropic transition (measured with increasing temperature)

(Voigt – notation). This seems to be correct, at least close to the mesomorphic transitions.

However, because of the fact that the quasistatic glass transition ($T_g \approx 290$ K) is found below the mesomorphic transitions, the latter are superimposed by the dynamic glass transition. For technical reasons, until now we could neither measure the quasistatic glass transition of PA6BOCH3 with Brillouin spectroscopy nor evaluate the hypersonic attenuation behavior with a sufficient accuracy. Therefore the classical way to detect the glass transition from acoustic measurements did not work.

A clear experimental hint for the influence of the dynamic glass transition on the elastic properties is the temperature dependence of special opto-acoustic functions, which we call D-functions [13a, b]. It turns out that the quantity D_i (see below) is a sensitive indicator for hypersonic dispersion. This probe for acoustic dispersion effects is of special interest if the phonon line widths are difficult to obtain, as in the case of the present measurements on a thin liquid crystal film.

We define only those D-functions, from which in the absence of hypersonic dispersion the refractive indices $n_i\,(i=1,3)$ of our samples can be calculated. Assuming $c'' \ll c'$, one has:

$$D_i^{90R}(\hat{q}) = \sqrt{\{c'^{90R}(\hat{q},p)/c'^{90A}(\hat{q},p)\}[n^2-0.5]+0.5} \tag{5}$$

c' and c'' are the real and imaginary part of the complex elastic stiffness modulus $c^* = c' - jc''$. p gives the polarization of the sound mode involved. Because x_1 and x_2 define symmetry equivalent directions as well for the smectic A as for the nematic state, the D_i^{90R}-functions can be evaluated from sound frequencies using the following scattering conditions:

$90A$-scattering geometry: q parallel to x_1 and the electric field vectors E_{inc} and E_{sc} of the incident and scattered laser light parallel to x_3.

The wavelength Λ^{90A} of the longitudinal sound mode along x_1 is then independent of the optical properties:

$$\Lambda^{90A} = \lambda_0/\sqrt{2} \tag{6}$$

$90R$-scattering geometry: q parallel to x_2 and $\mathbf{E}_{\text{inc}}$ and E_{sc} both parallel to x_i $(i=1,3)$. The wave length Λ^{90R} of the longitudinal sound mode involved in the $90R$-scattering experiment depends on n_i:

$$\Lambda^{90R} = \lambda_0/\sqrt{(4n_i^2-2)}\ , \quad (i=1,3)\ . \tag{7}$$

From Eqs. (5–7) we get the desired relation between D_i^{90R} and the measured hypersonic frequencies:

$$D_i^{90R}(x_{\hat{q}}) = \sqrt{\{[f_i^{90R}(x_2)/f^{90A}(x_1)]^2+1\}/2}\ , \quad i=1,3 \tag{8}$$

$x_{\hat{q}}$ represents the set of symmetry equivalent directions of type x_1, meaning that if the fiber axis is denoted by x_3, sound propagation along x_1 and x_2 contains the same information on the elastic stiffness constants. As mentioned above, we can measure the $f_i^{90R}(q\,||\,x_2)$ and $f^{90A}(q\,||\,x_1)$ simultaneously from the same scattering volume (Fig. 2). If hypersonic dispersion can be neglected, this experimental procedure yields in addition to the stiffness modulus $c_{11}=c_{22}$, the refractive indices $n_1=n_2$ and n_3. The quantity $(D_i^{90R}-n_i)$ is an especially sensitive sign for hypersonic dispersion. For normal dispersion, $(D_i^{90R}-n_i) \geqq 0$, holds.

Figure 7 shows the temperature dependences of the sound frequencies $f_i^{90R}\,(i=1,3)$ for the entire temperature range investigated. At fixed temperature the difference between these frequencies is mainly due to the difference of the refractive indices n_1 and n_3. In accordance with the behavior of $c_{11}(T)$ in Fig. 6, only a small acoustic anomaly is observed at T_{ni} and almost no anomalous behavior is found at $T_{s_A n}$.

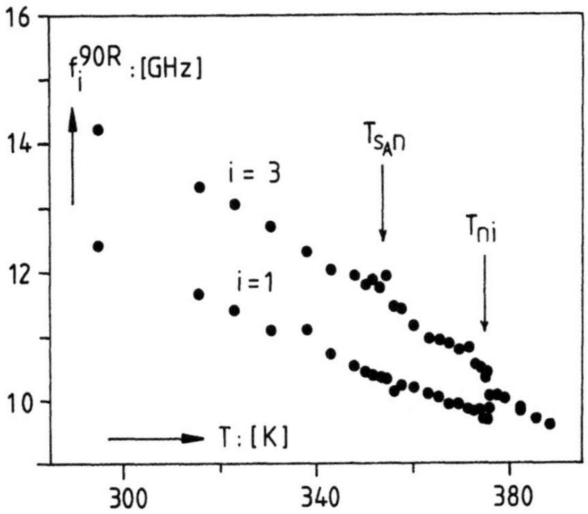

Fig. 7. Sound frequencies of the longitudinal polarized acoustic modes of PA6BOCH3 as a function of temperature T for propagation along x_2, using the $90R$-scattering geometry. The electric field vectors are chosen to be $E_i || E_s || x_1$ and $|| x_3$, respectively

Equation (8) has been used to determine $D_i^{90R}(i=1,3)$. Figure 8 shows $D_1^{90R}(T)$ and $D_3^{90R}(T)$ in comparison with $n_1(T)$ and $n_3(T)$. In the whole temperature range there is a deviation of the D^{90R}-functions from the corresponding n-functions which increases with temperature. This indicates the existence of an acoustic relaxation process at hypersonic frequencies. Since the increase becomes even stronger in the isotropic state, this relaxation process can be attributed to a glass relaxation process [12, 13a and b, 19]. The hypersonic loss maximum is expected above the highest measured temperature. This is supported by results recently obtained for PA6BCN [20], which shows the hypersonic loss maximum attributed to the α-process more than 60 K above the nematic-isotropic transition. It should be mentioned, that there is no evidence for order parameter fluctuations at hypersonic frequencies in the vicinity of T_{s_An} and T_{ni}, which would result in hypersonic loss maximum near the transition temperatures.

In addition to the n_i- and D_i^{90R}-values we present the orientation-averaged data $\langle n\rangle = (2n_1+n_3)/3$, (■, Fig. 8), and $\langle D^{90R}\rangle = (2D_1^{90R}+D_3^{90R})/3$, (●, Fig. 8), for the smectic A and the nematic phases. Both quantities show no anomalies at T_{s_An} and T_{ni}.

Finally the n-functions have been used to calculate the birefringence $\Delta n = n_3 - n_1$ as a measure for the static order parameter (Fig. 9). As expected, only small deviations appear between Δn, (□, Fig. 9) and $\Delta D = D_3^{90R} - D_1^{90R}$, (●, Fig. 9), indicating that in contrast to the D^{90R}-functions themselves, $\Delta D(T)$ is an interesting quantity to estimate the mesomorphic order purely from Brillouin measurements.

Comparing the temperature dependence of c_{33} within the smectic A phase with that of the ΔD-function, it turns out that $c_{33}(T)$ instead of reflecting the order parameter saturation with decreasing temperature, increases further, showing even an unexpected flex point between T_{s_An} and T_g, whereas the $\Delta D(T)$ curve shows the expected saturation behavior. Since the remarkable increase of c_{33} with decreasing temperature above T_g cannot be a consequence of the

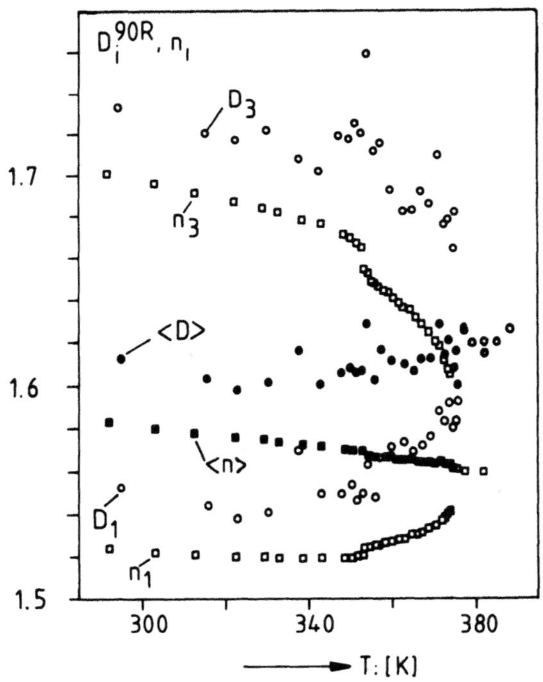

Fig. 8. D_i^{90R} (○) and $n_i(i=1,3)$ (□) of PA6BOCH3 as a function of temperature T. ●: $\langle D\rangle$, ■: $\langle n\rangle$ (explanation see text)

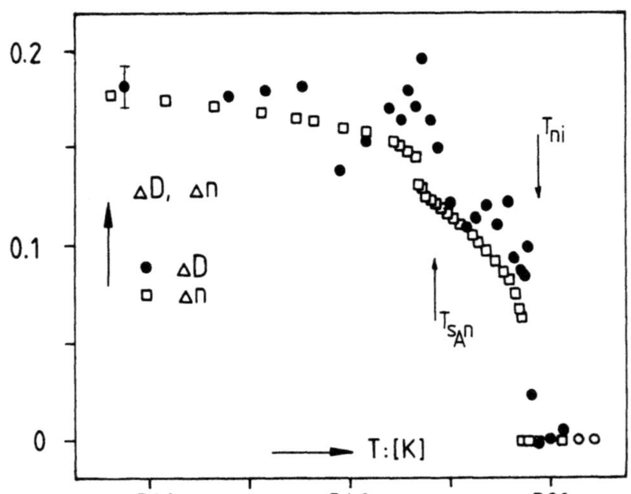

Fig. 9. ΔD and Δn as a function of temperature T of PA6BOCH3. ●: ΔD, □: Δn

order parameter saturation, we attribute it to the interference of the mesomorphic ordering with the dynamic glass transition. The difference in the elastic characteristics of monodomain CLCs on the one hand, and of monodomain PLCs on the other hand is then due to the fact that for CLCs the condition $\omega\tau \ll 1$ holds even at hypersonic frequencies.

Starting from the fact that the stiffness modulus c_{33} is mostly affected by the interference of the mesomorphic transitions and the dynamic glass transition, we have tried to describe the c_{33}-data of the smectic A phase by a simple two-phase model. Within this model the smectic PA6BOCH3 consists of the smectic layers and the interlayer regions, where the latter contain predominantly the polymer molecules and the spacers. For the sake of simplicity, we assume that only the interlayer regions reflect the dynamic glass transition, whereas the smectic layers reflect the mesomorphic ordering and its influence on c_{33}. Using a simplified Reuss-average for the calculation of c_{33} from the corresponding modulus of the components we find then the following relation:

$$L/c_{33} = L_i/c_i + L_m/c_m \ , \tag{9}$$

where c_m is the stiffness modulus c_{33} of the smectic layers and c_i is the corresponding modulus of the interlayer regions. L_m and L_i are the thicknesses of the smectic layers and interlamellar layers respectively. In our model calculations we have normalized $L = L_i + L_m$ to 1. c_m is believed to be independent of frequency and its temperature behavior is assumed to follow

$$c_m = c_m^0 + (c_m^{\max} - c_m^0) \tanh\{g(T_{s_A n} - T)\} \ , \tag{10}$$

where g is a constant which accounts for the order parameter's increase.

$\tanh\{g(T_{s_A n} - T)\}$ is a suitable function to describe the influence of the order parameter on the modulus c_m of the smectic layers.

c_m^0 and $c_m^{\max}$ are the elastic moduli at the order parameters $S = 0$ and $S = S^{\max}$, respectively. The frequency dependence of c_i is tentatively assumed to follow a Debye-law, and the Vogel-Fulcher-Tamman (VFT) law is assumed to be the temperature-time relation [21]:

$$c_i = c_i^\infty - (c_i^\infty - c_i^s)/ \{1 + \omega^2 \tau_0^2 \exp[2E_a/\{R(T - T_0)\}]\} \ , \tag{11}$$

Table 2. Fit parameters (+) and fixed parameters (*) of the series connection model given by Eq. (9)

*	τ_0	c_i^s	c_i^∞	c_m^0	$T_{s_A n}$	ω
	5×10^{-13} s	4 GPa	15 GPa	4 GPa	357 K	8π GHz
+	E_a	T_0	L_m	$c_m^{\max}$	g	
	6898 J/mol	113 K	0.245	815 GPa	0.00156 K^{-1}	

c_i^∞ is the limiting value of c_i for high frequencies, c_i^s is the low frequency value of c_i, τ_0^{-1} is the attempt frequency of the α-process, E_a is the activation energy of the α-process, R is the universal gas constant and T_0 is the ideal glass temperature of the interlamellar regions.

Equation (9) together with Eqs. (10) and (11) have been used to fit the c_{33}-data within the smectic A phase. However, some of the model parameters nearly known have been fixed (see Table 2) in order to avoid an instability of the fit because of correlation coefficients close to one. Figure 10 shows the measured c_{33}-data and the fit curve in quite good agreement. The fit parameters are also given in Table 2. The fit curve displays correctly the flex point in the $c_{33}(T)$ curve and the strong increase of c_{33} prior to the thermal glass transition.

Taking into account, that the model is of course strongly oversimplified, the fit parameters turn out to give at least the correct order of magnitude. However, the model considers neither the occurrence of the thermal glass transition nor a cooperative interaction between the dynamic glass transition with the mesomor-

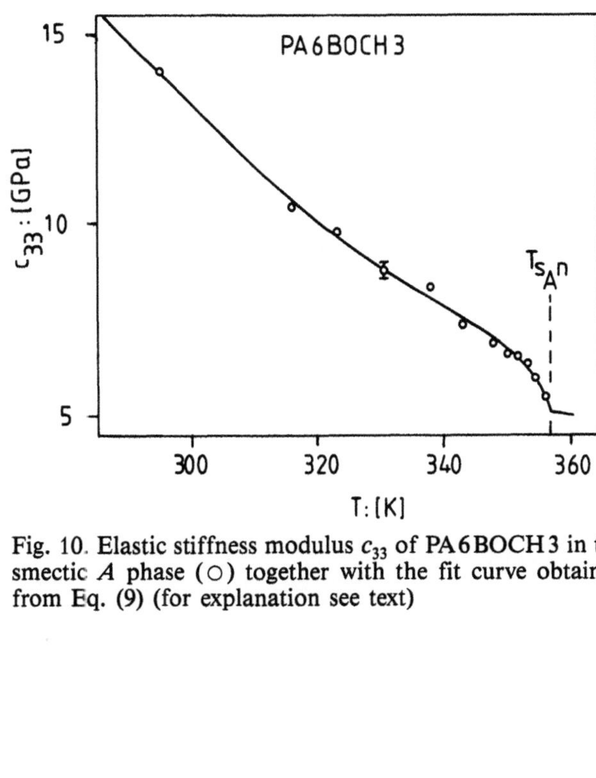

Fig. 10. Elastic stiffness modulus c_{33} of PA6BOCH3 in the smectic A phase (○) together with the fit curve obtained from Eq. (9) (for explanation see text)

phic ordering. A more general treatment in the frame of hydrodynamics is planned.

Acknowledgement

This work was kindly supported by the Deutsche Forschungsgemeinschaft, Schwerpunkt: "Thermotrope Flüssigkristalle". We thank Dr. M. Pietralla for interesting discussions and valuable comments on the manuscript.

References

1a. Kelker H, Hatz R (1980) Handbook of Liquid Crystals. Verlag Chemie, Weinheim
1b. de Gennes PG (1979) The Physics of Liquid Crystals. Clarendon Press, Oxford
1c. Chandrasekhar S (1980) Liquid Crystals. Cambridge Monographs on Physics, Cambridge
2. Miyano K, Ketterson JB (1979) in Physical Acoustics, Vol. XIV, p 93 ff
3. Wergin A, Krasser W, Stiller HH, Frischkorn CGB (1979) Phys Rev A20:1120
4. Copic M, Zgonik M, Lavrencic BB (1981) J Physique-Lett 42:L-405
5. Liao Y, Clark NA, Pershan PS (1973) Phys Rev Letters 30:639
6. Ward IM (1982) Developments in Oriented Polymers – 1. Appl Sci Pub, London
7. Kirste RG, Ohm HG (1985) Makromol Chem. Rapid Commun 6
8. Keller P, Carvalho B, Cotton JP, Lambert M, Moussa F, Pepy G (1985) J Physique Lett 46:L-1065
9. Kunchenko AB, Svetogorski DA (1986) J Physique 47:2015
10. Benguigui L, Ron P (1987) Mol Cryst Liq Cryst 153:241
11. Krüger JK, Peetz L, Zentel R, Claudy P (1986) Physics Letters 14A:51
12. Krüger JK, Peetz L, Siems R, Unruh H-G, Eich M, Hermann-Schönherr O, Wendorff JH (1988) Phys Rev A37:7
13a. Krüger JK, Marx A, Roberts R, Unruh H-G, Bitar MB, Nguyen-Trong H, Seliger H (1984) Makromol Chem 185:1469
13b. Krüger JK, Marx A, Peetz L, Roberts R, Unruh H-G (1986) Colloid & Polym Sci 264:403
14. Finkelmann H, Rehage G (1984) Adv Polym Sci 60–61:99
15. Wolf M, Wendorff JH (1987) Mol Cryst Liq Cryst 149:141
16. Krüger JK, Kimmich R, Sandercock J, Unruh H-G (1981) Polym Bulletin 5:615
17. Krüger JK, Peetz L, Pietralla M (1978) Polymer 19:1397
18. Krüger JK, Peetz L, Wildner W, Pietralla M (1980) Polymer 21:620
19. Krüger JK, Roberts R, Unruh H-G, Frühauf KP, Helwig J, Müser HE (1985) Progress Colloid & Polym Sci 71:77
20. Krüger JK, Grammes C, Wendorff JH (1989) In: Proceedings in Physics 37. Springer, Berlin
21. Donth E-J (1981) Der Glasübergang. Akademie Verlag, Berlin

Received February 27, 1989;
accepted March 6, 1989

Authors' address:

Dr. J.K. Krüger
Fachbereich Physik
Universität des Saarlandes
Bau 38
D-6600 Saarbrücken, FRG

Progress in Colloid & Polymer Science Progr Colloid & Polymer Sci 80:53–62 (1989)

Neutron spin-echo investigations on the dynamics of polymer systems*)

D. Richter[1]) and B. Ewen[2])

[1]) Institut Laue-Langevin, 156X, Grenoble, France; [2]) Max Planck Institut für Polymerforschung, Mainz, FRG

Abstract: Neutron spin-echo spectroscopy allows the observation of long-range internal relaxation mechanisms of macromolecules simultaneously in space and time. Thereby, it facilitates a microscopic study of molecular models applied for the explanation of macroscopic viscoelastic properties of polymer materials. After an outline of the method, we discuss experimental results on chain relaxation in polymer melts. We show that in the short time regime, the Rouse model describes perfectly well both the time dependences of the self and the pair correlation function. For longer times and stronger decay of the dynamic pair correlation function we observe distinct deviations from the Rouse model towards slower relaxation. However, other than predicted by the tube models the data do not reveal evidence for the existence of a microscopic length like the tube diameter or the entanglement distance but follow the time momentum transfer scaling behaviour of the Rouse-model. Thereafter we discuss aspects of the dynamics of polymer networks. We present neutron spin-echo results on the motion of labelled cross-links in a model network. These measurements reveal a surprisingly high mobility of the cross-links and are in qualitative agreement with the predictions of the phantom network theory.

Key words: Polymer dynamics, chain relaxation, cross-links, Reptation, Rouse motion, Rubbereleasticity, Neutron Spin Echo

I. Introduction

Viscoelasticity and transport properties of polymers in solution, melt and in the rubbery state are of basic physical interest and determine polymer applications to a large extent. An understanding of these properties on a molecular basis, therefore, is of great importance. Theoretically, the most simple approaches are based on Langevin equations with solutions comprising a spectrum of relaxational modes [1,2]. Spectial features are the incorporation of entropic forces (Rouse model) [3] which relax fluctuation of reduced entropy, hydrodynamic interactions which couple segment motions via the backflow field in solutions (Zimm model) [4] and the consideration of topological constraints which are mutually imposed within a dense ensemble of polymer chains [4–8]. These theories are used to describe rheological data like viscosity or shear moduli. In order to test these concepts on a molecular level, the space time evolution of the microscopic polymer motion has to be investigated. Here, scattering methods play a crucial role, since they allow a direct measurement of the corresponding correlation functions. In particular quasielastic neutron scattering (QNS) accesses the internal relaxation processes within one polymer chain on appropriate length and time scales simultaneously. Furthermore, the possibility of labeling by hydrogen-deuterium exchange allows the observation of single chain motion even in the melt.

On two examples, this paper discusses our experimental efforts to unravel characteristics of chain motion in a dense environment. After a brief outline of neutron spin-echo, we present a study on the dynamic structure factor for polymer motion in melts [9]. Then we report recent results on the dynamics of four-functional crosslinks in model networks [10].

II. The spin-echo method

The unique feature of neutron spin echo (NSE) is its ability to determine energy changes of neutrons occurring during a scattering process in a direct way

*) Dedicated to Prof. E. W. Fischer on the occasion of his 60th birthday

[11–13]. Unlike conventional techniques, NSE measures the neutron velocities of the incoming and scattered neutrons utilizing the Larmor precessions of the neutron spin in an external guide field. Since the neutron spin vector acts like the hand of an "internal clock" attached to each neutron which stores the result of the velocity measurement on the neutron itself, this measurement is performed for each neutron individually. Therefore, the incoming and outgoing velocities of one and the same neutron can be compared directly and a velocity difference measurement becomes possible. Thus, energy resolution and monochromatization of the primary beam or the proportional neutron intensity are decoupled and an energy resolution in the order of 10^{-5} can be achieved with an incident neutron spectrum $I(\lambda)$ of 20% bandwidth.

The basic experimental set up of a neutron spin-echo spectrometer is shown in Fig. 1. A velocity selector in the primary neutron beam selects a wavelength interval of about 20% full width half maximum. The spectrometer offers primary and secondary neutron flight paths where guide fields $\vec{H}$ and $\vec{H}'$ can be applied. At the beginning of the first flight path a supermirror polarizer polarizes the neutrons in direction of propagation. A first $\pi/2$ coil turns the neutron spins into the x-direction perpendicular to the neutron momentum $\hbar\vec{k}$. Starting with this well defined initial condition the neutrons commence to process in the applied guide fields. Without the action of the π-coil each neutron performs a phase angle $\phi \sim \lambda \int H ds$. Since the wavelengths are distributed over a wide range, in front of the second $\pi/2$-coil the phase angle will be different for each neutron and the beam will be completely depolarized. A π-coil positioned at the half value of the total field integral avoids this effect: on its way to the π-coil the neutron may pass an angle $\phi_1 = 2\pi n + \Delta\phi_1$. The action of the π-coil transforms the angle $\Delta\phi_1$ to $-\Delta\phi_1$. In a symmetric set up (both field integrals before and after the π-coil are identical)

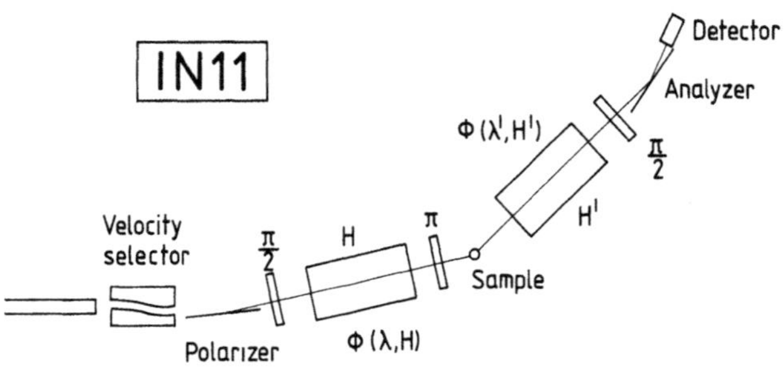

Fig. 1. Schematic of the neutron spin-echo spectrometer IN11 at the ILL Grenoble

the neutron spin turns by another phase angle $\pi_2 = \pi_1 = 2n\pi + \Delta\phi_1$. The spin transformation at the π-coil just compensates the residual angles $\Delta\phi_1$ and in front of the second $\pi/2$-coil and the neutron spin points again into the x-direction independent of its velocity. Finally, the second $\pi/2$-coil projects the x-component of the polarization in the z-direction and then at the supermirror analyser the total polarization is recovered. The experimental set up is spin focussing: similar to NMR spin echo methods in front of the second $\pi/2$ coil for each spin separately the phase is focussed to its initial value.

At the spin-echo spectrometer IN 11 realized at the Institut Laue-Langevin, Grenoble, the sample is positioned near the π-coil (Fig. 1). With exception of losses due to field inhomogeneities in the case of elastic scattering the polarization remains preserved. If the neutron energy is changed due to inelastic scattering at the sample, the neutron wavelength is modified from λ to $\lambda' = \lambda + \delta\lambda$. Then the phase angles ϕ_1 and ϕ_2 do not compensate each other and the second $\pi/2$-coil projects only the x-component of the polarization into the z-direction which passes afterwards through the analyzer. Apart from resolution corrections the final polarization P_f is then related to initial polarization P_i by:

$$P_f = P_i \int_{-\infty}^{+\infty} S(Q,\omega) \cos \omega t \, d\omega \qquad (1)$$

where the scattering function $S(Q,\omega)$ is the probability that during scattering at a certain momentum transfer $\hbar Q$ an energy change $\hbar\omega$ occurs. Furthermore, we have introduced the time variable $t \sim \lambda^3 H$. From Eq. (1) it is realized that the NSE is a Fourier method and essentially measures the real part of intermediate scattering function $S(Q,t)$. The Fourier time thereby is proportional to λ^3 and the applied guide field H. A spin echo scan is performed by varying the guide field and thereby studying the intermediate scattering function $S(Q,t)$ at different Fourier times. Finally, we notice that the use of a broad wavelength band introduces a further averaging process containing an integration over the incident wavelength distribution $I(\lambda)$

$$P_f = P_i \int_{0}^{\infty} I(\lambda) S(Q, t(\lambda)) d\lambda \ . \qquad (2)$$

This averaging process obscures somewhat the relationship between P_f and $S(Q,t)$. For many relaxation processes, however, where the quasielastic width varies with power laws in Q, the smearing of Eq (2) is of no practical importance. For example, for internal relaxation of polymers in dilute solution we have $S(Q,t) = S(Q^2 t^{2/3})$ [1]. Since Q varies with $1/\lambda$ and t with λ^3, the wavelength dependence drops out completely. The time range covered by the NSE-spectrometer at the Institut Laue-Langevin (ILL) in Grenoble depends on the neutron wavelength. At best, for $\lambda \simeq 12$ Å, times as long as 50 ns could be reached. The shortest times still detectable are in the 10^{-12} s regime. At longest wavelength an energy resolution of about 5 neV (1/e-decay) is achieved.

III. Polymer relaxation in melts

III.1 Theoretical considerations

The theoretical description of chain motion in polymer melts distinguishes three different time regimes separated by corresponding cross over times τ_s and τ_t.

i) At short times ($t<\tau_s$) the chains do not realize entanglement constraints and their motion is controlled by random thermal and entropic forces which are related to their conformational entropy. In this Rouse regime the mean square displacement of a chain segment increases proportionally to the square root of time. In Gaussian approximation the segment self-correlation function relates directly to the mean square segment displacement yielding a particularly simple intermediate scattering function [4].

$$S_{\text{self}}(Q,t) = \exp\left(-Q^2\sqrt{\frac{W\sigma^4 t}{9\pi}}\right) \tag{3}$$

Thereby σ denotes the segment length and $W = 3k_B T/\zeta_0\sigma^2$, where ζ_0 is a segment friction coefficient, k_B the Boltzmann constant, and T the temperature. The pair correlation function of a Rouse chain describing the interferences between the scattering from different segments is more complicated [14]. We note, however, that as S_{self} it scales with the universal "Rouse" variable $u = Q^2\sqrt{W\sigma^4 t}$. Besides the size of the chain R_g, there exists only *one* length scale, namely the length of the chain segment σ.

ii) At intermediate times ($\tau_s<t<\tau_t$) the viscoelastic data from long chain molecules show that the dynamic modulus exhibits a plateau regime indicating elastic behaviour. This elasticity may be understood in terms of a transient polymer network, the mesh size of which is given by the distance between entanglements or topological constraints. The most famous approach modeling these topological constraints is the reptation model. It assumes that these constraints restrict the chain motion to a tube along the chain profil [4]. At intermediate times in the so-called local reptation regime [15], the chain relaxes within the spatially fixed tube. Using the Doi-Edwards theory which generalizes the reptation concept to a theory of viscoelasticity the tube diameter can be evaluated from essentially the plateau modulus [16]

$$d_T = 4/5\left(\frac{6R_g^2}{M}\right)\left(\frac{\varrho L K_B T}{G_N^0}\right) \tag{4}$$

where R_g is the radius of gyration, G_N^0 is the plateau modulus, ϱ the polymer density, M is the molecular mass, and L the Avogadro number. For polydimethylsiloxan (PDMS), for example, this consideration yields $d_T \sim 50$ Å. Thus, there is considerable lateral freedom for the chain motion. The cross over time τ_s from the Rouse to the local reptation regime may be estimated by taking the slowest Rouse relaxation time for a chain, the end-to-end distance R_E of which corresponds to the tube diameter

$$\tau_s = d_T^4/(2W\sigma^4\pi^2)\ . \tag{5}$$

Concerning the scattering function the tube diameter – being a new dynamic length scale – is expected to invalidate the scaling with the Rouse variable u: In the local reptation regime the density fluctuations of a given chain equilibrate within the fixed tube. Consequently, the spatial segment positions stay correlated and the scattering function does not decay completely but contains a time-independent contribution related to the size of the tube. Neglecting the short time Rouse regime, de Gennes calculated the coherent scattering function for local reptation [15]

$$S(Q,t) = \left(1-\frac{Q^2 d_T^2}{36}\right) + \frac{Q^2 d_T^2}{36}\exp\left(\frac{u^2}{36}\right)\operatorname{erfc}\left(\frac{u}{6}\right)\ . \tag{6}$$

Equation (6) shows explicitly that $S(Q,t)$ contains a time-independent part which relates to the tube diameter and is calculated to second order in $d_T Q$. The presence of the intermediate length scale d_T destroys the scaling with u.

An alternative approach to treat topological constraints was recently proposed by Ronca [17]: he described the elastic interaction between different chains in dense systems by introducing a memory term into the Langevin equation for chain motion, the kernel of which is related to the relaxation modulus of the polymer ensemble. This ansatz is also able to reproduce the main experimental results and has the further virtue that it allows a calculation of the scattering function in the cross-over regime where the short time Rouse behavior transforms into entanglement controlled motion.

iii) For long times ($t > \tau_t$) where τ_t is the disentanglement time the polymer leaves its tube and the memory on the initial tube is lost. Then all inter-segment correlations vanish and $S(Q,t)$ decays to zero. This time regime is outside the reach of neutrons.

III.2 Neutron spin-echo experiments

Our neutron scattering investigations were performed on PDMS, a polymer which combines a very low monomeric friction coefficient ζ_0 with very high flexibility [2]. Two samples were prepared: i) a melt from deuterated PDMS molecules ($M_w = 10^5$; $M_w/M_n \simeq 1.6$) labelled at random positions along the chain. These polymers allowed us to study the self-correlation function since the labelled sections (8 protonated for on the average 80 deuterated monomers) were small enough to be seen as a single unit at the momentum transfer of the experiment, and on the other hand, large enough to create sufficient coherent scattering necessary for the spin-echo experiment; ii) a deuterated melt ($M_w = 1.5 \times 10^5, M_w/M_n \simeq 1.1$) containing 12% protonated polymers of the same molecular weight. The molecular weight of both samples was well above the entanglement molecular weight of $M_c = 21\,000$ for PDMS [18].

The experiments were carried out at the spin-echo spectrometer IN11 at the ILL, covering a Q-range $0.02 \leqslant Q \leqslant 0.15$ Å^{-1}. In a first experiment we measured at a neutron wavelength $\lambda = 8.56$ Å, achieving a resolution of 17 neV (1/e decay). Figure 2 presents spin-echo results for the self- and pair-correlation function from PDMS at 100 °C. The data are plotted vs the scaling variable of the Rouse model $u = Q^2 \cdot \sqrt{W\sigma^4 t}$. The solid lines present the results of a fit with the appropriate Rouse self- and pair-correlation functions [14] the only fit parameter being the "*Rouse parameter*" $W\sigma^4$. For both correlation functions we observe excellent agreement with Rouse dynamics. This concerns the Q as well as the time dependence: The characteristic frequencies of both sets of data follow the required Q^4 dependence, the lineshape of the self-correlation function agrees well with the predicted $t^{1/2}$ power law. For the case of the self-correlation function this is further illustrated in Fig. 3, which displays the characteristic frequency and the ex-

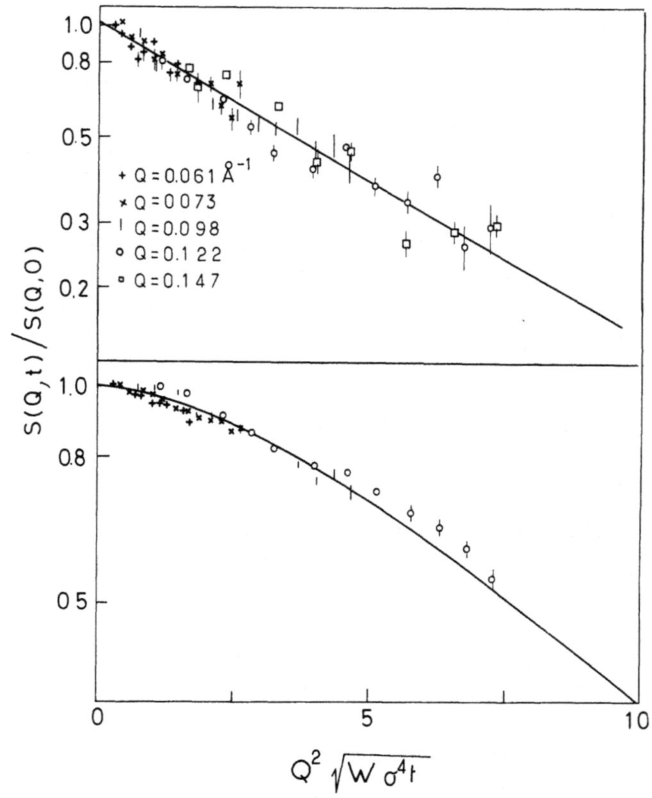

Fig. 2. Neutron spin-echo data for the self- (upper part) and pair-correlation functions obtained from PDMS melts at 100 °C. The data are scaled to the universal variable of the Rouse model. The symbols refer to the same Q-values in both parts of the figure

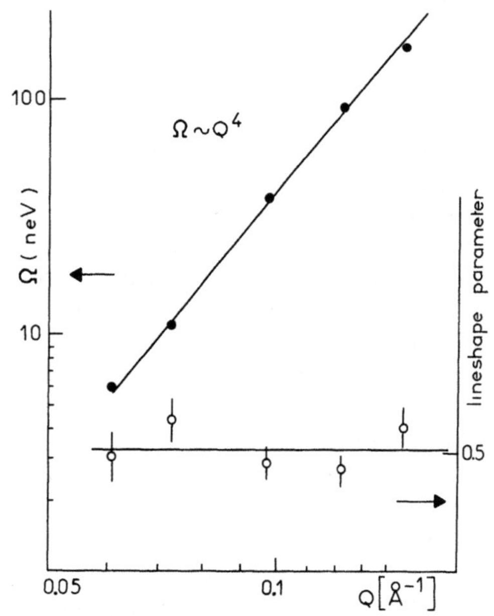

Fig. 3. The characteristic frequencies and time exponents in stretched exponential relaxation function obtained for the randomly labelled PDMS melt at 100 °C

perimental time exponent in the scattering function [Eq. (3)] as a function of Q. Within experimental accuracy the Rouse parameters obtained from the two measurements coincide (self-correlation: $W\sigma^4 = 1.75 + 0.15 \times 10^{13}$ Å^4 s^{-1}; pair-correlation: $W\sigma^4 = 1.85 \pm 0.05 \times 10^{13}$ Å^4 s^{-1}). The segmental friction coefficient $\zeta_0/\sigma^2 = 8.9 \pm 0.7 \times 10^5$ dynes/cm^3 derived from these microscopic data agrees well with the viscoelastic result $\zeta_0/\sigma^2 = 7.1 \times 10^5$ dynes/cm^3 [19].

In order to understand why these data do not show any influence of entanglement constraints, we estimate the cross-over time τ_s [Eq. (5)] using the viscoelastically determined tube diameter $d_T = 50$ Å and the measured Rouse parameter. We obtain $\tau_s = 18$ ns, somewhat larger than the experimentally covered time regime $0.35 \leqslant t \leqslant 14$ ns. Another estimate may be obtained from high frequency measurements of dynamic modulus of PDMS performed at room temperature [20]. Rescaling the measured cross-over frequency between the plateau modulus and the high frequency "Rouse regime" with the ratio of the temperatures and friction coefficients we find $\tau_s = 50$ ns also outside of our measuring range.

In order to improve experimentally, we made a second experiment by carefully tuning the spin-echo machine and increasing the wavelength to $\lambda = 10.94$ Å (Fourier time $t \sim \lambda^3 H$); we extended the experimental time range to $t_{max} = 40$ ns. Simultaneously we accelerated the polymer relaxation by heating to 200 °C. Both measures combined resulted in an increase of the time scale by one order of magnitude. From mechanical data τ_s is estimated to 20 ns (at 200 °C) while Eq. (5) gives 7 ns. Both values of τ_s are well within the extended experimental time range.

Figure 4 presents the measured relaxation curves in the short time domain ($t < 10$ ns). The solid lines present the fit with the $S(Q, t)$ derived from the pair correlation function of the Rouse model. Again we find very good agreement with this model. This can also be seen from the inset in Fig. 5 where the same data are

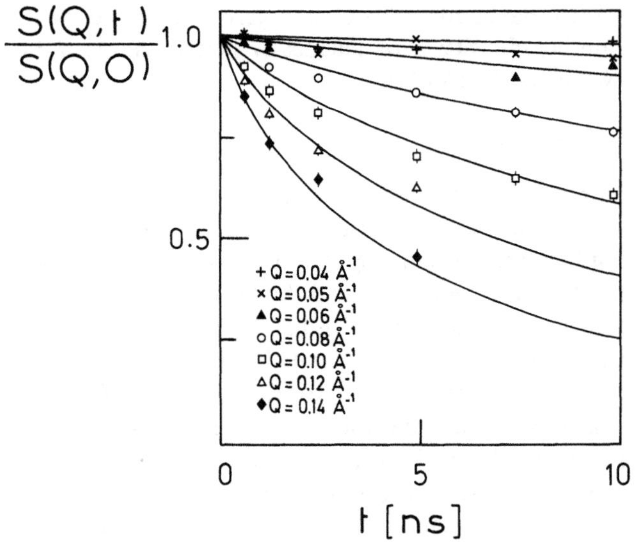

Fig. 4. Neutron spin-echo spectra from the PDMS-melt (pair correlation) at 200 °C in the short time regime ($t < 10$ ns). The solid lines are the relaxation curves predicted by the Rouse model

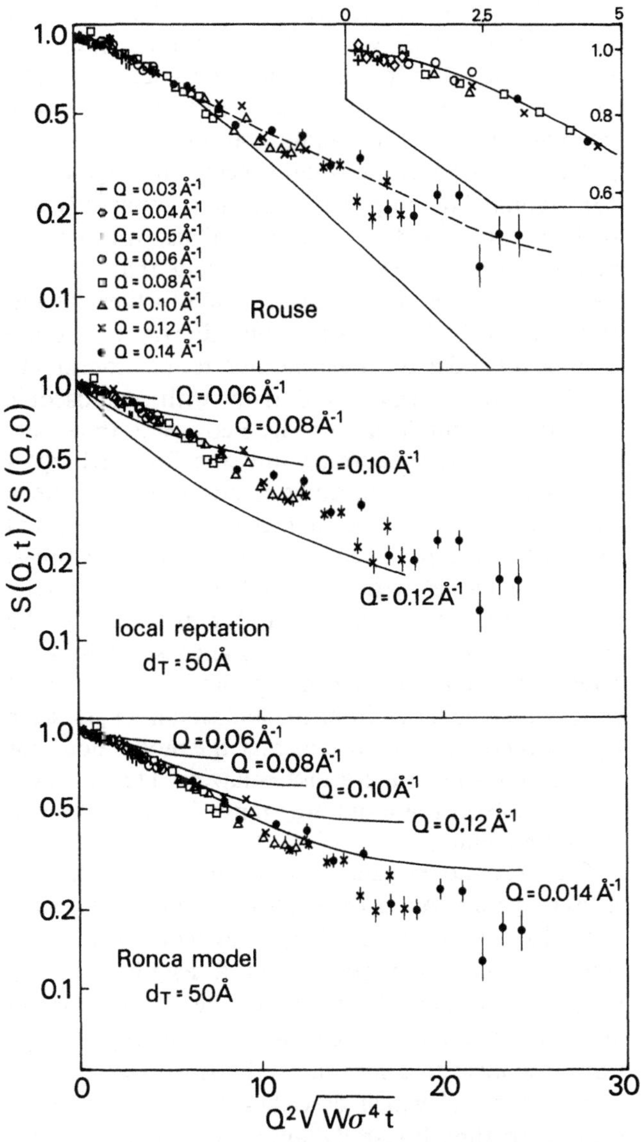

Fig. 5. Neutron spin-echo results from PDMS at 200 °C displayed in a scaling plot; Upper part: comparison with the Rouse model (solid line); the dashed line serves as a guide to the eye. Inset: behavior of the relaxation function at small values of the scaling variable. Middle section: comparison with the local reptation model [Eq. (6)]. Lower part: comparison with the generalized Rouse model [17]

displayed as a function of the scaling variable $u = Q^2\sqrt{W\sigma^4 t}$. For the Rouse parameter we find now $W\sigma^4 = 4.2 \pm 0.1 \times 10^{13}$ Å^4 s^{-1}. This value is the basis for the estimates of τ_s given above. The results further show that the experimental Q-range is well within the universal regime where deviation due to local potentials are of no importance. Figure 5 displays the obtained results over the full experimental time range, again using the scaling presentation. The solid line in the upper part of Fig. 5 shows the prediction of the Rouse model based on the short time behavior.

While for small values of the scaling variable ($u = Q^2\sqrt{W\sigma^4 t} \leqslant 7$) we again observe good agreement with the Rouse model, at higher u we see systematic deviations towards slower relaxation. Thus, qualitatively, we find additional hindrance to the polymer motion as we would expect from topological constraints. We note, however, that apparently the scaling with $Q^2\sqrt{W\sigma^4 t}$ prevails. Thus, without applying any further model, we may state that our data bear no evidence for the existence of a further dynamic length which necessarily must destroy the Rouse scaling. In order to underline this result more quantitatively we compare the data with microscopic models. The middle part of Fig. 5 compares the model prediction of the scattering function for local reptation [Eq. (6)] with the experimental results. Thereby we have fixed the tube diameter to $d_T = 50$ Å and used the Rouse parameter determined at short times. As we see, model predictions and experimental results are incompatible. In particular for different Qs the data do not split into different branches. As we have already mentioned earlier this Q-splitting in the model prediction originates from the additional length scale d_T. Furthermore, we observe that the model does not contain the short time Rouse behavior. This makes it inapplicable in the cross-over range between the two dynamic regimes.

The only model calculation treating explicitly the scattering function in the cross-over range is Ronca's generalized Rouse model [17]. This model leads to a temporal chain localization on a length scale d_T which may be identified with the tube diameter or the distance between entanglements. For each separate Q-value, the measured spectra can be fitted satifactorily with this model which describes well the deviations from short-time Rouse dynamics. However, then the tube diameter varies strongly with Q ($Q = 0.08$ Å^{-1}: $d_T = 110 \pm 42$ Å, $Q = 0.10$ Å^{-1}: $d_T = 75 \pm 5$ Å^{-1}; $Q = 0.12$ Å^{-1}: $d_T = 63 \pm 2.4$ Å and $Q = 0.14$ Å^{-1}: $d_T = 54 \pm 1.4$ Å). This is another manifestation of the observation that the experimental data do not split with Q but remain a function of the Rouse scaling variable. At the bottom of Fig. 5 we compare the prediction of the Ronca model with our data, fixing again the tube diameter to 50 Å and keeping the Rouse parameter obtained at short times. For small values of $Q^2\sqrt{W\sigma^4 t}$ the model describes well the Rouse dynamics which for higher values of the scaling variable slow down and cross over to a Q-dependent plateau level. Thus, the Ronca model depicts the cross-over from Rouse to entanglement controlled dynamics. However, again as a consequence of the additional dynamic length d_T, it predicts strong deviations from scaling with the Rouse variable – a prediction which experimentally is not confirmed.

Finally we remark on recent molecular dynamics simulations on polymer melts which exhibit a pronounced cross-over in the time dependence of the mean square displacement from a Rouse type $t^{1/2}$ to a reptation type$^{1/4}$ behavior [21]. These calculations indicate that finite size effects may obscure the reptation dynamics in a coherent scattering experiment. The chains end have much more freedom to move than the chain centers. Though we believe that such an effect is not very important for our chains which are built from ~2100 monomers, we consider a future experiment dealing with 3-block copolymers where only the inner block is labelled.

IV. Cross-link fluctuations in a rubber

If a polymer melt is chemically cross-linked, the liquid-like chain molecules are forced into a permanent three-dimensional network structure: a rubber is formed. Its elastic modulus G is largely determined by the number density v of elastically active chains in-between cross-links [1].

$$G = g v L k_B T \tag{7}$$

where L is the Avogadro number. As in the case of melt dynamics the elastic force originates from the conformational entropy of each chain. The proportionality factor g contains all the details of the network structure and dynamics. In particular it is determined by the local constraints, which are imposed on the cross-link motion. Most of the theories of rubber elasticity assume that the cross-links are embedded in the network medium and hence, they must affinely follow macroscopic displacements [23]. In this case, we have $g = 1$. On the other extreme and in contrast to the affine models, the so-called phantom network model [24–26] assumes a network consisting of Gaussian chains, whose only property is to transfer forces between cross-links. They are devoid of other material properties like, e.g., a specific volume. They even may cut through one another freely. In such a network the cross-links fluctuate within a fluctuation range approximately as large as the end-to-end distance of the mesh. The constraints on the cross-links are reduced and for a four-functional-network g becomes 1/2. Real networks will behave somewhere between the two extremes. Unfortunately, the cross-link fluctuation is not accessible by macroscopic experiments since the prefactor of Eq. (1) cannot be determined independently – it always contains the unknown number of elastically active chains and is affected by the network structure, network defects, etc.

Here we review first dynamic experiments which revealed information on the time scale and spatial extent of the cross link motion.

The experiments were performed on PDMS model networks. The polydispersity of the precursor polymer was $M_w/M_n = 1.8$ with $M_n = 3100$. Each cross-link was labelled by 50 protons. In order to look for correlations between the cross-links, SANS experiments were performed. In the Q-range $0.01 < Q < 0.3$ Å^{-1} the intensity profiles exhibit nearly no Q-dependence. A slight decay towards larger Q is probably due to the form factor of the individual cross-links which have a radius of gyration of about 10 Å. Since we do not observe any correlations between different cross-links, their scattering can be treated as originating from a *single junction, revealing the self-correlation function* of the cross-link. Furthermore, the radius of gyration of the precursor polymers was determined to be $R_g = 25$ Å, defining the mesh size of the network. For comparison also a melt of end-labelled deuterated PDMS of the same molecular weight as the precursor polymers was prepared.

The quasielastic neutron scattering experiments were performed at 100 °C using the spin-echo spectrometer IN 11 at the ILL, thereby, covering a Q-range $0.04 \leqslant Q \leqslant 0.130$ Å^{-1}. Figure 6 compares spectra obtained for the labelled cross-links and the end-labelled chains at two different Q-values. Without any analysis a strong reduction of the cross-link mobility compared to that of the chain ends is obvious. A closer inspection also shows that the lineshape of both curves differs. While the dynamic structure factor $S(Q,t)$ from the chain ends decays continuously, $S(Q,t)$ from the cross-links appears to decay faster at short times than for longer times. This difference is quantified in Fig. 7 where the time exponent of the stretched exponential relaxation function for both samples is plotted vs Q. The measured exponent for the end-labelled chain is in close agreement with the $n = 1/2$ prediction for the self-correlation function of a Rouse chain [see Eq. (3)]. The observation of an initially faster decaying relaxation curve for the cross-links translates into a significantly smaller time exponent n. A fit with Eq. (3) yields the Rouse parameter for the end-labelled chain. We found $W\sigma^4 = 3.2 \pm 0.4 \times 10^{13}$ Å^4 s^{-1}, about twice as large as that obtained from labels at random positions along the chain ($1.75 \pm 0.15 \times 10^{13}$ Å^4 s^{-1}; see above) and in good agreement with a prediction of Warner [27], according to which the characteristic frequency doubles for the chain ends while for f-fold connected segments it is reduced by $2/f$. For the Q-exponent of the characteristic frequency a value of 3.6 ± 0.4 in close accord with Rouse model is found.

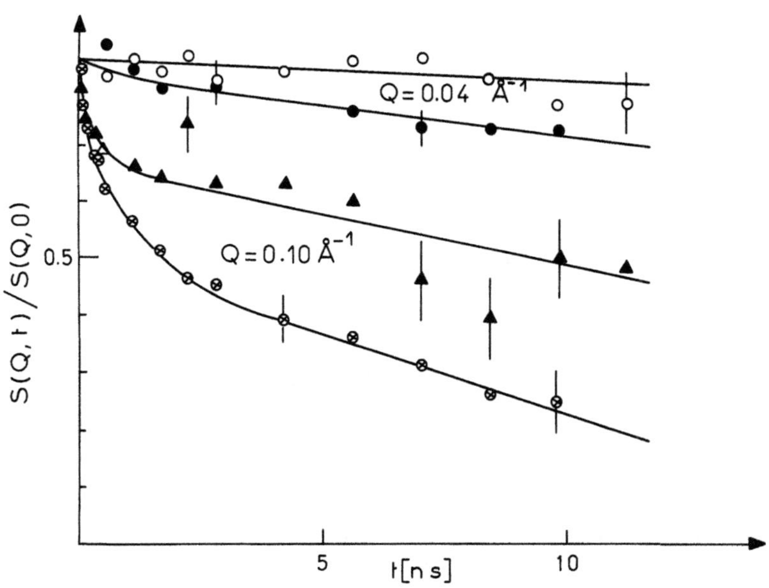

Fig. 6. NSE spectra from end-labelled PDMS chains in the melt (○, ⊗) ($M_w = 5500$) and labelled cross-links in a network with a mesh of $M_w = 5500$ at $T = 100$ °C. The solid lines are guides to the eyes

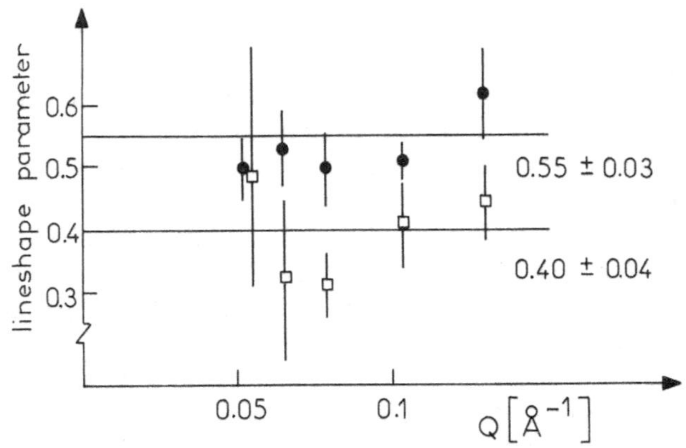

Fig. 7. Time exponents in the stretched exponential relaxation functions obtained from the NSE spectra of the end-labelled PDMS chains and the labelled cross-links

The same analysis of the network data yields again a Q-exponent of 4 but a strongly reduced Rouse parameter $W\sigma^4 = 0.54 \pm 0.06 \times 10^{13}$ Å^4 s^{-1}, 40% below the prediction for a four functional cross-link.

In a second step wo now consider spatial restrictions on the cross-link motion: other than the motion of a labelled segment in a linear chain which may essentially probe the whole sample volume, the spatial fluctuation of a cross-link being part of a macroscopic network is confined in space. We assume that for asymptotic times the probability to find a junction at a distance r from its equilibrium position is given by a Gaussian profile $\exp(-r^2/l^2)$. Then, besides a component decaying in time the self-correlation function contains a time independent part, the Fourier transform of which appears as a constant part in the intermediate scattering function. This contribution is also termed elastic incoherent structure factor (EISF). Assuming the the dynamics of a cross-link still remain Rouse-like, Eq. (3) transforms to:

$$S(Q,t) = \exp\left(-\frac{Q^2 l^2}{4}\right) + \left[1 - \exp\left(\frac{Q^2 l^2}{4}\right)\right] \times \exp\left(-Q^2 \sqrt{\frac{W\sigma^4 t}{9\pi}}\right) . \quad (8)$$

Figure 8 presents the fit of Eq. (8) to the cross-link data. The shaded areas display the constant EISF contribution to $S(Q,t)$. A combined fit to all spectra yields $W\sigma^4 = 0.84 \pm 0.08 \times 10^{13}$ Å^4 s^{-1} and $l = 24.5 \pm 1.5$ Å. Displayed are fits to single spectra with the rate fixed to the value obtained from the joint fit. In this way l-values are determined for each momentum transfer separately. The results fluctuate between 22.6

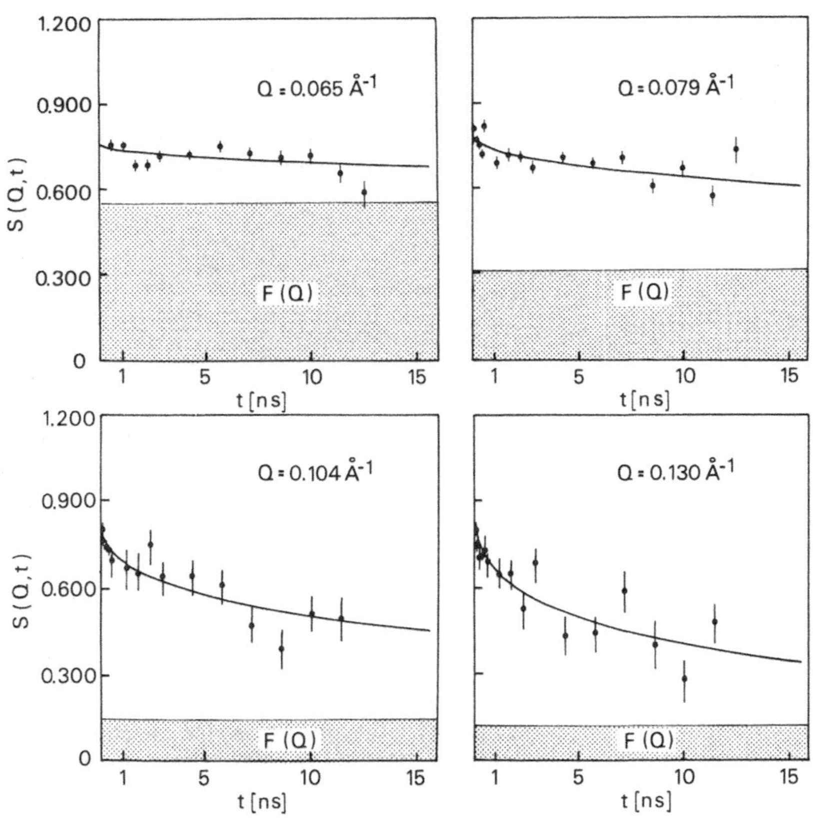

Fig. 8. NSE-spectra for labelled cross-links at four different Q-values. Included is the result of a fit with a sum of a constant EISF contribution $F(Q)$ and a Rouse relaxation spectrum. Note that the NSE spectra do not approach 1 for $t \to 0$. This effect is related to fast relaxation processes of the deuterated mesh which have not been subtracted

and 28.1 Å and enclose the value obtained from the joint fit, thereby demonstrating the internal consistency of the description. Thus, the fluctuation range of the cross-links is rather extended with $l \simeq 24.5$ Å nearly equal to the radius of gyration of a mesh ($R_g = 25$ Å). The obtained Rouse parameter amounts to closely half the value of the randomly labelled melt in very good agreement with the prediction of a $2/f$ reduction.

We now compare the experimentally obtained fluctuation range with the prediction for the phantom network. In this model the width of the Gaussian distribution of cross-link fluctuations is given by

$$\langle l^2 \rangle = \frac{4}{3f} \langle R_E^2 \rangle \tag{9}$$

where f is the functionality of the network. With our experimental value for $R_g^{\mathrm{mesh}} = 25$ Å we obtain $l^{\mathrm{theo}} = 35$ Å, somewhat larger than the experimental value of 24.5 Å. Thus, qualitatively, theoretical prediction and experimental results agree; quantitatively the smaller experimental fluctuation ranges suggest that in a real network additional constraints are active which reduce the fluctuation range below the phantom network prediction.

V. Conclusion

In this short review we surveyed neutron spin-echo investigations on the intramolecular motion in polymer melts and rubbers: i) We presented results on the time-dependent self- and pair-correlation function of a highly entangled polymer chain in a melt. In the short time regime both correlation functions are in very good agreement with the prediction of the Rouse model. For longer times or larger values of the Rouse scaling variable strong deviations from Rouse dynamics towards slower relaxation are found. Thereby, and importantly, the scaling with the Rouse variable is preserved. Thus the experiments bear no evidence for the existence of a second microscopic length like the tube diameter or the entanglement distance. With this respect the data are at variance with the model of local reptation of de Gennes or the generalized Rouse model by Ronca [17]. The second part of the paper dealt with the motion of labelled cross-links in a rubber. The NSE experiments show that the diffusive motions of the junctions are Rouse-like and are limited in space. For the first time the spatial fluctuation range which is covered by the junctions was determined experimentally. The results are in satisfactory agreement with phantom network predictions.

Acknowledgement

The autors are indebted to R. Oeser and T. Wagner (MPI für Polymerforschung, Mainz) for the preparation of the samples and to B. Farago (ILL) for his contributions to the experiment.

References

1. De Gennes PG (1979) Scaling concepts in polymer physics. Cornell University Press, Ithaca, London
2. Ferry JD (1980) Viscoelastic properties of polymers. Wiley & Sons, New York Chichester Brisbane Toronto
3. Rouse PE (1953) J Chem Phys 21:1272–1280
4. De Gennes PG (1971) J Chem Phys 55:572–579
5. Edwards SF, Grant JMV (1973) J Phys A6:1169–1185
6. Doi M, Edwards SF (1978) J Chem Soc Faraday Trans 274:1789–1832
7. Doi M (1980) J Polymer Sci Phys Ed 18:1005–1020
8. Doi M (1983) J Polymer Sci Phys Ed 21:667–687
9. Richter D, Ewen B, Farago B, Wagner T (1989) preprint
10. Oeser R, Ewen B, Richter D, Farago B (1988) Phys Rev Lett 60:1041–1044
11. Mezei F (1972) Z Phys 255:146–160
12. Mezei F (1980) (ed) Neutron spin echo. Springer Verlag, Heidelberg Berlin New York
13. Hayter JB, Penfold J (1979) Z Phys B35:199–205
14. De Gennes PG (1967) Physics (USA) 3:37–45
15. De Gennes PG (1981) J Physique 42:735–740
16. Graessly WW (1980) J Polymer Sci Phys Ed 18:27–34
17. Ronca G (1983) J Chem Phys 79:1031–1043
18. Graessly WW (1977) Advances in Polymer Science 16
19. Barlow AJ, Harrison G, Lamb J (1964) Proc Roy Soc London A282:228–251
20. Rahalker RR, Lamb J, Harrison G, Barlow AJ, Hawthorne W, Semlyen JA, North AM, Pethrick RA (1983) Faraday Symp Chem Soc 18:103–114
21. Kremer K, Grest GS, Carmesin I (1988) Phys Rev Lett 61:566–569
22. For a recent overview see e.g. Pietralla M, Kilian HG (eds) (1987) Permanent and transient networks. Progress in Colloid and Polymer Science 75
23. Flory PJ (1976) Proc Roy Soc A351:351–378
24. James HM (1947) J Chem Phys 15:651–668
25. James HM, Guth E (1947) J Chem Phys 15:669–683
26. James HM, Guth E (1953) J Chem Phys 21:1039–1049
27. Warner M (1981) J Phys C Solid State 14:4985–4994

Received January 30, 1989;
accepted February 22, 1989

Authors' address:

Dr. D. Richter
Institut Laue-Langevin
156X
38042 Grenoble Cedex, France

Discussion

KIMMICH:

You identify the reptation model with the mean square displacement behavior $\langle r^2\rangle \sim t^{1/4}(t < T_d)$. Previously we have concluded from NMR T_1-dispersion data that rather the proportionality $\langle r^2\rangle \sim t^{1/2}(T_d < t < \tau_d)$ is valid (Kimmich R (1975) Polymer 16:851 and subsequent papers). Nevertheless, we speak of reptational displacements which are one-dimensional per definition. This reduced dimensionality reveals itself in different dependences of diverse NMR parameters in spite of the superposition of other degrees of freedom.

RICHTER:

Since in a neutron scattering experiment we are sensitive to the space time development of a diffusing entity, we have direct access to the mean square displacement of a polymer segment. In the case of Rouse-type dynamics, this mean square displacement already increases much more slowly than for normal diffusion, where $\langle r^2\rangle \sim t$ holds. The prediction is $\langle r^2\rangle \sim t^{1/2}$. If in addition, due to topological constraints the polymer motion is confined to a tube, this power law changes to $\langle r^2\rangle \sim t^{1/4}$ [1].

[1] De Gennes PG (1971) J Chem Phys 55:572–579

WARTEWIG:

I am surprised that the Rouse model works well in the time scale $10^{-9} \ldots 10^{-10}$ s. Is that a consequence of the very high flexibility of the PDMS? Our ultrasonic studies on polybutadiene and poly(phenylmethylsiloxane) have shown that the minimum relaxation time of Rouse-like motions is in the range of $10^{-7} \ldots 10^{-8}$ s.

RICHTER:

Without doubt our experimental data show that in the time range between a few times 10^{-10} to 10^{-8} s in all its aspects, the polymer relaxation of PDMS in a melt is well described by the Rouse model. This concerns the time and the Q-dependence of the self- as well as the pair-correlation function. This result must be related to the very high flexibility and the very low monomeric friction coefficient of PDMS. According to the tables in Ferry's book there is nearly a two order of magnitude difference between the friction coefficient of polybutadiene at room temperature and PDMS at 100 °C. This difference is the answer to your question.

Progress in Colloid & Polymer Science Progr Colloid & Polymer Sci 80:63–68 (1989)

Optical measurements of the deformation behavior of thin metallized viscoelastic layers for light valves

R. Gerhard-Multhaupt, W. Brinker and R. Tepe

Heinrich-Hertz-Institut für Nachrichtentechnik Berlin GmbH, Berlin, FRG

Abstract: The investigation of thin viscoelastic layers coated with a highly reflective top electrode is surveyed. Such layers find applications in active-matrix-addressed spatial light modulators that are of potential use for the light-valve projection of high-resolution images, for input devices of optical information-processing systems, for adaptive optical components, for optical interconnections, for wavelength conversion of images, etc. Spatially and temporally varying electric fields generated by active matrices lead to a deformation of the top mirror electrode. Phase-shift interferometry is employed in order to gain quantitative experimental information on the deformation behavior. From a comparison of experimental results with the predictions of an extensive theoretical analysis, some material properties of the thin viscoelastic layers may be estimated.

Key words: Elastomer layer, silicon gel, light valve, phase-shift interferometry, linear viscoelasticity, diffraction, spatial light modulator

1. Introduction

The basic principle of a light valve with a deformable layer was conceived some 50 years ago: In what was to become the commercial "Eidophor" ("image carrier") projector [1,2], the light of an arc lamp is diffracted by electron-beam-generated deformations of an oil layer that represents a phase grating; the phase modulation of the light is transformed into the required amplitude modulation by means of a schlieren-optical system that blocks all undiffracted light. Since the coexistence of an oil layer and an electron source inside a high-vacuum vessel poses some problems, the basic idea of the "Eidophor" was modified by use of a gel instead of the oil and a photoconductor layer for the generation of the deforming electric field [3]. Several optical arrangements and addressing schemes were suggested [4,5] for this "image amplifier" which can also be employed in optical-information-processing systems [6].

In the devices discussed so far, the read-out light is reflected from a mirror behind the oil layer (Eidophor) or from the interface between the gel layer and air (total internal reflection). Thus, the deformable layer has to be optically transparent and extremely homogeneous. These requirements can be avoided if the gel layer is covered by a mirror electrode as in the so-called "gamma-Ruticon" devices [7–15] proposed for image intensification [7], holographic recording [7], image storage [7,10], optical processing [10], and light-valve projection [13]. The performance of the photoconductor-addressed "Ruticon" ("wrinkle image") was investigated in detail, both experimentally [8,9,12,14,15] and theoretically [8,9,11,12,15]. It did, however, not find its way into commercial applications.

Another commercial device based on the same principle is the "Talaria" projector in which a deformable oil film covers a glass disk with a transparent electrode [16]; the electron-beam-addressed spatial light modulator can thus be used in transmission instead of the reflection found with all previously discussed cases. Again, further development of the principle led to a device with a viscoelastic gel layer [17–21]; the mirror electrode is now supported by a thin pellicle [17–20], which improves its durability and optical quality, and an active matrix containing electronic switches at each picture element is suggested for addressing the solid-state spatial light modulator [19,21].

In the present survey, recent theoretical [22–26] and experimental [22,25,26] investigations of this active-matrix-addressed viscoelastic spatial light modulator

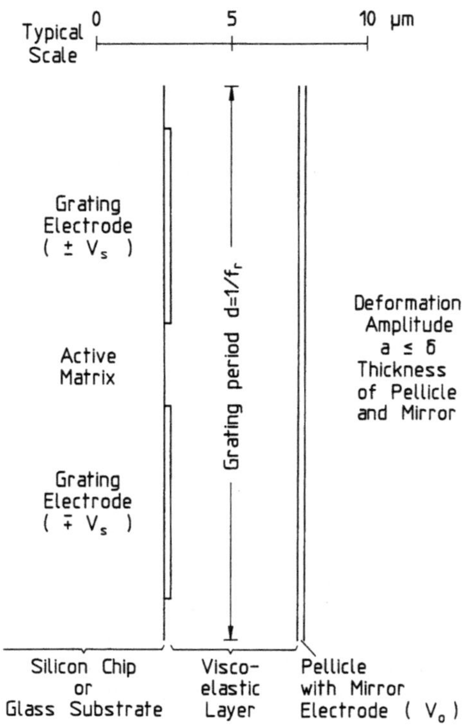

Fig. 1. Schematic of an active-matrix-addressed viscoelastic spatial light modulator with pellicle and mirror electrode; all dimensions are shown in approximately the same scale as indicated

are briefly summarized and discussed in view of their potential use for the measurement of the viscoelastic properties of thin layers. Further details of the methods and the results cited here can be found in the original publications [22–26].

2. Theoretical and experimental studies

Figure 1 shows the geometry of a viscoelastic spatial light modulator: The silicon wafer or glass substrate carrying an active matrix with peripheral driving electronics, row and column electrodes, switching transistors, and grating electrodes are spin-coated with a silicon-gel layer. On top of this viscoelastic layer a thin pellicle and a highly reflective mirror electrode are deposited. A bias voltage V_0 is applied between the mirror electrode and the grating electrodes which are in turn driven by the active matrix with a spatially alternating signal voltage V_s that varies from picture element to picture element. The resulting periodic potential distribution can be approximated by its first harmonic: $V_e(x) = V_0 + V_s \cos(\omega_r x)$ where $\omega_r = 2\pi f_r$ is the spatial angular frequency, $f_r = 1/d$ is the spatial frequency and d is the grating period of the electrodes. It leads to a periodic deformation of the mirror electrode with an amplitude a that is typically about equal to the thickness δ of the pellicle and its metallization.

2.1 Theoretical analysis of the deformation behavior

The theoretical analysis [23, 24, 26] of the metallized viscoelastic spatial light modulator consists of three parts as indicated in the block diagram of Fig. 2. Electrostatics is employed in order to determine the electrostatic stress distribution acting on the mirror electrode from the exciting potential distribution $V_e(x)$ given above; the reaction of the silicon-gel layer is derived from the theory of linear viscoelasticity (with the assumption of frequency-independent shear modulus and viscosity); and the reaction of the metallization is determined by means of the theory of strong bending of plates. Equilibrium of the exciting [electrical transfer function $T_e(\omega_r)$] and reacting [mechanical transfer function $T_m(\omega_r)$] stresses finally yields the required surface deformation [22, 24, 26]. Major theoretical analyses of metallized viscoelastic spatial light modulators are listed in Table 1 which shows that the most complete treatment including the viscous and the elastic properties of the silicon gel, as well as the bending resistance of the metal electrode was published only recently [24, 26].

2.2 Experimental investigation of the deformation behavior

Initial deformation measurements [22] were carried out by the qualitative method of differential-interference-contrast microscopy, by relative-diffraction-efficiency experiments, which lead to correct results only if the deformations are sinusoidal, and by Mirau two-beam interferometry, which requires a time-consuming photometric evaluation for accurate results. A fully quantitative technique that yields not only the amplitude but also the shape of the surface deformation is the so-called phase-shift interferometry [27]; the experimental set-up for this method is schematically depicted in Fig. 3.

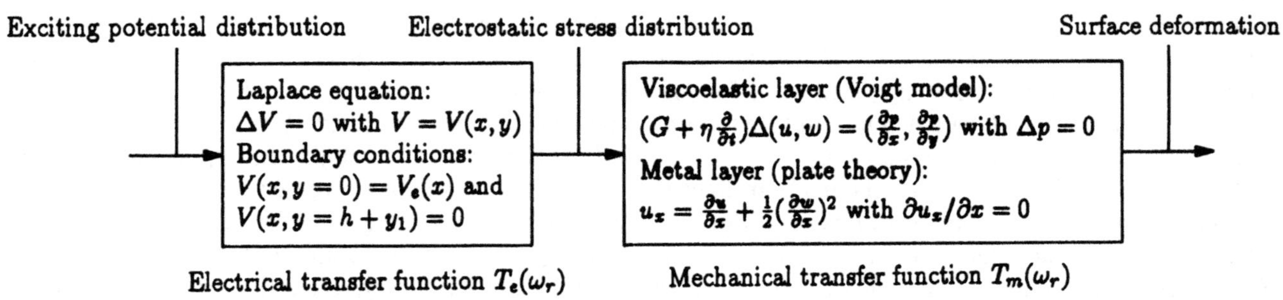

Fig. 2. Block diagram of the physical processes governing the deformation behavior of the metallized viscoelastic spatial light modulator and of the models used for its analysis

Table 1. Physical models employed for the description of the viscoelastic and metal layers in theoretical treatments of metallized viscoelastic spatial light modulators [11, 15, 23, 24]

Reference	Viscolelastic layer	Metal layer
[11]	Elastic shear modulus	Effective surface tension
[15]	Elastic shear modulus	Bending modulus of plate
[23]	Linear viscoelasticity	Effective surface tension
[24]	Linear viscoelasticity	Bending modulus of plate

A microscope interferometer is used in order to determine the phase distribution of the light reflected from the sample under investigation. If this phase distribution is measured for three different vertical positions of the sample surface, whose separation by a phase shift of $\pi/4$ is controlled by a piezoelectric actuator, then the required height profile across the sample surface can be calculated from the three recorded interference patterns. Details of the implementation and further references are given in a previous paper [25].

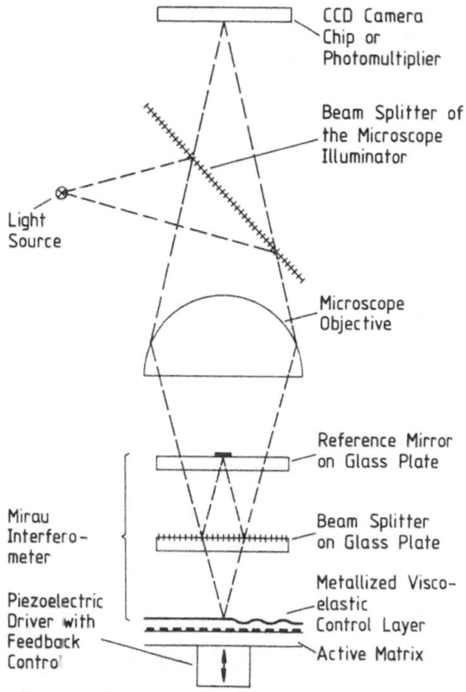

Fig. 3. Schematic diagram of the experimental arrangement for phase-shift interferometry consisting of a piezoelectrically positioned sample stage, a two-beam interferometer, a microscope, and a CCD camera or a photomultiplier tube

3. Typical results

Figure 4 depicts the surface-deformation profile of an 8.3 µm-thick viscoelastic layer at a grating period of 100 µm. As seen in the lower part of the figure, the bias voltage $V_0 = 175$ V already leads to a deformation because of the periodical structure of electrodes and spacings. Application of an additional signal voltage $V_s = \pm 10$ V, whose polarity alternates from electrode to electrode, yields a deformation with twice the period of that found without the signal (upper part of the figure); furthermore, the shape of the surface deformation is not purely sinusoidal, since the grating period is much larger than the layer thickness (see below).

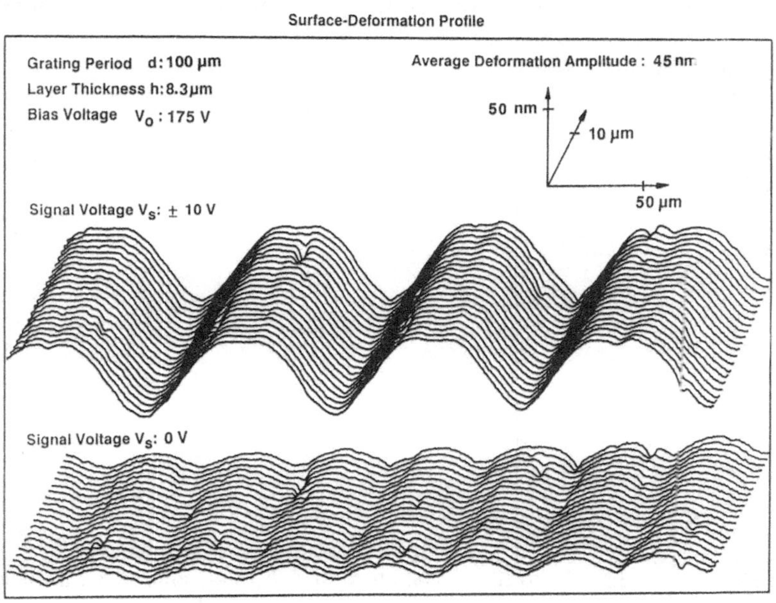

Fig. 4. Measured surface-deformation profiles of an 8.3 µm-thick metallized viscoelastic layer with and without a voltage signal of 10 V amplitude and 100 µm grating period

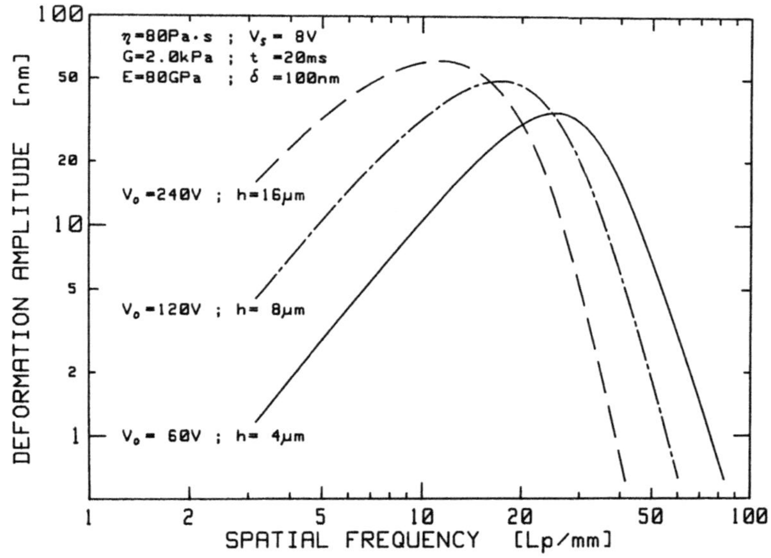

Fig. 5. Calculated spatial-frequency responses of 4-, 8-, and 16 μm-thick metallized viscoelastic layers

3.1 Time response

Previously reported theoretical and experimental results [22–26] demonstrate that the rise and fall times of the deformation are on the order of a few ms. For television and other video applications with their typical frame rates of 50 or 60 Hz and corresponding image periods of 20 or 16.6 ms, this time response is sufficient, since the active matrix provides almost constant signal voltages at every picture element for one image period, which also leads to very low flicker.

3.2 Spatial-frequency response

The resolution of an image generated by the viscoelastic spatial light modulator is limited by the modulator's spatial-frequency response. Theoretical and experimental spatial-frequency-response curves for three different gel-layer thicknesses h are shown in Figs. 5 and 6, respectively. In order to avoid destructive electrical breakdown, the bias fields have to be kept approximately constant for the different thicknesses; therefore, the bias voltages were adjusted accordingly for the experiments (Fig. 6) and also for the numerical evaluations (Fig. 5). In both cases, the deformation time was chosen as $t = 20$ ms and the signal-voltage amplitude as $V_s \approx 8$ V. For the numerical evaluation of the theoretical spatial-frequency response, the material parameters $G = 2$ kPa (shear modulus of the viscoelastic layer), $\eta = 80$ Pa s (viscosity of the gel), $E = 80$ GPa (elastic modulus of the metallization), and $\delta = 100$ nm (mirror-electrode thickness) were chosen. In view of the non-sinusoidal

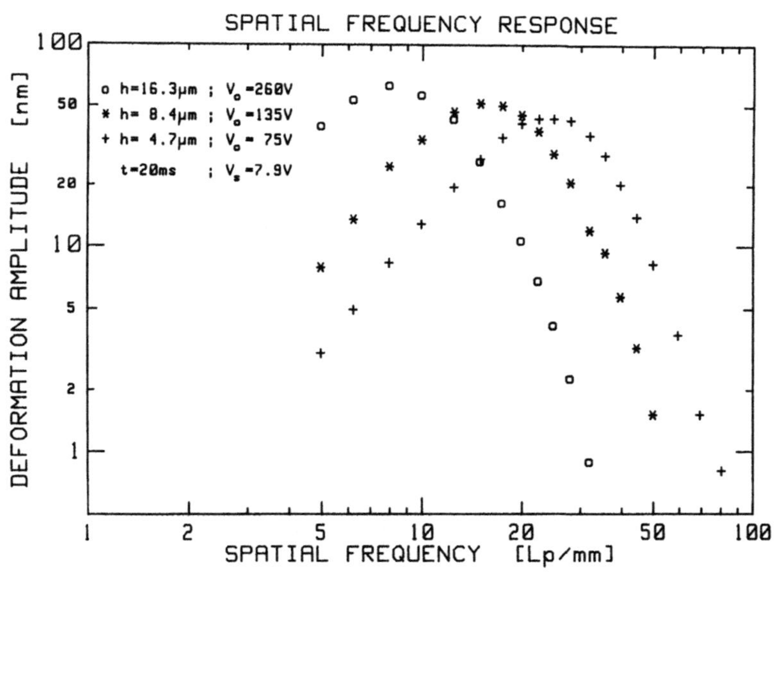

Fig. 6. Measured spatial-frequency responses of 4.7-, 8.4-, and 16.3 μm-thick metallized viscoelastic layers

deformations found for large grating periods (see Fig. 4), Fourier transformation, suppression of higher spatial-frequency components by filtering, and inverse Fourier transformation were employed [25] in order to obtain the experimental results shown in Fig. 6.

A comparison of Figs. 5 and 6 reveals that the band-pass character of the spatial-frequency response is very well predicted by the theoretical analysis. As discussed in previous papers [22–26], the sensitivity decrease at lower spatial frequencies is mainly caused by the incompressibility of the silicon gel, while the steep decrease at higher spatial frequencies is a consequence of a field homogenization for grating periods d equal to or smaller than the layer thicknesses h and of the metal-layer's bending stiffness, which strongly depends on the radius of curvature. Further agreement between theory and experiment is found insofar as the maximum of the band pass shifts towards higher spatial frequencies and lower deformation amplitudes if the thickness of the viscoelastic gel is decreased.

4. Discussion

By comparing the theoretically derived and the experimentally determined fall times of the deformation, the viscosity of the viscoelastic sample layer may be estimated. A comparison between the slopes of the measured and the calculated spatial-frequency responses at low spatial frequencies permits an estimate of the gel-layer's shear modulus. Both parameters were derived from the measurements whose spatial-frequency results are given in Fig. 6; the estimated values for a 4.7 µm thick experimental silicon-gel layer are 2–7 kPa for the shear modulus and approximately 10 Pa s for the dynamic viscosity. Bulk data for some commercial silicon gels, all of which are addition-crosslinked two-component systems with a catalyst, are listed in Table 2.

Table 2. Viscoelastic properties of commercial addition-crosslinked two-component silicon gels as given by the manufacturers

Material (manufacturer)	Penetration after curing (test method)	Dynamic viscosity before curing (Pa s)
RTV 6157 (General Electric)	5.7 mm (CTM 0155, at room temperature)	0.6
Sylgard 527 (Dow Corning)	8 mm (69.5 g shaft)	0.33
SilGel 612 (Wacker-Chemie)	30 mm (DIN ISO 2137, 150 g hollow cone)	0.8 – 1.2

In the investigations briefly described above, the frequency dependencies of the shear modulus and the viscosity were neglected, since the time window of 20 ms used in the experiments and also in possible applications corresponds to a rather narrow frequency range. However, measurements of the frequency dependence can be carried out if a passive matrix of interdigital electrodes is used and if the time dependence of the signal voltages is not chosen to be a square function, but rather a sine function. Then the resulting deformations represent forced standing waves whose spatial frequency (or wave vector) is determined by the electrode grating, while their frequency can be varied easily over several decades. Such measurements, which are planned for the future, will allow for a nondestructive investigation of the relaxation behavior of thin viscoelastic layers; additionally, they will permit a further optimization of the silicon-gel materials, and they will demonstrate whether the assumption of frequency-independent quantities is justified.

Acknowledgements

The authors are indebted to Professor Dr. G. Mahler, Dr. W. E. Glenn, and W.-D. Molzow for stimulating discussions and to W. Wirges, R. Schardin, L. Emich, and R. Woyzella for sample preparation and numerical evaluation. Financial support by the Bundesminister für Forschung und Technologie (BMFT) within the framework of the research project No. TK 0444 4 is gratefully acknowledged; the responsibility for the contents of this paper rests, however, solely with the authors.

References

1. Baumann E (1952) J Brit Inst Radio Eng 12:69
2. Mol JC (1962) Photogr J 1962:128
3. Baumgartner W (1967) Z Angew Math Phys 18:31
4. Mast F, Waser R (1983) Europ Pat 0029006
5. Rhomberg B (1986) Swiss Pat Appl No 9-15914
6. Hess K, Dändliker R, Thalmann R (1987) Opt Eng 26:418
7. Sheridon NK (1972) IEEE Trans Electron Dev ED-19:1003
8. Lakatos AI (1974) J Appl Phys 45:4857
9. Lakatos AI (1975) J Appl Phys 46:1744
10. Sheridon NK, Berkovitz MA (1976) SPIE Proc 83:68
11. Kermisch D (1976) Appl Opt 15:1775
12. Lakatos AI (1977) J Appl Phys 48:2346
13. Lakatos AI, Bergen RF (1977) IEEE Trans Electron Dev ED-24:930
14. Wysocki JJ (1982) Appl Opt 21:2205
15. Bernstein H (1986) M Sc Thesis, Weizmann Institute of Science, Rehovot, Israel
16. Glenn WE (1970) J SMPTE 79:788

17. Glenn WE (1985) US Pat 4529620
18. Glenn WE (1985) US Pat 4626920
19. Glenn WE (1985) Europ Pat 0184378
20. Glenn WE (1985) Deut Pat 3503079
21. Glenn WE (1987) SID Digest 87:72
22. Tepe R, Gerhard-Multhaupt R, Brinker W (1986) Proc SPIE 684:20
23. Tepe R (1987) J Opt Soc Am A 7:1273
24. Tepe R (1988) ntz-Archiv 10:269 and 10:295
25. Brinker W, Gerhard-Multhaupt R, Molzow W-D, Tepe R (1989) Proc SPIE 1018:79
26. Tepe R, Gerhard-Multhaupt R, Brinker W, Molzow W-D (1989) Appl Opt 28: in press
27. Wyant JC, Koliopoulos CL, Bhushan B, George OE (1983) ASLE Transactions 27:101

Received March 10, 1989;
accepted March 30, 1989

Authors' address:

Dr. R. Gerhard-Multhaupt
Heinrich-Hertz-Institut für Nachrichtentechnik
Einsteinufer 37
1000 Berlin 10, FRG

Two-dimensional NMR studies of structure and dynamics of solid polymers

H. W. Spiess

Max-Planck-Institut für Polymerforschung, Mainz

Two-dimensional (2D) NMR offers unique possibilities for a detailed characterization of molecular order and dynamics in solids and solid polymers. This holds in particular for ultraslow dynamic processes occurring on a timescale above 1 ms [1].

Two-dimensional exchange spectra of static samples display characteristic ridge patterns if the reorientational motion occurs around well-defined angles [2], e.g., in crystalline polymers. In amorphous polymers, however, rotational motions of both polymer chains and side groups typically involve rotations about ill-defined angles, leading to angular distributions [3]. These are directly projected into broad exchange patterns of the 2D-NMR spectra [4].

The *angular* resolution is highest for static samples, e.g., for ^{2}H-NMR of isotopically labelled samples or ^{13}C-NMR of simple polymers [5]. The *spectral* resolution can be increased at the expense of the angular resolution ($\pm 10°$ instead of $\pm 1°$) by invoking magic angle spinning (MAS) [5]. Rotor-synchronized MAS together with rotations in spin space, e.g., by the TOSS-sequence, not only allows the detection of molecular order and dynamics separately but also offers a means to *correlate* these two molecular parameters [6] which are often supposed to govern the macroscopic properties of solid polymers.

The techniques have been applied to study helical jump motions in polyoxymethylene [5], the chain motion of polystyrene in the vicinity of the glass transition [4], the molecular order in drawn fibers of polyethyleneterephthalate [7] and in side-group liquid crystalline polymers [8]. The correlation between order and mobility was demonstrated in drawn fibers of polyoxymethylene [9].

References

1. Blümich B, Spiess HW (1988) Angew Chem 100:1716, Int Ed Engl 27:1655
2. Schmidt C, Wefing S, Blümich B, Spiess HW (1986) Chem Phys Letters 130:84
3. Wefing S, Spiess HW (1988) J Chem Phys 89:1219
4. Wefing S, Kaufmann S, Spiess HW (1988) J Chem Phys 89:1234
5. Hagemeyer A, Schmidt-Rohr K, Spiess HW (1989) Adv Magn Res 13, in press
6. Yang Y, Hagemeyer A, Blümich B, Spiess HW (1988) Chem Phys Letters 150:1
7. Harbison GS, Vogt V-D, Spiess HW (1987) J Chem Phys 86:1206
8. Boeffel C, Spiess HW (1989) in "Side Chain Liquid Crystal Polymers" (ed) by C. B. McArdle. The Blackie Publ Group, Glasgow, p 224
9. Yang Y, Hagemeyer A, Spiess HW (1989) Macromolecules 22:1004

Received March 16, 1989;
accepted March 30, 1989

Author's address:

H. W. Spiess
Max-Planck-Institut für Polymerforschung
Postfach 3148
6500 Mainz, FRG

The influence of the molecular weight distribution on the viscosity and elastic properties of polymer melts

H. Schuch

Kunststofflaboratorium, Abt. Polymerphysik, Festkörperphysik, BASF Aktiengesellschaft, Ludwigshafen/Rhein, FRG

The molecular weight distribution of unbranched polymers controls their viscoelastic properties. In order to find out this dependency, extreme distributions represented by binary blends of narrowly distributed polymers were studied.

A generalization of the blending rule for the zero shear viscosity gives a blending rule for the *components of the complex viscosity in oscillatory shear*, which is known to be valid for some polystyrene melts (Montfort et al. (1979) Rheol. Acta 18:623). This rule reads:

$$\overset{*}{\eta}{}_b^a = w_1 \overset{*}{\eta}{}_1^a + w_2 \overset{*}{\eta}{}_2^a \;;$$

with $a = 1/3.4$; w_1, w_2 weight fractions of the components ($w_1 + w_2 = 1$); $\overset{*}{\eta}_1, \overset{*}{\eta}_2, \overset{*}{\eta}_b$ complex viscosities of the components and the blend for a given frequency ω, the viscosities being in complex notation.

An examination of the rule using literature data and our own for melts of *unbranched* polymers like PS, HDPE, polybutadiene yields:

All experimental results including those for the components of the dynamic viscosity are well described (For details see Schuch H. (1988) Rheol Acta 27:384).

The *elongational viscosity* at finite stress obeys the same blending rule as the zero shear viscosity at least in the case studied here:

A linear polystyrene ($M_w = 197\,000$) does not show any maximum of the elongational viscosity if blended with another linear polystyrene of $M_w = 1\,300\,000$. The result supports the hypothesis that long chain branching (as in LDPE) is much more effective than the molecular weight distribution in causing a pronounced maximum in the steady-state elongational viscosity vs stress.

The addition of high molecular weight components may drastically enhance *the steady state compliance* in shear and in elongation.

Details of the experiments and the results are represented in the publication: Laun HM, Schuch H (1989) J Rheology 33:119.

Received February 13, 1989;
accepted March 13, 1989

Author's address:

H. Schuch
Kunststofflaboratorium
Abt. Polymerphysik/Festkörperphysik
BASF Aktiengesellschaft
6700 Ludwigshafen/Rhein, FRG

Progress in Colloid & Polymer Science Progr Colloid & Polymer Sci 80:71–77 (1989)

Molecular motion and orientational distribution in polymer electrets as detected by nuclear magnetic resonance and relaxation

D. Geschke and P. Holstein

Karl-Marx-Universität Leipzig, Sektion Physik

Abstract: Solid state NMR is used to study molecular motion and orientational distribution in thin films of PVDF and PA 11 which are of particular interest in technical application. The variation of angle between the static magnetic B_0-field and preferred directions given by drawing and poling, respectively, leads to characteristic angular dependencies of the effective transverse relaxation time $T_{2\mathrm{eff}}$. This allows one to obtain information on the effect of external mechanical and electrical fields on *c*- and *b*-axes distribution. A good *c*-axes alignment could be reached by drawing. The tendency for *b*-axes to orient towards the direction of the applied electric field is evident. This effect is stable over a longer period of time but can be reversed at high temperatures. The results of first in-situ measurements are given showing the the *b*-axes alignment in electric fields up to 2 MV cm^{-1} can be "switched" periodically).

Key words: Solid state NMR, polymer electrets, orientational distributions, external fields, PVDF, PA 11

Introduction

Today, electret research is very active and the wealth of publication becomes more and more difficult to assess (cf. reviews [1,2]). This is elucidated especially by the steep increases of published work in field of piezo-, pyro-, and ferroelectric polymer electrets. Besides other non-destructive techniques Nuclear Magnetic Resonance (NMR) became a valuable tool for studying molecular motions, conformations, orientational distributions, etc. (e.g., [3–5]).

The poor sensitivity of NMR, which is a disadvantage in investigations of thin films, is in contrast to the advantage of uncomplicate inclusion of large temperature ranges and the possibility to detect separately different morphological regions of macromolecular systems.

NMR investigations in polymer electret research have been done up to now especially on fluoropolymers as poly(vinylidene fluoride) (PVDF) which is of particular technical interest [6–14] (see also [15]).

This paper is concerned with 1H-NMR investigations on thin films of PVDF and PA11 (polyamide 11). Both polymers are known to exist in different crystallographic phases causing polar and non-polar structures, respectively. The extent to which macroscopic polarization can be developed strongly depends on the thermal history of the samples as well as on the effect of external mechanical (drawing) and electrical (poling) fields. PVDF generally exists in an intimate mixture of tiny crystalline regions surrounded by an amorphous matrix. It normally crystallizes from the melt in the α or II phase, in which there are portions of two $TGT\bar{G}$ chains in the unit cell, arranged so that their individual electric dipole moments cancel, leaving a non-polar structure. The technically most important polar and piezoelectric structure is the β or I phase, in which all chains are oriented similarly and in which each chain has a planar trans arrangement with negative fluorines extending in one direction from the chain backbone and positive hydrogens in the other. It can be produced by uniaxial stretching and poling, respectively.

For PA 11 also several polymorphic forms have been found. The α crystal structure is characterized by a triclinic unit cell containing one polymer repeat. The planar zig-zag conformations produce large dipole moments perpendicular to the chain. Like other nylons PA 11 undergoes a gradual structural transformation with increasing temperature from the α to the γ form. The γ form is termed pseudohexagonal. When the temperature is above 95 °C the hydrogen bonds are distributed isotropically in the direction perpendicular to the chain axis.

In both polymers structural transformations and changes of dipolar orientations can be enforced by external mechanical or electrical fields.

Experiments

Industrial PVDF films of different provenience (thickness approximately 10 ... 25 μm) have been investigated. The unoriented material contained modification II in crystalline regions in contrast to the uniaxially drawn films (drawing parameter λ) which were of type I.

For the Thermal Electrode Poling (TEP) process evaporated aluminium was used as electrodes. In order to guarantee large polarizations the following poling conditions have been choosen: $E_p = 0.5 \ldots 2.0\,\mathrm{MV\,cm^{-1}}$, $t_p = 30$ min, $T_p = 80\,°\mathrm{C}$. PA 11 films were pressed from molten pellets which have been vacuum-dried at 80 °C for 3 days. Relatively low pressures up to 0.5 MPa were employed. Drawing of films has been performed at 120 °C with $\lambda = 1 \ldots 4$. The NMR measurements were performed at a proton resonance frequency of 60 MHz using a FT-spectrometer of type FKS 178 from the Zentrum für Wissenschaftlichen Gerätebau of the Academy of Sciences of GDR. For applying strong electric fields up to 2 MV cm^{-1} (in situ measurements) a special probe head with a modified goniometer device has been constructed [13]. In order to detect separately the effect of drawing and poling two different types of sections were cut from the films (cf. Fig. 1) allowing angle variation between the static magnetic field $\boldsymbol{B}_0$ and the drawing direction (angle γ) and between $\boldsymbol{B}_0$ and the normal of the film (poling direction, angle β).

For different reasons (cf. [12]) multi-pulse measurements have been favored.

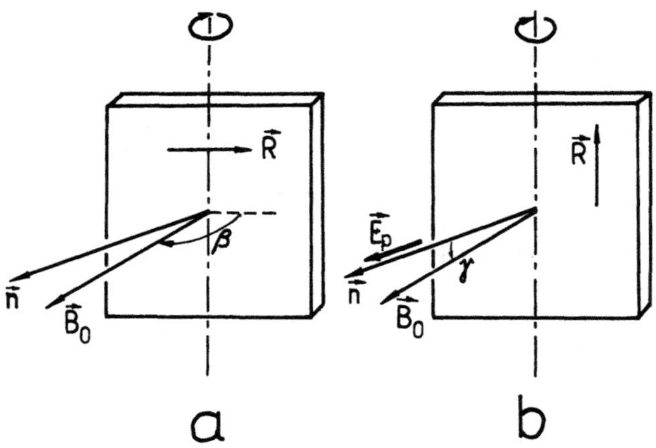

Fig. 1. Arrangement of film sections in the magnetic field for separate detection of drawing (a) and poling (b) effect; $\boldsymbol{R}$: drawing direction, $\boldsymbol{n}$: poling direction (normal of the film)

Theory

Different attempts have been made in the past to reduce the NMR linewidth of solid materials (cf., e.g., [16]). By means of multi-pulse sequences, especially, an artificial time-dependence of the dipolar interaction can be realized, leading to narrower resonance lines. In our investigations the MW 4 sequence proposed by Mansfield and Ware [17] has been applied. It can be used successfully for studies of slow molecular motions [18, 19].

According to theory the following relation holds for $\tau_c \gg \tau$

$$T_{2\mathrm{eff}} \simeq 3\tau_c / \langle \Delta B_p^2 \rangle \tau^2 \tag{1}$$

giving a direct connection between $T_{2\mathrm{eff}}$ and that part of the total second moment of the resonance line of the crystalline region $\langle \Delta B_p^2 \rangle$ which is averaged out by the α-relaxation process ($\tau_c < \tau$) or is not influenced by it ($\tau_c \geqslant \tau$). τ_c denotes the correlation time of the molecular process considered (e.g., α-relaxation) and 2τ the pulse distance in the MW 4 experiment. Therefore, the interpretation of the angular dependencies of $T_{2\mathrm{eff}}$ can be based on theoretical relations which originally have been worked out for the second moment in order to connect NMR typical parameters with those of the molecular system (e.g., distribution parameters of c- and b-axes). Distributions in general can be described by the following type of functions [20]:

$$N(\alpha', \beta', \gamma') = \sum_{l=0}^{\infty} \sum_{m=-l}^{l} \sum_{n=-l}^{l} \langle P_{lmn} \rangle Z_{lmn} \times (\cos\alpha') e^{-im\beta'} e^{-in\gamma'} \tag{2}$$

where $N(\alpha', \beta', \gamma') \sin\alpha' d\alpha' d\beta' d\gamma'$ is the fraction of structural units whose axes lie in the space angle in $\alpha' d\alpha' d\beta' d\gamma'$. The Z_{lmn} are a generalization of the Legendre function (cf., e.g., [21]). The quantities $\langle P_{lmn} \rangle$, the coefficients of the expansion, are averages of functions $P_{lmn}(\alpha', \beta', \gamma')$ taken over the aggregate of units. The assumption of at least orthorhombic symmetry, together with the fact that NMR measurements will be considered where l is even (as in the case of infrared and Raman spectroscopy) means that the $\langle P_{lmn} \rangle$ are only non-zero when l, m, n are even, ensuring that P_{lmn} are real.

Very often special cases of physically relevant symmetries are discussed in the literature for which relatively simple expressions can be derived [20]. For the second moment of the NMR resonance line the following equation holds in the case of rigid lattice and equal nuclei (cf., e.g., [22])

$$\langle \Delta B_0^2 \rangle = \frac{3}{4} K N^{-1} \sum_{\substack{j,k=1 \\ j \neq k}}^{N} [(3\cos^2\theta_{jk} - 1)/r_{jk}^3]^2 \quad (3)$$

with

$$K = \gamma_p^2 \hbar^2 I(I+1) \; . \quad (4)$$

The symbols denote:

γ_p: gyromagnetic ratio of protons
I: spin quantum number of protons
$\hbar$: Planck's constant, divided by 2π
θ_{jk}: angle between r_{jk} and B_0
r_{jk}: internuclear vector
N: number of nuclei

In the case of PVDF the proton-fluorine interaction is neglected. If the correlation frequency (τ_c^{-1}) of a possible motion is larger than the width of the resonance line of the rigid lattice, the magnetic dipole interaction must be averaged over the time and one obtains in analogy to Eq. (3) the relation

$$\langle \Delta B_{0,t}^2 \rangle = \frac{3}{4} K N^{-1} \sum_{\substack{j,k=1 \\ j \neq k}}^{N} \langle (3\cos^2\theta_{jk} - 1)/r_{jk}^3 \rangle_t^2 \; . \quad (5)$$

In the following some special cases will discussed.

Case 1

The partially oriented polymer consists of an aggregate of units of structure. Uniaxial drawing produces an oriented system characterized by a c-axes alignment along a perferred axis R together with an isotropic b- and a-axes distribution in a plane perpendicular to it.

In this case one gets for the second moment the well known equation (cf. [23, 11])

$$\langle \Delta B_0^2 \rangle = \frac{3}{4} K N^{-1} \sum_{l=0,2,4} a_l S_l P_l(\cos\beta) \langle P_l(\cos\beta') \rangle \quad (6)$$

with the lattice sums

$$S_l = \sum_{\substack{j,k=l \\ j \neq k}}^{N} P_l(\cos\lambda_{jk})/r_{jk}^6 \quad (7)$$

and $a_0 = 1/5$, $a_2 = 2/7$ and $a_4 = 18/35$.

λ_{jk} means the angle between the internuclear vector r_{jk} and the c-axis and P_l denote the Legendre polynomials.

Using known values for the lattice sums a simple relation can be derived for the dependence of the second moment on the angle β between $\boldsymbol{R}$ and $\boldsymbol{B}_0$ (cf. Fig. 1).

$$\langle \Delta B_0^2 \rangle = A + B[\tfrac{1}{2}(\cos^2\beta - 1)] + C[\tfrac{1}{8}(35\cos^4\beta - 30\cos^2\beta + 3)] \quad (8)$$

or

$$\langle \Delta B_0^2 \rangle = A' + B'\cos^2\beta + C'\cos^4\beta \quad (9)$$

where the coefficients A, B, C and A', B', C' are related to $\langle P_0 \rangle, \langle P_2 \rangle, \langle P_4 \rangle$.

For a full c-axes orientation one obtains $\langle P_2 \rangle = \langle P_4 \rangle = 1$ and therefore, a minimum value of the second moment for $\beta = 45°$.

Case 2

The system possesses the same (molecular) symmetry as in case 1 but additionally there is a fast reorientation of macromolecules around their long axes (as a model for the α-relaxation).

In this case one has to start from Eq. (5) leading to

$$\langle \Delta B_{0,t}^2 \rangle = \frac{3}{4} K N^{-1} \sum_{\substack{j,k=l \\ j \neq k}}^{N} [(3\cos^2\lambda_{jk} - 1)/r_{jk}^3]^2 \cdot S_\theta^2 \quad (10)$$

with

$$S_\theta = \tfrac{1}{2}(3\cos^2\theta - 1) \quad (11)$$

and

$$S_\theta^2 = \frac{1}{4}\Bigg\{\langle\cos^4\beta'\rangle\left[9\cos^4\beta + \frac{27}{8}\sin^4\beta - 27\cos^2\beta\sin^2\right] + \langle\cos^2\beta'\rangle \times \left[-\frac{27}{4}\sin^4\beta + 27\cos^2\beta\sin^2\beta - 6\cos^2\beta + 3\sin^2\beta\right] + \left[\frac{27}{8}\sin^2\beta - 3\sin^2\beta + 1\right]\Bigg\} \; . \quad (12)$$

θ is the angle between the magnetic field $\boldsymbol{B}_0$ and the c-axis. For full alignment one gets $\langle\cos^2(\beta')\rangle = \langle\cos^4(\beta')\rangle = 1$ and thereby

$$S_\theta^2 = [\tfrac{1}{2}(3\cos^2\beta - 1)]^2 \quad (13)$$

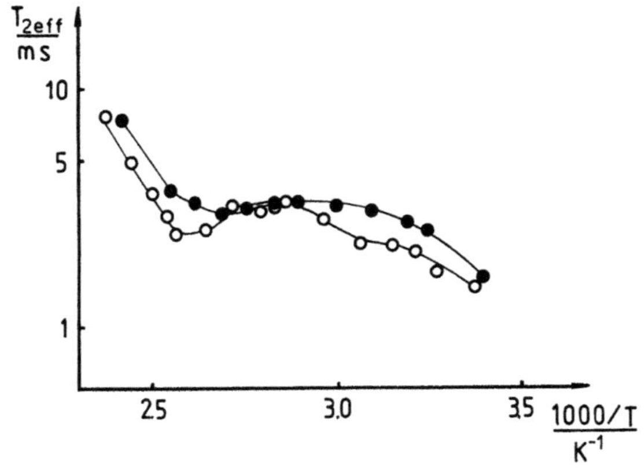

Fig. 2. Temperature dependence of $T_{2\mathrm{eff}}$ of poled (●) and unpoled (○) PVDF films

with a minimum value of the second moment at $\beta = 54.7°$.

Case 3

The assumption of isotropic distribution of *b*- and *a*-axes no longer holds. An electric field acting along the normal ***n*** of the film affects a preferred orientation of *b*-axes in the direction of E_p (cf. Fig. 1).

This case has been treated carefully by Douglass et al. [6], and in a similar way be Ishii and coworkers [8,9]. It leads for $\langle P_2\rangle = \langle P_4\rangle \simeq 1$ in analogy to Eq. (9) to the expression

$$\langle B_0^2\rangle = A'' + B'' \cos 2\gamma + C'' \cos 4\gamma \qquad (14)$$

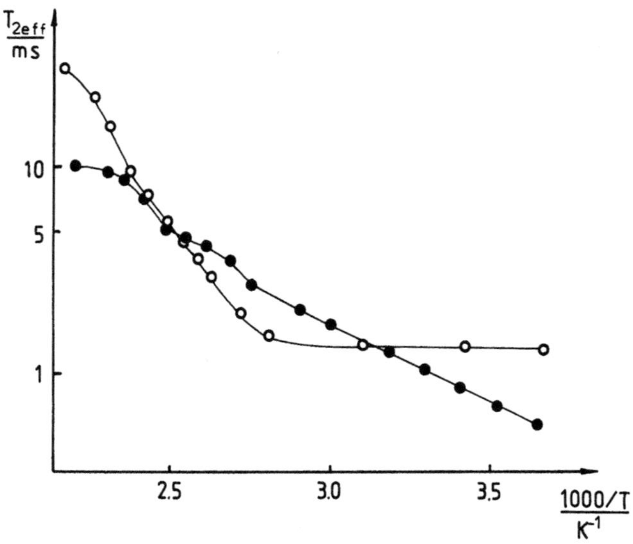

Fig. 3. Temperature dependence of $T_{2\mathrm{eff}}$ of drawn (●) and undrawn (○) PA 11 films

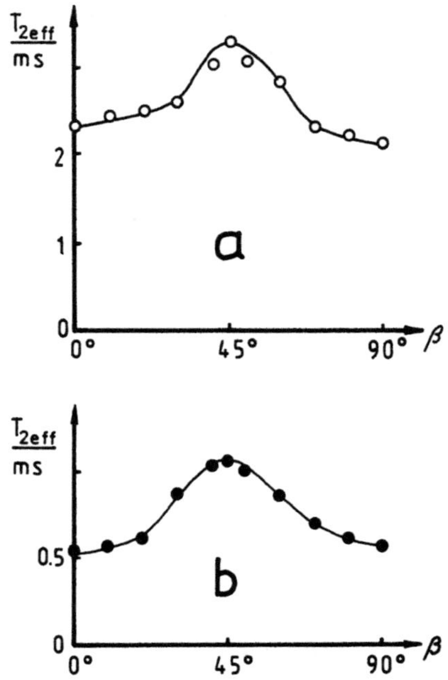

Fig. 4. Angular dependence $T_{2\mathrm{eff}}(\beta)$ of drawn PVDF (*a*) and PA11 (*b*) films, $T = 25\,°C$

(Note: $\beta = 90°$), where the parameters A'', B'', C'' are connected with the Fourier coefficients $\langle \cos 2m\delta\rangle$ and $\langle \cos 4m\delta\rangle$ of the *b*-axes distribution function [6].

For small β' ($\beta' \leqslant 20°$) δ gives the angle between ***n*** and the *b*-axis, assuming

$$\gamma' = (3\pi/2) - \alpha' + \delta \; . \qquad (15)$$

Results and discussion

In Figs. 2–7 typical results of measurements obtained on PVDF and PA 11 are presented.

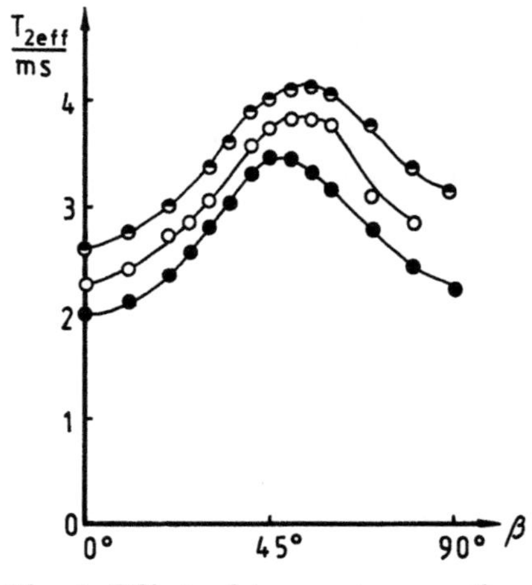

Fig. 5. Effect of temperature on the angular dependence $T_{2\mathrm{eff}}(\beta)$ of a drawn PVDF film (sample 2): ● $T = 25\,°C$; ○ $T = 100\,°C$; ◓ $T = 120\,°C$

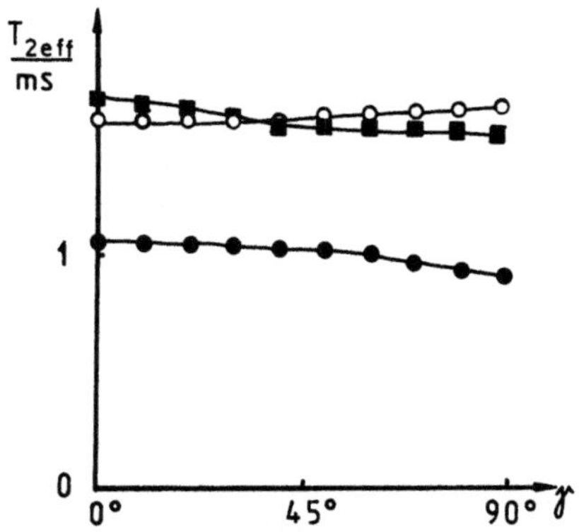

Fig. 6. Effect of poling on the angular dependence $T_{2eff}(\gamma)$ of a PVDF film ($T_p = 80\,°C, E_p = 1$ MV cm^{-1}): ○ unpoled, $T = 25\,°C$ (sample 4); ● poled, $T = -25\,°C$ (sample 5); ■ poled, $T = 25°$ (sample 5*) (measured 2 years after the poling)

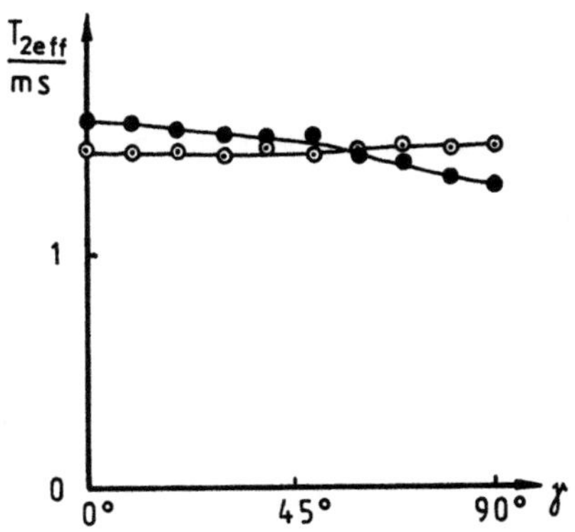

Fig. 7. Effect of annealing temperature T_e on the angular dependence of $T_{2eff}(\gamma)$ of a poled ($E_p = 1.5$ MV/cm^{-1}) PVDF film (measuring temperature $T = 20\,°C$); ● $T_e = 20\,°C$ (sample 6); ⊙ $T_e = 120\,°C$ (sample 6*)

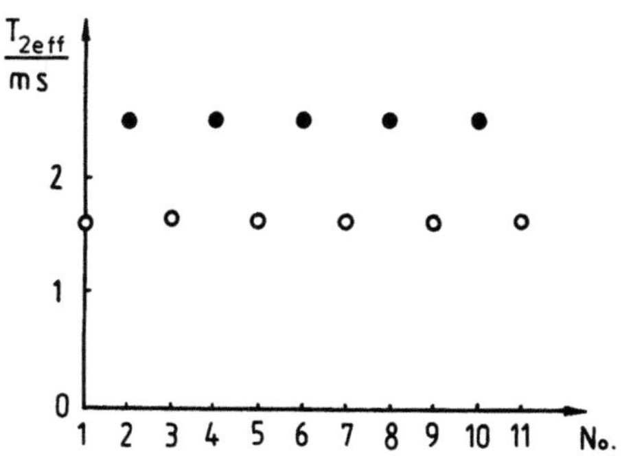

Fig. 8. Electric field effect on a PVDF film during an in-situ experiment (● on, ○ off) with $T_p = 25\,°C$, $E_p = 1$ MV cm^{-1} and time interval $\Delta t = 3$ min. N_0, denotes the number of the repeated MW 4 measurements with and without electric field, respectively

In most cases a non-exponential relaxation behavior has been observed. For this reason the values given for T_{2eff} in the following are really time constants for the decay of the effective transverse magnetization from 1 to $1/e$.

For PVDF the temperature dependence of T_{2eff} shows a pronounced minimum at $T = -25\,°C$ and a second one, weekly indicated at $T = 90\,°C$.

The former should be due to the glass transition, the latter may be explained by a hindered rotation of macromolecules around their long axes [12]. For the correlation time τ_c of this process which preferentially takes place in the crystalline regions (α-relaxation) up to $T \simeq 90\,°C\, \tau_c \gtrsim \tau$ may be assumed.

From the minimum condition for T_{2eff} (cf. [18])

$$\omega_{eff}\tau_c = \frac{\pi\tau_c}{2\tau} = 1 \ . \tag{16}$$

$\tau_c = 3.8\,\mu s$ can be estimated (with $\tau = 6\,\mu s$). It is interesting to note that this time is in the order of the dielectric relaxation time which dominantly determines the ferroelectric switching time at the Curie-temperature, as recently could be shown by Tajitsu et al. [24] for fluorine copolymers. At the present stage, however, there are many unresolved problems about ferroelectric reorientation and residual polarization which are due to the complicated structure of drawn PVDF-films. The purpose of the investigations of the temperature dependence is to clarify the influence of molecular motion on the reorientation process caused by poling. In this direction detailed investigations have been made by Ishii and coworkers [10], especially concerning the contribution of the non-crystalline regions.

In the temperature dependence of T_{2eff} in PA 11 minima are only indicated.

In analogy to studies of dielectric relaxation [25] they should be connected with motions of the amide groups (β-relaxation), the α-relaxation in the amorphous regions, and the transition at the Currie-temperature $T_c = 95\,°C$ between the α and the γ modification. There are, however, additional measurements necessary.

Figure 4 gives typical dependencies of T_{2eff} on β for PVDF and PA 11 at room temperature.

Figure 5 shows the expected effect of the temperature on the angular dependence of T_{2eff}, i.e., the shift of the maximum towards the "magic angle" due to the α-relaxation.

Figures 6 and 7 demonstrate the effect of poling.

From the experiments the following conclusions can be drawn:

1) The dipole alignment caused by an electric field is stable over a longer period of time (e.g., 2 years), if the material is kept at room temperature (cf. Fig. 6).

2) The angular dependencies as typical for $T_{2\mathrm{eff}}$ after poling (same results are obtained for T_2) can be reversed, when the temperature is increased; i.e., the thermally stimulated depolarization which begins at 60 ° ... 80 °C is finally stopped at 120 °C (cf. Fig. 7).

Figure 8 shows the result of the in-situ measurement at angle $\gamma = 35$ °.

Some quantitative results which could be obtained using Eqs. (10) and (15) are collected in Tables 1 and 2. From the values of $\langle P_2 \rangle$ and $\langle P_4 \rangle$ one can conclude that a high *c*-axes alignment could be reached by the drawing process.

The values preserved for the coefficients $\langle \cos 2\delta \rangle$ and $\langle \cos 4\delta \rangle$ can be discussed in the following way (see also [6]):

1) for the unpoled samples the absolute values are relatively small, but opposite in sign ($\langle \cos 2\delta \rangle < 0$, $\langle \cos 4\delta \rangle > 0$). The negative sign of $\langle \cos (2\delta) \rangle$ shows that mechanical processing leaves the film in a state wherein the *b*-axes have a larger component in the plane of the film than normal to it (see also [14]). However, the deviation from the isotropic case ($\langle \cos (2\delta) \rangle = \langle \cos (4\delta) \rangle = 0$) is small.

2) The tendency for *b*-axes to orient towards the direction of the applied electric field is evident but ultimate alignment is by no means achieved; both coefficients remain small in their absolute values.

3) The alignment of the *b*-axes towards the electric field can be "switched" periodically and monitored in an MW 4 experiment. This is in accordance with the fact that the electric dipoles relax to an equilibrium when the poling field is switched off (reversible polarization).

A detailed interpretation of the in-situ measurements, however, requires further experiments.

Table 1. Numerical values obtained from $T_{2\mathrm{eff}}(\beta)$

Sample	$T/°C$	$\langle P_2 \rangle$	$\langle P_4 \rangle$
1	−25	1.1	0.53
2	25	0.88	0.35
3*	25	0.00	0.00

* undrawn film

Table 2. Numerical values obtained from $T_{2\mathrm{eff}}(\gamma)$

Sample	$E_p/MV\,\mathrm{cm}^{-1}$	$T/°C$	$\langle \cos 2\delta \rangle$	$\langle \cos 4\delta \rangle$
4	–	20	−0.027	0.002
5	1	−25	0.079	0.046
5*	1	20	0.068	−0.027
6	1.5	20	0.09	0.018
6*	1.5	20	−0.037	−0.013

5* sample 5 measured 2 years after the poling
6* sample 6 measured after annealing at 120 °C

References

1. Gerhard-Multhaupt R (1987) IEEE Transactions on Electrical Insulation E 1 – 22
2. Gerhard-Multhaupt R, Gross B, Sessler GM (1987) In: Sessler GM (ed) Electrets. Springer, pp 383-431
3. McBrierty VJ (1974) Polymer 15:503 – 520
4. McBrierty VJ, Douglass DC (1981) J Polym Sci, Macromolec Rev 16:295 – 366
5. Sillescu H (1986) Makromolek Chem, Macromolec Symp 1:39 – 49
6. Douglass DC, McBrierty VJ, Wang TT (1982) J Chem Phys 77:5826 – 5835
7. McBrierty VJ, Douglass DC, Wang TT (1982) Appl Phys Lett 41:1051 – 1053
8. Ishii F, Sawatari T, Odajima A (1981) Rep on Progr in Polym Phys of Jap 24:493 – 494
9. Ishii F, Sawatari T, Odajima A (1982) Rep on Progr in Polym Phys of Jap 25:607 – 608
10. Ishii F, Sawartari T, Odajima A (1982) Jap Journ Appl Phys 21:L 251 – L 253
11. Clements J, Davies GR, Ward IM (1985) Polymer 26:208 – 216
12. Holstein P, Geschke D, Petzsche Th (1986) Acta Polymerica 37:60 – 62
13. Geschke D, Holstein P, Mendler M (1988) Acta Polymerica 39:206 – 207
14. Frigge K, Geiss D, Jancke A, Künstler W (1988) Acta Polymerica 39:168 – 171
15. Schmidt VH (1987) Ferroelectrics 73:5 – 14
16. Mehring M (1976) in: Diehl P, Fluck E, Kosfeld R (eds) NMR Basic Principles and Progress, Vol. 11, 1 – 246. Springer, Berlin Heidelberg New York
17. Mansfield P, Ware D (1966) Phys Lett 23:421 – 422
18. Gründer W (1974) Wiss Zschr Karl-Marx-Universität Leipzig. Naturwiss R 23:466 – 478
19. Müller R, Willsch R (1976) J Magn Res 21:135 – 142
20. Ward IM (1985) Adv Polym Sci 66:81 – 115
21. Dawydow AS (1974) Quantenmechanik. Deutscher Verl der Wiss, Berlin
22. Abragam A (1961) The principles of Nuclear Magnetism, Oxford Univ Press
23. McBrierty VJ, Ward IM (1968) J Appl Phys (J Phys D) 1:1525 – 1542
24. Tajitsu Y, Masuda T, Furukawa T (1987) Jap Journ Appl Phys 26:1749 – 1753
25. Wu G, Yano O, Soen T (1986) Polym Journ 18:51 – 61

Received January 14, 1989;
accepted April 14, 1989

Authors' address:

Doz. Dr. sc. D. Geschke
Karl-Marx-Universität Leipzig
Sektion Physik
Linnéstraße 5
7010-Leipzig, GDR

Discussion

GEISS:

Your findings that the dipoles are preferentially oriented in the sample plane before poling confirm results we have obtained by WAXS investigations. Furthermore we found that the change in the orientation of the dipoles caused by poling is measured only as field strength when a considerable amount of polarization is reached in the sample (above 100 MV/m). Do you have any results which say something about the poling mechanism at lower field strength?

GESCHKE:

The effect of poling on the angular dependence of T_{2eff} we observed also with lower fields, approximately 500 MV m^{-1}.

FUHRMANN:

According to the large $P2$-values the question arises whether you have information about the type of ODF, e.g. drawn from the P_2 versus P_4 relation?

GESCHKE:

The $P2$-values given in the table were calculated from the T_{2eff} angular dependence of protons within the crystalline regions of the material. Because of the limited number of experimental results up to now we did not try to use the $P4$ versus $P2$ relation for getting further information about the type of the ODF.

FUHRMANN:

You have shown that the poling field dependent value of T_{2eff} is in the time region around 1 ms, i.e., around 1 KHz. For a deeper understanding of the dynamics of a β-PVDF electret property it seems to be important to know whether there is any loss in the 1 KHz region in some PVDF properties.

GESCHKE:

One has to consider that the absolute T_{2eff}-values depend on both the magnetic interaction energy and molecular motions which modulate it. It is known that T_{2eff} reflects motions in the time region around 1 μs which can be used to explain typical (motional) processes characterizing special β-PVDF electret properties

Progress in Colloid & Polymer Science Progr Colloid & Polymer Sci 80:78–82 (1989)

Nonlinear dynamics of polymer crystals. Soliton models of structural defects in the polyethylene crystals

L. I. Manevitch and N. G. Ryvkina

Institute of Chemicals Physics, the USSR Academy of Sciences, Moscow, USSR
Institute of Synthetic Polymeric Materials, the USSR Academy of Sciences, Moscow, USSR

Abstract: The asymptotics of the nonlinear equations of motion corresponding to bends and twists of polymer chains in an orthorhombic polyethylene crystal are obtained analytically. Using these asymptotics, soliton models describing structural defects are constructed. The present results are compared with the results of a numerical simulation.

Key words: Nonlinear dynamics, soliton model, structural defect, polyethylene crystal

The investigation of the linear dynamics of polymer crystals was initiated as far back as 1939 in the well-known paper by Kirkwood [1] where the problem of plane vibrations of the single chain of a polyethylene crystal was solved analytically. Later this solution was furnished with the analysis of the bend-twist out-of-plane trans zigzag vibrations [2]. The influence of the intermolecular interaction was taken into account in the numerical calculations of the harmonic oscillation spectrum [3, 4] and analytically in [5] where the asymptotic approach was used. Because of the difficulties involved in the analytical calculations, the allowance for the nonlinear effects was made only recently in the molecular dynamics method, i.e., in the numerical integration of the equations of motion of a polymer crystal. However, the interpretation of the results of these numerical calculations presents many difficulties because the ordinary spectral language used to describe the calculated and experimental data in the linear theory is unacceptable when the anharmonicity is strong.

The recent developments in nonlinear physics permit the inference of specific localized motions (the solitons and the soliton-like solitary waves) of essentially nonlinear systems. The existence of these solutions makes it possible to study the anharmonic effects and to interpret the numerical data with the maximum possible simplicity. But it is difficult to apply these results to the physics of polymer crystals because the corresponding equations of motion are complex. However, polymer crystals possess qualitative features which are a consequence of the hierarchy of interactions typical of these crystals (the intermolecular interaction is substantially weaker then the intramolecular interaction, the change of the angles of internal rotation requires substantially less energy then does the comparable change of the valence angles, the even more so, of the valence bond lengths). Using the corresponding small parameters one can construct approximate nonlinear equations describing different deformations that can be used effectively in the analytic study. At low energies these equations should reduce to the linear equations suggested in [5].

The present paper deals with nonlinear effects under bend-and-twist deformations of polymer chains. We consider the case of cooperative motion of mass-centers of crystallographic cell of orthorhombic polyethylene (PE). In this case the Bravais lattices move either in phase or in antiphase. In view of cooperation, we may restrict ourselves to the one-dimensional equations of motion. The static case has been analyzed in detail, which makes it possible to model the structural defects in a polyethylene crystal.

For an isolated chain in the trans state the twist-bend motions are completely determined by the change of the conformational angles. In [6] these motions were considered in the model of fixed bonds and valence angles. It was shown that the conformational vibrations can be represented as a combination of twist and bend components, the long-wave twist (around the axis of the molecule) is accompanied by the short-wave small-amplitude bend and the long-wave bend (out of the planar trans zigzag), by the short-wave weak twist. The smallness of the accompa-

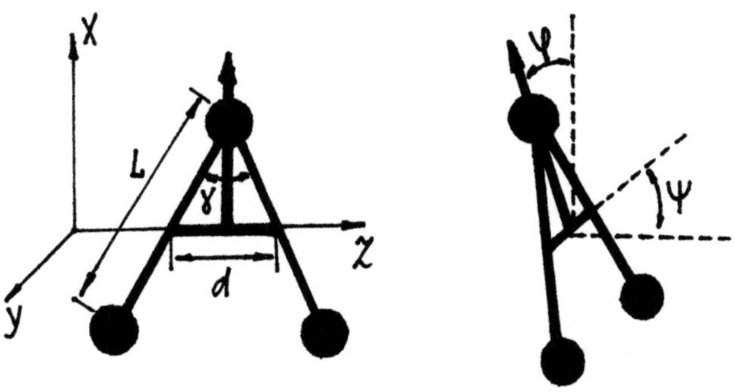

Fig. 1. Fragments of a polyethylene molecule. The XZ-plane of the trans zigzag of the undistorted chain. The twist and bend components of the bend-twist deformation can be described by the variables φ and ψ, respectively

nying deformation is the more obvious, the greater is the characteristic wavelength of the main deformation. As a limiting case, one can discard the accompanying deformation and consider the total deformation as pure tiwst or pure bend. As is shown in [5], the model of fixed valence bonds and angles is directly applicable in the limit of twists and for bends it is to be modified because the deformation is not purely conformational.

In the present paper the twist deformation, which is treated first, is considered as a twist of the PE molecule around the long axis that is described by the variable φ (see Fig. 1). The intermolecular energy of the crystal is allowed for as a sum of pairwise interactions of adjacent molecules. Each CH_2-group of the molecule interacts with the three nearest CH_2-groups of an adjacent molecule in accordance with the Lenard-Jones potential

$$V = 2\varepsilon(r^*/r)^9 - 3\varepsilon(r^*/r)^6 , \quad \text{where} \quad r^* = 4.4\ \text{Å} . \tag{1}$$

The effective mass of the CH_2-group is $m_{\text{eff}} = 33.38$ a.u., the effective energy parameter of the potential $\varepsilon_{\text{eff}} = 5.72$ kJ/mol. In [5] $m_{\text{eff}} = 23.5$ a.u., $\varepsilon_{\text{eff}} = 3.08$ kJ/mol. The effective parameters were calculated in the present work following the procedure of [5]. However, the deformation was described, not as a displacement in the variable Y, but as a twist in φ that leads to differing displacement ratios of carbon and hydrogen atoms in the CH_2-group. A comparison between the spectral gaps of the harmonic torsional vibrations that were obtained in the description of motion as Y-vibrations and φ-vibrations and the parameters given in [5] and in the present paper, respectively, is made in Table 1. The energy of the intermolecular interaction can be represented as a series [6]:

$$V_i = -A_1 \cos 2\varphi_i - A_2 \cos 4\varphi_i - A_3 \cos 6\varphi_i - A_4 \cos 8\varphi_i - \dots$$

where i denotes the position of the CH_2-group in the chain.

For the in-phase motion of Bravais lattices the following values of the parameters of the potential (2) are obtained:

$$A_1 = 0.58\ \text{kJ/mol} , \quad A_2 = 1.45\ \text{kJ/mol}$$
$$A_3 = -3\times10^{-3}\ \text{kJ/mol} , \quad A_4 = -4\times10^{-5}\ \text{kJ/mol} , \tag{3}$$

Thus, $A_1 \approx A_2$, and the equation of motion takes the form

$$I\ddot{\varphi} - C\varphi'' + 2A_1 \sin 2\varphi + 4A_2 \sin 4\varphi = 0 , \tag{4}$$

where $I = 10.14\times10^{-27}$ kg Å^2 is the inertia moment, $C = 115.94$ kJ Å^2/mol is the torsional rigidity [6].

This equation gives a better description of the situation than the one given in [6] where in calculating the coefficients of the series (2) the motion was considered as a displacement in the Y-axis (in the nonlinear case this approximation leads to substantial error).

The static solution of Eq. (4), which describes the structural defect, is of the form

$$\varphi = \text{across} \left[\sqrt{\xi}\, \text{sh arth} \left(\frac{1}{\sqrt{1+\xi}}\, \text{th}\,(-\eta z) \right) \right] , \tag{5}$$

where $\begin{cases} \xi = A_1/4A_2 = 0.10 ; \\ \eta = 4\sqrt{(1+\xi)A_2/C} = 0.469\ \text{Å}^{-1} . \end{cases}$

This solution describes the chain twist by 180°. The energy of the defect is

$$E = 4\sqrt{A_2 C} \left(\lambda + \frac{\xi}{2} \ln \frac{\lambda+1}{\lambda-1} \right) / d , \quad \lambda = \sqrt{1+\xi} . \tag{6}$$

Computer simulation of structural defects of orthorombic PE was made in [7, 8]. One of the simulated defects can be compared with the defect described by Eq. (5). The lengths and the energies of these defects are in good agreement (see Fig. 2 and Table 2).

For the antiphase motion of Bravais lattices, the coefficients of series (2) are

$$A_1 = 0 , \quad A_2 = 2.41\ \text{kJ/mol} , \quad A_3 = 0$$
$$A_4 = -3\times10^{-3}\ \text{kJ/mol} . \tag{7}$$

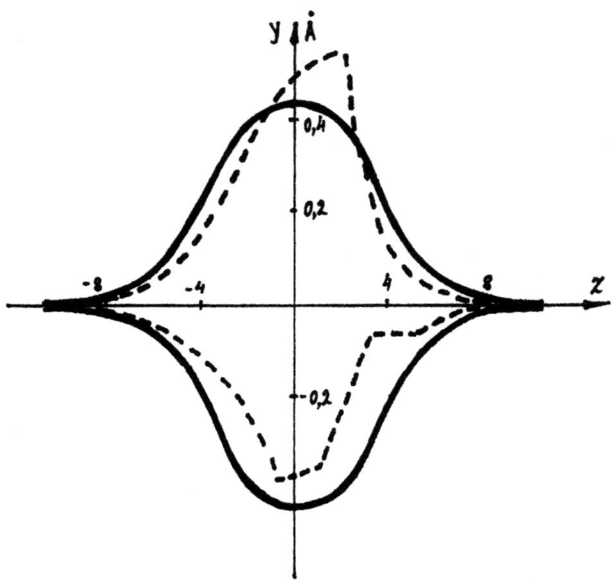

Fig. 2. The chain twist by 180°. The solid curve corresponds to the present results. The broken-line curve represents the results of [8]. The scale in the Z-axis is $d = 1.27$ Å

The equation of motion takes the form

$$I\ddot{\varphi} - C\varphi'' + 4A_2 \sin 4\varphi = 0 \ . \tag{8}$$

The static solution of this equation is

$$\varphi = \text{arctg} \exp (z/\sigma)$$
$$\text{where} \quad \sigma = \tfrac{1}{4}\sqrt{C/A_2} = 1.74 \text{ Å} \ . \tag{9}$$

The energy of the defect is $E = \sqrt{CA_2}/d$. This defect describes the cooperative chain twist by 90°. It can also be compared with one of the simulated defects from [7, 8]. The profiles and the energies of these defects are compared in Fig. 3 and in Table 2, respectively.

According to the results of [7, 8], the defects of type (5) and (9) can exist in crystals with extended chains. The conjectured defect formation mechanism in the preparation of crystals with the extended chains from the folded crystals is described in [7].

Table 1. The spectrum gap values (cm^{-1})

Variables	In-phase vibrations	Antiphase vibrations
Y	115	131
φ	108	133
Experimental values [3]	108	136

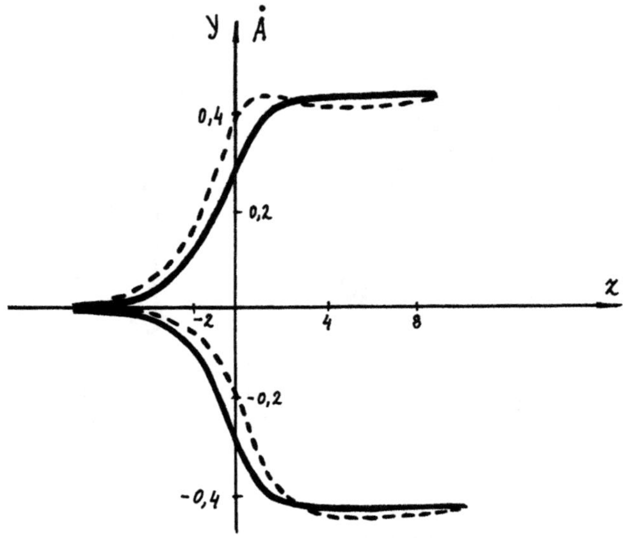

Fig. 3. Same as in Fig. 2 but for the twist of 90°

Above we obtained the static solutions corresponding to structural defects in crystalline PE. However, the equations of motions are invariant under the Lorentz transformation group and so each of these solutions can be associated with the dynamic wave process:

$$\varphi_1 = \arccos \left[\sqrt{\xi} \, \text{sh arth} \left(\frac{1}{\sqrt{1+\xi}} \times \text{th} \left(- \frac{\eta(z \pm vt)}{\sqrt{1 - v^2/c_0^2}} \right) \right) \right] \tag{10}$$

$$\varphi_2 = \text{arctg} \exp \left(\frac{z \pm vt}{\sigma \sqrt{1 - v^2/c_0^2}} \right) , \quad c_0^2 = C/I \ . \tag{11}$$

Now we proceed to the consideration of the bend deformation. As follows from group theory [3] and from the results of the asymptotic linear theory [5], this deformation corresponds to the bend in the a and b direction of the PE crystal. Owing to this deformation structural defects caused by the conformational transitions can enter into the crystal. Consider as an example the defect $g + tg^-$. When this defect arises in the undistorted chain, the direction of its axis changes, and the defect cannot enter into the crystal

Table 2. The energy of the defect (kJ/mol)

	Twist by 180°	Twist by 90°
Our results	50.49	26.33
The results of [8]	62.17	32.70

without additional distortions in the defect-free region. We suggest to model these distortions using the static solutions corresponding to the bend-twist deformation of the nonlinear equations in the continuum approximation. The twists cannot give sufficient displacement and, consequently, the peripheral regions should be deformed in the bend-like manner.

As is seen from Fig. 4, in the regions A and B, the chain is close to trans zigzag. This is the periphery of the defect and region C is its "core". The CH_2-group designated as "0" and "−1" are the connections of the core and the periphery.

The linear equation of bends [5] is

$$m\ddot{u}+Gu''+Du^{IV}+Bu=0 \ , \tag{12}$$

where u is a or b displacement. Using the periodicity of the lattices in the a and b directions, we can write an expression for the intermolecular energy as

$$V_i=-B_1\cos \varkappa u-B_2\cos 2\varkappa u-\dots \tag{13}$$

where $\varkappa=2\pi/a$ is for the bend in the a direction and $\varkappa=2\pi/b$ is for the bend in the b direction. If we use only the first term of this series, we get the nonlinear equation

$$m\ddot{u}+Gu''+Du^{IV}+Bu-Qu^3=0 \ . \tag{14}$$

We are interested in its solution at positive z. At negative z Eq. (14) can be solved considering the condition of antisymmetry with respect to the point $z=0.5\,d$.

The boundary conditions of Eq. (14) are the decay of the deformation and of its derivative at infinity and the equality of forces and moments at the connection of the core and the periphery of the defect. Before writing these conditions in an explicit form, we make some transformations of the conformational potential, determining the behavior of the defect core. The conformational potential used is of a simple form

$$V=A(1-\cos 3\theta) \ , \tag{15}$$

where θ is an angle of internal rotation. Consider the joint "0" (see Fig. 4). From geometrical considerations we get

$$\begin{aligned}\cos\theta=1-2\delta^2[&(x_0-x_{-1}-x_{-2}+x_{-3})^2\\&+(y_0-y_{-1}-y_{-2}+y_{-3})^2\\&+(z_0-z_{-1}-z_{-2}+z_{-3})^2] \ ,\end{aligned} \tag{16}$$

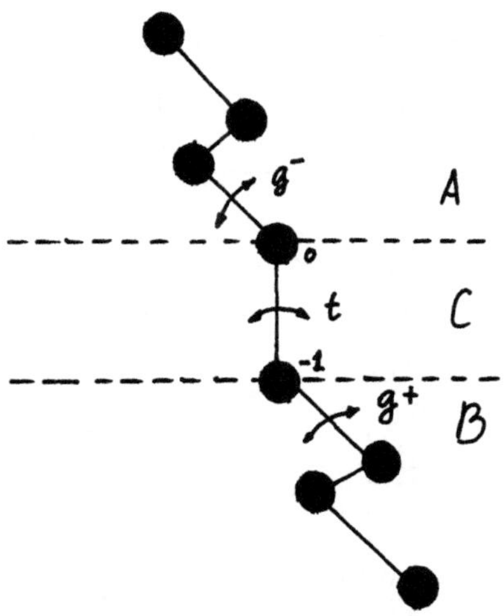

Fig. 4. Fragment of a polyethylene molecule containing the defect g^+tg^-

where $\delta=1/(2L\sin\gamma)$, L is the bond length, γ is the bond angle. Orientation of X,Y,Z axes is seen from Fig. 1, numbering of CH_2-groups – from Fig. 4. Using the antisymmetry condition and assuming that in the periphery a bend occurs and that the continuum description is acceptable, we get

$$\begin{aligned}\cos\theta&\approx 1-2\delta^2(u_0-u_{-1}-u_{-2}+u_{-3})^2\\&=1-2\delta^2\left(2u-du'+\frac{3d^2}{2}u''\right)^2\Bigg|_0\end{aligned} \tag{17}$$

$$\sin\frac{\theta}{2}\approx\left(2u-du'+\frac{3d^2}{2}u''\right) \ . \tag{18}$$

Then the boundary conditions of Eq. (14) are

$$\begin{aligned}&\frac{\partial H}{\partial u'}-\frac{\partial}{\partial z}\left(\frac{\partial H}{\partial u''}\right)\Bigg|_0=\frac{\partial V}{\partial u'}-\frac{\partial}{\partial z}\left(\frac{\partial V}{\partial u''}\right)\Bigg|_0 \ ,\\&\frac{\partial H}{\partial u''}\Bigg|_0=\frac{\partial V}{\partial u''}\Bigg|_0\\&u_{z\to\infty}=0 \ , \quad u'_{z\to\infty}=0\end{aligned} \tag{19}$$

where H is the Hamiltonian corresponding to (14); $\dfrac{\partial V}{\partial u'}$ and $\dfrac{\partial V}{\partial u''}$ are derivatives of the functional

$$V=\int dz\left\{\frac{P}{d}\left[\delta\left(2u-du'+\frac{3d^2}{2}u''\right)-\sin 60^\circ\right]^2\right\} \ . \tag{20}$$

Table 3. The parameters of Eqs. (14), (20), (22)

Parameters	Bend in the b direction	Bend in the a direction
B(kJ/mol Å^2)	27.46	60.34
D (kJ Å^2/mol)	413.62	413.62
G (kJ/mol)	10.56	22.01
Q (kJ/mol Å^4)	7.63	7.74
P(kJ/mol Å^2)	391.7	391.7
a_3 (Å)	1.22	1.22
a_4 (Å)	1.00	8.3×10^{-1}
b_1 (Å)	2.0×10^{-2}	9.1×10^{-3}
b_2 (Å)	-1.9×10^{-2}	-8.5×10^{-3}
c_1 (Å)	5.0×10^{-4}	5.0×10^{-5}
c_2 (Å)	-3.0×10^{-3}	-1.2×10^{-3}
α (Å^{-1})	0.350	0.442
β (Å^{-1})	0.368	0.453

The static solution of Eq. (12) is

$$u = a_1 \exp(-\alpha z)\cos(\beta z) + a_2 \exp(-\alpha z)\sin(\beta z)\ . \tag{21}$$

The static solution of Eq. (14) represented as a series in powers of (21) is

$$u = \exp(-\alpha z)(a_3 \cos\beta z + a_4 \sin\beta z) + \exp(-3\alpha z)(b_1 \cos\beta z + b_2 \sin\beta z + c_1 \cos 3\beta z + c_2 \sin 3\beta z) + \ldots \tag{22}$$

The parameters of Eq. (14) and the coefficients of its solutions (22) with the boundary conditions (19) for bends in the direction of the a and b axes are given in Table 3.

The present data for the structural defect associated with the bend in the b direction can be compared with the result of the computer simulation [8]. The comparison between the profiles of the defects is made in Fig. 5.

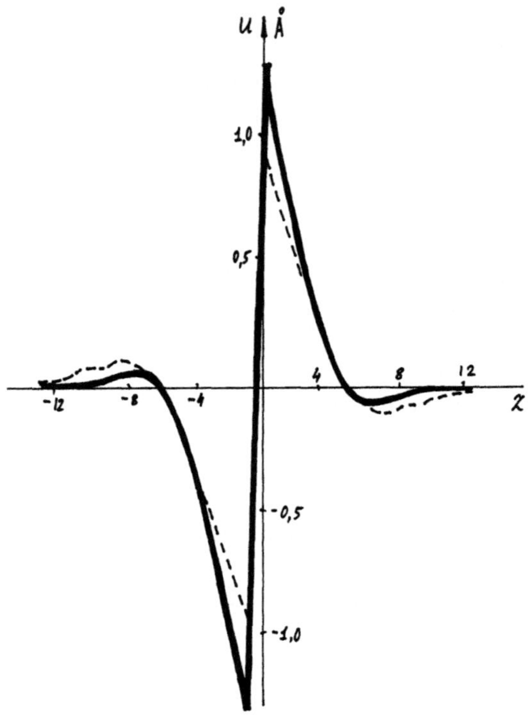

Fig. 5. The defect with the core g^+tg^- and the bend-like periphery. The solid-line curve shows the present results. The broken-line curve is for the results of ref. [8]. The scale in the Z-axis is $d = 1.27$ Å

References

1. Kirkwood JG (1939) J Chem Phys 7:506–509
2. Pitzer KS (1940) J Chem Phys 8:711–720
3. Painter PC, Coleman MM, Koening JL (1986) The theory of vibrational spectroscopy and its application of polymeric materials. Wiley, New York
4. Wunderlich B, Baur H (1970) Heat capacities of linear high polymers. Springer, Berlin
5. Ginzburg VV, Manevitch LI, Ryvkina NG (1989) Vysokomoleculyarnye Soedineniya 31:1149–1155
6. Manevitch LI, Ryvkina NG (1988) Khim Fiz 7: 1620–1628
7. Reneker DH, Mazur J (1983) Polymer 24:1387–1400
8. Reneker DH, Mazur J (1988) Polymer 29:3–13

Received December 12, 1988;
accepted April 20, 1989

Authors' address:

Prof. Dr. L. Manevitch
Institute of Chemical Physics
USSR Academy of Sciences
u. Kosygina 4
117977 Moscow V-334, USSR

Progress in Colloid & Polymer Science Progr Colloid & Polymer Sci 80:83–92 (1989)

Applications of "Forced Rayleigh Scattering" and "Photon Correlation Spectroscopy" for the examination of transport properties in polymer melts

M. Antonietti, D. Ehlich, H. Sillescu and M. Wesselmann

Institut für Physikalische Chemie der Universität Mainz, Mainz, FRG

Abstract: The importance of optical techniques for the characterization of the dynamics in polymers has increased in the last 10 years. This was caused by the rapid development in optics and microelectronics which has made these techniques easily available and extremely precise. However, in case of the examination of transport properties in bulk polymers, new difficulties arise due to the comparably slow dynamics underlying these phenomena. The topic of this paper is a small survey of two optical techniques which can be applied to solve the arising problems and a presentation of some results of our working group concerning slow translational transport processes in polymers. In a first part, the technique of "Forced Rayleigh Scattering" will be discussed according to two selected examples, the self-diffusion of polymers and the mobility of small molecules in glassy polymers. In the second part, some peculiarities "Photon Correlation Spectroscopy" applied in the bulk will be treated.

Key words: Forced rayleigh scattering; polymer self-diffusion; reptation; photon correlation spectroscopy; slow mode

I. "Forced Rayleigh Scattering"

It is just 15 years ago since the Swiss scientist H. J. Eichler developed the "Forced Rayleigh Scattering" (FRS) for application in heat conduction experiments [1]. Five years later FRS was applied to the problem of particle diffusion in liquid crystals [2]. Afterwards the number of referred experiments has enormously increased, especially in the field of polymer science. The experiments which are referred to here started in 1982. Two particular improvements, the introduction of irreversible gratings and the use of reflection holography, allowed our working group to extend the method by another five orders of magnitude to smaller diffusion coefficients.

The technique is also called "Holographic Grating Technique" because we use the simplest possible hologram that is formed by recording in a light sensitive material the interference pattern of two coherent plane wave fields of equal intensity. The resulting interference pattern (Fig. 1a) has a grid spacing d, the distance between two maxima, given by the Bragg-equation:

$$d = \lambda / 2 \cdot \sin(\Theta/2) \; . \qquad (1)$$

λ is the wavelength, Θ the angle between the two interfering light beams. Since we use scattering from a planar film geometry, the situation is different for the formation of a reflection hologram where the two light beams come from above and below the sample (Fig. 1b). In this case, the grating is parallel to the sample surface, and d is given by

$$d = \lambda / 2 \cdot \{n^2 - \sin^2 [(180 - \Theta)/2]\}^{0.5} \qquad (2)$$

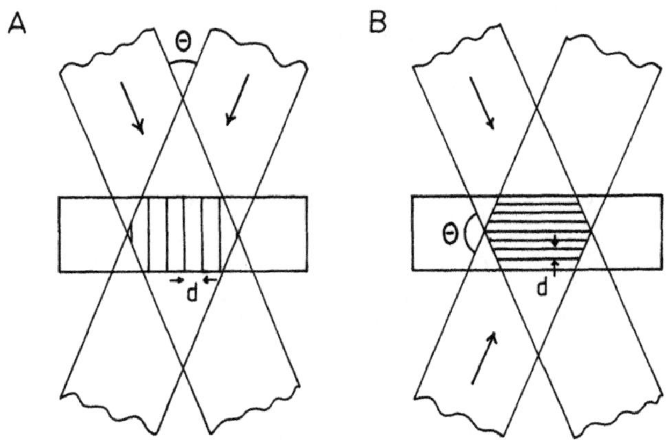

Fig. 1. Geometries of transmission (A) and reflection (B) hologram

where n is the index of refraction of the medium. It is especially this technique that allows the application of scattering angles close to 180°, resulting in grid spacings of 150 nm (with $\lambda = 458$ nm). With that we have experimentally realized the minimal spatial resolution which can be obtained with light scattering techniques. Therefore, the range accessible with our set-up covers the region between 150 nm and 20 µm.

Due to the light sensitivity of the material, the pattern is transferred to the sample: at places of high light intensity a large amount of photoproduct is formed and vice versa. We call this process *writing* of the hologram.

The photoreaction is accompanied by a change of the absorption or the refractive index of the medium. Thus the written pattern can be used as a grid for light scattering. For a sinusoidal pattern the Fourier-transformation results in a δ-function in Q space (Q = scattering vector) at the Bragg-angle. The scattered light is "forced" into one direction, all our information about the amplitude or the time-dependence of the grid is contained in one spot of scattered light where it can be easily detected. This process is called *reading* the hologram.

For the determination of dynamic phenomena, the time evolution of the written hologram is of importance. In case of a thermal irreversible photo-chemistry, the grid can just vanish when the photochromic molecules diffuse. By solving the Fick law for the given spatial distribution one obtains for the time-dependent concentration c (see also Fig. 2):

$$c(x,t) = C_0 \cdot \sin(2\pi x/d) \cdot \exp(-t/\tau) + C_1 \quad (3)$$

with $$\tau = d^2/4\pi^2 D \quad (4)$$

where D is the diffusion coefficient. One important fact should be noted here: the relaxation time depends on the square of the grid spacing which is also an unambiguous test for diffusional processes. The use of

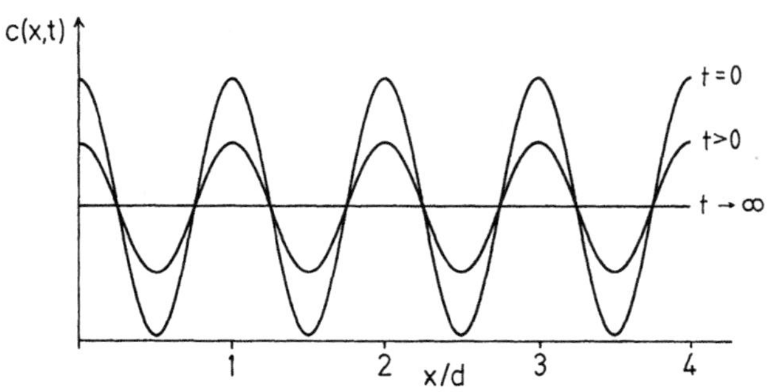

Fig. 2. Time-evolution of the concentration profile

small grids allows the determination of very slow diffusion coefficients.

With d-values of $150\,\text{nm} < d < 20000\,\text{nm}$ and experimentally manageable waiting times of less than one day we can cover directly a span of diffusion coefficients between $10^{-7} < D < 5 \cdot 10^{-17}\,\text{cm}^2\,\text{s}^{-1}$ or 10 orders of magnitude with just one technique. For a more practical understanding it should be stated that the smallest diffusion coefficient corresponds to diffusive displacements of only about 2 Å per s.

On the other hand, the application of FRS is restricted to very special diffusion problems: the samples have to be optically transparent and photoreactive. Practically this reduces the use to the observation of dye-molecules or polymers with covalently bound photolabels. The necessity of labeling the polymers requires some polymer chemistry which makes sample preparation inconvenient. However, this can be turned into an advantage because any technique working with labels allows the diversification between a probe (the labeled species) and the matrix (unlabeled environment).

One has also to ensure that the diffusion behavior of the polymer is not severely influenced by the label. In case of the smallest applied label, the o-nitrostilbene (see text below), the diffusion coefficients obtained with FRS are in excellent agreement with data from the "forward-recoil-technique" which uses isotope-labeling (16).

We wish to demonstrate the possibilities of FRS at the example of diffusion of linear polymer chains where the independent variation of probe and matrix allows the description of a complex transport problem.

I.1 Diffusion of linear polymer chains in the bulk

The structure of a polymer melt is very complicated in terms of intermolecular interactions. Due to their Gaussian character, any chain is interpenetrated by many others, and a description of polymer dynamics has to consider these topological constraints and entanglements. The idea of reptation, first introduced by de Gennes [3] and Edwards [4], approximates the geometrical influence of the surrounding chains by a tube with also Gaussian statistics in which the observed chain has to move. It results in a "curvilinear", snake-like motion where the chain has to follow its own contour.

From this model, characteristic parameter dependencies can be calculated. For instance, the zero shear viscosity η_0 is expected to vary with the cube of the molecular weight of the polymer chains, and the self

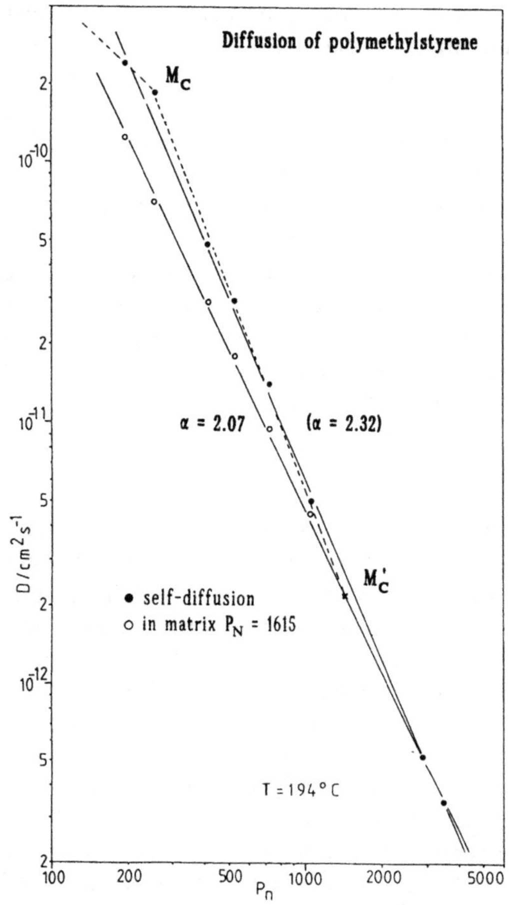

Fig. 3. Diffusion coefficients D of polymethylstyrenes with different degrees of polymerization P_n. Filled circles denote self-diffusion data whereas open circles represent data measured in a linear matrix of $P_n = 1615$

diffusion coefficient should be inversely proportional to the square of the molecular weight. The scaling exponent for the viscosity, which has been known since 1953 ([5], η_0 (exp.) $\sim M^{3.4}$) differs from this prediction and the difference has led to numerous controversies. In addition to the scaling predictions, a certain invariance has to be fulfilled. If the picture of a fixed tube is correct, the mobility of the probe chain remains the same when the matrix consists of very large chains or even of a cross-linked polymer network.

This can be checked with FRS diffusion experiments too, in contrast to viscosity experiments.

Figure 3 shows the molecular weight dependence of the diffusion coefficient of polymethylstyrene (a copolymer of 60% of the meta- and 40% of the paraderivative) at 194 °C [6]. Typical experimental errors and also reproducibility are within 10% of the presented values. In this double-logarithmic plot we observe in the case of diffusion of labeled chains in a high molecular weight matrix a clearly linear dependence; the slope is close to two, which is predicted by reptation.

Apparently the situation is more complex in the case of self-diffusion when diffusant and matrix molecules have the same molecular weight. A force-fit of the data points with one exponent leads to moderate agreement and an exponent higher than two. However, the agreement is much better when the description uses three different regimes of mobility which are separated by two critical molecular weights corresponding to M_c and M'_c, as determined by mechanical experiments:

$$M<M_c:\ D \sim M^{-1.5} \quad \text{Rouse mobility} \tag{5a}$$

$$M_c<M<M'_c:\ D \sim M^{-2.5} \quad \text{regime of tube formation} \tag{5b}$$

$$M'_c<M:\ D \sim M^{-2.0} \quad \text{regime of reptation} \tag{5c}$$

Below M_c (for polystyrenes ≈ 32000 g/mol) the chains behave unentangled; the exponent of −1.5 is slightly lower than predicted from the Rouse-theory. At molecular weights larger than M'_c self-diffusion and diffusion in high molecular weight matrix is approximately the same, and reptation describes the observed mobility. Between the two critical molecular weights, the exponent of −2.5 indicates that simple reptation is not the complete truth. The comparison of self- and matrix diffusion indicates that this change is not due to the probe chain which behaves correctly in the large chain environment; the difference results from the increased mobility of the surrounding chains which should build up the tube. Apparently small chains with these intermediate molecular weights do not produce the same entanglement density as large linear polymers. We call this region the regime of "tube formation".

The sharp transition from Rouse-behavior to "tube formation" looks quite arbitrary within the presented data-set, but the behavior of small chains has been analyzed in more detail in our polystyrene experiments [7], where the sketched behavior has been found. Apart from that, these experiments coincide with the polymethylstyrene-data so they will be not considered in this context.

The nature of the additional mobility can be better examined with experiments where the molecular weight of the probe is fixed and the matrix is varied. Here, the postulated invariance of a reptative movement to cross-linking can also be tested.

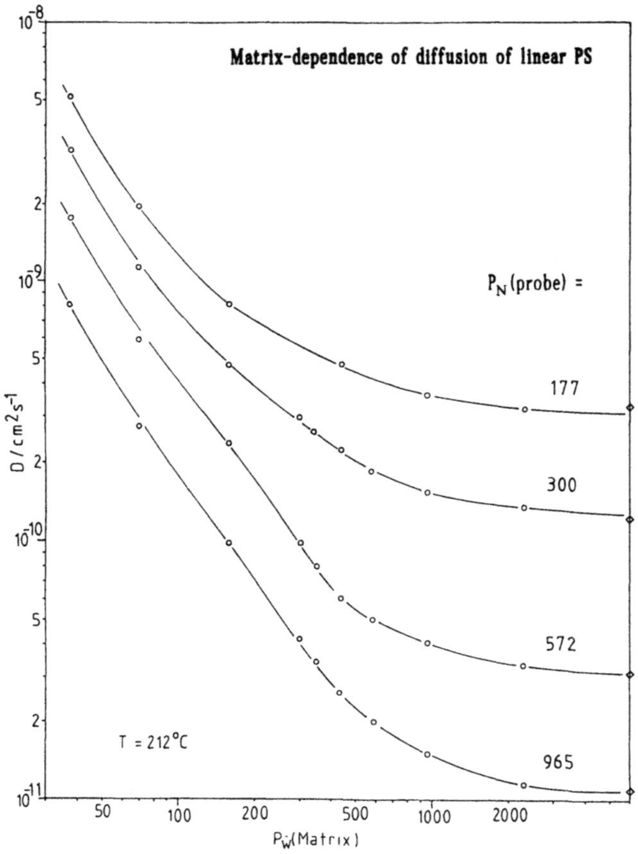

Fig. 4. Dependence of the diffusion coefficient of probe polymers with four different degrees of polymerization upon the length of the matrix chains. The diamonds at the rightside denote the data obtained in a cross-linked matrix. The solid lines are a guide for the eye

Figure 4 shows the results of these experiments for four different polystyrene probes in a wide range of matrices [6]. The data are already shifted due to their difference in T_g with a standard free volume correction. The diamonds at the right limit of the curves indicate the diffusion coefficients in a network. The identity of the values observed in high molecular weight matrix with those of the network experiments clearly prove the validity of reptation in this limit. The mobility begins to increase when the matrix molecular weight drops below M_c'. This increase is not given by simple scaling laws; different probes are influenced in a different way. The observed behavior cannot be explained by a simple "tube renewal" (8) which predicts that D is inversely proportional to the cube of the matrix chain length. Even the diffusion coefficient of the smallest probe chain which is too short to be entangled exhibits a molecular weight dependence.

At present there is no theory which explains all three scalings (matrix-diffusion, self-diffusion, *and* matrix-dependence of the mobility). We believe that the data can be qualitatively explained be the conception that the tube is not immediately created at the cross-over molecular weight M_c but is continuously formed, beginning with small entanglement densities and a correspondingly large tube. This process of tube formation is completed at M_c' where the maximal entanglement density is reached.

It should be noted that a similar cross-over regime has to exist also within the tube-model, though it has never been explicitly discussed. Comparing the expression for diffusion coefficients from the Rouse-model and the tube-model we obtain [8]:

$$D_{REP} = D_{ROUSE}/3N \ . \qquad (6)$$

The advantage of this expression is that it is free of all physical or geometrical parameters; N simply denotes the mass of chain in units of M_e. We see that this formula contradicts the existence of one critical point which separates the regimes of validity of both theories, because this would result in a discontinuous jump of mobility at M_e by a factor of three. Rather, it is physically more meaningful when a transition zone exists, as experimentally observed. It is worth mentioning that the extrapolated values of self-diffusion and diffusion in a long-chain matrix of a chain with the mass M_e do indeed differ by a factor of three, as seen in Fig. 3.

Figure 5 concludes the observations in one schematic drawing where D is plotted in dependence on the mass of probe chain M and the mass of the matrix chains P. The scalings of self-diffusion ($M=P$) and diffusion in a long-chain matrix (P constant) result from different derivations of the plane. The transition from isotropic to curvilinear motion has become a M_c-line which intersects the self diffusion at M_c and diffusion in a long-chain matrix at M_e.

For the direct observation of additional types of mobility besides the dynamically dominant reptation we can choose a polymer geometry where the movement by reptation is topologically forbidden: ring polymers or spherical particles for instance.

We use the examination of the diffusion of μ-gels, small network particles which are made by cross-linking of linear polymers in high dilution. By taking a precursor polymer with a narrow molecular weight distribution, well-defined μ-gels are easily available. Since these particles represent highly branched structures with cyclic subelements, the only observable mobility should be a cooperative one.

Figure 6 shows the self-diffusion data of μ-gels with different cross-linking densities in dependence on their

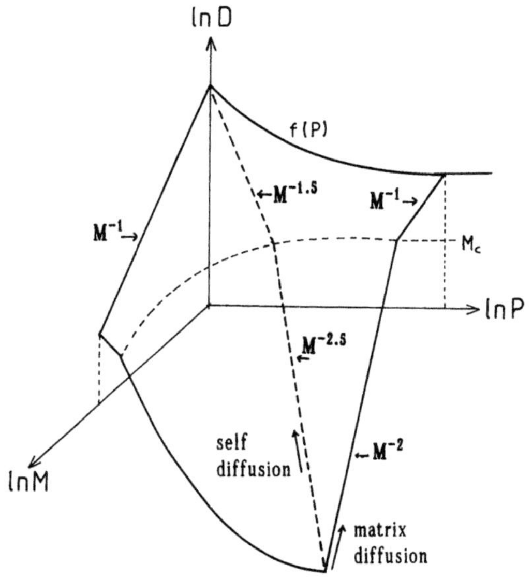

Fig. 5. Scheme of the dependence of the polymer diffusion coefficient on the mass of the probe chain (*M*) and the mass of matrix chain (*P*). The M_c-line separates the regimes of isotropic and curvilinear movement of the probe. Note that the scalings of self- and matrix-diffusion in a long-chain result from different deviations of the plane

molecular weight [9]. We observe that the mobility of small μ-gels with a molecular mass of ≈ 20000 g/mol is close to that of linear chains; apparently, domains of approximately 200 monomer units are highly mobile. This mobility drops strongly until it becomes unobservable with our set-up at $M_w \approx 200000$ g/mol. Therefore, one can obtain with these experiments a rough estimate of a "cooperative length" in bulk polymers. Such a cooperativity of movements is also indicated by the relatively high activation energy of flow properties of polymers close to the glass transition. Further, we have demonstrated that there are – in addition to reptation – secondary types of mobilities which help the system to rearrange in situations which are critical in a topological sense, for instance in case of polymeric rings, stars, or branched systems. Of course, this mobility is also present in linear polymers, but due to its smaller influence it should be harder to detect. In case of "tube renewal", some cooperativity has been postulated to explain deviations of the experimentally observed exponents from theory [8].

I.2 Mobility of low molecular weight compounds in glassy polymers

Another problem where FRS is used favorably is the examination of the translatory mobility of low molecular weight compounds in polymers above and below the glass transition temperature T_g. In addition to the scientific importance of these experiments for the study of the glass transition there are also technological implications. The ageing of polymer materials or food poisoning by the release of plasticizers and colorants or the stability of optical storage media are just three obvious applications.

Since a labeled species is necessary for FRS, we have to model these systems using photochemically-active dye molecules of different sizes. In the following, data from two different types of dye will be reported: TTI represents a thiopheneindigo-derivative which can be involved in a photochemical cis-trans-isomerization, whereas o-nitrostilbene-derivatives (ONS) react with light irreversibly to isatogenes.

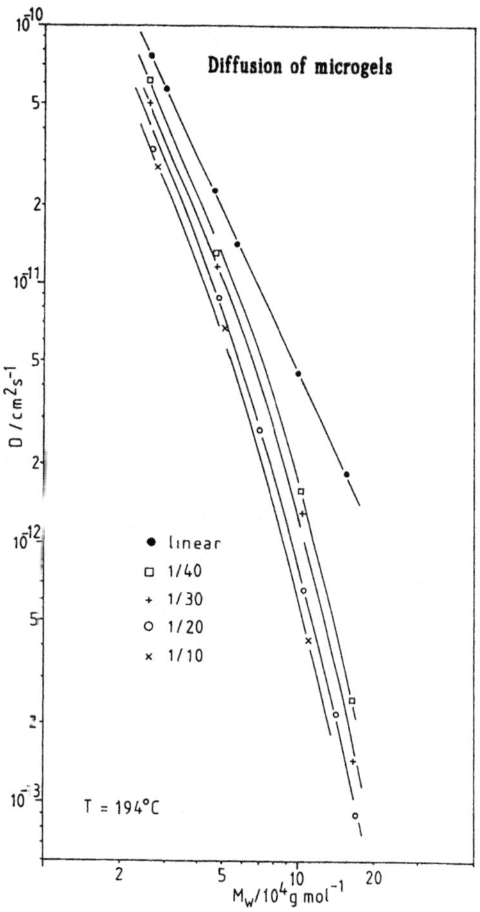

Fig. 6. Diffusion coefficients of microgels with four different cross-linking densities in dependence of their molecular weight. A cross-linking density of 1/10 means that in average every 10th monomer is crosslinked. For comparison the filled circles denote the behavior of linear chains. Solid lines are a guide for the eye

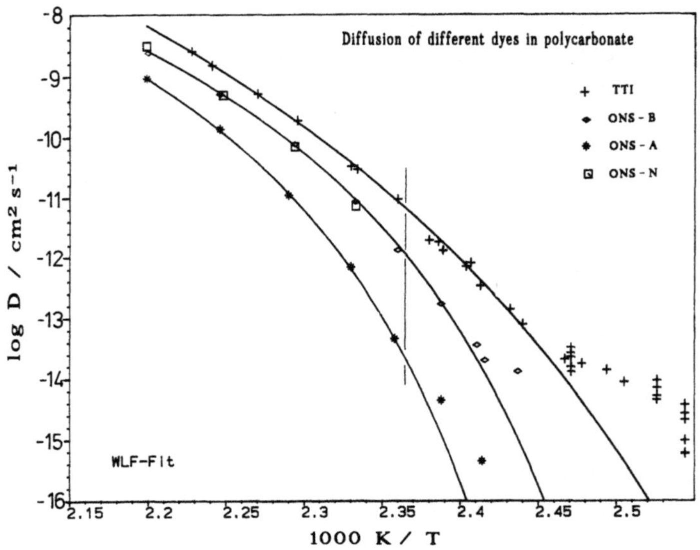

Fig. 7. Temperature dependence of diffusion coefficients of four different dye molecules in polycarbonate: (+) TTI, (◇) ONS-benzylester, (□) ONS-nitrile and (*) ONS-azobenzeneester

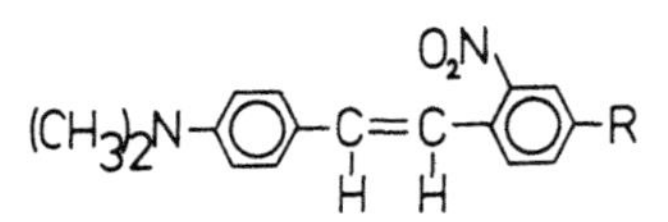

R = CN, COOCH3, ...

ONS

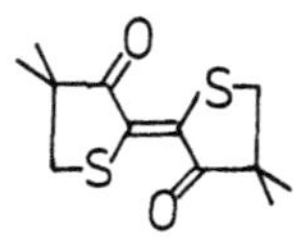

TTI

Figure 7 presents diffusion data with dye molecules of different sizes in dependence on the inverse of the temperature [10]. The concentration of dye molecules is typically between 0.2 and 0.5%, which should be a weak perturbation of the matrix polymer. In all cases, the glass transition equals that of the neat polycarbonate. It should be noted that these measurements may cover a range of diffusion coefficients of 10 orders of magnitude which are directly accessible. The vertical bars represent the glass transition temperatures determined with DSC. Above T_g, all curves can be described by the WLF-equation:

$$\log \frac{D}{D_{ref}} = \frac{C_1 \cdot (T - T_{ref})}{C_2 + T - T_{ref}} \quad (7)$$

The calculated C_1 and C_2-values which result from a fit of Eq. (7) to the experimental data are physically reasonable. Fixing C_2 at the value obtained from mechanical relaxation of the polymer matrix leads to smaller C_1-values for the probe due to the smaller jumping units (10).

Table 1 presents the resulting C_1 and C_2-values for the different dye-mobilities; for illustration, the length l of the dye molecules is given. At the smallest probe molecule the difference in C_1 is most pronounced, whereas the data of the largest dye are closer to those of the matrix. At T_g, there is a notable mobility for all dye molecules. However, this mobility is strongly molecular-weight-dependent: an increase of the dye size by a factor of three decreases the mobility by more than two orders of magnitude. Obviously we cannot describe the data with the Stokes-Einstein law which would result in a probe-size independent viscosity of the polymeric environment at T_g. Below the static glass transition, the temperature behavior changes and can be described by a straight line corresponding to a single activation energy which is lower than above T_g. The larger the size of the probe the closer to T_g is the occurrence of this discontinuity. In the case of the TTI experiments, we observed the ageing of polycarbonate glass. The six values at the rightside show the ageing of polymer glasses by means of the probe diffusion behavior. The fastest value has been obtained by

Table 1. C_1 and C_2-values for dye mobilities

	C_1	C_2	l/nm
mech. relaxation (from Ref. [17])	16.1	56	–
TTI	8.1	56	0.8
ONS-nitrile	9.5	56	1.5
ONS-benzylester	9.5	56	2.0
ONS-methylazobenzeneester	12.6	56	2.6

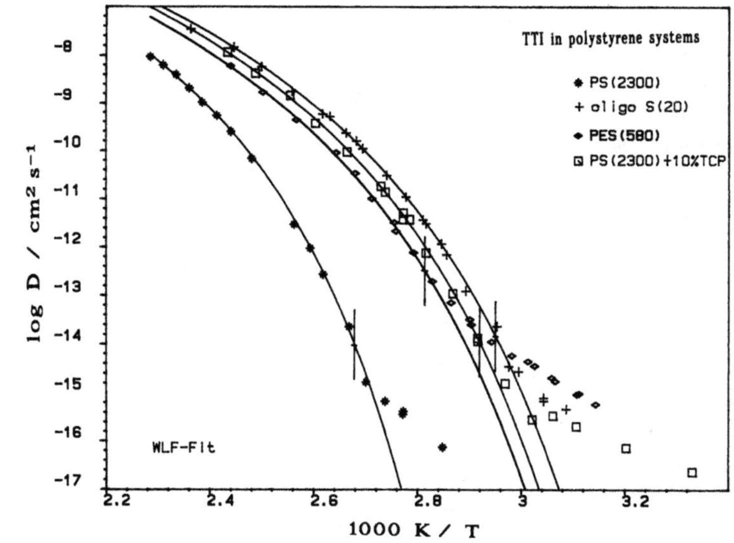

Fig. 8. Temperature dependence of the TTI diffusion coefficient in four different polystyrene systems: (*) PS with $M_w \approx 240000$ g/mol; (+) PS with $M_w \approx 2000$ g/mol; (□) high molecular mass PS with 10% tricresylphosphate; (◇) Polyethylstyrene with $M_w \approx 77000$ g/mol

measuring directly after temperature equilibration of the sample, whereas the lowest probe mobility resulted after 55 days of tempering at 393 K.

Figure 8 shows data of the TTI mobility in different polystyrene (PS) matrices, again in dependence on the inverse of the temperature [10]. In this set of experiments, the influence of different ways of plasticization on the probe mobility has been examined. One can observe that the mobility of TTI in polystyrene melt is again characterized by a WLF-behavior, but two main differences to the behavior in polycarbonate exist. On the one hand, the mobility at T_g in PS is about two orders of magnitude smaller; on the other, the change of activation energy occurs closer to T_g. The difference of about 5 K found in all polystyrene systems can unambiguously be attributed to the different time scales of probe diffusion measurements and of the determination of T_g with DSC. Changing T_g in polystyrene by changing the molecular weight of the polymer does not alter the behavior of probe diffusion relative to T_g; it is possible to construct a master curve for $(T-T_g)$. The same holds for a system consisting of high molecular weight PS containing 10% of a commercial plasticizer, tricresylphosphate (TCP).

But the third way of lowering T_g by flexible side chains ("internal" plasticization), which is actually discussed for different applications, exhibits a different behavior. Although the amount of T_g-reduction is smaller, the diffusion curve intersects the T_g line at much higher values of probe mobility and the residual mobility in the glassy state stays considerably above the corresponding values in PS and PS/TCP. The crossover from WLF to Arrhenius-type behavior extends over a broader temperature range. The nature of this phenomenon is still not completely understood. We can speculate that this decoupling is due to the additional free volume introduced by the flexible side chains, which in this case is *homogeneously* distributed along the chains. Additional experiments using other glasses and probes will surely aid our understanding of the relationship between glass structure and translatory probe mobility.

II. Photon correlation spectroscopy (PCS)

Since the theory of PCS or "quasielastic light scattering" is treated in detail in two other contributions of this issue [11, 12], we discuss only the peculiarities of dynamic light scattering in highly viscous systems.

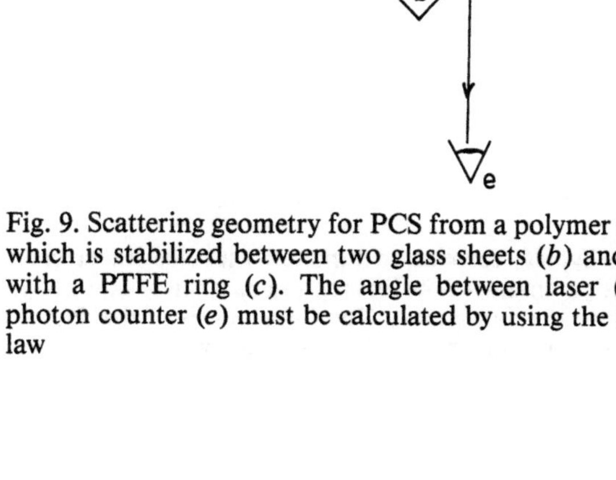

Fig. 9. Scattering geometry for PCS from a polymer film (*a*) which is stabilized between two glass sheets (*b*) and sealed with a PTFE ring (*c*). The angle between laser (*d*) and photon counter (*e*) must be calculated by using the Snellius law

Due to the relatively high temperature which is necessary for these measurements, the tendency to include small bubbles and the extremely slow setting of an equilibrium state, it appeared that scattering from a film geometry might be favorable. Figure 9 presents the geometry which is inclined to the incoming beam in order to separate parasitic scattering from reflected beams.

The disadvantage of this set-up consists in a restricted Q-range that is due to the Snellius-law and is indirectly measurable.

In the examination of polymer-polymer mixtures, the predominant scattering intensity results from concentration fluctuations, the relaxation of which is determined by a "mutual" or "inter"-diffusion coefficient. Therefore, PCS should represent an access to the characterization of polymer bulk dynamics [13]. Because the problem of interdiffusion is complicated and not fully clarified [14], we decided to examine the most simple case, which is given by a mixture of two distinguishable, but thermodynamically very similar polymers.

For this purpose, isotope labeling has been introduced in neutron scattering or NMR. It was not recognized that the same technique can be used for light scattering since protonated and deuterated polystyrene, for instance, exhibit a difference in the index of refraction of $\Delta n \approx 0.040$. This small but significant value is sufficient for application of well known optical techniques. In the case of equal molecular masses of the polymers, PCS should result in the determination of self-diffusion instead of interdiffusion coefficients.

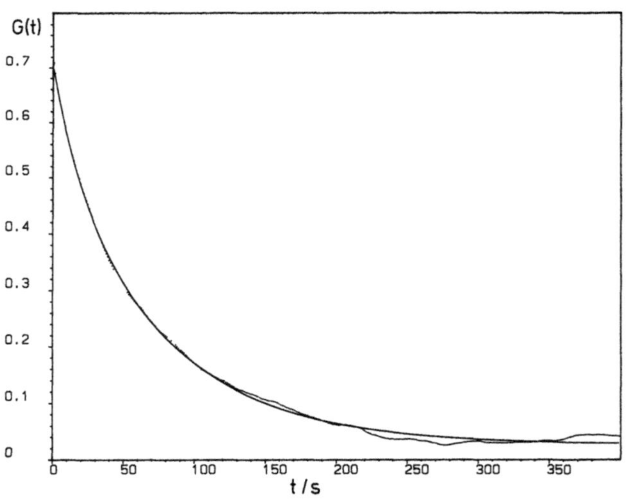

Fig. 10. Typical correlation function with 400 linear channels and 1 s/channel. The line represents a single exponential force-fit to the data points. The fluctuations of the data are typical in this time range of the measurement

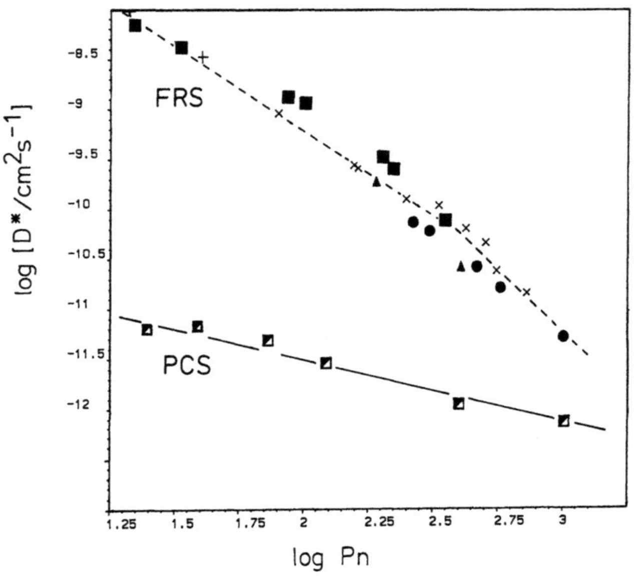

Fig. 11. Comparison of the apparent diffusion coefficients determined by PCS (◪) with those obtained by different FRS experiments (●, ■, ×, ▲, +)

A second problem of PCS of polymers in the bulk state results from the very slow processes observed thus leading to relatively long correlation times. Correlators with a signal width of usually 1–256 have to use prescaling techniques in this time range which implies a waste of signal.

In our set-up, a homemade software correlator based on an ATARI 1040ST computer was used, it covers a dynamic range from 1.4 μs to 10 s sampling time with a signal width of 16 bits. In addition to the direct treatment of high counting rates this system has the advantage of being low-priced.

Figure 10 shows a typical correlation function of a mixture of protonated and deuterated polystyrene at 467 K. The scattering intensity, due to concentration fluctuations, is more than one order of magnitude larger than that caused by density fluctuations which are too fast to be included in Fig. 10. The decay can be described with a single exponential function. The examination of the Q-variation results in a Q^{-2}-dependence, as expected for diffusional processes [15].

If we compare the calculated diffusion coefficient from this experiment with the ones determined with FRS [6, 7] as carried out in Fig. 11 for a reference temperature of $T_g + 94$ K, we observe that neither the magnitude nor the molecular weight dependence coincide. We have to conclude that the relaxation of the optically detectable coherent structure factor in PS is *not* given by the diffusional mobility of polymers; apparently we observe with PCS much slower modes, which are possibly due to the movement of some kind of clusters. This could explain the weak molecular weight dependence. In very careful experiments [15] we have shown that this mobility is not caused by dust or impurities, that the relaxation time is independent on the volume fraction of protonated or deuterated polymer, and that it is thermally activated as the usual self-diffusion. The insensitivity of this process concerning concentration variation leads us to the conclusion that it must also exist in the "pure" states, although in this case the missing contrast hinders an optical detection. It should also be stressed that this "slow-mode" is probably not directly detectable in mechanical experiments since these data are in harmony with the FRS results.

At this stage of the examinations we cannot relate this process definitely to a theoretical description, although different possibilities exist. However, other authors in this issue [11, 12] have presented similar observations in different systems, even in low molecular weight compounds close to T_g, and they have interpreted them in the same sense. Especially it is the "clusters" found in the experiments of the Fischer-group that behave in the same characteristic way, so that both sets of data can possibly be attributed to the same physical phenomenon. Within the framework of our procedure additional experiments, especially concerning the static structure factor, have to be carried out for a better understanding of this slow, but possibly very predicative part of polymer bulk dynamics.

Acknowledgement

Support by the "Deutsche Forschungsgemeinschaft", the "Fonds der Chemischen Industrie" und the "Arbeitsgemeinschaft Industrieller Forschungsvereinigungen" is gratefully acknowledged.

References

1. Eichler HJ, Salje G, Stahl H (1973) J Appl Phys 44:5383
2. Hervet H, Leger L, Rondelez F (1979) J Chem Phys 68:2725
3. de Gennes PG (1971) J Chem Phys 52:572
4. Doi M, Edwards SF (1978) J Chem Soc Faraday Trans II 74:1789
5. Fox TG, Flory PJ (1954) J Polym Sci 14:315
6. Antonietti M, Coutandin J. Sillescu H (1986) Macromolecules 19:793
7. Antonietti M, Fölsch KJ, Sillescu H (1987) Makromol Chem 188:2317
8. Klein J (1986) Macromolecules 19:103
9. Antonietti M, Sillescu H (1986) Macromolecules 19:798
10. Ehlich D (1989) Dissertation Mainz
11. Fischer EW, Becker C, Hagenah JU, Meier G (1989) This issue
12. Burchard W (1989) This issue
13. Brereton MG, Fischer EW, Fytas G, Murschall U (1987) J Chem Phys 86:5174
14. Hess W, Akcasu AZ (1988) J Pys (France) 49:1261 and refs. cited therein
15. Wesselmann M (1988) Dissertation Mainz
16. Green PF, Mills PJ, Palmstrom CJ, Mayer JW, Kramer EJ (1984) Phys Rev Letters 53:2145
17. Mercier JP, Aklonis JJ, Litt M, Tobolsky AV (1965) J Appl Polym Sci 9:447

Received December 29, 1988;
accepted April 18, 1989

Authors' address:

M. Antonietti
Institut für Physikalische Chemie
Universität Mainz
Welder-Weg 15
6500 Mainz, FRG

Discussion

DUSEK:

The positive derivation of D of the probe polymer in networks from the expected dependence on crosslinking density can hardly be explained by some cooperative displacement of crosslinks with probe diffusion. Would not fluctuations in crosslinking density be an alternative explanation?

ANTONIETTI:

We cannot exclude heterogeneities of the crosslinking density which would result – as you have said – also in a positive deviation of D to the theoretically expected behavior. However, the observed difference is too large to be related with fluctuations, especially in the regime of high crosslinking densities. Cooperative movements of neighboring chains are just a simple trial to explain the observed behavior, but possible masks the real situation which can be far more complicated

RICHTER:

I would like to recall the observation of a translational diffusion coefficient which is proportional to $1/M^2$. It is a necessary condition but by no means a proof of the reptation concept. The $D \approx 1/M^2$ behavior is, for example, exhibited also by generalized Rouse models. The essential feature of reptation is the existence of a new intermediate length scale, namely the tube diameter, governing microscopic motion. By a diffusion experiment alone this feature cannot be tested

ANTONIETTI:

I agree with your opinion that a simple scaling relationship is not a proof for a reptative movement. But in addition, we have demonstrated a "topological" invariance, namely, the independence of probe chain mobility on matrix crosslinking. This fact can only be explained by a curvilinear movement which is the basic idea of all reptation concepts, but strictly contradicts all generalized Rouse-theories. I also believe that the existence of a distinct tube diameter simplifies the theoretical treatment but should not be overestimated from an experimental point of view. Simple cooperative movements would "soften" the topological constraints and shift the (possibly broad) transition of local dynamics out of the present range of neutron scattering.

KIMMICH:

Did you investigate whether the evaluated diffusion coefficient depends on the grid spacing? In other words, is the diffusion coefficient a constant with respect to the mean square displacement in the accessible range?

ANTONIETTI:

We have checked the dependence of the diffusion coefficients on the grid spacing *d*. In the accessible range of our experiments which covers the range of some hundreds of nanometers to 20 µm, we have not observed any size-dependence. It is my feeling that such experiments are of fundamental importance in the range of the so-called tube diameter, say 10–100 nm, but for these non-optical experiments, NMR is necessary and most favorable.

Progress in Colloid & Polymer Science Progr Colloid & Polymer Sci 80:93–100 (1989)

TSD-α-peak and non-equilibrium processes

S. Schrader and A. Schönhals

Akademie der Wissenschaften der DDR, Zentralinstitut für Organische Chemie, Bereich Makromolekulare Verbindungen, Berlin, DDR

Abstract: The influence of physical aging on the electrical properties of poly(vinylacetate) in the glass transition region were investigated using the thermally stimulated depolarization (TSD) technique. The position, intensity, and shape of TSD-α-peaks were studied in dependence on cooling and heating rates. Furthermore, a model for TSD measurements in the glass transition region is presented based on the phenomenological approach, as proposed by Narayanaswamy. The theoretical predictions are in qualitative agreement with experimental findings.

Key words: Thermally stimulated depolarization, glass transition, physical aging, shape of TSD-peaks, theoretical model, poly(vinylacetate)

Introduction

The method of thermally stimulated depolarization (TSD) is a powerful tool for the study of molecular mobility of polymeric systems. In recent years this method has also been applied to the study of the glass transition region (cf. [1]). Often during TSD-experiments polymeric samples are cooled down below the glass transition temperature T_g. However, it is a well-known fact that in experiments where the glass transition is passed a non-equilibrium state occurs, dependent on thermal pre-history. This non-equilibrium causes a structural relaxation which is responsible for time-dependent changes of the thermal, mechanical, electrical, and other properties of the sample. Because no chemical reactions occur these processes are denoted as "physical aging" [2]. In the majority of publications concerning TSD measurements the influence of physical aging on the detected current signals has been neglected. This can be the reason for discrepancies in the interpretation of TSD measurements carried out in order to study the main relaxation process (α-peak) which is coupled to the glass transition of polymers. Only a few publications discuss the influence of physical aging, i.e., of structural relaxation on the TSD signals (cf. [3–6]).

In contradiction, the influence of physical aging on other properties was extensively studied [2]. Figure 1 gives an example of how physical aging changes the enthalpy $H[t, T(t)]$ of a polymer with time t and temperature T. After cooling down with constant rate from a high temperature where the sample is in a molten or rubbery state to a temperature below glass transition vitrification of the specimen occurs. Dependent on cooling rate enthalpy deviates from the equilibrium isobar during vitrification. After having reached a distinct value of T below T_g the tempera-

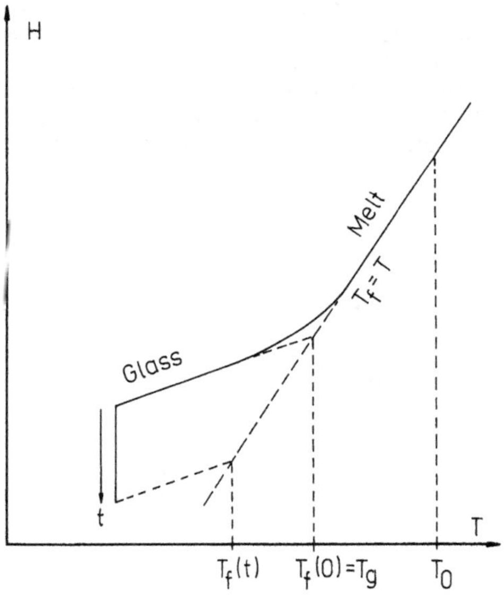

Fig. 1. Schematic plot of enthalpy H of a polymeric system during cooling through glass transition and subsequent annealing below T_g (T_g = glass transition temperature, T_f = fictive temperature, T_0 = reference temperature, t = time)

ture is kept constant. During this isothermal annealing process enthalpy decreases with time towards equilibrium isobar (dashed line in Fig. 1) due to structural relaxation.

In the present paper a phenomenological approach is developed in order to describe this influence on TSD-peaks theoretically. The predictions are compared with experimental results measured on poly(vinylacetate) (PVAc).

Experimental

Figure 2 presents the scheme of a TSD measurement. The sample is polarized at a high temperature and then cooled down to a deep temperature where the investigated polarization process is frozen-in. Subsequently, the sample is short-circuited and heated with constant heating rate while the relaxation or TSD current $J[t, T(t)]$ is recorded. The TSD current increases with rising temperature due to the increase of relaxation rate. When the number of relaxing dipoles is exhausted a drop of detected current occurs. As a consequence of superposition of these two tendencies a current peak is recorded during experiment. This current peak bears essential information about the related relaxation processes.

The polymer selected for the study was a commercial PVAc of VEB Buna (GDR) synthesized by suspension polymerization ($M_n = 67445$, $M_w = 407530$, $T_g = 312$ K, determined by DSC with a heating rate of 10 K/min) [7]. The sample was cast from a 5 vol% acetone solution on a 1.5-mm-thick copper plate serving as the bottom electrode. The 77.0-μm-thick film was dried under high vacuum using an annealing program where the sample was heated in steps of 10 K from room temperature up to 423 K. At each temperature step the sample was annealed for about 1 h. After drying of the sample a silver electrode was evaporated which had guard-ring configuration (diameter of the hot electrode 1.740 cm, active electrode area 2.378 cm²).

In our experiment the PVAc sample was short-circuited and heated to 393 K in order to depolarize all stored space charges. Then the sample was cooled to 343 K and at this temperature a polarization voltage of $U_p = 500$ V was applied for a time $t_p = 10$ min. This formation conditions guaranteed an almost complete filling of dipolar polarization and a low amount of injected space charges. After formation the sample was cooled with applied field at a constant cooling rate down to liquid nitrogen temperature. Then the sample was short-circuited and the TSD experiment was carried out with constant heating rate.

In our experiments the influence of different cooling and heating rates on the measured TSD current was studied.

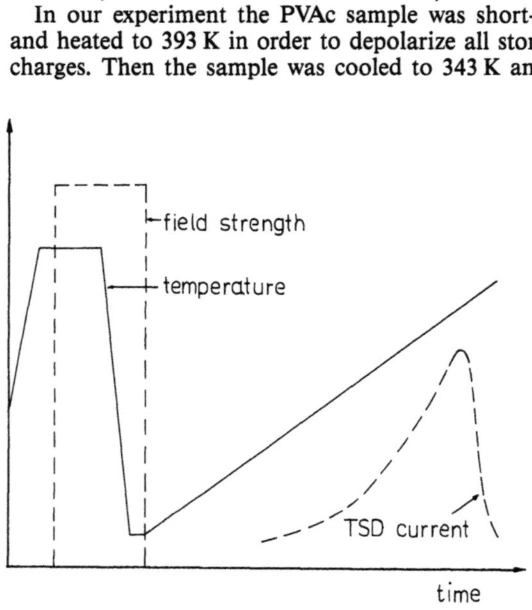

Fig. 2. Scheme of a TSD experiment

Model

The model presented in the following section is based on the concept that the measured TSD response around T_g is strongly influenced by physical aging. In order to predict the TSD current in the glass-transition region it is, therefore, necessary to use information about physical aging. To take this information into consideration in a practicable manner some assumptions have to be made:

1) In a first approximation, a perturbation (e.g., the electrical field) in the low-level range does not influence the spectrum of molecular motion and, hence, the structural relaxation [8]. Therefore, in the model the influence of the electrical field on the spectrum of relaxation times is neglected.
2) The actual state of the system is described by one state property only, the time-dependent value of enthalpy $H(T, t)$.
3) For very long annealing times below T_g the thermodynamic equilibrium should be attainable, and for the same reason the relaxation current of long-time annealed samples should reflect the equilibrium behavior too.

In order to model structural relaxation the concept first proposed by Narayanaswamy [9] is used. In the following only some essential ideas of this concept are given. For further details see [9, 10].

In the framework of this model it is convenient to characterize the actual state of the system by a fictive temperature $T_f(t)$ which is defined according to Tools construction [11] by

$$H(T) = H_e(T_f) - \int_T^{T_f} c_{pg} dT' \quad . \tag{1}$$

$H_e[T_f(t)]$ is the equilibrium value of H at temperature T_f. c_{pg} is the heat capacity in the glassy state as measured by DSC. For temperatures above T_g where the system is in equilibrium the equation $T = T_f$ is fulfilled. Below T_g when a non-equilibrium state exists $T_f(t)$ is the abscissa of the projected value of

$H[T(t)]$ along a line with slope c_{pg} onto the equilibrium (melt) curve of enthalpy (see Fig. 1).

Treating a change of temperature as perturbation in the framework of linear response theory the system follows a small step in temperature with a non-exponential response described by a dimensionless response function $K(t)$ [9, 10]. Additionally, the kinetics of structural relaxation is non-linear in the sense that the non-exponential response function $K(t)$ depends both on the actual temperature and on the deviation from equilibrium expressed in the present formulation by the difference (T_f-T). This difference represents the relaxing part of enthalpy measured in units of temperature.

Using a material time or reduced time $\Delta\xi$ the influence of both T and T_f on structural relaxation can by taken into account [9]. $\Delta\xi$ is given by

$$\Delta\xi(t,t_0)=\int_{t_0}^{t} a[T(t'),T_f(t')]^{-1}dt' \tag{2}$$

where $a[T(t),T_f(t)]$ is a dimensionless shift factor depending on T and T_f and, therefore on t. Note that the material time has the unit of time. Physically spoken, $\Delta\xi$ measures the time in units of the mean relaxation time of physical aging depending on temperature and fictive temperature. The Boltzmann superposition principle is valid for the material time scale. Hence, the change of enthalpy of the system, expressed by T_f (as a reaction to arbitrary temperature programs) follows the well-known convolution integral wherein the time differences in the response-function have to be replaced by material time differences [7, 9, 10]

$$T_f(t)=T_0-\int_{0}^{t}\dot{T}(t')K[\Delta\xi(t,t')/\tau_{0H}]dt' \ , \tag{3}$$

where $\dot{T}=dT/dt$ and τ_{0H} is the mean relaxation time of enthalpy at the starting temperature T_0 (see Fig. 1) of experiment ($T_0>T_g$). Narayanaswamy [9] suggested the mixing rule for T and T_f with respect to their influence on $a(t)$:

$$\ln a(t) = -c\{x[T(t)-T_0]+y[T_f(t)-T_0]\} \tag{4}$$

$$x+y=1$$

y is a mixing parameter weighting the influence of T_f with respect to T. Note that in Eq. (4) the empirical WLF-dependence is replaced by a linear approximation which is valid in narrow temperature ranges. c is the slope of the WLF-plot at $T=T_0$. Using a fitting procedure the response function $K(t)$ and the parameters c, τ_{0H}, and y can be obtained from DSC-measurements [7]. For the mathematical representation of $K(t)$ the Havriliak-Negami-function [12] is used. The present model for structural relaxation describes the experimental heat capacity curves both qualitatively and quantitatively in a correct manner [7]. In principle the above-mentioned parameters can be obtained by a non-linear fitting procedure to experimental c_p-curves [13]. However, in [7] the width, the asymmetry, and the height of the over-heating peak are taken from a series of DSC-curves which were measured with different cooling and heating rates, and these quantities were compared with the respective calculated quantities. The parameter set which provided the best fit to the experimentally obtained quantities was chosen for further calculations.

Because the material time $\Delta\xi$ contains all information about the thermal history this quantity is suitable to connect structural relaxation and electrical behavior. There is experimental evidence that different quantities like enthalpy, dielectric permittivity or mechanical compliance have different mean relaxation times but equal shift factors [7]. Thus for structural and dielectric relaxation the same material time scale can be defined. Therefore, the relaxation current $J(T)$ is expressed by the following relation [5]

$$J(T)=A\,\varepsilon_a E\dot{\varepsilon}_{eq}[\Delta\xi(T,T_f)/\tau_{0D}] \ . \tag{5}$$

A is a geometrical constant, ε_a is the absolute dielectric constant, E is the electric field, $\dot{\varepsilon}_{eq}$ is the time derivative of isothermal time dependent permittivity ($\dot{\varepsilon}_{eq}=d\varepsilon_{eq}/dt$) in the equilibrium, measured at $T=T_0$. τ_{0D} is the mean dielectric relaxation time at T_0. $\dot{\varepsilon}_{eq}(t)$ can be determined by isothermal depolarization measurements and for the mathematical representation of $\dot{\varepsilon}_{eq}(t)$ also the model function of Havriliak-Negami is used [14, 15]. In order to simplify

Table 1. Parameters used for calculation of TSD-α-peaks

Thermal behavior [7]	Dielectric behavior
β_T = 0.65	β_D = 0.95
γ_T = 0.9	γ_D = 0.49
τ_{0T} = 2.9 E −4 s	τ_{0D} = 2.54 E −3 s
c = 0.78 K^{-1}	$\Delta\varepsilon$ = 6.2
y = 0.65	
T_0 = 330 K	

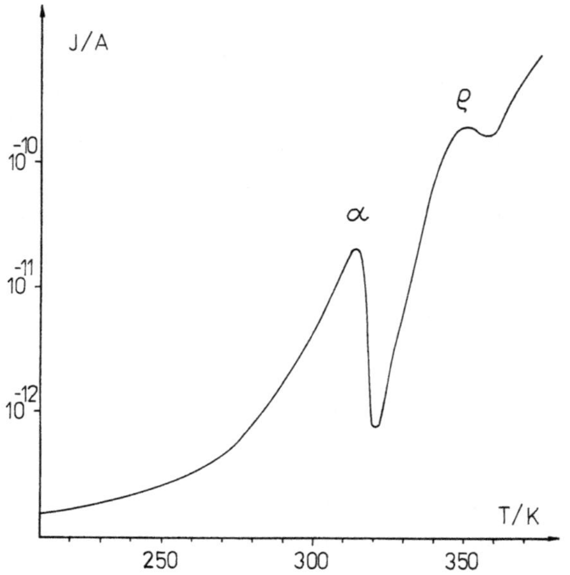

Fig. 3. TSD thermogram of PVAc measured with a heating rate of 1.9 K/min after cooling with 6.9 K/min

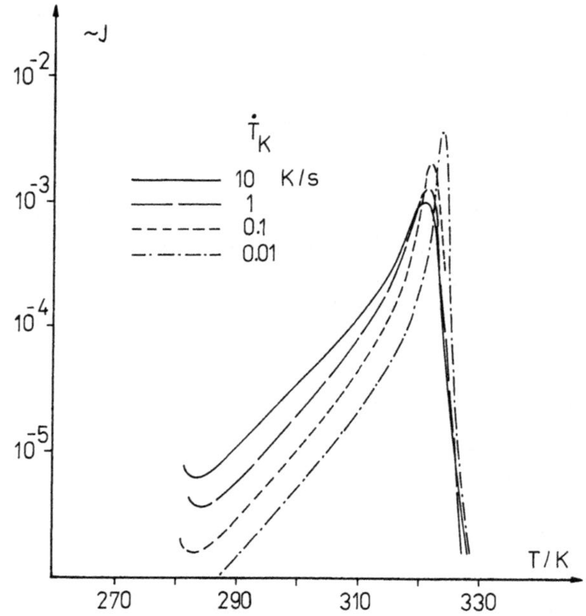

Fig. 5. α-peaks calculated for various cooling rates but for fixed heating rate (1 K/s)

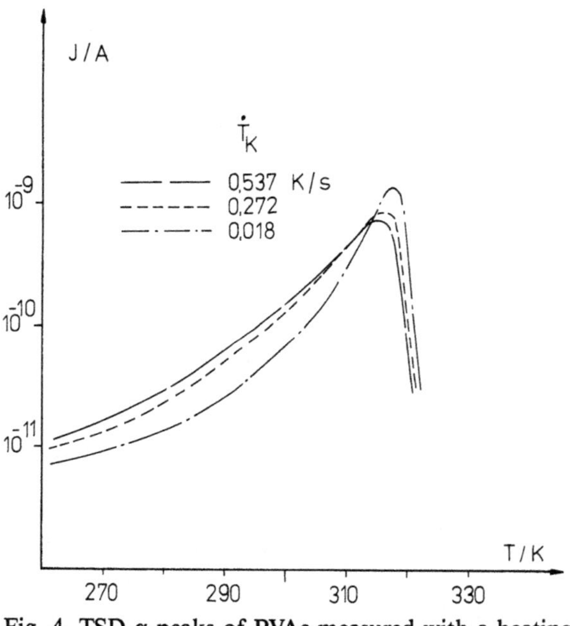

Fig. 4. TSD-α-peaks of PVAc measured with a heating rate of 6 K/min after cooling with different cooling rates $\dot{T}_K$

the approach the model is based on the time-temperature-shift principle [cf. Eq. (5)] and so the small variations in shape and intensity of $\dot{\varepsilon}_{eq}(t)$ with temperature are neglected. After solving Eq. (3) numerically by means of an iteration procedure $J(T)$ is calculated according to Eq. (5). Further expressions and the model parameters (see Table 1) used for the calculation of TSD current are given in the appendix.

Results

The TSD thermogram of PVAc (Fig. 3) shows two pronounced peaks, the α-peak connected with the cooperative movement of chain segments near 315 K (glass transition), and a ϱ-peak at about 353 K which is connected with the decay of an accumulated space charge cloud. In the following the α-peak is analyzed in more detail.

Figure 4 shows α-peaks of PVAc measured with the same heating rate of 6 K/min but after cooling with different rates. Peak width decreases, maximum current increases and maximum temperature shifts to higher values for slower cooling.

Calculated TSD peaks for the same situation after cooling down with different rates but measured with the same heating rate are shown in Fig. 5. It is obvious that the experimental behavior as described above is reproduced by the presented model.

The influence of heating rate on the α-peak at fixed cooling rate is demonstrated in Figs. 6 and 7. Figure 6 shows measured thermograms and Fig. 7 shows calculated thermograms. The TSD current increases and the maximum temperature shifts to higher values when heating rate increases. A comparison of Figs. 6 and 7 also demonstrates a good qualitative agreement between experiment and theory. In order to characterize the shape of TSD α-peaks the half-width $\delta_{1/2}$ and the tenth-value-width $\delta_{1/10}$ of measured and calculated thermograms are given in Figs. 8 and 9,

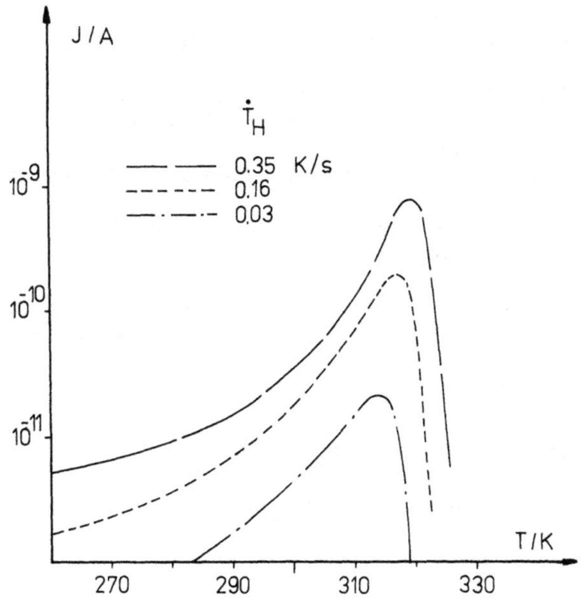

Fig. 6. α-peaks of PVAc measured with different heating rates $\dot{T}_H$ (cooling rate 6.9 K/min)

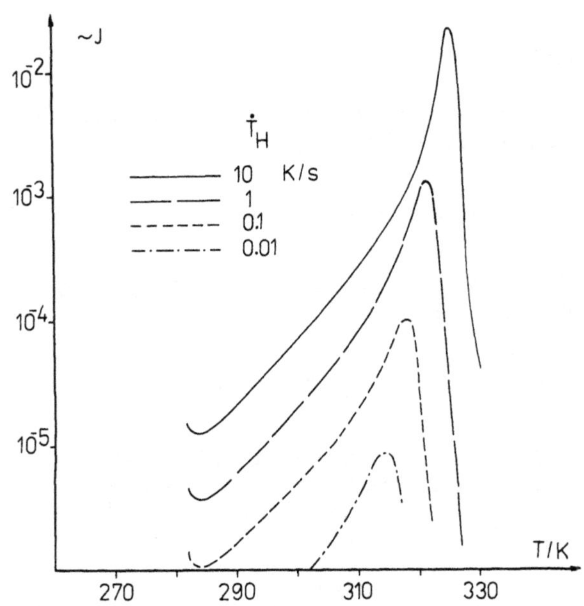

Fig. 7. α-peaks calculated for different heating rates and for fixed cooling rate (1 K/s)

respectively. Although there is a deviation between measured and calculated peak parameters, both representations reveal an almost linear dependence of both quantities on the logarithm of cooling rate.

Discussion

It was demonstrated that the structural relaxation has a considerable influence on the polarization state

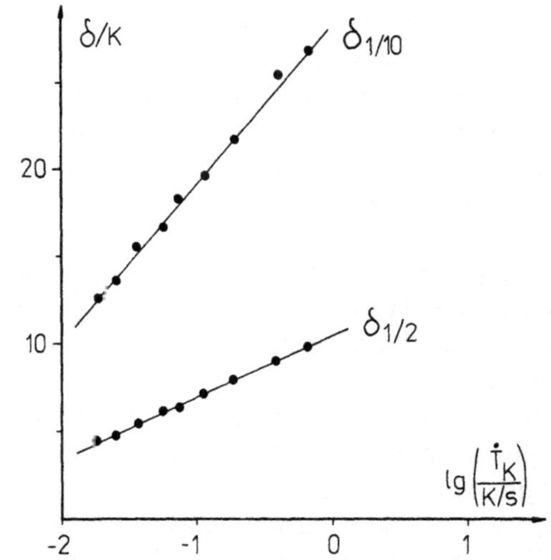

Fig. 8. Half-width $\delta_{1/2}$ and tenth-value-width $\delta_{1/10}$ of measured TSD-α-peaks as a function of logarithm of cooling rate

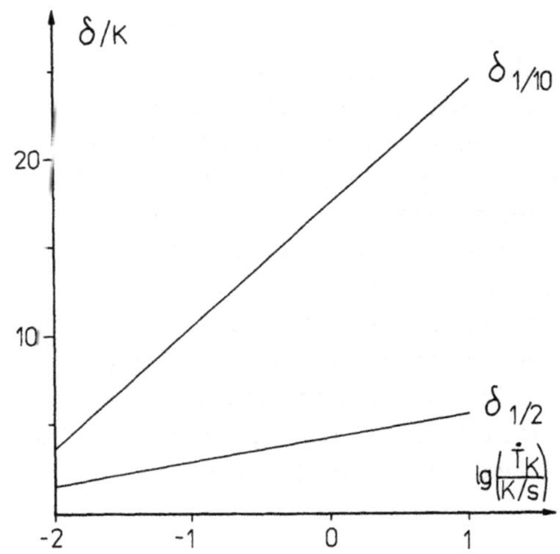

Fig. 9. Half-width $\delta_{1/2}$ and tenth-value-width $\delta_{1/10}$ of calculated TSD-α-peaks as a function of logarithm of cooling rate

reached during the formation period. Especially by variation of cooling rate or of time of annealing below glass transition temperature considerable changes in the detected TSD signals occur. This is due to the fact that the system relaxes more towards equilibrium at low cooling rates or during annealing below T_g. These findings were also confirmed by Kubon et al. [6], who carried out TSD investigations on annealed PMMA electrets. For theoretical description of their results they used the phenomenological aging theory of Kovacs et al. [16], combined with the three-parameter Williams-Watts-type representation of dielectric response. This approach is physically equivalent to the model used in [5] and in the present paper. But in the latter the dielectric response is described by the four-parameter formula of Havriliak-Negami [12]. It was shown, for example, by Schlosser et al. [14] that a four-parameter representation is necessary in order to reflect completely the dielectric relaxation behavior of polymers.

From the experimental results and from the presented model calculations it has to be concluded that the analysis of TSD signals of the main relaxation process has to take into account the circumstances of thermal pre-history. When the so-called thermosampling technique is applied for the analysis of main relaxation region the formation procedure is carried out in a temperature window around glass transition. Therefore, effects of physical aging should especially be taken into account in the interpretation of these sampling thermograms. The presented model is able to predict the peculiar effects of physical aging on the sampling curves.

Furthermore, the theory predicts that the initial rise of the α-peak is connected with a factor which depends on the actual deviation from equilibrium state. Hence, the initial rise alone is not sufficient for the estimation of apparent activation energy of the main relaxation process.

The shift of the maximum position of the TSD α-peak with heating rate depends slightly on the deviation from the equilibrium state which was frozen-in during cooling. Therefore, the widely used analysis of the maximum shift with heating rate provides only a rough approximation of the apparent activation energy of the main relaxation process.

References

1. Belana J, Colomer P, Pujal M, Montserrat S (1984–1985) J Macromol Sci Phys B23:467
2. Struik LCE (1978) Physical aging in amorphous polymers and other materials. Elsevier, Amsterdam
3. van Turnhout J (1975) Thermally Stimulated Discharge of Polymer Electrets. Elsevier, Amsterdam, pp 33–34
4. van Turnhout J (1980) in: Sessler GM (ed) Electrets. Springer, Berlin, Heidelberg, New York, p 160
5. Schrader S, Schönhals A, Schlosser E (1987) Poster IV/124. 31. IUPAC Macromolecular Symposium, Microsymposium IV, June 30–July 4, Merseburg, GDR
6. Kubon U, Schilling R, Wendorff JH (1988) Colloid Polym Sci 266:123
7. Schneider K (1984) Dissertation A, Technische Hochschule "Carl Schorlemmer" Leuna-Merseburg
8. Struik LCE (1980) Polymer 21:962
9. Narayanaswamy OS (1971) J Amer Ceram Soc 54:491
10. Sasabe H, Moynihan CT (1978) J Polym Sci. Polym Phys Ed 16:1447
11. Tool AQ (1946) J Res Nat Bur Standards 37:73
12. Havriliak S, Negami S (1966) J Polym Sci C Polym Symp 14:59
13. Hodge IM, Berens AR (1982) Macromolecules 15:762
14. Schlosser E, Kästner S, Friedland KJ (1981) Plaste u. Kautschuk 28:77
15. Schlosser E, Schönhals A (1987) Acta Polymerica 38:357
16. Kovacs AJ, Aklonis JJ, Hutchinson JM, Ramos AR (1979) J Polym Sci Polym Phys Ed 17:1097

Received January 13, 1989;
accepted May 12, 1989

Authors' address:

Dr. S. Schrader
Akademie der Wissenschaften der DDR
Zentralinstitut für Organische Chemie
Bereich Makromolekulare Verbindungen
Rudower Chaussee 5
1199 Berlin, DDR

Appendix

The response function $K(t/\tau_0)$ used for description of the thermal relaxation is defined by

$$K(t/\tau_0) = \int_{-\infty}^{\infty} L_T(s, \beta_T, \gamma_T)[1-\exp(-ts/\tau_0)]\, d\ln s \;. \qquad (A1)$$

The derivation of time-dependent permittivity $\dot{\varepsilon}(t/\tau_0)$ of equilibrium state is given by

$$\dot{\varepsilon}(t/\tau_0) = \int_{-\infty}^{\infty} L_D(s, \beta_D, \gamma_D)\, s \exp(-ts/\tau_0)\, d\ln s \qquad (A2)$$

L_T and L_D are thermal and dielectric distribution functions of relaxation times, respectively. Both are given by the same mathematical expression [14, 15] which was derived from the Havriliak-Negami-formula [12]

$$L(s) = \frac{1}{\pi} \frac{\sin[\gamma \psi(s)]}{[1+2s^\beta \cos(\beta\pi)+s^{2\beta}]^{\gamma/2}} \qquad (A3)$$

with

$$\psi(s) = \frac{\pi}{2} - \arctan\left[\frac{s-\beta+\cos(\beta\pi)}{\sin(\beta\pi)}\right] \qquad (A4)$$

and the normalized relaxation frequency

$$s = \tau_0/\tau$$

where τ is a relaxation time and τ_0 represents the mean relaxation time of the process.

In Table 1 the parameters used for calculations are given.

Discussion

OLEYNIK:

Comments: 1. Different rates of cooling of polymeric samples during polarisation introduce not only shape changes of alpha-relaxation peak (or peaks), but also change the shape of gamma- and beta-relaxation peaks. We have found that with epoxy polymers.

2. Of course one should expect the change of relaxation peak broadness as a function of cooling rate within of the T_g region. But there is another way of glass-preparation. That is polymerisation or polycondensation of monomeres at constant temperature. If electric field is applied during polycondensation process, the final glassy state becomes polarized and one may measure TSD during heating. Again for epoxies we found that the broadness of the alpha-peak becomes extremely small (about 4–5 °C). That is practically impossible to reach this state of the glass by changing of cooling rate.

It means in our understanding, that the state of chemically identic glasses, which were prepared by different ways may seriously differ from each other.

SCHRADER:

1. It is a well-known fact that physical aging influences also to a small extent secondary relaxation processes. This is understandable if one imagines that the surrounding of groups which carry out local motions responsible for these relaxation processes is coupled to the frozen-in structure, e.g. to the amount of free volume. This structure is dependent on cooling or annealing conditions.

2. If one assumes that the chemical process of curing is not influenced by electrical field the polarization state reached after a slow polycondensation reaction with applied electrical field corresponds to that of a well-aged sample which is near thermal equilibrium. It is questionable if such a state can be reached during cooling or annealing experiments in a finite time scale. In that sense the two epoxies have to be regarded as two different glasses. That corresponds to the concept of M. Goldstein. On the other hand the ultimate state approached by the sample during annealing or during polarization period below T_g depends on the annealing temperature. So the equilibrium state at each temperature is connected with a an other motional spectrum. Therefore, one can ask which annealing temperature corresponds to the state reached after slow, complete curing.

BÄSSLER:

Can the profile of the alpha-relaxation peak be translated into a distribution of relaxation times reflecting the thermal history of the sample?

SCHRADER:

This question touches two unsolved problems of relaxation spectroscopy.

The first problem results from the temperature dependence of equilibrium spectra of relaxation times. Assuming equilibrium state as initial condition of experiment then, roughly spoken, the TSD alpha-peak reflects at each temperature a distinct fraction of the relaxation spectrum which corresponds to the equilibrium state of this temperature. It is already very difficult to deconvolute the alpha-peak into a series of equilibrium spectra, because their temperature dependence is unknown. Furthermore, only a very high accuracy of measuring would allow such an deconvolution. But this accuracy is hardly to reach. Therefore, a combination of isothermal and non-isothermal dielectric measurements is necessary to solve this problem. The second problem concerns the influence of non-equilibrium processes on relaxation spectra. In addition to the mentioned temperature dependence a time dependence of the spectra occurs which is due to physical aging. Of course this dependence reflects the thermal pre-history, but it is not investigated satisfactorily until now.

KRÜGER:

Which is the physical significance of T_0 you have used in your calculations?

SCHRADER:

The state of the system at temperature T_0 should be a state in thermal equilibrium.

KRÜGER:

I would like to mention that at least some part of the observed enthalpy relaxations may be due to quenched in mechanical constraints.

SCHRADER:

The mechanical constraints can, in fact, contribute to a great extent to the excess enthalpy. However, two kinds of constraints should be distinguished: macroscopic constraints which can be visualized by optical means, and constraints on a microscopic or molecular level. The latter are probably the main driving force of structural relaxation.

HOLZMÜLLER:

There are two different possibilities to describe the behavior near the glass transition temperature:

i) Cooling from a liquid state and freeze in holes in the structure.
ii) Heating up from an ideal solid state and forming holes above the glass transition temperature.

I think the second way will better describe viscous flow and molecular mobility and the first way better describes electrets.

SCHRADER:

The first mentioned way to describe the behaviour near glass transition is similar to the way used in the presented model. It was assumed that the equilibrium state of system below T_g is expressed by an equilibrium isobar of a liquid.

PECHHOLD:

Have you tried to perform similar TSC-measurements and analyses on samples cooled under static pressure?

Calorimetry and mechanical compliance on such samples show a strong shift of T_g to lower temperatures. This may be due to an increased order under pressure, causing a strong driving force towards equilibrium when pressure is released.

SCHRADER:

No, we did not carry out such measurements under hydrostatic pressure. The variation of pressure as additional external parameter will obviously give more information about the thermodynamics of the glassy state.

DUSEK:
What are the factors the time dependence of enthalpy you measure is determined by? One can assume that $(dH(t)/dt)_T$ will be proportional to the difference in free energies of the system at time t and $t \to \infty$ with a proportionality factor also depending on the state of glass. Is there any correction between such an approach and the time temperature shift of the retardation spectra you are using?

SCHRADER:
The time dependence of enthalpy is determined by the factors temperature, the deviation from equilibrium, e.g. characterized by a fictive temperature or an excess free volume, and the underlying relaxation time spectrum. It was shown by Kovacs or others that the description of enthalpy relaxation by one ordinary differential equation with proportionality factor also depending on the state is possible, but gives only a rough approximation of the experimental time dependence. A better description gives the multiparameter model proposed by Kovacs et al. This model is mathematically and physically equivalent to the presented approach of enthalpy relaxation.

Deterministic chaos in the motion of chain molecules

P. Reineker and R. G. Winkler

Abteilung Theoretische Physik, Universität Ulm, Ulm, FRG

Abstract: The dynamics of a molecular chain is investigated numerically. The chain consists of a one-dimensional arrangement of mass points which are connected by massless rigid rods. The dynamical behavior of the chain is determined by a set of nonlinear classical equations of motion. These equations are solved numerically for chains of various lengths and for motion of the chain in two and three dimensions under the restriction that its endpoints are fixed. The representation of the solution using Poincaré maps shows regions of chaotic motion of the chain whose size depends on its end-to-end distance. In addition, the time-dependent force needed to fix the end points of the chain is calculated. Its time average is determined as a function of the end-to-end distance of the chain. A discussion of the various cases mentioned above and of the connection with earlier models is given.

Key words: Networks, rubber elasticity, deterministic chaos, entropy, elastic forces, molecular dynamics simulation

1. Introduction

Various properties of polymers are not yet completely understood from a microscopic point of view, some of them are connected to entropy elasticity of polymeric networks. The stress-strain relation of a network is an example for the deviation between microscopic theories and experimental results. The difference is most clearly seen in the Mooney plot [1].

Calculations of the stress-strain relation for molecular chains and networks are based usually on statistical treatments [2–8]. By counting the number of possible static configurations for a given extension, the distribution function of the end-to-end distance of a single chain is determined. This distribution function is used to calculate the free energy and from this quantity finally the stress-strain relation is obtained. Treloar [9] gave an exact expression for the probability distribution of the end points of a chain; however, much more often used is the Gaussian distribution [10]. Such a distribution is obtained if the distance between successive points is much smaller than the end-to-end distance and the latter quantity is much smaller than the length of the chain. If, in addition, the finite length of the chain is taken into account the distribution function is characterized by a Langevin function [11]. Starting from these single chain approaches several authors [2–4, 6, 12] calculated the stress-strain relation for polymeric networks assuming an affine deformation down to microscopic length scales. These phantom chain models neglect the inter- and intramolecular interactions. By comparing the theories with experimental results one finds large deviations. There are several models which partly remove these discrepancies including the inter- and intramolecular interaction in the form of entanglements [13, 14], tubes [15], or constrained junction fluctuations [16]. Besides these semi-phenomenological models there exists some purely phenomenological approaches that well describe the stress-strain behavior of networks [17–19].

The dynamical treatment of molecular chains is considered within two classical models: the flexible and the rigid chain models. Both types of models consider the chain as a one-dimensional arrangement of mass points which are connected in different ways. In the flexible chain model the mass points are connected by harmonic or anharmonic springs. These models give, in the limit of an infinite spring constant, the well known results described above [6]. In the rigid chain model the mass points are connected by massless rigid rods. Due to these constraints there is a coupling between the conformations of the chain in space and its kinetic energy. As a consequence the equilibrium distribution in phase space is different for these two types of models [20].

Our calculations address the investigation of entropy elastic properties of networks using a dynamical approach. In this connection we also work out the differences between the rigid and flexible chain models. The purpose of this paper is to present results of the dynamic investigations for the motion of short chains.

2. Dynamical model

We consider a polymeric chain consisting of $N+2$ mass points which are connected by massless rigid rods of length l (see Fig. 1). The endpoints of the chain are fixed at a distance $\vec{d}$. The other points can move freely and they only have to satisfy the constraints

$$|\vec{r}_i-\vec{r}_{i-1}|-l=0\ ,\quad i=1,\dots,N+1\ . \tag{1}$$

These constraints are taken into account in the equations of motion by Lagrangian multipliers. The forces stemming from these constraints are the only forces acting on the mass points of the chain. The equations of motion are then given by

$$m_i\ddot{\vec{r}}_i=\bar{\lambda}_{i+1}(\vec{r}_i-\vec{r}_{i+1})+\bar{\lambda}_i(\vec{r}_i-\vec{r}_{i-1})\ ,\quad i=1,\dots,N \tag{2}$$

where the $\bar{\lambda}_i$'s are the Lagrangian multipliers. It is convenient to normalize position vectors, time, Lagrangian multipliers, end-to-end distance, and velocities in the following way:

$$\vec{r}_i=l\vec{\varrho}_i \tag{3a}$$

$$t=\left(\frac{ml^2}{E}\right)^{1/2}\tau \tag{3b}$$

$$\bar{\lambda}_i(t)=\frac{E}{l^2}\lambda_i(\tau) \tag{3c}$$

$$\vec{d}=l\vec{\alpha} \tag{3d}$$

$$\dot{\vec{r}}_i=\sqrt{\frac{E}{m}}\,\dot{\vec{\varrho}}_i(\tau) \tag{3e}$$

E is the (conserved) kinetic energy of the system. For simplicity, equal masses for all points are assumed. In terms of these quantities the equations of motion read:

$$\ddot{\vec{\varrho}}_i(\tau)=\lambda_{i+1}(\vec{\varrho}_i-\vec{\varrho}_{i+1})+\lambda_i(\vec{\varrho}_i-\vec{\varrho}_{i-1})\ . \tag{4}$$

For the total energy we obtain

$$\frac{1}{2}\sum_{i=1}^{N}\dot{\vec{\varrho}}_i^{\,2}=1\ , \tag{5}$$

i.e., the energy is a constant of the motion and normalized to one. It is obvious that the end-to-end distance α and the number of links $(N+1)$ are the only specific quantities of the chain which appear in its equations of motion. With the help of the holonomic constraints [Eq. (1)] we derive a linear system of equations for the Lagrangian multipliers by differentiating the constraints twice with respect to τ and inserting Eq. (4). Solving this system of equations and inserting the solution in the equations of motion we can calculate the orbits $\vec{\varrho}_i(\tau)$ numerically.

The Lagrangian multipliers allow to calculate the forces of constraint, especially the forces needed to fix the end points of the chain. For the righthand end point, $\vec{r}_{N+1}=\vec{d}$, the force in dimensionless quantities is given by

$$\vec{f}_{N+1}=\lambda_{N+1}(\vec{\varrho}_n-\vec{\alpha})\ . \tag{6}$$

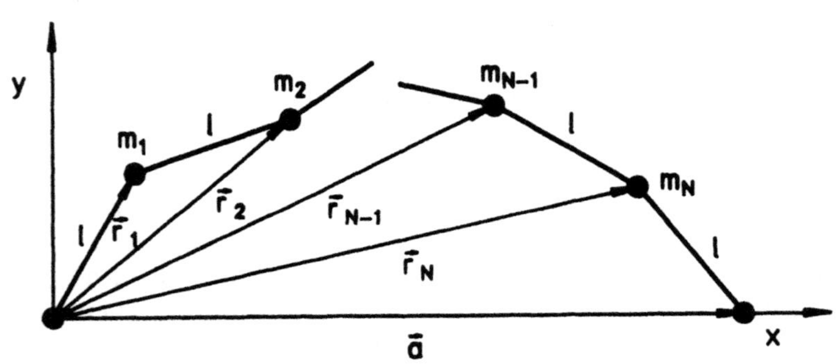

Fig. 1. Model of the polymeric chain

We are not only interested in the force as a function of time but also in its time average. This average is calculated from

$$\langle \vec{F}_{N+1}(t)\rangle = \frac{E}{l}\lim_{t\to\infty}\frac{1}{t}\int_0^t \lambda_{N+1}(\tau)\times(\vec{\varrho}_n-\vec{a})\,d\tau = \frac{E}{l}\langle \vec{f}_{N+1}\rangle\ . \tag{7}$$

The only quantity characterizing the chain which appears in this average is the length of a segment l. Especially the equation is independent of the masses m_i because of the assumption of equal masses for all points. The use of the canonical ensemble average of the kinetic energy gives

$$\bar{E} = \tfrac{1}{2}([d-1]N-1)kT\ , \tag{8}$$

where d is the dimension of the motion. With this property the averaged force of Eq. (7) is proportional to the temperature ($\langle \vec{F}_{N+1}\rangle \sim T$), a relation which is well known in polymer physics [4]. In the following section we consider the two-dimensional motion of a chain with $N=3$ mobile mass points (or four segments) in more detail.

3. Chaotic motion of chains

For three mobile mass points we have the following equations of motion:

$$\ddot{\vec{\varrho}}_1 = \lambda_1\vec{\varrho}_1+\lambda_2(\vec{\varrho}_1-\vec{\varrho}_2) \tag{9a}$$

$$\ddot{\vec{\varrho}}_2 = \lambda_2(\vec{\varrho}_2-\vec{\varrho}_1)+\lambda_3(\vec{\varrho}_2-\vec{\varrho}_3) \tag{9b}$$

$$\ddot{\vec{\varrho}}_3 = \lambda_3(\vec{\varrho}_3-\vec{\varrho}_2)+\lambda_4(\vec{\varrho}_3-\vec{a})\ . \tag{9c}$$

The λ_i's are obtained from a linear system of equations as described above. For the purpose of illustration, in this way we get for λ_1

$$\lambda_1 = \frac{A}{B} \tag{10a}$$

$$\begin{aligned} A = {} & \dot{\vec{\varrho}}_1^{\,2}\{[(\vec{\varrho}_1-\vec{\varrho}_2)(\vec{\varrho}_3-\vec{\varrho}_2)]^2 \\ & -2(2-[(\vec{\varrho}_3-\vec{\varrho}_2)(\vec{\varrho}_3-\vec{a})]^2)\} \\ & +(\dot{\vec{\varrho}}_1-\dot{\vec{\varrho}}_2)^2[\vec{\varrho}_1(\vec{\varrho}_1-\vec{\varrho}_2)](2-[(\vec{\varrho}_3-\vec{\varrho}_2)(\vec{\varrho}_3-\vec{a})]^2) \\ & -(\dot{\vec{\varrho}}_2-\dot{\vec{\varrho}}_3)^2[\vec{\varrho}_1(\vec{\varrho}_1-\vec{\varrho}_2)][(\vec{\varrho}_1-\vec{\varrho}_2)(\vec{\varrho}_3-\vec{\varrho}_2)] \\ & +\dot{\vec{\varrho}}_3^{\,2}[(\vec{\varrho}_3-\vec{\varrho}_2)(\vec{\varrho}_3-\vec{a})][\vec{\varrho}_1(\vec{\varrho}_1-\vec{\varrho}_2)] \\ & \times[(\vec{\varrho}_1-\vec{\varrho}_2)(\vec{\varrho}_3-\vec{\varrho}_2)] \end{aligned} \tag{10b}$$

$$\begin{aligned} B = {} & 4-2[(\vec{\varrho}_3-\vec{\varrho}_2)(\vec{\varrho}_3-\vec{a})]^2-[(\vec{\varrho}_1-\vec{\varrho}_2)(\vec{\varrho}_3-\vec{\varrho}_2)]^2 \\ & -2[\vec{\varrho}_1(\vec{\varrho}_1-\vec{\varrho}_3)]^2 \\ & +[\vec{\varrho}_1(\vec{\varrho}_1-\vec{\varrho}_2)]^2[(\vec{\varrho}_3-\vec{\varrho}_2)(\vec{\varrho}_3-\vec{a})]^2\ . \end{aligned} \tag{10c}$$

These expressions show that the $\{\lambda_i\}$ depend in a nonlinear manner on the scaled velocities and coordinates. Because of the constraints the simple problem of the motion of mass points in free space becomes highly nonlinear.

The system described by Eq. (9) has two degrees of freedom and one integral of motion (energy). It is well known [23] that in such a situation deterministic chaos can occur [21]. From Eqs. (9) and (10) we also see that the only parameter occurring in the equations is the extension a of the chain, i.e., the normalized distance of the initial and final points of the chain. Therefore, this parameter has to determine the degree of chaotic motion. This is in contrast to systems with potential energy in which the degree of irregular motion is governed by the energy.

The numerical solution of the set of nonlinear differential equations is obtained by using a fourth order predictor corrector method [22]. Solutions of differential equations with more than two degrees of freedom are usually presented using Poincaré sections [23]. In our problem we have chosen the following conditions for the surface of section (see Fig. 2): the angle η be-

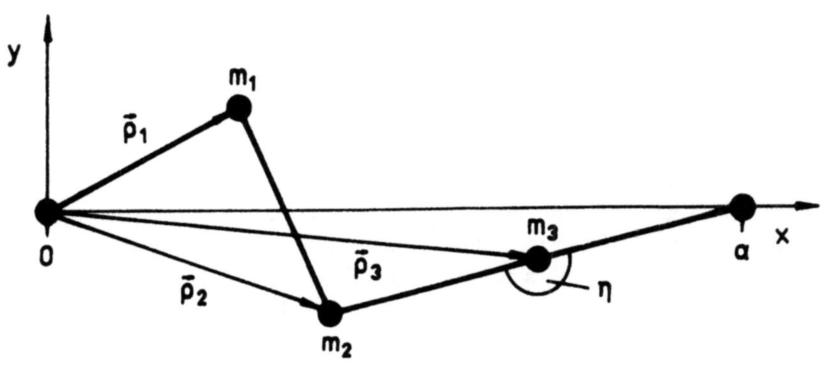

Fig. 2. Definition of the Poincaré surface of section

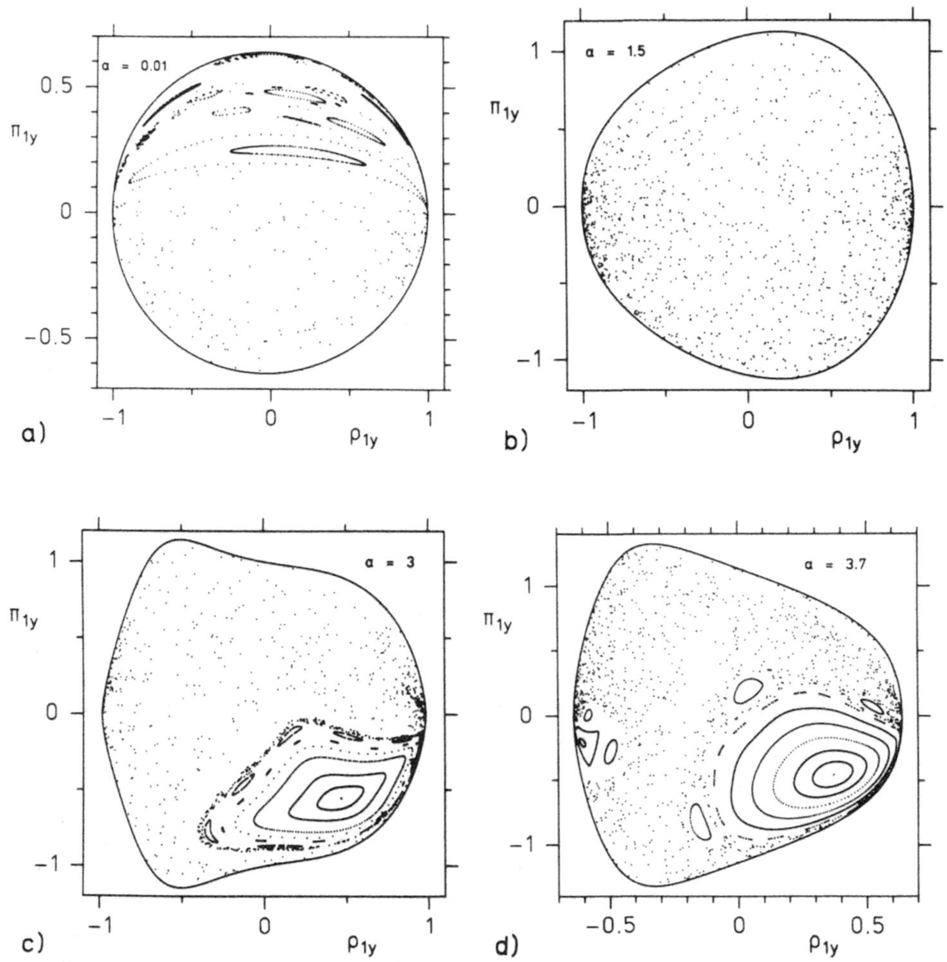

Fig. 3. Surface of section as a function of the extension for $\alpha = 0.01$ (a), $\alpha = 1.5$ (b), $\alpha = 3.0$ (c), $\alpha = 3.7$ (d)

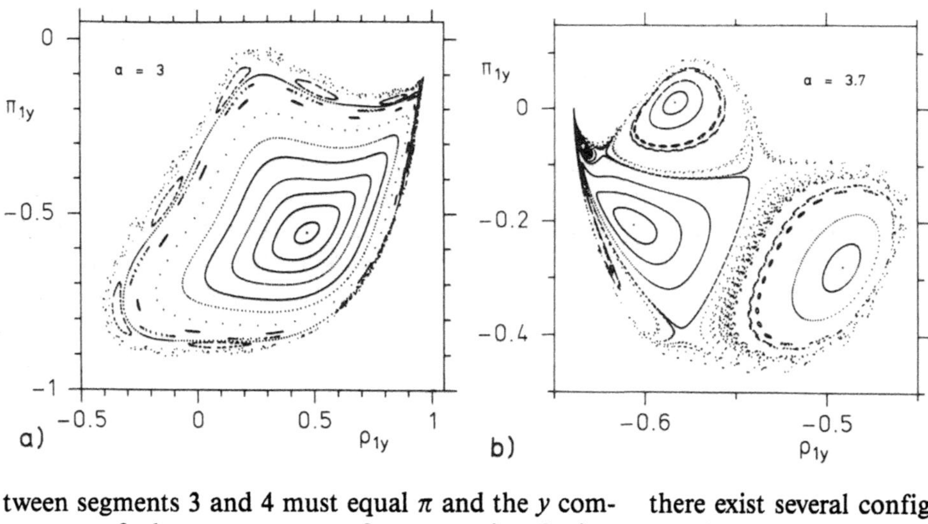

Fig. 4. Regular region in the surface of section for different extensions of the chain. $\alpha = 3.0$ (a), $\alpha = 3.7$ (b) (enlarged parts of Figs. 3c and 3d)

tween segments 3 and 4 must equal π and the y component of the momentum of mass point 3, i.e., $\Pi_{3y} = \dot{\varrho}_{3y}$, has to be positive. Any other condition for the angle η restricts the surface of section and we do not obtain a complete representation of the motion of the chain. Depending on the end-to-end distance α, there exist several configurations for a given value of ϱ_{1y} that satisfy the above conditions for the surface of section. If $\alpha > \sqrt{8}$ we have two configurations of the mass points 2 and 3 with $\varrho_{1y} > 0$. They are symmetrical with respect to the line connecting point 1 and the fixed point at α. In the case $\alpha < \sqrt{8}$ two ad-

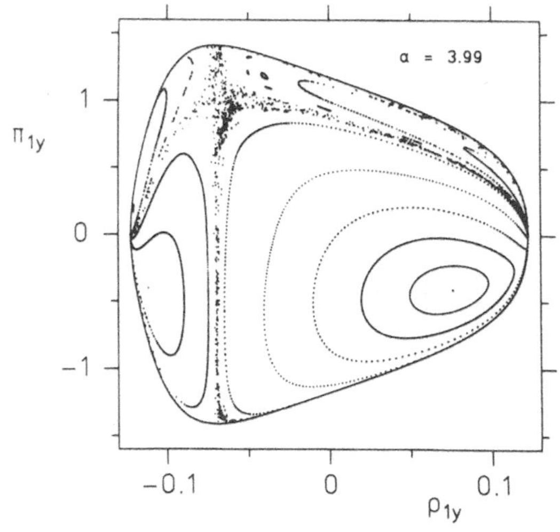

Fig. 5. Surface of section for $\alpha = 3.99$

ditional configurations occur for which $\varrho_{1x}<0$. In order to get a well defined representation, we choose for the surface of section the lower position for the mass points 2 and 3 for all values of α. If $\alpha>2$, only positions are represented with $\varrho_{1x}>0$, in the case $\alpha<2$, only those with $\varrho_{1x}<0$.

Figure 3 shows Poincaré sections for different extensions α, chosen in the way discussed above. We have represented the y components of the position coordinate of mass point 1, ϱ_{1y}, and its momentum, Π_{1y}. In Fig. 3a we have $\alpha = 0.01$, i.e., a very small extension with very close end points. In the upper part of the figure we see the intersections of regular orbits with the Poincaré sections. The lower part seems to consist only of intersections with a chaotic trajectory. All the irregular dots belong to the same trajectory in phase space, whereas in the case of the regular motion almost every "curve" belongs to a different trajectory. The proof of existence of such periodic or quasi-periodic orbits (in the environment of an elliptic fixed point) is given by the KAM-theorem [24–26]. In Fig. 3b ($\alpha = 1.5$) the perturbation of the system is stronger, all regular orbits seem to have disappeared. Quite probably, there exist some regular trajectories also in this case, but the extension of the regular regions in the surface of section is unobservably small on the scale shown in the figure. For $\alpha = 3$ there are again periodic or quasi-periodic orbits as shown in Fig. 3c. At the transition to the chaotic region, we observe an island structure as expected according to the Poincaré-Birkhoff theorem [27]. This region is represented in more detail in Fig. 4a. A further increase in the extension to $\alpha = 3.7$ leads to an additional regime of regular motion (cf. Fig. 3d) which is represented in detail in Fig. 4b. This figure shows several regions of periodic or quasi-periodic orbits, around elliptic fixed points, and also several hyperbolic fixed points which are connected to the chaotic properties.

By further increasing the end-to-end distance of the chain new regular regions occur. However, even for $\alpha = 3.99$, i.e., in the case of an almost stretched chain, there is still a chaotic region in the surface of section as shown by Fig. 5. This indicates that a separation of the system into an integrable problem and a small perturbation – which is usually possible for systems in which the chaotic properties are determined by the variation of the energy – is not possible.

4. Numerical results for forces

The end points of the considered chain model are fixed and the required forces [see Eq. (6)] fluctuate in time. The nature of these fluctuations depends on whether the trajectories are chaotic or regular and thus on the initial conditions. For a chaotic trajectory these fluctuations are completely irregular. For a regular trajectory they are periodic or quasi-periodic. Figure 6 shows the force of constraint at the righthand end point, i.e., $\vec{f}_4$, as a function of time τ for a chain of three mobile mass points which move in the x-y plane [28]. For a chaotic trajectory we plotted the x and y components of $\vec{f}_4$ pointing, respectively, along the direction of the connection of the end points and perpendicular to it. Obviously the forces are not periodic in the time window shown in the figure. The Fourier transform of these data shows that there is really no regular behavior. Figures 6a and 6c demonstrate that the fluctuations of f_x are mostly in the positive direction, whereas f_y in Figs. 6b and 6d fluctuates symmetrically around zero. Therefore, the time average of f_x should be positive and the one of f_y is expected to vanish. Comparing the forces for $\alpha = 3$ and $\alpha = 3.9$, we observe that in the second case the amplitude of the fluctuations and the number of peaks per time unit is larger. Both effects increase the average value of the force.

In [28] we also discussed the time dependence of the force for a chain consisting of two mobile mass points, e.g., $N=2$, moving in a plane. This system has one degree of freedom and the energy is an integral of motion. Therefore, the system is integrable [23] and chaotic motion is not possible. In addition, the time average for this system is independent of the initial condition of the motion.

For the chain with $N=3$ mass points the time average depends on the initial conditions, because of the

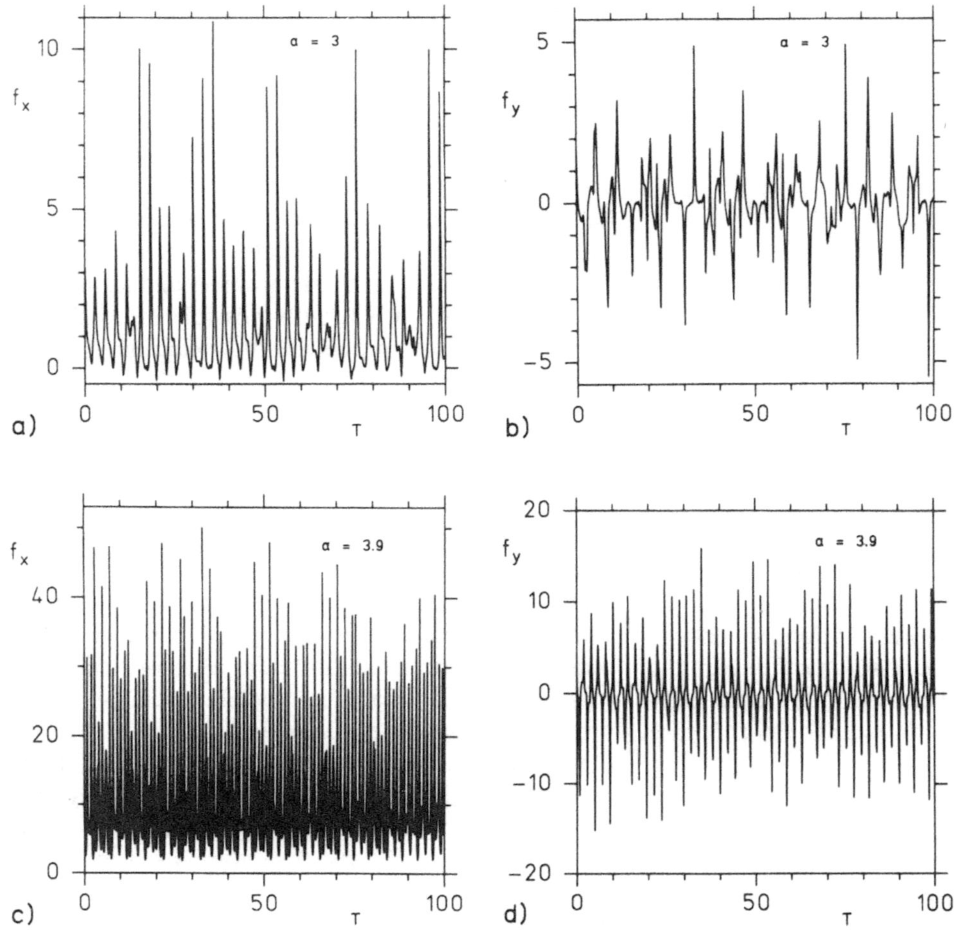

Fig. 6. X- and y-component of the force at the righthand fixation point as a function of time $\tau(N = 3)$. $\alpha = 3.0$ (a), (b); $\alpha = 3.9$ (c), (d)

existence of regular trajectories. As a consequence the system is usually not ergodic. Only for α values close to $\alpha = 1.5$ can we expect a nearly ergodic behavior. Averaging the forces along a chaotic trajectory we get – within the accuracy of our calculations – the same value for different initial conditions with deviations less than 10%. Due to the fact that for the interesting range of α values (see Fig. 7) most of the surface of section is covered by dots corresponding to an irregular orbit, we calculate the averaged force only for chaotic trajectories, assuming that the existing regular regions in phase space do not influence the average very much. A check of this assumption will be done in another simulation procedure.

Figure 7a shows the time average of the x-component of the force for the chain with $N = 3$ mobile mass points moving in two dimensions. Generally $\langle f_x \rangle$ increases with increasing α. The singularities for $\alpha = 0$ and $\alpha = 2$ are connected to special conformations of the short chain for which the segments become almost parallel. The singularity for $\alpha \to 4$ is due to the finite length of the chain.

In Fig. 7b the force-strain relation for a chain with one more segment is shown, i.e., $N = 4$. This increase in the chain length smooths the force vs extension curve. For $\alpha = 1$ and $\alpha = 3$ the segments again have the possibility of becoming parallel. $\alpha = 5$ corresponds in this case to the stretched chain and the force diverges when this extension is approached. The last figure presents the time-averaged force for $N = 5$ mobile mass points. In this case the $\langle f_x \rangle$ vs α curve looks rather smooth. By further increasing the degrees of freedom we expect a smooth relation between $\langle f_x \rangle$ and α. This increase can be achieved in two ways: either by further increasing the number of points in the chain and/or by allowing for three-dimensional motions.

Figure 8 shows the averaged force for the motion of a chain with $N = 5$ mobile mass points in three dimensions. As expected, the curves are rather smooth. In contrast to the two-dimensional motion we now have a second integral of motion for our system. Due to the rotational symmetry about the x-axis of the problem the angular momentum is conserved for this direction.

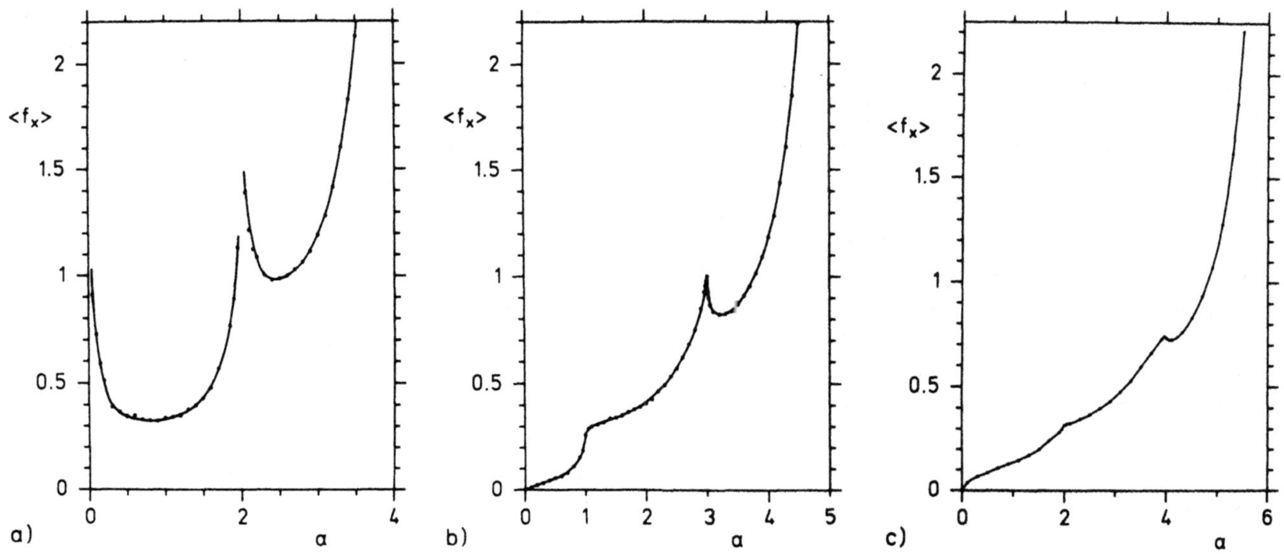

Fig. 7. Time-averaged force $\langle f_x \rangle$ as a function of the extension α for several numbers of mobile mass points. $N=3$ (a), $N=4$ (b), $N=5$ (c)

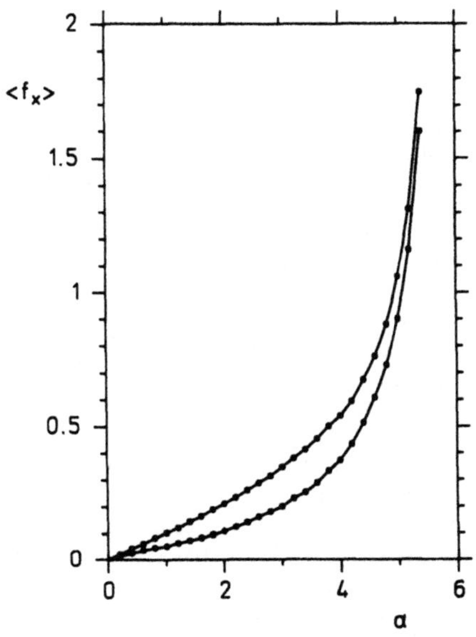

Fig. 8. Time averaged force $\langle f_x \rangle$ as function of the extension for $N=5$ mobile mass points moving in three dimensions. The upper and lower curves correspond to different angular momenta

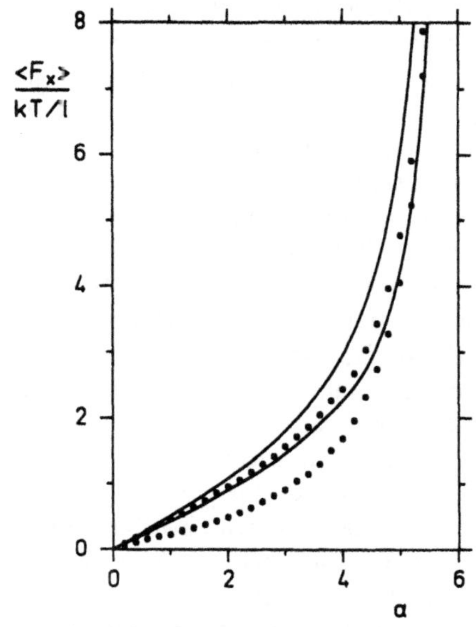

Fig. 9. Comparison of the numerical results for $\langle F_x \rangle$ with results of other theories. Upper solid curve: inverse Langevin function. Lower solid curve: force according the formula of Treloar [4]

Therefore, in Fig. 8 two curves are represented: the curve with the higher force values corresponds to the angular momentum $L_x = 0$, the lower curve is calculated for a very large angular momentum, which is different for different extensions α. In order to get a single curve corresponding to a microcanonical or canonical ensemble of chains we should average the force-strain curve over the distribution of angular momenta. However, this distribution function is not known and therefore, the average cannot be calculated in this way. To find out the canonical average (lying between the two curves in Fig. 8) we shall use a different simulation procedure.

Figure 9 compares our numerical results with forces calculated on the basis of other theories [4]. The upper solid curve is the inverse Langevin function and the lower one is the force according to the exact formula given by Treloar [9]. In contrast to Fig. 8, we represent the unnormalized force $\langle F_x \rangle$ in units of kT/l instead of the normalized quantity $\langle f_x \rangle$. The picture shows that there are differences between the flexible model (which corresponds to the expression of Treloar) and the rigid model. The force calculated with help of the formula of Treloar approaches the inverse Langevin function for $N \to \infty$. The reason for the differences between the flexible and the rigid model is the different number of degrees of freedom. For the flexible model the motion in the direction of the connection between neighboring points is not frozen as in the rigid model [20]. This difference results in different equilibrium properties of the models.

5. Conclusion

We investigated the dynamical behavior of a single polymeric chain, which is considered as an arrangement of mass points with constant distances between successive points. Furthermore, the end points of the chain are fixed. These constraints are included in the equations of motion by Lagrangian multipliers. The corresponding equations of motion are then solved for two- and three-dimensional motion of the chain and for different chain lengths.

For a chain with $N = 3$ mobile mass points moving in two dimensions we find that its motion may be chaotic. The representation of the trajectories by Poincaré sections shows the dependence of the size of the chaotic region in the surface of section from the end-to-end distance of the chain. For small end-to-end distances we find both regular and chaotic behavior. An increase of the extension destroys more and more regular trajectories, but for extensions larger than $\alpha = 2$ regular trajectories again appear. For the almost stretched chain most of the surface of section is covered by the intersections of regular trajectories.

Furthermore, we calculated the time dependence of the force of constraint at one end point of the chain and showed the irregular fluctuations of this force.

Finally, we determined the time-averaged force component along the direction connecting the end points of the chain. This average was calculated for chains with $N = 3$, 4, and 5 mobile masses moving in two dimensions and for a chain with $N = 5$ mobile masses moving in three dimensions. The averaged force for the short chain ($N = 3$) shows a singular behavior which is connected to some special conformations and which is removed with increasing number of mass points.

For chains which move in three dimensions the force is also a function of the angular momentum about the line connecting the end points. Therefore, we calculated the force for two values of the angular momentum, zero and a large angular momentum. The force calculated as an ensemble average should then be located between these two curves. The comparison of our numerical results with forces obtained by earlier theories shows differences which are due to the different number of degrees of freedom in the two models. Further investigations consider the influence of different masses and of side chains on the force-strain behavior. The results will be published in a following paper.

Acknowledgement

This project is part of SFB 239 and has been supported by the Deutsche Forschungsgemeinschaft.

References

1. Enderle HF, Kilian H-G (1987) In: Kilian H-G, Lagaly G (eds) Progress in Colloid & Polymer Science. Steinkopff, Darmstadt 75:55
2. Volkenstein MV (1963) Configurational Statistics of Polymeric Chains. John Wiley & Sons, New York
3. Flory PJ (1969) Statistical Mechanics of Chain Molecules. John Wiley & Sons, New York
4. Treloar LRG (1975) The Physics of Rubber Elasticity, 3rd ed. Clarendon Press, Oxford
5. De Gennes P-G (1979) Scaling Concepts in Polymer Physics. Cornell University Press, Ithaca
6. Weiner JH (1983) Statistical Mechanics of Elasticity. John Wiley & Sons, New York
7. Doi M, Edwards SF (1986) The Theory of Polymer Dynamics. Clarendon Press, Oxford
8. Des Cloizeaux J, Jannink G (1987) Les Polymères en Solution: leur Modélisation et leur Structure. Les Editions de Physique, Les Ulis
9. Trelocar LRG (1946) Trans Faraday Soc 42:77
10. Kuhn W (1934) Kolloid Z 68:2
11. Kuhn W, Grün F (1942) Kolloid Z 101:248
12. James HM, Guth E (1943) J Chem Phys 11:445
13. Ball RC, Doi M, Edwards SF, Warner M (1981) Polymer 22:1010
14. Edwards SF, Vilgis Th (1986) Polymer 27:483
15. Gottlieb M, Gaylord RJ (1983) Polymer 24:1644
16. Flory PJ, Erman B (1982) Macromolecules 15:800
17. Mooney M (1940) J Appl Phys 11:582
18. Rivlin RS (1948) Trans R Soc (Lond) A240:459, 491, 509
19. Kilian H-G (1981) Polymer 22:209
20. Van Kampen NG, Lodder JJ (1984) Am J Phys 52:419
21. Reineker P, Winkler RG (1989) Phys Lett A in print
22. Conte CD, de Boor D (1965) Elementary Numerical Analysis 2nd ed. McGraw-Hill Kogakusha, Ltd, Tokyo
23. Lichtenberg AJ, Liebermann MA (1983) Regular and Stochastic Motion, Appl Math Sci 38. Springer, New York

24. Kolmogorov AW (1954) Dokl Akad Nauk SSSR 98:527
25. Arnold VI (1963) Sov Math Dokt 3:136
26. Moser J (1962) Nachr Akad Wiss Göttingen, Math-Phys Kl IIa 1:1
27. Arnold VI, Avez A (1968) Ergodic Problems of Classical Mechanics. Benjamin, New York
28. Winkler, RG, Reineker P, Schreiber M (1989) Europhys Lett 8:493

Authors' address:

Prof. Dr. P. Reineker
Dipl.-Phys. R. G. Winkler
Universität Ulm
Abt. Theoretische Physik
Oberer Eselsberg
D-7900 Ulm, FRG

Received February 14, 1989;
accepted April 25, 1989

Discussion

GOTLIB:

Because the problem discussed in your lecture is connected with the problem of influence of inertial effects (mass distribution) on the brownian motion and relaxation the question arises: Can you say how the introduction of different masses can effect on the conformation properties or the distribution of α's at a given force, or on the average force at a given α. Could such effects be really observed?

WINKLER:

The model we presented in this lecture is not yet a model of a network. We only investigated the behavior of a single chain. If this chain is considered in connection with a network it may represent only a chain between two (fixed) crosslinks.

Because of the fact, that the endpoints of the chain are fixed in our model, we can not quantitatively estimate the effect of a mass distribution in the case of a constant force and a fluctuating endpoint. But it is easy to show that for a model like ours there is a coupling between conformations in space and kinetic energy independent of the constraints at the endpoints.

For our model we are able to calculate the force as a function of momentum about x-axis, the end to end distance and a mass distribution. If we change only one mass we find a slow increase of force with increasing mass at a constant end to end distance. This increase is a function of angular momentum and is zero within the accuracy of our calculations for vanishing angular momentum and increases with increasing momentum. The observation of this effect is difficult because mainly the modulus is changed in a stress strain experiment and there are usually chains of different masses in a polymer. But it should be observable in a relaxation experiment at high frequencies.

Interrelation of phase and relaxation behavior in polymer blends and block copolymers with crystallizable components

Yu. D. Shibanov and Yu. K. Godovsky

Karpov Institute of Physical Chemistry, Moscow, USSR

Abstract: Coupling of demixing and glass transition in polymer blends and block copolymers having the liquid-liquid separation curves with both the upper and the lower critical separation temperatures is discussed. Particularly, in the latter case a transition from liquid to glassy state at increased temperature was found. Crystallization kinetics and morphology of polymer blends and block copolymers under superposition of demixing and crystallization was studied, and their interaction and coupling was found manifested in formation of various non-equilibrium morphologies and abnormal increase of crystallization rate at the separation boundary. The morphology is determined by the ratio of demixing and crystallization rates.

Key words: Demixing, glass transition, crystallization, polymer blends, critical temperatures

Introduction

Binary mixtures can have various types of phase transitions such as liquid-liquid separation, crystallization, liquid crystallization, etc., as well as the relaxation-glass transition. Each transition is an individual object of a thorough study. For that purpose the conditions are selected under which these transitions do not affect each other, for instance, demixing is studied apart from glass transition and crystallization. In the general case, however, phase diagrams of the mixtures possess intersections of curves of various phase and relaxation transitions, and as a consequence, their interaction. This interaction is manifested under sharp transfer of the mixture from one representative point at the phase diagram to another, where a coexistence of various types of phase equilibria is possible. In this case, coupling (interaction) of various transitions should occur at the kinetic stage of transition from one phase to another. At the intermediate stages of phase separation, coupling leads to formation of non-equilibrium supramolecular structures that become sufficiently stable due to setting at the final stage of phase separation. By altering the conditions of transfer from one phase diagram point to another it is possible to get various finite morphologies of the mixtures that should possess different physical properties. Consequently, the study of interrelations of different phase and relaxation transitions has not only scientific but also practical interest.

Of all the types of transitions superposition, the present paper discuss two types: the demixing + glass transition and the demixing + crystallization. The effects originating under this superposition are illustrated by the polymer blends and block copolymers, though, generally speaking, similar phenomena can occur also in polymer solutions, low-molecular mixtures – and in particular, in binary glasses, metal alloys, etc.

Demixing and glass transition

By now, two types of the liquid-liquid separation curves (binodales) have been found in polymer blends with the upper critical solution temperature (UCST) and with the lower critical solution temperature (LCST). For experimental study of superposition of transitions it is essential to have a complete phase diagram in the experimentally attainable temperature range (bound above by destruction temperatures). Of the known homopolymer blends the oligomer blends have UCST in the permissiable range; LCST is mostly typical of the polymer blends with specific interactions. Let us consider each blend separately.

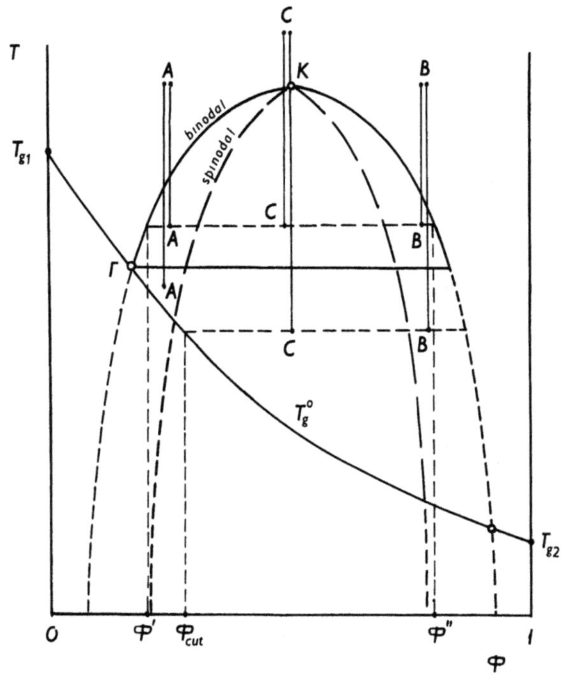

Fig. 1. Schematic constitutional diagram with intersection of the binodal and glass transition curve of the one-phase blend, T_g^0, *at* Γ-point. K is the UCST; T_{g1}, T_{g2} are the glass transition temperatures of pure components; AA', BB', CC' are the blend transfer paths from one representative point to the other. The binodal and the spinodal are "dot-projected" to the glassy-state region. Φ_{cut} is the cutting concentration. Φ', Φ'' are the concentrations of existing phases

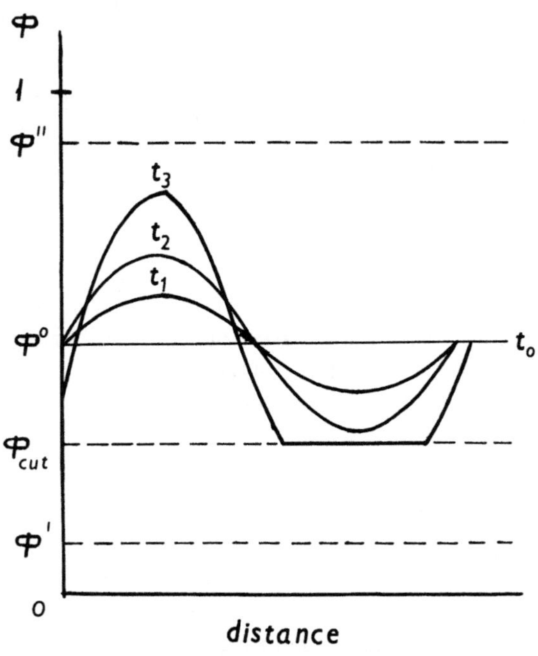

Fig. 2. Development of the predominant spinodal concentration fluctuation in time below Γ point (CC' path in Fig. 1). $t_0 = 0$; t_1 t_2 t_3. Φ', Φ'' are the concentrations of existing phases; Φ^0 is the blend-initial concentration; Φ_{cut} is the cutting concentration

Blends with UCST

The examples of the blends on which apparently superposition of demixing and glass transition has been first observed and studied are the oligomer blends and block copolymers of polyacrylate (PA) and polyethyleneoxide (PEO) [1–4] and the oligomer blends of polystyrene (PS) and polymethylphenylsiloxane (PMPS) [5, 6]. Recently, studies have begun on the coupling of demixing and glass transition in the atactic PS solutions [7–9].

A block diagram of the phase-aggregate state, which is the binodal and the glass transition curve of one-phase system, $T_g^0(\Phi)$ intersecting in point Γ, is shown in Fig. 1. The T_g^0 curve in the region of stable two-phase state is plotted via determination of the glass transition temperatures of blends rapidly quenched to temperatures below T_{g2}, or by interpolation of the experimental curves obtained for the one-phase region onto the two-phase region by the semi-empiric glass-transition formulas of one-phase blends [2, 3].

Quenching of the blend from a homogeneous state above the binodal to the two-phase zone leads to different phase-separation mechanisms depending on the relative position of representative point to point Γ. If the representative point is above the Γ point, then we have a common liquid-liquid separation either by the spinodal mechanism (path $C-C'$) [5, 6, 10] or the one by the nucleation mechanism (paths $A-A'$ and $B-B'$). (Since the nucleation mechanism in polymer blends seems to have never been studied we will discuss below the demixing process illustrating it by the spinodal decomposition.) Compared to common unmixing, it is necessary to account for a low mobility of macromolecules related to the nearness of T_g, as well as a large difference of viscosities of the forming phases in case of a large difference between T_{g1} and T_{g2}.

When the representative point is below the Γ point (although above the T_g^0 curve), we have coupling of demixing and glass transition related to transition of one of the phases, formed in the process of demixing, into the glasy state [1–5]. In the initial moment of time, development of concentration fluctuations occurs by Cahn mechanism with increasing fluctuation amplitude (Fig. 2). In this region of the constitution diagram, however, at the fluctuations half-cycles, corresponding to low Φ values (large content of the component with high T_g, i.e., T_{g1}), the amplitude grows until it reaches the value of cutting concentration, $\Phi_{cut}(T)$, which is determined by the intersection point between the node, corresponding to the liquid-liquid separation temperature and the $T_g^0(\Phi)$ curve (Fig. 1). (Here, as in [2, 3, 5], we believe that glass transition occurs in a point rather than in a range, as is described below.) Further concentration growth at these half-cycles is retarded due to the substance transition from the liquid to glassy state. Thus, coupling of demixing and glass transition means that demixing causes glass transition, which in turn arrests mobility of one of the phases thereby excluding this phase from separation process.

The experimental proof of glass transition in the process of demixing was obtained on block copolymers PA-PEO (Fig. 3a, b). After quenching from the one-phase region to the two-phase one, below Γ-point there forms a heat capacity jump characterizing transition of one of the phases in time from the liquid to glassy state (Fig. 3) [3]. The time of half-transition into the glassy state is around 3 h, and the full transformation time is ca. 25 h. The T_g corresponding to the middle of heat-capacity jump grows in time and reaches a steady-state value, which is approximately 35 K higher, at which the phase separation occurs [3]. Demixing study at various temperatures has shown that unmixing with subsequent glass transition occurs also at temperatures approx. 40 K lower than T_g^0, determined by the glass transition of the quenched blend and by computations. The latter two facts testify to retained molecular mobility in the glassy state to the temperatures approx. 40 K lower than T_g^0.

The glassed phase removal from demixing process is shown by the fact that upon incomplete phase separation at T_1 temperature, subsequent annealing at lower temperature T_2 ($T_2 < T_1$) leads again to liquid phase unmixing and formation of the second (low-temperature) heat-capacity jump, the first one remaining practically intact [3].

Coupling of demixing and glass transition leads, under falling temperature, to narrowing concentration range of demixing due to impossibility to reach equilibrium concentrations of coexisting phases determined by the binodal, and decreased T_g of glass-transition phase, and can also lead to fixation of the spinodal structure [11]. Fixation of spinodal structure of the interpenetrating type, due to glass transition of one of them, excludes to a great extent transition to further decomposition of the spinodal structure and subsequent ripening. The fixed spinodal structure reinforces blends at formation stage and, as [7–9] is supposed, imparts gel properties to polymer solutions.

Besides the above narrowing of demixing range under falling temperature due to impact of vitrifying phase on the blend, [5, 8] predicted narrowing also due to glass transition effect on the liquid phase composition. It should be borne in mind, however, that this deduction was made based on the equilibrium thermodynamics. The kinetic stage of separation requires the use of the irreversible processes thermodynamics.

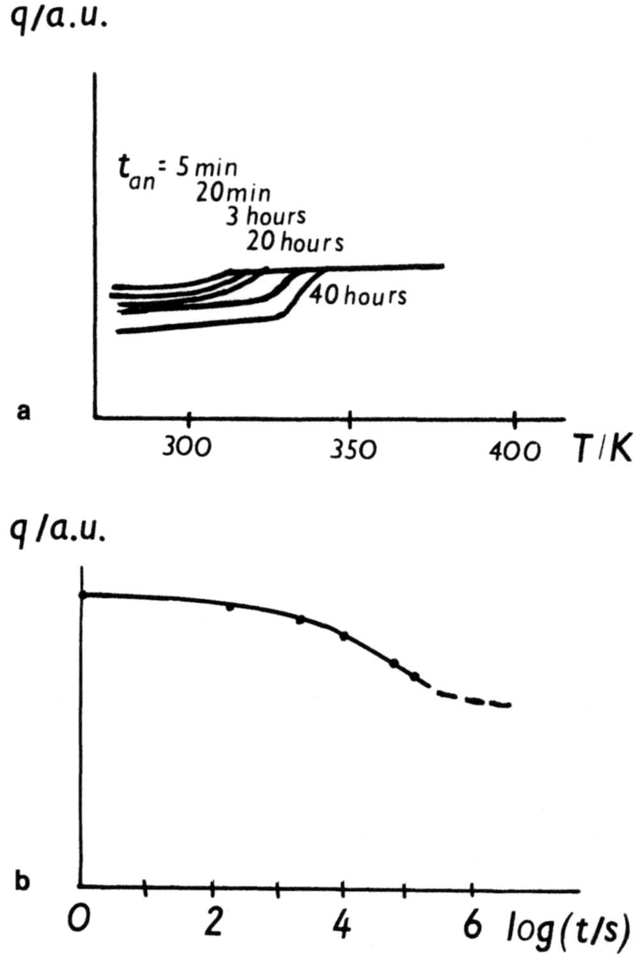

Fig. 3. Heat capacity jump formation in time in the PA+PEO block copolymer after quenching from the one-phase region to 290 K. $x_{PEO} = 0.29$ (wt fraction); M(PEO) = 1000; M(PA) = 2500; T_g(PEO) = 260 K; T_g(PA) = 516 K [2, 3]. g: heat flux. a) DSC-thermograms upon different exposures; b) Isothermal heat-capacity jump (cross-section of Fig. 3a at 290 K)

Blends with LCST

Intersection of separation curve having LCST and the glass transition curve of the one-phase system, $T_g^0(\Phi)$ was found in the blends of polysulfones with PEO [11, 12]. The phase diagram of the blend of poly(acrylene sulfoneoxide) (PASO)+PEO is shown in Fig. 4. Transferred from the one-phase region of the melt into the two-phase one above the cloud point curve (CPC) the blend below the Γ-point separates in-

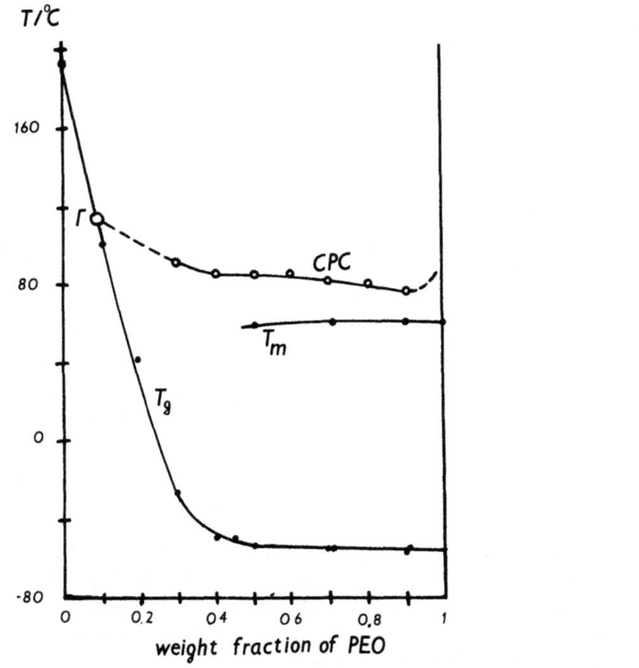

Fig. 4. Phase diagram of the PASO+PEO blend. M_n(PASO) = 44000; M_w/M_n = 2.28; M(PEO) = 75000. The CPC is the cloud-point curve; T_m is the melting-point curve; T_g is the glass transition curve; Γ is the CPC intersection point with the T_g curve [11]

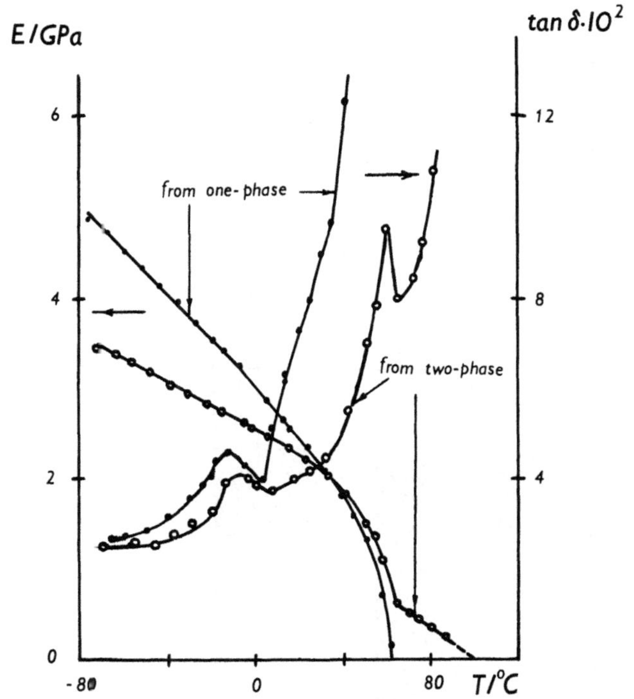

Fig. 6. Modulus E and tan δ as functions of temperature of the PASO+PEO blend (40 wt% PEO) cooled from the one-phase (75 °C) and two-phase (120 °C) states [11]

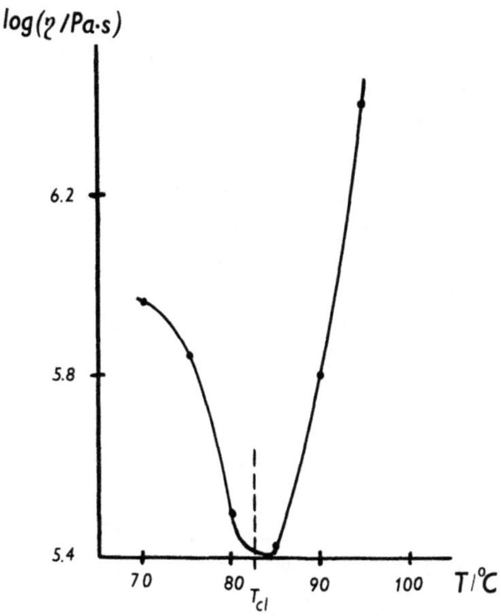

Fig. 5. Temperature relationship of the initial shear viscosity of the PASO+PEO blend (40 wt% PEO). T_{cl} is the cloud point [11]

to two liquid phases, and above the Γ-point one of the phases (rich in PASO) undergoes glass transition.

Glass transition at elevated temperature is confirmed by the temperature relation of shear viscosity (Fig. 5). In the one-phase region at rising temperature the viscosity, as a rule, drops and under phase separation it starts to rapidly rise, reaching the values typically close to T_g. At temperatures above 100 °C the sample becomes an optically clouded glass.

A sharp transition of blend from the homogeneous melt into the two-phase region above the Γ-point leads to demixing accompanied with glass transition of one of the phases. If overheating relative to CPC is sufficiently high, demixing of blend with PASO concentration of 60 wt% goes apparently by the spinodal mechanism. It means that the blend is beyond the percolation threshold, and the glassy net that is formed and permeates through the sample reinforces it. This is clearly demonstrated by the comparison of the temperature relationships of the elasticity moduli of this blend and the one cooled from the homogeneous melt (Fig. 6). The separated blend retains rigidity and finite values of elasticity modulus above PEO melting temperature (up to 100 °C), while the sample prepared from the one-phase state loses the modulus and rigidity in PEO melting region (ca. 60 °C), losing shape under a small load.

The glassy state of the PASO-rich phase is also indicated by the glass-transition region at the losses curve lying above PEO melting temperature, as well as the lower temperature coefficient of the elasticity modulus at lower temperatures.

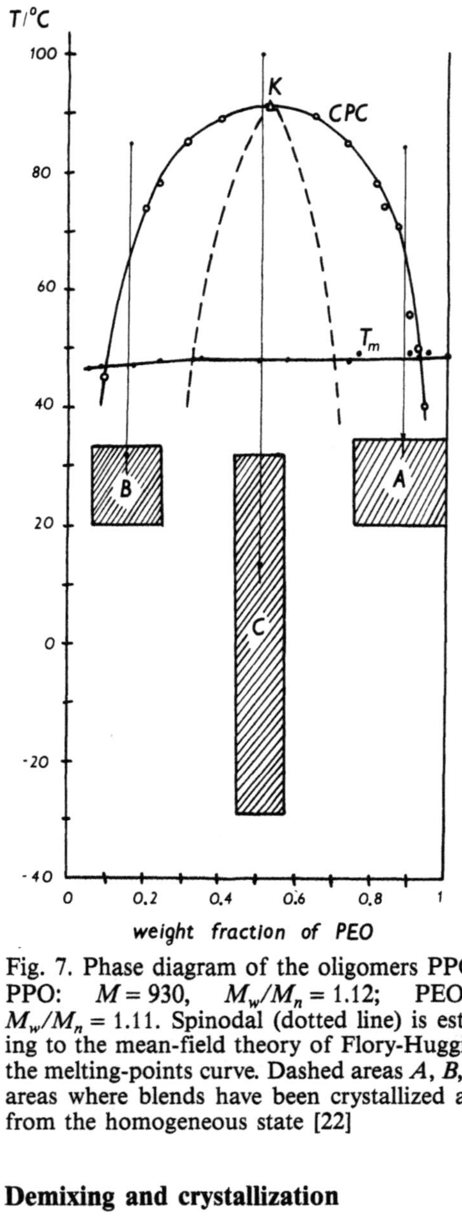

Fig. 7. Phase diagram of the oligomers PPO+PEO blend. PPO: $M = 930$, $M_w/M_n = 1.12$; PEO: $M_n = 1240$, $M_w/M_n = 1.11$. Spinodal (dotted line) is estimated according to the mean-field theory of Flory-Huggins-Scott. T_m is the melting-points curve. Dashed areas *A*, *B*, and *C* indicate areas where blends have been crystallized after quenching from the homogeneous state [22]

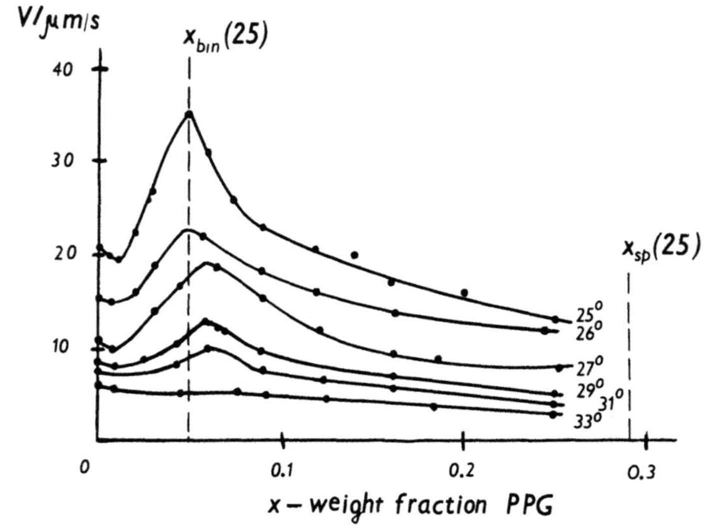

Fig. 8. Growth rate of PEO spherulites as a function of PPO concentration in the area *A* of phase diagram (see Fig. 7). Figures near curves indicate the crystallization temperature $T_{cr} \cdot x_{bin}$ (25) corresponds to the decomposition border at 25 °C, x_{sp} is the estimated spinodal concentration at 25 °C [16]

Demixing and crystallization

Coupling of liquid-liquid separation and crystallization is better studied in the blends of amorphous and crystalline polymers with UCST, and in block copolymers. For blends with LCST there are as of yet only preliminary data.

Blends with UCST

Interaction of liquid-liquid separation and crystallization was carried out on the blends and block copolymers of PA-PEO [2, 13], as well as on the blends of polystyrene (PS)+polycaprolactone (PCL) [14], polypropylene glycol (PPG)+polyethylene glycol (PEG) [15–18], and poly(ethylene-*co*-propylene)+ polypropylene [19–21].

Figure 7 shows a phase diagram of the PPG+PEG oligomers blend [22] consisting of the binodal with UCST and the melting-point curve. For analysis the blend from the homogeneous melt above the binodal was quenched below the melting-point curve. In the A-region (high concentrations of crystallizing PEG component), an abnormal rate of spherulite growth was found under changing amorphous PPG concentration (Fig. 8). The abnormality in the form of a peak is observed in the region of considerable overcooling (over 17 K), and the peak's height decreases with overcooling. The peak's position within the experimental error coincides with the separation boundary at each crystallization temperature [16].

The nature of this peak is related to the liquid-liquid separation processes occurring in the diffusion zone near the boundary of growing spherulite. A possible mechanism of this phenomenon can be PPG liquid phase nucleation and nuclei growth in the tangential direction, causing the amorphous component drainage and crystallization front instability [16]. At the binodal and out of it, nucleation of drops in the bulk sample far from the crystallization front is excepted due to the infinite critical size of the nucleus. Crystal formation proceeds from the homogeneous solution, but in its growth in the diffusion zone, due to depletion of crystallizable component, a metastable (in relation to demixing) layer is formed in which the PPG phase liquid nuclei can form (Fig. 9). (Generally

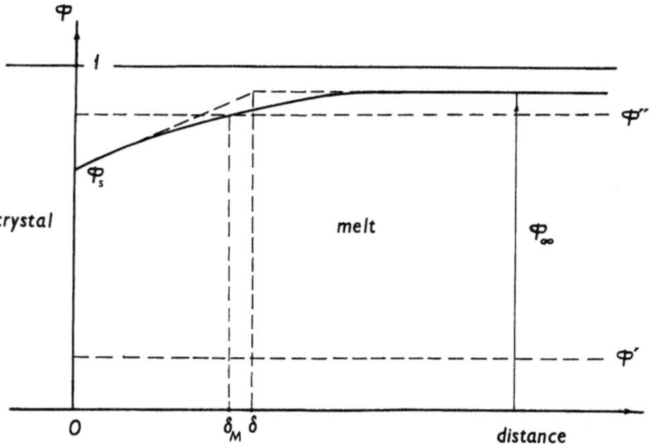

Fig. 9. The profile of the concentration of the crystallizable component ϕ near the border of a growing crystal. ϕ' and ϕ'' correspond to the concentrations of the coexisting phases, ϕ_∞ is the concentration at the infinite distance from the crystal (an initial concentration of the mixture), δ is thickness of the diffusion layer, δ_M is the thickness of the metastable with reference to decomposition diffusion layer, and ϕ_s is the concentration at the crystalline surface

speaking, a binodal decomposition is also possible in the metastable diffusion layer if the concentration at the crystal surface is under the spinodal. It is possible under extreme overcooling.) Formation on the growing crystal surface of a thin layer in which the concentration of crystallizable component is much lower (as determined by the spinodal left branch) than that beyond the layer can lead to unstable growth and accelerated crystallization.

Besides this mechanism, redistribution of H-bonds between the end hydroxyl groups of both component's chains can have its effect, as well as collapse of the chains of amorphous component found near the binodal with the Monte-Carlo modeling method [23]. This phenomenon requires further study.

The morphology of blends crystallized in the A-region below the binodal is that of PEG spherulites with PPG phase inclusions stretching along the spherulite radius (Fig. 10a). It differs from that found in [14] by the shape of amorphous inclusions (there they were spherical). This difference is related to the high viscosity of amorphous phase drops (the glass transition temperature of the PS oligomer used is close to crystallization temperature) that suppresses drop deformation by the crystallization front in the PS+PCL blend. In the PPG-PEG blend the viscosity of amorphous phase is low, which enables the spherulite front to deform the originally spheric drops (Fig. 10b).

An important consequence of the morphology study in the A-region under the binodal, but at a significant distance from it, is the rate of liquid phase nucleation exceeding that of crystalline nucleation, at least in the studied temperature range. For this reason the spherulites start growing in the demixed melt containing the amorphous component phase droplike in-

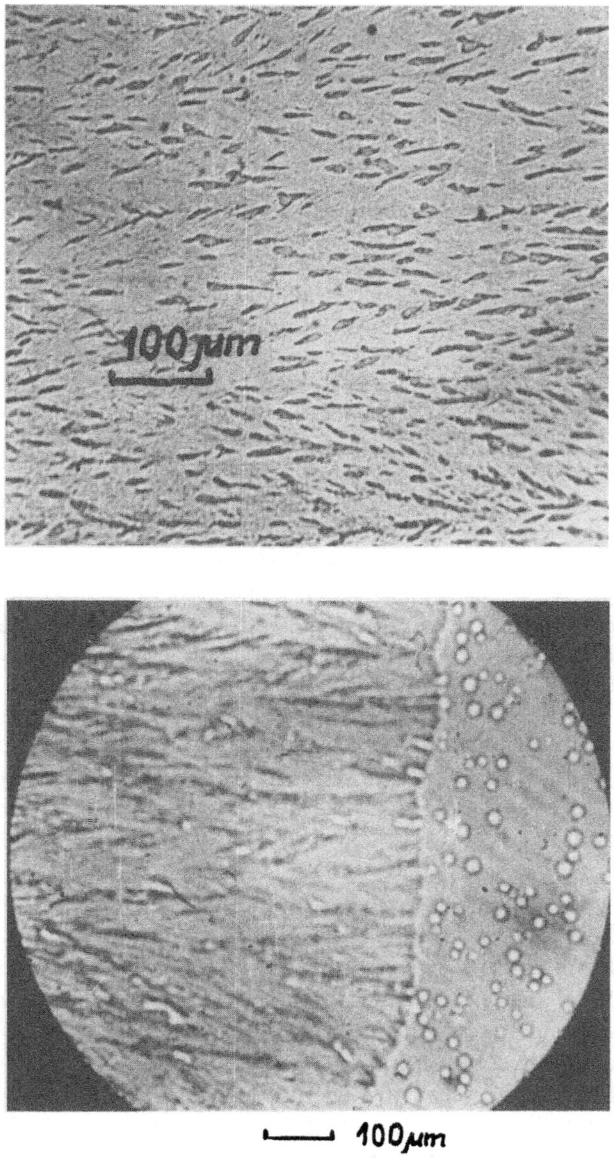

Fig. 10. Nonpolarized-light micrograph of blends PPO + PEO crystallized in the area *A* under binodal (see Fig. 7). a) Final morphology, $x = 87.9$ wt% PEO, $T = 301.9$ K. Only a small part of a spherulite situated far from its center is shown (the center of the spherulite is outside the photo). The particles of the PPO phase are elongated along the spherulite radius. b) The growth process. To the right: melt with the spherical droplets of the PPO phase. To the left: a spherulite with engulfed and deformed droplets. At the border one can see the process of engulfing and deforming of the droplets (large droplets are engulfed by the border of the spherulite, small ones are pushing away)

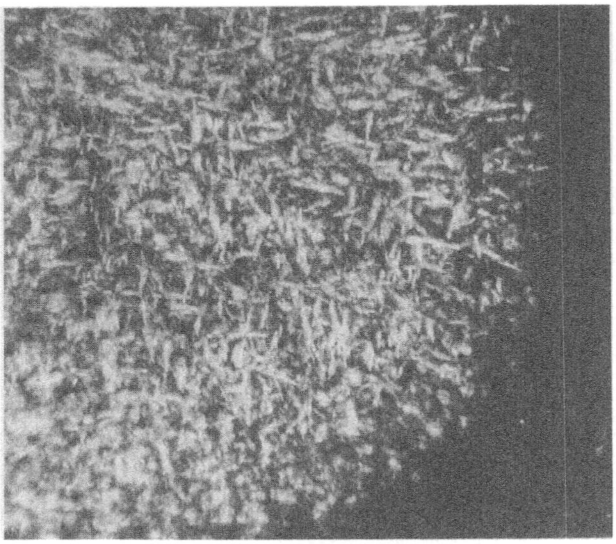

100 μm

Fig. 11. Polarized-light micrograph for the border of the growing structure of PPO+PEO mixture in the area *B* (see Fig. 7). $x = 6.3$ wt% PEO, $T_{cr} = 297.7$ K. The center of the growing spherulite is outside the picture

clusions. It should be borne in mind, however, that the presence of impurities seriously affects the nucleation rate. Therefore, the nucleation rates ratio can be altered by dopants selectively adsorbing the desired component.

In the B-region below the binodal (Fig. 7), i.e., at low concentration of the crystallizable component, PEG phase before crystallization is distributed in the form of drops in PPG phase. However, crystallization occurs not via formation of separate crystallites, as could have been expected, but in the form of a structure successively growing from one center by inducing crystallization in the neighboring drops (Fig. 11) [17]. In spite of friable morphology and a seriously eroded "crystallization front", the crystalline phase is the one-bond dendritic (similar to fractal) structure.

In the C-region (medium concentration) (Fig. 7), liquid-liquid separation in absence of crystallization occur by the spinodal mechanism. Under rapid quenching of homogeneous solution to relatively high temperatures crystallization goes slowly and unmixing gets to advanced stages, forming huge inclusions into spherulites, often of irregular shapes (Fig. 12a). At lower temperatures, i.e., under rapid crystallization, an unusual structure is formed by as spherulites with concentric-ring inclusions of PPG amorphous phase (Fig. 12b), and the distance between the neighboring rings increases with distance to the spherulite center. At a considerable distance from the spherulite center the amorphous rings dissociate into drops. This structure

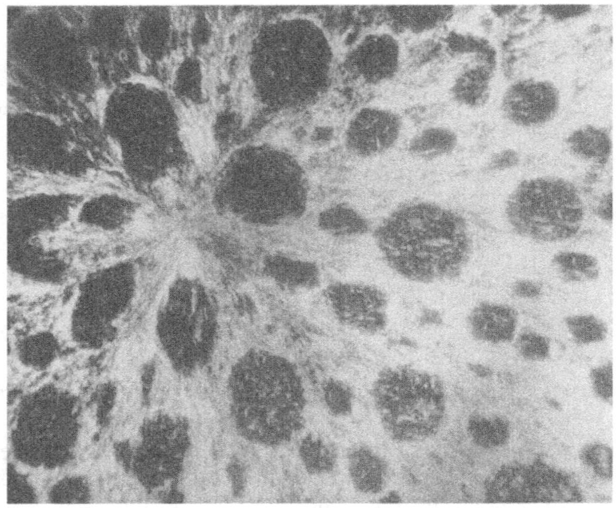

100 μm

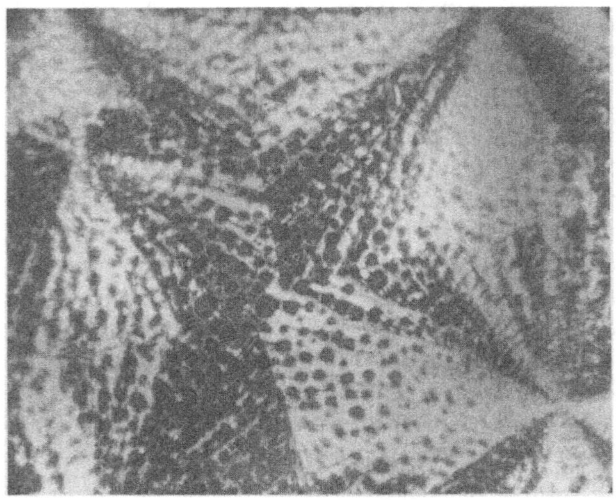

100 μm

Fig. 12. Polarized (crossed polars) micrographs of final structures in PPO+PEO blend, crystallized in the area *C* (see Fig. 7). Inclusions correspond to the PPO phase. $x = 54.0$ wt% PEO. a) $T_{cr} = 295$ K; b) $T_{cr} = 273.2$ K

is a consequence of the competition between demixing and crystallization.

The difference in morphologies, obtained at different temperatures, is related with the different spinodal decomposition and crystallization rates ratio. The spinodal decomposition rate is determined by overcooling with respect to the binodal and by diffusion coefficient, which in turn depends on the polymer chain length and proximity to glass transition temperature (GTT). With increased overcooling relative to the binodal, the spinodal decomposition rate should increase from zero (at the spinodal) to the maximum, and then again decrease to zero near GTT. At high temperatures the demixing rate is determined by the thermodynamic driving force of demixing – and at low ones, by the molecular mobility, determined by Williams-Lendel-Ferry mechanism. Crystallization rate changes similarly with changing overcooling relative to melting temperature*).

At relatively high temperatures the spinodal decomposition rate exceeds crystallization rate manifold, leading to the final morphology shown in Fig. 12a. At lower temperatures the rates are converging, leading to crystalline nucleation at the spinodal decomposition stage. Growing spherulite deforms the continuous spinodal structure leading to ring structure of amorphous phase at the early growth stages (Fig. 12b). At later growth stages the spinodal structure has time to dissociate into drops which are observed far from the spherulite center, their size increasing with distance. Thus, changing shape, size, and distribution of particles within the spherulite prove competition and coupling of demixing and crystallization.

At temperatures intermediate between the two described above we observe morphology similar to that described in [14].

In the described PPG+PEG oligomer blends, as well as in PS+PCL [14] the unentangled (Rouseian) chain dynamics regime is realized, since the molecular masses of the components are lower than the entanglement critical molecular mass. Polymer blends with high molecular masses demonstrated crystallization without any changes in the pre-formed spinodal structure [19, 20]. It is related to the low interdiffusion coefficients apparently realized in the entangled regime, and high viscosity of the amorphous phase.

Interaction of demixing and crystallization has been found for the first time on the PA+PEO block copolymers and blends [13] by decelerated crystallization under quenching from the one-phase state, compared to that under quenching from the two-phase state (Fig. 13). In that case, however, demixing and crystallization can be affected by glass transition (see above). This should be taken into consideration also for other blends with greatly different GTT of components, e.g., in PS+PCL blends.

Thus, morphology formation under coupling of demixing and crystallization is determined by the ratio of demixing and crystallization rates, as well as by the formed amorphous phase viscosity. The formed inhomogeneous structures do not seem to be the dissipative ones "according to Prigogine" [25], since they are not stable. However, due to solidification and, consequently, very low rates of subsequent processes, they are rather stable and remain unchanged for quite longe. Therefore, they can be termed as the metadissipative structures.

Blends with LCST

In this blend type, having a phase diagram similar to that shown in Fig. 4, superposition of demixing and crystallization is possible under system transfer from

*) An important question requiring separate consideration is a possibility of the maximum crystallization rate exceeding that of separation. If this is possible, then the crystalline nuclei should form without pre-unmixing of the bulk sample, although formation of drops around the nucleus in the diffusion zone is possible

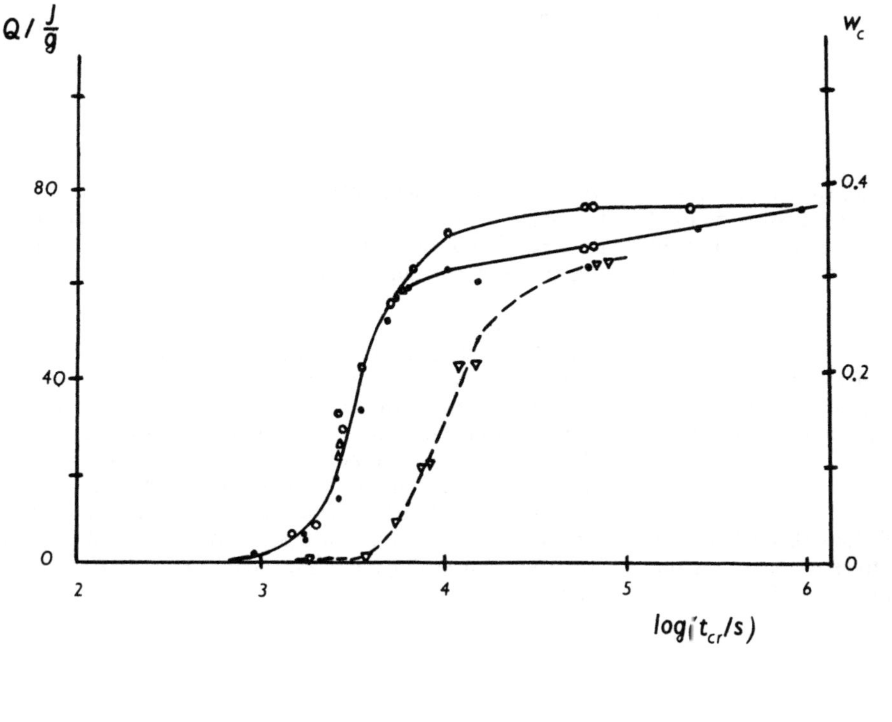

Fig. 13. Crystallization isotherms of PAr-PEO block copolymer. M (PAr) = 2500, M(PEO) = 4000, x_{PEO} = 0.62. T_{cr} = 291.0 K. The predetermined annealing temperature: 353 K (●), 400 K (○), 450 K (△), and 500 K (▽). Annealing time – 10 min. Q: heat of crystallization, w_c: degree of crystallinity

the one-phase state, first to the two-phase one by rising the melt temperature, and then by sharply cooling to the crystallization temperature. This blend type is practically unstudied. So far, only preliminary results on morphology changes and accelerated crystallization – compared to direct crystallization of the one-phase melt – of the PASO+PEO blend under the above described thermal treatment have been obtained.

Conclusion

A research into coupling of phase and relaxation transitions is very important for understanding of kinetics of superpositioned phase transitions and formation of inhomogeneous supramolecular structures in various blends. These phenomena, however, can also find practical applications, since different finite structures produced under different thermal treatment conditions should possess different physical properties.

References

1. Shibanov YD, Godovsky YK (1982) Preprints of II Sov Conference on Thermodynamics of Organic Compounds. Gorkii p 159
2. Shibanov YD (1982) Thesis. The Karpov Institute of Physical Chemistry Moscow
3. Shibanov YD, Godovsky YK (1985) Colloid Polym Sci 263:202–216
4. Shibanov YD, Godovsky YK (1988) Uspechi Chimii 57:1713–1741
5. Nojima S, Nose T (1982) Polymer J 14:269–275
6. Murschall U, Fischer EW, Herkt-Maetzky C, Fykes G (1986) J Polym Sci, Polymer Lett 24:191
7. Arnauts J, Berghmans H (1987) Polymer Commun 28:66–68
8. Frank FC, Keller A (1988) Polymer Commun 29: 186–188
9. Hikmet RM, Callister S, Keller A (1988) Polymer 29:1378–1388
10. Meier H, Strobl GR (1987) Macromolecules 20:649–657
11. Shibanov YD, Rajabov TM (1988) Vysokomol Soed, B30:281–283
12. Walsh DI, Singh BV (1984) Makromol Chem B185:1979
13. Shibanov YD, Godovsky YK (1983) Vysokomol Soed, A25:339–345; (1983) Polymer Science USSR, A25: 395–404
14. Tanaka N, Nishi T (1985) Phys Rev Lett 55:1102–1106
15. Shibanov YD, Komaricheva LI (1986) Preprints of I Sov Conference "Polymer Blends". Ivanovo, p 64
16. Shibanov YD, Komaricheva LI (1988) Vysokomol Soed, B30:135–137
17. Komaricheva LI, Shibanov YD (1989) Vysokomol Soed, A31:1367–1372
18. Philippov AV, Smirnov VS, Gimatdinov RS, Shibanov YD (1988) Vysokomol Soed, B30:552–553
19. Inaba N, Sato K, Suzuki S, Hashimoto T (1986) Macromolecules 19:1690
20. Inaba N, Yamada T, Suzuki S, Hasimoto T (1988) Macromolecules 21:407–414
21. Nakai A, Shiwaku T, Hasegawa H, Hashimoto T (1986) Macromolecules 19:3008–3010
22. Komaricheva LI, Shibanov YD (1988) Vysokomol Soed, A30:1436–1441
23. Sariban A, Binder K (1988) Makromol Chem 189: 2357–2365
24. Binder K (1983) Chem Phys 79:6387–6409
25. Nicolis G, Prigogine I (1978) Self-Organization in Nonequilibrium Systems. Wiley, New York, Haken H (1978) Synergetics. Springer Verlag, Berlin; Haken H (1983) Advanced Synergetics. Springer Verlag, Berlin

Received February 24, 1989;
accepted June 6, 1989

Authors' address:

Yu. D. Shibanov
Karpov Institute of Physical Chemistry
ul. Obukha 10
Moscow 103064, USSR

Discussion

KILIAN:

Can you please comment on the coexistence conditions when crystallisation happens in one of the existing phases? I think that for three coexisting phases in a two-component system under constant pressure conditions the phase equilibrium should be indifferent.

GODOVSKY:

I agree with you that for three coexistence phases in a two-component system under constant pressure conditions the phase equilibrium should be indifferent. But i speak on the non-equilibrium processes.

DUSEK:

I think that the binodal curve is only hypothetical because it cannot be reached in finite time. However, phase separation or crystallization can still occur below the T_g curve, however, with reduced rate given by the distance from T_g. The phase separation rate obviously depends on whether the T_g curve is concave or convex. Do data exist on phase separation or crystallization below T_g?

GODOVSKY:

Really the cloud point curve and the binodal life differ in some respect. But this difference is not very important for our consideration. In principle, the phase separation or crystallisation can still occur below T_g because of the existence of some molecular motion even below T_g ut in real situations the rate of the processes below T_g is close to zero. Therefore, if one of the phases transforms to the glassy state the crystallization in it practically stops. I have no information concerning the phase separation or crystallisation below T_g.

Progress in Colloid & Polymer Science Progr Colloid & Polymer Sci 80:119–128 (1989)

Field-induced structure conversions in PVDF

D. Geiss and C. Ruscher [1)]

Institute of Polymer Chemistry "Erich Correns", Academy of Science of the GDR, Teltow-Seehof, GDR
[1)] "Karl Marx"-University, Department of Physics, Leipzig, GDR

Abstract: Poly(vinylidene fluoride) undergoes several structural conversions during electrical poling which contribute in different way to the pyro- and piezoelectrical properties. These conversions originate at the molecular dipole of the $-CF_2CH_2-$ monomer unit and its interaction with the poling field. Therefore orientation effects play an important role in these structure formation processes.

The mechanism of structural conversions of material having different conformations and crystalline structure are investigated in dependence on the poling conditions. By means of modified wide angle x-ray methods it is possible to separate individual effects from each other.

Key words: Poly(vinylidene fluoride), WAXS, texture, structure conversion

Introduction

Numerous investigations during the last decade on poly(vinylidene fluoride) (PVDF) confirmed that the ferroelectric behavior of this polymer can be attributed above all to molecular properties. It was found that the molecular dipoles of the CF-bonds may effect a polarization as a volume property of several polar PVDF modifications [1]. In this connection polymorphism leads to a rather complicated structure formation during mechanical treatment and electrical poling.

PVDF can crystallize in at least four different modifications (α-, β-, γ- and α_p-phase, also called

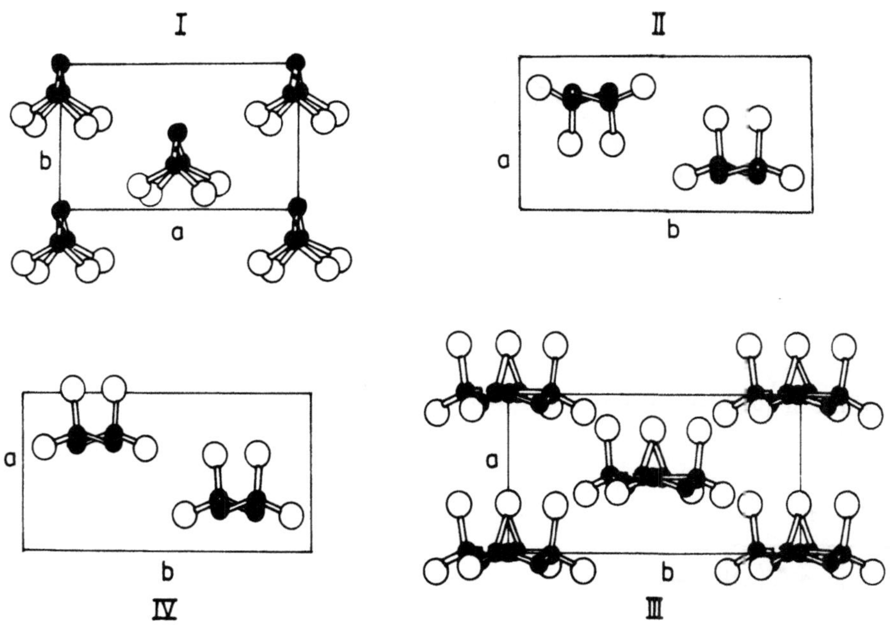

Fig. 1. The crystalline modifications of PVDF (projections along the crystalline c-axes)

Table 1. Crystallographic data of the four PVDF-modifications

Modification	Unit cell	Space gr.	Unit cell parameter	$\beta/°$	Conformation	Literature
β	orthorhomb.	Cm2m	0.858 0.491 0.256		*TT*	[34]
α	monoclinic	$P2_1/c$	0.496 0.964 0.462	90	*TGTG'*	[2]
α	orthorhomb.	P2cm	0.496 0.964 0.462		*TGTG'*	[1]
γ	monoclinic	Cc	0.496 0.958 0.923	92.9	*TTTG* *TTTG*	[35]
γ	orthorhomb.	C2cm	0.497 0.966 0.918		*TTTG* *TTTG'*	[6]
α_p	orthorhomb.	$P2_1cn$	0.496 0.964 0.462		*TGTG'*	[10]

forms II, I, III, and IV), the structure and symmetry elements of which are shown in Fig. 1 and Table 1.

In the β-phase (form I) the molecular chains are arranged in an orthorhombic lattice [2]. The unit cell consists of two all-trans chains packed with their dipoles pointing in the same direction (b-axis) and therefore resulting in a dipole moment of each single crystallite. Nevertheless, ordinary poly-crystalline PVDF material containing β-phase crystallites does not show any macroscopically acting polarization since the spontaneous polarization of oppositely oriented crystallites will neutralize each other.

Unlike the β-phase the monoclinic α-phase (form II) consists of two chains in the *tg tg'* conformation, whose dipole components (parallel to the b-direction and smaller than that of the β-phase) are antiparallel, thus neutralizing each other [2–5].

The polar α_p-phase (also often called form IV or δ-phase) can be obtained by applying a high electric poling field to films of the α-phase. The α- and the α_p-phases have identical sized unit cells and both contain chains in the *tg tg'* conformation. For the polar γ-phase (form III) of PVDF, Weinhold [6] detected an orthorhombic crystalline lattice.

Figure 2 shows a schematic survey of the preparation conditions for the formation of the different modifications of PVDF and the conditions for transformations between them. Some details concerning this survey will be discussed in the following.

I. Field-induced changes of the structure in the α-phase

Spherulites in which the ordered regions consist of the α-phase are formed during cooling of a PVDF melt. Such material can be transformed into a well-oriented state by uniaxial drawing at temperatures above 373 K to medium draw ratios <5 without generating any considerable amount of the β-phase [7]. In this procedure the spherulitic structure vanishes

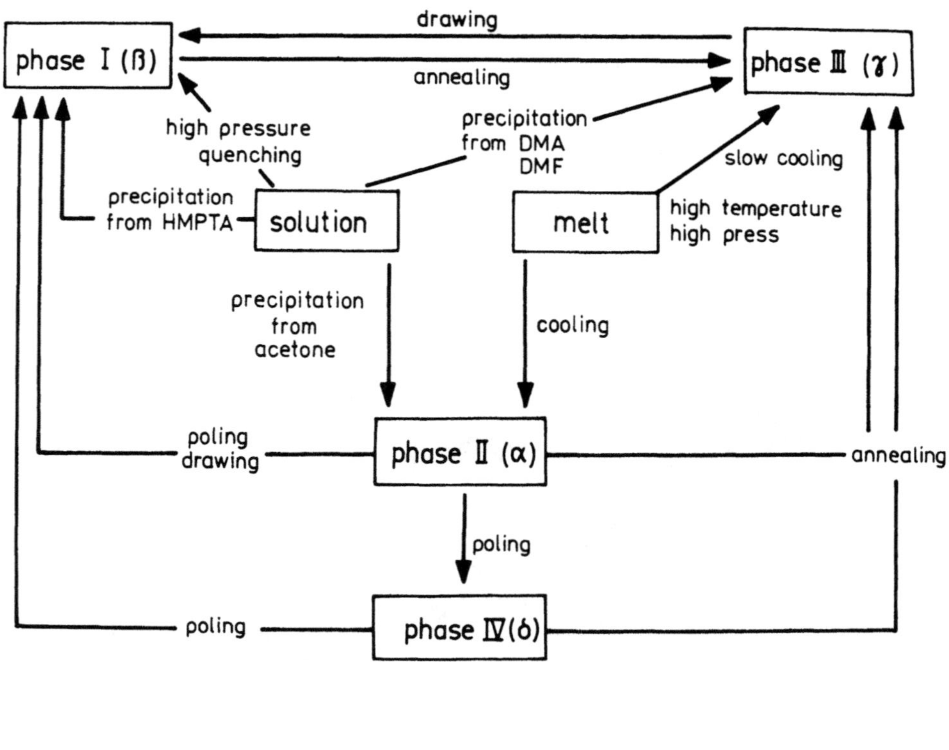

Fig. 2. Conversion scheme for PVF_2 (after [34] with some supplements)

and an extensive alignment of the chains is achieved along the draw direction. Samples produced in this way and to which a strong electrical poling field is applied exhibit several conversions of chain conformation and crystalline modifications [8–11]. At poling field strengths between 150 and 200 MV/m a certain amount of crystallites can be converted to the polar α_p-phase.

The authors are divided on the question of which of the crystallites with different orientations to the poling field are involved in these conversions [12, 13].

Experiments

Unoriented 100-µm-thick film material from Kureha was used for our investigations of the α-phase. This material was drawn (draw direction parallel to the machine direction M) with a draw ratio of $\lambda = 6$ at 423 K in a homogeneous temperature field by means of a tensile testing device (Instron). After drawing the samples were poled at a field strength of $E = 180$ MV/m (field direction parallel to the sample normal direction N).

x-Ray film photographs were taken in three directions $\vec{s}_0$ (M, N, T; T perpendicular to N and M), to estimate the content of different crystal forms and to make a rough estimation of the orientation state.

After drawing and poling, pole figures were taken by means of a Siemens texture goniometer working in the Schulz-reflection mode [14].

The frequency-distribution of the orientations of particular lattice planes was determined from the pole figures. The frequency distribution of lattice planes (hkl) with $l \neq 0$ was projected on the equator $N-T$ of the pole figure in order to compare several distributions having different l.

Results and discussion

Figure 3 shows the flat film photographs taken before and after poling ($\vec{s}_0 = N$). The distribution of the reflected intensities is similar to that of a fiber texture.

The portion of the β-phase in the ordered content of the material was determined to be 10% to 20%.

The pole figures in Fig. 4 obtained at a scattering angle of $\vartheta = 13°$ may be considered as a representative example of the state of orientation in the investigated material. In them the orientation distributions of the normals of the lattice planes (120), (021), and (111) of the α-phase are overlapping. The comparison of the pole figures before (below) and after (above) poling reveals considerable differences. It follows from the flat film, photographs in Fig. 3 that these differences cannot be interpreted as changes of the orientation state without any further explanation. The changes in the intensity in Fig. 4 represent the entire structure changes induced by the electric field, i.e., changes to

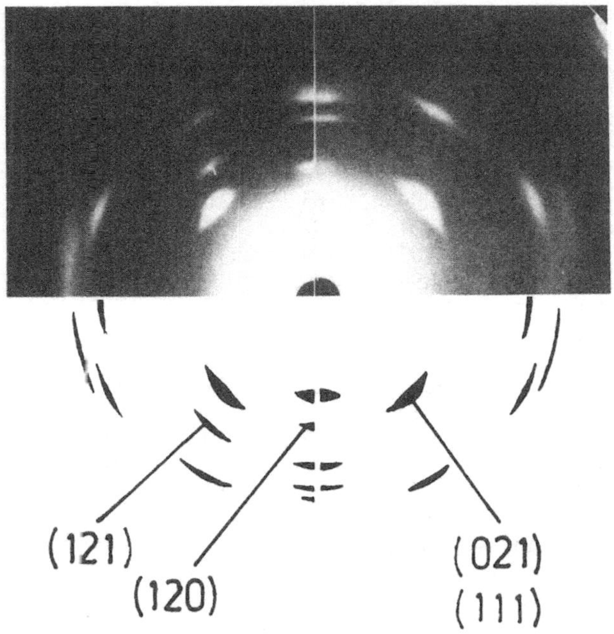

Fig. 3. Flat film photographs of a drawn PVDF-film-sample ($T = 423$ K, $\lambda = 6$): left = unpoled; right = poled at 180 MV/m

the orientation state and crystalline structure. The reflected integral intensities of the individual lattice planes correspond to the calculated x-ray form factors of the α-phase (before poling, on the left) and the α_p-phase (after poling, on the right) as shown in Table 2.

To a first approximation the intensities displayed in the pole figure on the circle at an angle of $\theta = 44°$ to

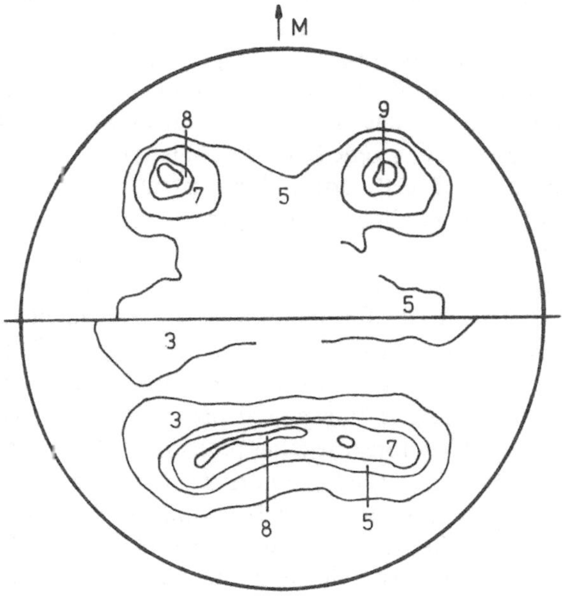

Fig. 4. Pole figure of drawn PVDF-film-sample ($T = 423$ K, $\lambda = 6$, $\vartheta = 13°$); above = after poling at 180 MV/m; bottom = before poling

Table 2. Comparison of x-ray form factors of the α- and the α_p-phase (arbitrary units)

Reflection (hkl)	$\vartheta/°$	[10]		[36]	
		II	IV	II	IV
100	8.94	65	0	135	0
020	9.21	93	93	130	130
110	10.07	149	179	175	215
120*	12.98	100	0	100	0
021	13.19	129	129		
111	13.99	85	92		
121	16.18	71	0		
130	16.60	81	101	95	105
200	18.10	74	115	135	170
210	18.72	88	0	70	0
040	18.66	15	15	70	70
002	19.50	84	84		

* The intensity of the (120)-reflex of the α-phase was assumed to be 100

the fiber axis M can be regarded as belonging to the (021)-lattice planes. Changes in the reflected intensity caused by the conversion of the α-phase into the α_p-phase can be neglected in this case (cf. Table 2). The scattered intensity of the lattice planes (120) will become zero in the case of the conversion of α into α_p. Figures 5 and 6 give the distributions of the normals of the (021)- and (120)-planes, respectively, in terms of the sample deflection angle α along $N-T$ [(021) normals distribution projected on NT, thus giving the (020) normals distribution).

Figure 5 represents only that part of the whole structure conversion which is caused by the change of the state of orientation. Provided [001] direction and draw direction M coincide, the weak maximum occurring in this figure shows that prior to poling α-phase dipoles are preferentially aligned in the T-direction. (Note: The normals of (020) lattice planes are at an angle of 90° to the dipole direction [100] in the crystal lattice of the α_p-phase).

There is a significant shifting of the (020) normals' intensity towards T after poling which is equivalent to an improved alignment of the dipoles towards the field direction N. One result of the reflection mode used is a typical decrease in intensity at high sample deflection angles. Therefore, the exact position of the points of concentration of the (021)-normals as well as of the dipole direction is not detectable from the measured intensity distributions. However, the observed difference in the intensity at the sample deflection angle $\alpha = 0°$ gives evidence that crystallites located with their dipole directions in the film plane before poling have changed their positions during poling. Figure 6

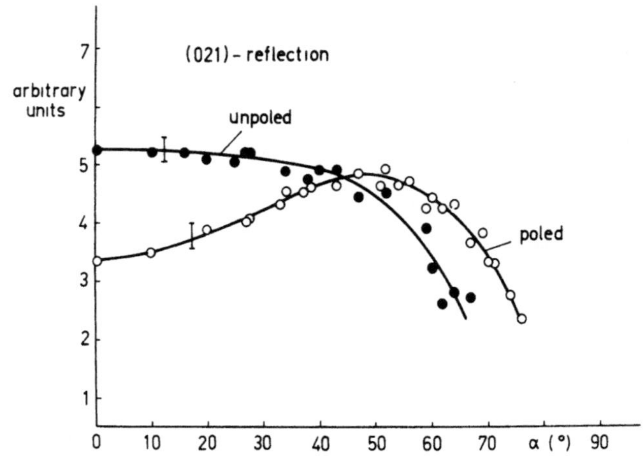

Fig. 5. Frequency-distribution of the $N-T$ projection of the normals of the (021)-lattice phase as a function of the sample deflecting angle α along $N-T$ before and after poling

gives the distribution of the normals of the (120)-lattice planes obtained from the same pole figure as the (021)-distribution. There is a considerable difference in the intensity levels between the unpoled and the poled samples in comparison with Fig. 5. Since the x-ray form factor of the lattice plane (120) of the α-phase becomes zero for the α_p-phase the decrease in intensity observed for the lattice plane (120) over a wide range of sample deflection angles confirms that a partial conversion of the α-phase into the α_p-phase took place during poling. The content of ordered regions of the α-phase converted into the α_p-phase obviously involves crystallites with a braod spectrum of orientations.

Summarizing, it can be stated that the structure conversions of the α-phase due to poling consists of at least two different components.

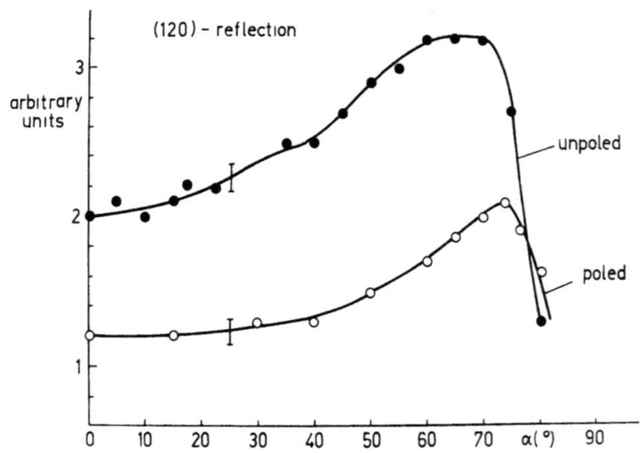

Fig. 6. Frequency-distribution of the normals of the (120)-lattice planes as a function of the sample deflection angle α along $N-T$ before and after poling

1. Crystallites of the α-phase are converted to the α_p-phase

Crystallites of the α-phase involved in this conversion have a broad orientation distribution of the crystallographic *a*-axes to the poling field direction *N*.

According to the changes of dipole orientation during poling the dipole axes of the α_p-phase-crystallites have a changed orientation distribution after poling, as compared to their orientation state prior to poling. Before poling the *a*-axes of the α-phase-crystallites show a slightly preferred orientation parallel to the sample direction *T*, whereas after poling the *a*-axes of the α_p-phase show a lie significantly preferred orientation parallel to the field direction *N*.

2. Field-induced structural changes in the β-phase

The all-trans conformation of PVDF has a dipole moment of 2.1 D (7.0×10^{-30} cm per repeat unit [15]. All other conformations and modifications of PVDF show smaller dipole moments. This suggests that the β-phase might be best suited for achieving a high macroscopically acting polarization. As a matter of fact there are some studies reporting polar coefficients rising with increasing content of β-phase [1, 16]. Conformation and modification changes due to poling are assumed to be negligible here.

The usual way of obtaining PVDF samples with a high amount of β-phase consists in uniaxial drawing of non-oriented α-phase films. Typically temperatures above 373 K and medium drawing ratios ($\lambda\approx4-6$) should be chosen [17–19]. This produces material which has only a small amount of α-phase (<5%). The chain axes are highly aligned in drawing direction *M* [20]. The orientation distribution of the crystalline dipole axes ([010]-direction) of such films is especially clear in the *N-T*-plane because of the nearly perfect orientation of crystalline chain axes in *M*-direction. The state of orientation of such materials has been studied in detail using for instance WAXS-texture methods [21, 22]. It was found that after drawing the dipoles are slightly preferentially aligned in the film plane along the transverse direction *T*. As in the case of the α-phase samples, only strong poling fields may induce a macroscopically acting polarization in the β-phase [23–25]. In this case the polarization may arise from changes in the dipole orientation distribution as well as from changes in the degree of order. Both processes have been used in the literature to interpret the experimental results [21, 22, 26–32]. To get a deeper insight into the molecular mechanisms of polarization of PVDF β-phase, x-ray pole figure measurements were performed under applied poling fields with varying field strengths.

Experimental

The starting material for a modified zone drawing procedure [18, 20] was 50-μm-thick Bemberg (FRG) films. Applying a drawing stress of 2.45 MPa, a drawing velocity of 30 mm/min, a drawing temperature of 383 K, and a drawing ratio of $\lambda = 6$, the drawn samples contained less then 5% α-phase.

The poling field strenghts were between 0 and 340 MV/m. Poling lasted between 10 and 60 s. For poling experiments and x-ray investigations 7-nm-thick aluminum electrodes were evaporated on both sides of the samples. It could be shown that these electrodes did not significantly influence the WAXS results [33]. The texture in the starting material, in the drawn unpoled samples and the drawn poled samples, was investigated using reflection- and transmission techniques. Additionally, WAXS plane film photographs were taken after every significant step of preparation. The measurements with applied poling field were performed by making use of a transmission mode texture goniometer with a steady electric field (0–250 MV/m) applied to the sample as described by Janke [33].

The interpretation of β-phase pole figures in general is even more difficult due to overlap between neighboring (hkl)-WAXS-reflections [2]. We have chosen the reflection group (111/021) for our experiments. In this case the contribution of the (021)-intensity to the whole integral intensity of the (111/021)-combination is less than 5% [2, 21]. Because of the nearly perfect orientation of the crystalline *c*-axes along *M* and since the normals on the (111)-lattice planes form an angle of $\theta = 31°$ relative to the *c*-axes, the (111)-lattice plane normals will be preferentially oriented along a circle with an angular distance of 31° to *M* in the pole figure. In this case, therefore, the transmission technique is sufficient for recording the complete orientation distribution of the (111) lattice plane normals. Under these conditions the dipole orientation along the *N–T*-direction in the pole figure can be determined from the measured orientation distribution of the (111)-lattice plane normals. Provided that the *c*-axes are perfectly aligned along *M*, the projection of the (111)-lattice plane normals orientation distribution on *N–T* is identical to the orientation distribution of the (110) lattice plane normals on *N–T* from which we can get the distribution of the dipolar direction using the crystal symmetry.

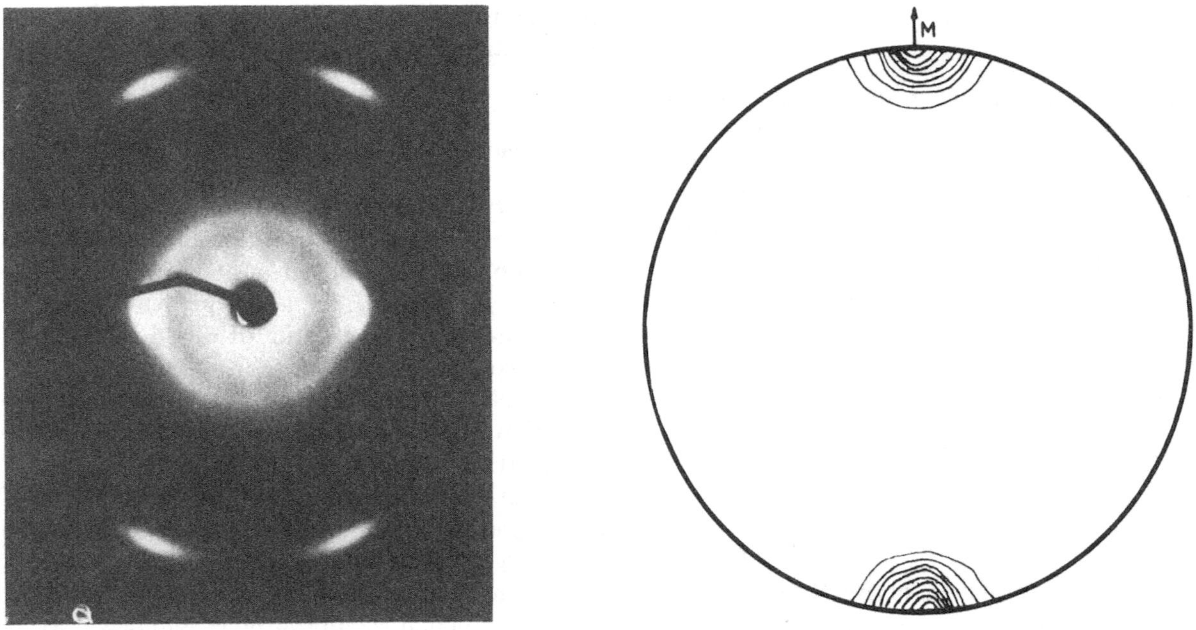

Fig. 7. WAXS-pattern (a) and transmission mode (001) pole figure (b) of a PVDF sample drawn with $\lambda = 6$ at 383 K

Results and discussion

The state of orientation before poling is shown by the x-ray diffraction pattern (Fig. 7 a) taken by a flat film camera and the (001) pole figure (Fig. 7 b). Both figures confirm that the crystalline chains are nearly perfectly aligned in the drawing direction *M*.

Ordinary x-ray methods cannot be used for the determination of changes in crystallinity of highly oriented drawn PVDF films during poling, since these methods require isotropic samples. Here, therefore, changes in crystallinity for different poling field strengths were estimated by integration of x-ray intensity over the whole (111)-pole sphere. This is assumed to be a good approximation for the number of lattice cells present in a sample, if the average crystallite shape does not change extremely [33], which is the case here.

The change in the orientation distribution due to poling is shown in Fig. 8 ($E = 150$ MV/m) as a representative example. Based on such pole figures the orientation distribution of the uniplanar-axial oriented dipoles along $N-T$ was determined. Here the dipole orientation distribution in $N-M$-direction is assumed not to depend on the poling field strength. The results are collected in Fig. 9. The remaining dipole orientation distributions for field strengths of 0 – 250 MV/m are shown together with the respective distribution at each foregoing poling field strength stage. Up to 100 MV/m the dipole distribution under poling field is also represented. Above 100 MV/m such curves cannot be measured, because the films become wrinkled during poling.

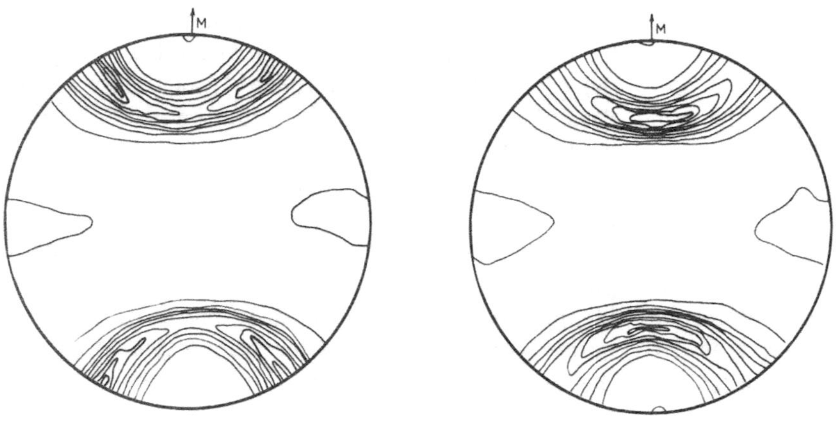

Fig. 8. (111)-pole figures of an a) unpoled sample, and b) poled at 250 MV/m

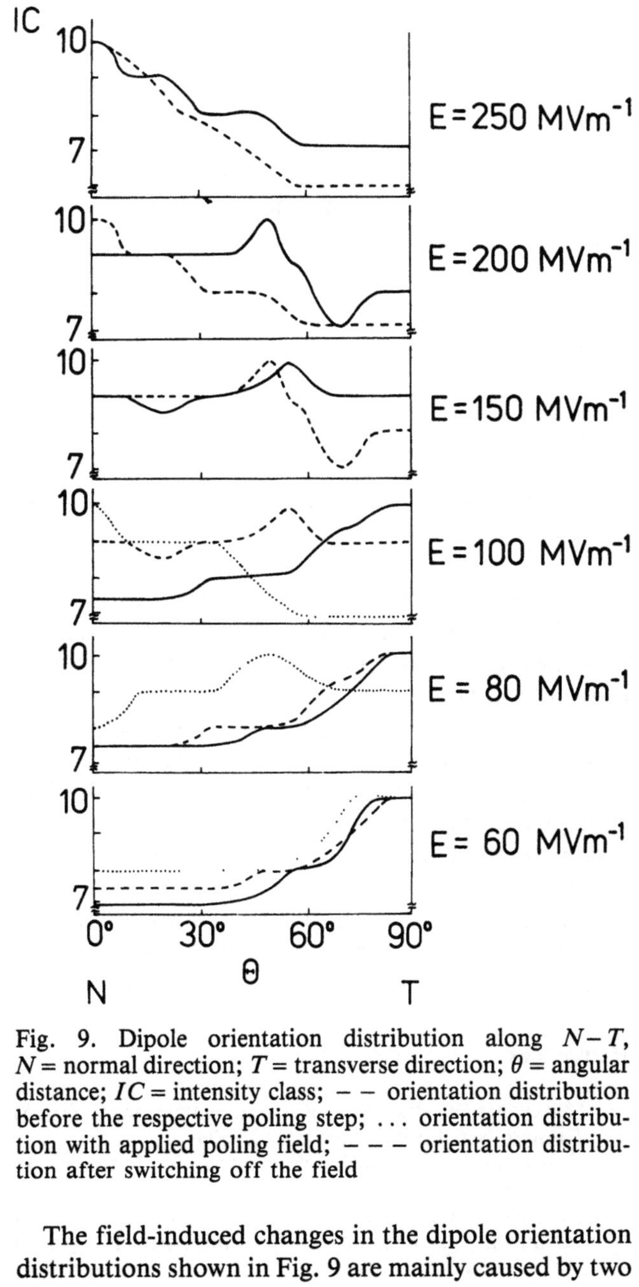

Fig. 9. Dipole orientation distribution along $N-T$, N = normal direction; T = transverse direction; θ = angular distance; IC = intensity class; – – orientation distribution before the respective poling step; ... orientation distribution with applied poling field; – – – orientation distribution after switching off the field

The field-induced changes in the dipole orientation distributions shown in Fig. 9 are mainly caused by two mechanisms:

1) Changes in orientation of the crystallites. In this case the area under the dipole orientation distribution curve in Fig. 9 remains constant.
2) Changes in crystallinity, i.e., changes in the number of crystalline dipoles. In this case the area under the dipole orientation distribution curve in Fig. 9 changes.

The dipole orientation distribution of the drawn unpoled starting material (solid line in the 60 MV/m part of Fig. 9) quantifies the deviation of the dipole orientation in this material from a pure fiber texture around M.

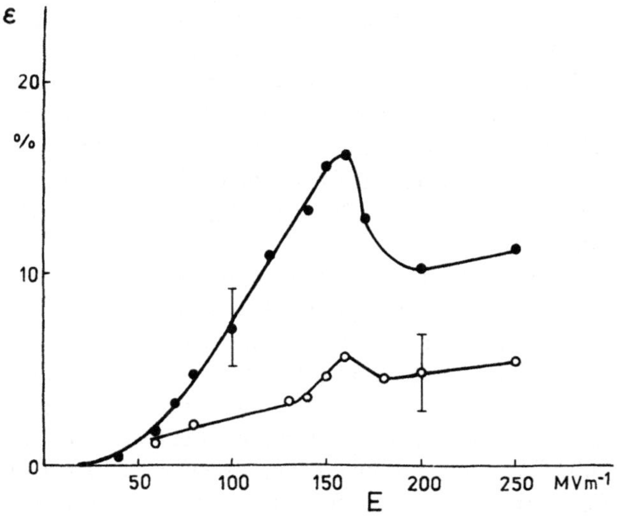

Fig. 10. Change in crystallinity ε as a function of poling field strength: ● = under field; ○ = after switching off the field

Below $E = 50$ MV/m no significant changes in the dipole orientation distribution have been observed. Such changes can be found for field strengths >60 MV/m. It has to be noticed that already at relatively small field strengths ($E < 80$ MV/m) a broad orientation range is involved in temporary changes in dipole orientation distribution. Comparing the areas below the distribution curves it follows that there are minor changes in reorientation of the crystalline dipoles up to a poling field strength of 80 MV/m. An increase in crystallinity has to be stated in both cases during poling, and to a significantly lower extent, after poling.

During poling at 100 MV/m the uniplanar-axial material portion shows substantial changes in orientation. The dipole axes are now preferentially aligned along the poling direction N (dotted line at 100 MV/m), while the amount of dipoles oriented in T-direction is strongly reduced. On removal of the field, however, a considerable relaxation is observed, with the relaxed dipoles showing a pronounced maximum at an angular distance of 55° to N. This maximum, as well as that observed for the 80 MV/m field suggests that stepwise changes in dipole orientation are caused by the field. This conclusion is also confirmed by the changes in the remanent dipole orientation distribution for field strengths >100 MV/m. With increasing field strength the maximum of the distribution changes its position step-by-step towards the normal direction N of the sample. At 200 MV/m or more, the preferred orientation of the uniplanar-axial-oriented crystalline dipoles is parallel to N, but with a broad distribution around this position.

It has to be noticed, however, that all the changes in orientation of dipoles caused by poling are small.

The relative change in crystallinity ε was determined as a function of poling field strength using the method of (111)-reflection intensity summation over the whole pole figure. The results for applied poling field and for poling field switched off after poling are shown in Fig. 10.

A significant change in crystallinity is seen to begin at a field strength of about $E = 60$ MV/m. Then the crystallinity increases with rising field strength up to 160 MV/m. At 160 MV/m the increase in crystallinity under field amounts 16% related to the unpoled material. For poling field strengths between 160 and 200 MV/m, crystallinity drops down with increasing field strength. This decrease reaches about 6% at 200 MV/m. Above 200 MV/m there is again a certain but lower increase in crystallinity, with no saturation visible in the investigated range of field strength.

The course of the change in crystallinity ε in the case of a field switched off is similar to that with applied poling field – but here the maximum increase in crystallinity is considerably smaller and reaches only 6% at 160 MV/m. Several repeated measurements with different samples ensured the reliability of this result.

Taking into account the special course of the parameter ε, as shown in Fig. 10, it can be assumed that poling leads to two counteracting processes, in addition to the changes in dipole orientation.

On the one hand, new or existing crystallites will grow due to the insertion of chains from the non-crystalline regions into the crystalline lattices. On the other hand, crystallites, oriented unfavorably in relation to field direction before poling, are destroyed by the partial rotation of dipoles.

At low field strengths (up to 160 MV/m) the first effect is predominant, while with increasing poling field strength the second effect becomes stronger and predominant at a range of $E = 160 \ldots 200$ MV/m.

For a series of samples some piezoelectric constants and mechanical parameters have also been determined as functions of poling field strength. The piezoelectric strain coefficient d_{31} shows a relative minimum at about 160 MV/m (cf. Fig. 11). Above 200 MV/m, d_{31} reaches a certain saturation. The determined d_{31} minimum correlates well with the maximum of crystallinity at 160 MV m.

Summarizing, it can be concluded that changes in crystalline orientation, which are caused by poling of PVDF β-phase films, take part in a very complex process of structural changes under electric poling fields.

For low poling field strengths (< 100 MV/m) the polarization is mainly related to an increase in

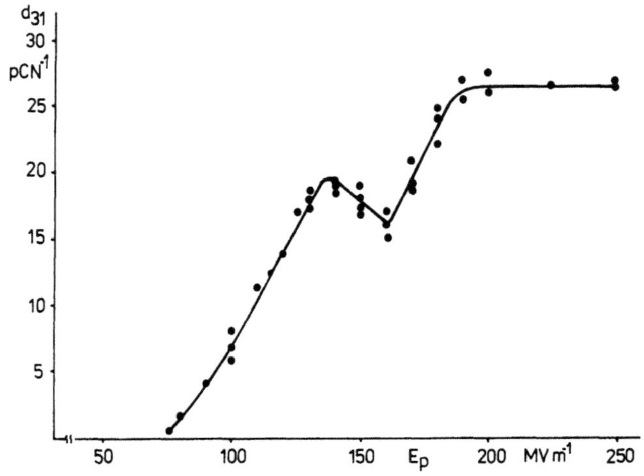

Fig. 11. Piezoelectric strain coefficient d_{31} vs poling field strength

crystalline order, 180° changes in the orientation of crystalline dipoles and remanent poling field dependent changes in the orientation of non-crystalline dipoles. Major changes in texture, which would be related to considerable changes in orientation of the crystallites as a whole, are of lower importance for this range of poling field strengths. For higher poling field strengths (> 100 MV/m) changes in crystalline texture become more and more significant for the structural changes observed during or after poling.

Conversions between crystalline and non-crystalline material portions as a result of poling have a major influence on crystallinity and on material properties.

References

1. Murayama N, Oikawa T, Katto T, Nakamura K (1975) J Polym Sci: Polym Phys Ed 13:1633
2. Hasegawa R, Takahashi Y, Chatani Y, Tadokoro H (1972) Polym J 3:600
3. Bachmann MA, Lando JB (1981) Macromolecules 14:40
4. Takahashi Y, Matsubara Y, Tadokoro H (1983) Macromolecules 16:1588
5. Takahashi Y, Tadokoro H (1984) Ferroelectrics 87:187
6. Weinhold R, Litt MH, Lando JB (1979) J Polym Sci: Polym Lett Ed 17:585
7. Kosmynin BP, Galperin EL, Cvankin DJ (1970) Vysokomol Soed A12:1254
8. Southgate PD (1976) Appl Phys Lett 28:250
9. Das Gupta DK, Doughty K (1977) Appl Phys Lett 31:585
10. Davis GT, McKinney JE, Broadhurst MG, Roth SC (1978) J Appl Phys 49:4998
11. Newman BA, Yoon CH, Pae KD, Scheinbeim JI (1979) J Appl Phys 50:6095
12. Newman BA, Scheinbeim JI (1983) Macromolecules 16:60
13. Weinhold R (1982) Thesis. Case Western Reserve Univ Cleveland, Ohio, USA

14. Schulz LG (1949) J Appl Phys 20:1030
15. Nakamura K, Wadda Y (1971) J Polym Sci A2:161
16. Shuford R, Wilke AF, Ricca JJ, Thomas GR (1976) Polym Eg Sc 16:25
17. Kosmynin BP, Galperin EL, Cvankin DJ (1970) Vysokomol Soed A12:1254
18. Kofer U (1980) Thesis A. Academy of Sciences of the GDR, Teltow-Seehof
19. Matsushige K, Nagawa K, Imada S, Takemura T (1980) Polymer 21:1391
20. Kuenstler W, Stark W, Danz R, Geiss D (1983) unpublished report. Academy of Sciences of the GDR, Teltow-Seehof
21. Takahashi N, Odajima A (1981) Jpn J Appl Phys 20:59
22. Geiss D, Ruscher C (1984) Acta Polym 35:299
23. Kawai H (1969) J Appl Phys 8:975
24. Bergmann JG, McFee JH, Grane GR (1971) Appl Phys Lett 18:203
25. Broadhurst MG, Davis GT (1980) In: Sessler GM (ed) Top Appl Phys, Electrets 33:285
26. McFee JH, Bergmann JG, Crane GR (1972) Ferroelectrics 3:305
27. Ohigashi H (1976) J Appl Phys 47:949
28. Tamura M, Ogasawora K, Oho N, Hegiwora S (1974) J Appl Phys 45:3768
29. Buchmann P (1973) Ferroelectrics 5:39
30. Kepler RG, Anderson RA (1978) J Appl Phys 49:1232
31. Kepler RG, Anderson RA, Lagasse RR (1984) Ferroelectrics 57:151
32. Latour M, Abo Dorra H, Galigne JL (1984) J Polym Sci: Polym Phys Ed 22:345
33. Janke A (1989) Thesis A. Academy of Sciences of the GDR, Teltow-Seehof
34. Takahashi Y, Matsubara Y, Tadokoro H (1982) Macromolecules 15:334
35. Takahashi T, Date M, Fukada E (1980) Appl Phys Lett 37:791
36. Davies GR, Singh H (1979) Polymer 20:772

Received February 6, 1989;
accepted June 13, 1989

Authors' address:

Prof. Dr. D. Geiss
Akademie der Wissenschaften der DDR
Institut für Polymerchemie "Erich Correns"
Kantstraße 55
DDR-1530 Teltow-Seehof, GDR

Discussion

KILIAN:

Have you found indication whether the transformation induced by the field depends on the crystal orientation? If considered as a field-induced phase transition such a dependence should indeed occur.

GEISS:

The views of the authors are devided in the question which crystallites having different orientations to the poling field are involved in the transition α into α_p. We favour that crystallites from a wide range of orientations of the dipole direction take part in this conversion. Especially such crystallites are converted whose dipoles lie in the sample plane prior to poling. The largest number of crystallites transformed is at an angular distance of about 45° (dipole direction to the poling field).

PIETRALLA:

After the phase transition took place under the influence of the electric field the α_p-crystals are stabilized under the action of their own internal field. Is it possible then to observe the phase transition back to the α-phase by heating or at least to measure a change in melt enthalpy?

GEISS:

The probability of reorientation the chains to come back to the α_p-phase is rather small as lattice energy calculations show.

GERHARD-MULKTHAUPT:

i) With thermal poling of PVDF, the generated polarization profiles are often very non-uniform; this can lead to erroneous interpretations if uniform profiles are assumed. Did you check the polarization profiles after thermal poling?

ii) Is there any indication of a ferroelectric to paraelectric phase transition for the α_p-phase in PVDF?

iii) In a simplified concept, the piezoelectricity of PVDF arises from the combination of the crystalline polarization acting as a bias and of the amorphous phase acting as an easily deformable medium. Consequently, there should be a certain crystallinity for which the piezoelectric activity has a maximum. Is such a maximum found experimentally?

GEISS:

i) You are right in saying that we can expect a non-uniform polarization distribution in our samples along the poling field direction. This polarization distribution should correspond to a non uniform structure conversion in the material. But up to now we have no experimental results which confirm the latter assumption and therefore we have not yet measured the polarization distribution in the sample.

ii) We have no indication that such a transition takes place. We assume that the situation is here similar to that we know from the β-phase where the ferroelectric-paraelectric transition lies above the melting point.

iii) In a general view you are right, but it is not so easy to change the crystallinity in PVDF systematically in a wide range to measure such a dependence. What we have observed is the change in the crystallinity in a very small range. Due to poling 10% – 15% under field influence with only a few percent remaining after poling. This increase amounts in a significant change of the piezo- and pyroelectrical response but no maximum has been observed.

KRÜGER:

Has the α_p-process been observed in statistic copolymers of P(VDF-TrFE)?

GEISS:

As far as I know the α to α_p conversion has not been observed in the copolymers of PVDF till now.

OLEYNIK:

Can you tell something about the thermodynamics of the α_p-state of PVDF? Is it metastable in comparison with non-polarized state? Could you find any differences in the thermodynamics of melting of the polar state of PVDF?

GEISS:

The state of the α_p-phase can be assumed to be thermodynamically very similar to that of the α-phase because the chain conformation as well as the dimensions of the crystalline lattice remain unchanged during the conversion. Furthermore there exist calculations which show that the lattice energy is very similar in both cases.

GOTLIB:

What can you say about the time scale of transitions from one to another type modification of PVDF?

The rearrangement of local order at the scale of the monomeric units if repeated at many steps can lead to a long scale variations of chain conformation which could be accompanied (or must?) by long relaxation times which could be very great in solid polymeric state.

GEISS:

According to the different mechanisms to get polarization which we have observed, there is a broad distribution of transition times. In other words, the transition times depend strongly on the amounts of different crystalline modifications in the sample, on the supermolecular structure – especially on the orientation state – and on the probing field strength.

Progress in Colloid & Polymer Science Progr Colloid & Polymer Sci 80:129–139 (1989)

Broadband dielectric spectroscopy on the molecular dynamics in polymer model-systems: polyphenylmethylsiloxane and linear and star-branched poly(cis-1,4-isoprene)

F. Kremer, D. Boese, G. Meier, and E.W. Fischer

Max-Planck Institut für Polymerforschung, Mainz, FRG

Abstract: Broadband dielectric spectroscopy (10^{-1} Hz – 10^9 Hz) is employed to study the molecular dynamics in polymer model-systems. For poly(methylphenylsiloxane) the measured molecular dynamics is compared with the collective dynamics determined by quasielastic light scattering. Quantitative agreement in the mean relaxation time and its shape parameters β ("stretched exponential") is found with both methods. It is concluded that the collective motions are the rate-limiting quantity and hence determine the molecular dynamics. In the case of poly(cis-1,4-isoprene) the molecular dynamics of the linear and the star-branched polymer are compared. Two relaxation processes, a segmental and a normal mode, are observed. Its dependence on temperature, molecular weight, and the chemical structure is discussed.

Key words: Dielectric relaxation; photon correlation; collective dynamics; spectroscopy; poly(phenylmethylsiloxan); poly(cis-1,4-isoprene)

Introduction

Dielectric spectroscopy is extensively used to study the molecular dynamics in polymers [1–4]. It is an interesting question of how these dynamics compare with the collective dynamics measured by quasielastic light scattering. In this paper such a comparison is carried out for the example of poly(methylphenylsiloxane), a polymer having an especially high flexibility [5]. It is of further interest of how the chemical structure of a polymer influences its molecular dynamics, and therefore, we chose to study poly(cis-1,4-isoprene). This polymer can be synthesized as a linear or a star-branched polymer with narrow molecular weight distributions [6–10]. Due to the lack of symmetry in its chemical structure poly(cis-1,4-isoprene) has non-zero components of the dipole moment, both perpendicular and parallel to the chain backbone. This gives rise to two dielectrically active relaxation processes, a segmental and a normal mode [6–10]. The former is a local process, while the latter corresponds to extended motions of the polymer chain. This peculiarity makes cis-1,4-polyisoprene an interesting candidate for studying the influence of the molecular structure on the molecular dynamics in polymers.

Experimental

To cover the frequency range from 10^{-4} Hz up to 10^9 Hz three different measurement systems were employed: i) a frequency response analyzer (Solartron Schlumberger 1254), ii) an automatic low-frequency bridge (Hewlett-Packard 4192A), and iii) an impedance analyzer (Hewlett-Packard 4191A) based on the principle of a reflectometer.

The frequency-response analysis was accomplished by a high-impedance buffer amplifier of variable gain [11], whose generator (Fig. 1) applies a digitally synthesized sine-wave signal to the dielectric sample (or a reference capacitor)

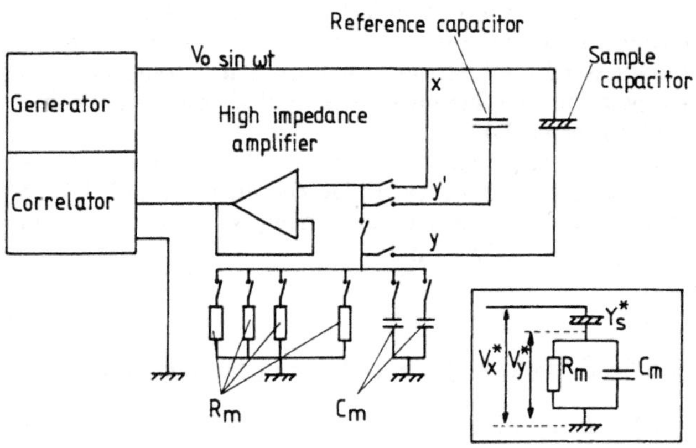

Fig. 1. Scheme of the frequency response analyzer and the high impedance buffer amplifier

of up to 9.99 volts in the frequency range from $5\cdot 10^{-5}$ Hz to 65 kHz. Using the buffer amplifier the potentials at point X and $Y(V_x^*$ and V_y^*, see inset in Fig. 1) are measured by the correlator in its in-phase and quadrature component with respect to the generator signal. For that a digital integration technique is employed which yields complete harmonic rejection [4, 12]. From the measured complex voltages V_x^* and V_y^* the complex admittance Y_s^* of the sample can be calculated:

$$Y_s^* = Y_m^*/[(V_x^*/V_y^*)-1] \tag{1}$$

where

$$Y_s^* = G_s + i\omega C_s \quad \text{and} \quad Y_m^* = 1/R_m + i\omega C_m \; . \tag{2}$$

Using these expressions the dielectric function $\varepsilon^* = \varepsilon' + i\varepsilon''$ may be calculated according to $\varepsilon^* = Y_s^*/i\omega C_0$, where C_0 is the capacitance of the empty sample capacitor. In order to improve the accuracy of the measurement system especially for low-loss materials a low-loss air capacitor (General Radio) can be used as continuously available reference during the course of the measurement [11]. The amplifier is now switched to point Y' and a measurement of $V_{y'}^*$ is made, retaining the same value of Y_m^*. Then V_x^* is measured and in analogy to Eq. (1):

$$Y_r^* = Y_m^*/[(V_x^*/V_{y'}^*)-1] \; , \tag{3}$$

where $Y_r^* = G_r + i\omega C_r$. Combining Eqs. (1) and (3) yields for the admittance of the sample Y_s^*,

$$Y_s^* = Y_r^*[(V_x^*/V_{y'}^*)-1]/[(V_x^*/V_y^*)-1] \; . \tag{4}$$

Now the conductance Y_m^* of the measuring circuit and associated errors are eliminated. All measurement functions, e.g., the switching from the sample capacitor to the reference capacitor and the selection of the measuring resistors R_m and capacitors C_m are fully computer-controlled.

The sample capacitor was made of two gold-plated polished stainless steel electrodes (diameter 40 mm), having a separation of 150 μm, maintained by three small, fused silica pieces (area = 2 mm²). The capacitor plates were pressed together by a micrometer screw and the whole arrangement was placed in a custom-made cryostat, which will be described below. The sample temperature was measured with a platinum resistor (PT 100) that was inserted in one of the electrodes. (Diameter of the PT 100 resistor = 1.8 mm; thickness of the electrode = 4 mm). The resolution of the temperature measurement was 0.01 K using a Keithley Multimeter (195 A).

In the frequency range from 10 Hz to 10^7 Hz an automatic low-frequency bridge (Hewlett-Packard 4192 A) was used (Fig. 2). At zero adjustment, where the bridge current $I_d = 0$, it follows that

$$I_x = I_r \tag{5}$$

and hence the complex impedance of the sample Z_x^* is

$$Z_x^* = R_r \cdot e_s^*/e_r^* \; , \tag{6}$$

where e_s^* and e_r^* are complex voltages (Fig. 2). From the complex impedance the complex dielectric function can be calculated if the sample geometry is known. The sample capacitor arrangement was the same as used for the measurements with the frequency response analyzer. The whole measurement system is fully automatic as well.

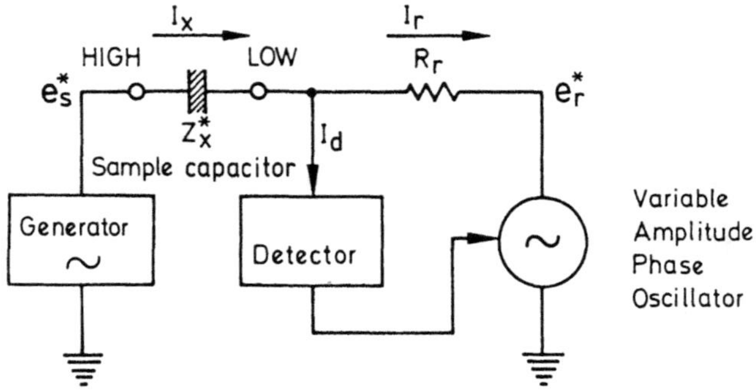

Fig. 2. Scheme of the automatic low frequency bridge (10 Hz to 10^7 Hz)

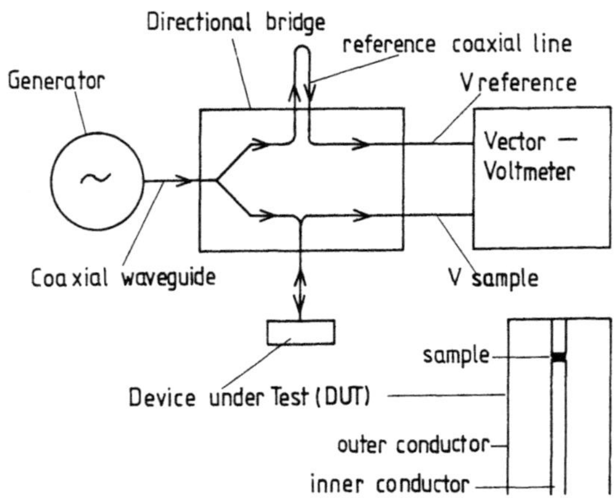

Fig. 3. Scheme of the radio frequency impedance analyzer (10^6 Hz to 10^9 Hz)

For the frequency range from 10^6 Hz to 10^9 Hz an impedance analyzer (Hewlett-Packard 4191 A) was employed, which is, in principle, a coaxial line reflectometer [13, 14]. The signal which is reflected by the device under test is lead to a vector voltmeter by use of a directional bridge, where it is compared with the signal from a reference channel. Thus the complex reflection coefficient,

$$\Gamma^* = V_{\text{sample}}^*/V_{\text{reference}}^* \; , \tag{7}$$

is measured. From that the complex impedance Z_x^* can be deduced.

$$Z_x^* = Z_0(1+\Gamma^*)/(1-\Gamma^*) \tag{8}$$

where Z_0 denotes a 50 Ω reference. The sample is mounted as part of the inner conductor (diameter: 3 mm) of a coaxial short (see inset in Fig. 3). A separation of 50±2 μm for the sample is maintained by the use of two fused silica glass fibers. Their contribution to the volume in the sample capacitor is negligible. The sample cell is mounted in a custom-made cryostat which is designed for the temperature range from 100 K to 500 K. The sample temperature is measured with a platinum resistor (PT 100), which is inserted in the outer conductor of the coaxial cell. The resolution of the temperature measurement is 0.01 using a Keithly 195 A.

The sample temperature could be adjusted between 100 K and 500 K, by a jet of temperature-controlled nitrogen gas (Fig. 4), produced by mounting a heater in a liquid nitrogen container. The stability of the temperature adjustment was better than ±0.02 K. Temperature gradients within the sample cell were e.g. less than 0.2 K within a sample of 40-mm diameter.

The synthesis and characterization of polymethylphenylsiloxane and of linear and star-branched poly(cis-1,4-isoprene) is described elsewhere [15–17].

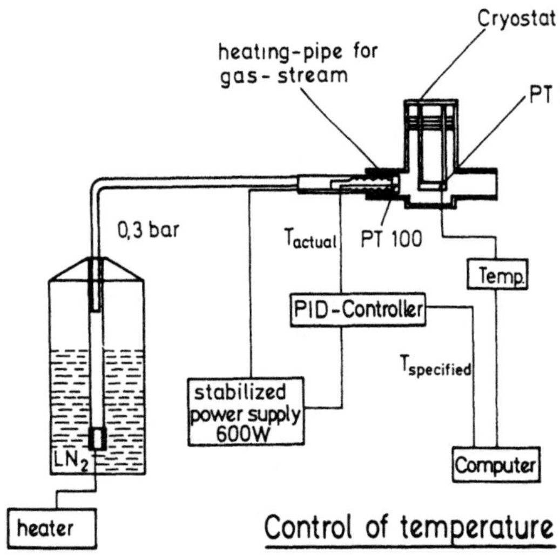

Fig. 4. Schematic of the automatic nitrogen gas temperature control system. It covers the temperature range from 100 K to 500 K

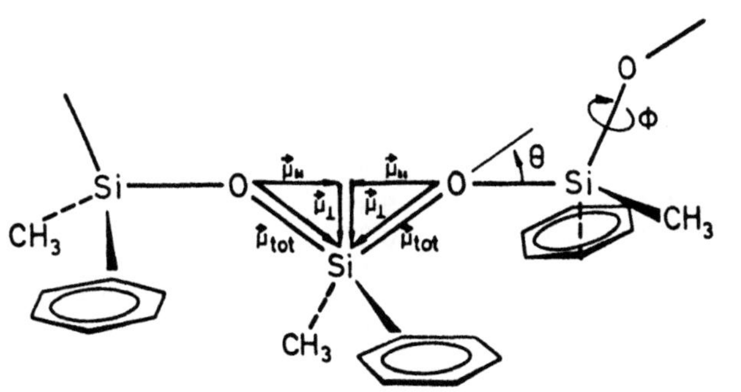

Fig. 5. Chemical structure of poly(methylphenylsiloxane)

Molecular and collective dynamics in polymethylphenylsiloxane

The fact that poly(methylphenylsiloxane) (PMPS) has a dielectrically active relaxation process originates from the dipole moment of the Si-O bond [18–20]. Due to the chemical structure of PMPS (Fig. 5) the dipole components in the main axis of the polymer cancel each other, while they add up in the perpendicular direction. Thus, one has to expect one dielectric relaxation process, assigned to a local motion of the Si-O bond. It corresponds to the dynamic glass transition (α-relaxation) of the bulk polymer [9, 20]. Measured from temperatures between −25 °C to 33 °C this relaxation process shifts from about 1 Hz to about 10^8 Hz (Fig. 6). On the low-frequency side the measurement is limited by a conductivity contribution, on the high frequency side by resonance effects of the sample cell. In order to deduce the mean relaxation rate and the shape parameter of the relaxation time distribution from the measured data the ansatz of Havriliak-Negami [21] is used:

$$\varepsilon^*(\omega) = \varepsilon_\infty + \frac{\varepsilon_s - \varepsilon_\infty}{[1+(i\omega\tau)^\alpha]^\gamma} \tag{9}$$

with $0 \le \alpha \le 1$ and $0 \le \gamma \le 1$, ε_s, respectively ε_∞ describes the low frequency, respectively the high frequency value of the real part of the dielectric function. The parameter α, respectively γ characterize a symmetrical, respectively asymmetrical broadening of the relaxation time distribution. From Eq. (9) the imaginary part $\varepsilon''(\omega)$ of the complex dielectric function can be deduced [21]:

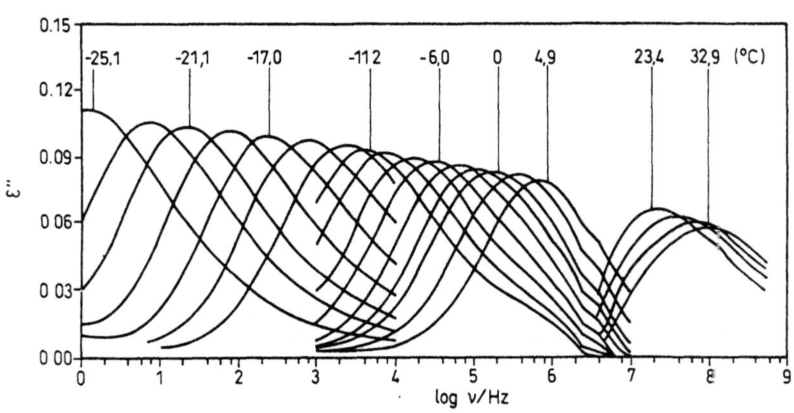

Fig. 6. Dielectric loss $\varepsilon''(\nu)$ vs frequency for PMPS ($M_n = 28500$ with $T_g = -26\,°C$) at different temperatures as indicated. At temperatures below $T = -12.9\,°C$ the frequency response analyzer system was used, for very high frequencies the reflectometer method was used. Accuracy of the measurement of the dielectric loss is ±5% in the low frequency region; above 1 MHz the measurement accuracy is ±10%

$$\varepsilon''(\omega) = (\varepsilon_s - \varepsilon_\infty)\tau^{-\gamma} \sin \gamma\psi \quad (10)$$

$$r^2 = 1 + 2(\omega\tau)^\alpha \cos(\alpha\pi/2) + (\psi\tau)^{2\alpha} \quad (11)$$

$$\tan\psi = \frac{(\omega\tau)^\alpha \sin(\alpha\pi/2)}{1+(\omega\tau)^\alpha \cos(\alpha\pi/2)} \,. \quad (12)$$

The experimental data can be described by this Ansatz within the experimental accuracy – over the entire frequency range (Fig. 7) leading to the fit parameters listed in Table 1.

In order to compare the data measured in the frequency domain (with dielectric spectroscopy) with the results of photon correlation spectroscopy measured in the time domain, one has either to transfer the frequency dependent data in the time domain or vice versa. Because of the fact that the dielectric measurements cover a much wider frequency and temperature range than the photon correlation measurements and because of the authors' interest in a possible temperature dependence of the relaxation time distribution parameters, it was decided to transform the dielectric data into the time domain. The total dipole moment correlation function $\phi(t)$ is given by:

$$\phi(t) = \frac{\sum_i^N \sum_j^N \langle \mu_i(0)\mu_j(t)\rangle}{\sum_i^N \sum_i^N \langle \mu_i(0)\mu_j(0)\rangle} \,, \quad (13)$$

where $\mu_j(t)$ denotes the elementary dipole moment in the chain at time t [22–25]. Since the real and the imaginary part of the dielectric function are connected by the Kramers-Kronig relation, it is sufficient to use only $\varepsilon''(\omega)$ in order to determine the autocorrelation function.

$$\phi(t) = \frac{2}{\pi}\int_0^\infty \frac{\varepsilon''(\omega)}{(\varepsilon_s - \varepsilon_\infty)} \frac{\cos\omega t}{\omega} d\omega \,. \quad (14)$$

Thus the analytical form of Eq. (10) for the dielectric loss (with the Havriliak-Negami fit parameter from Table 1) is used to calculate numerically the autocorrelation function $\phi(t)$ for different temperatures. The resulting function $\phi(t)$ is asymmetric with respect to time (Fig. 8); it can be within the experimental accuracy, described by the well-known Ansatz of Kohlrausch-Williams-Watts (KWW):

$$g(t) = a \cdot \exp\left[-\left(\frac{t}{\tau_{KWW}}\right)^\beta\right] \quad (15)$$

with the parameters a, τ_{KWW} and β ("stretched exponential") with $(0 \leq \beta \leq 1)$ [26]. It is also possible to describe the experimental data by using the Laplace

Table 1. Temperature dependence of the fit parameters $\Delta\varepsilon_{HN}, \alpha, \gamma, \tau_{HN}$, from the Havriliak-Negami equation [Eq. (9)]. To describe the data within experimental accuracy a variation of the fit parameters in $\Delta\varepsilon_{HN}$: ±0.005, α: ±0.05, γ: ±0.06 and τ_{HN}: 10% is possible

T/°C	$\Delta\varepsilon_{HN}$	α	γ	τ_{HN}/s
−22.8	0.42	0.76	0.56	$4.7\cdot10^{-2}$
−21.1	0.41	0.80	0.46	$1.7\cdot10^{-2}$
−19.1	0.40	0.82	0.43	$5.1\cdot10^{-3}$
−17.0	0.40	0.84	0.39	$1.7\cdot10^{-3}$
−11.2	0.35	0.76	0.56	$7.6\cdot10^{-5}$
−9.7	0.35	0.77	0.55	$4.1\cdot10^{-5}$
−7.8	0.34	0.83	0.46	$2.3\cdot10^{-5}$
−6.0	0.34	0.85	0.43	$1.3\cdot10^{-5}$
−4.0	0.34	0.89	0.36	$7.1\cdot10^{-6}$
0	0.31	0.85	0.46	$2.1\cdot10^{-6}$
23.4	0.25	0.96	0.34	$1.7\cdot10^{-8}$
28.7	0.25	0.99	0.29	$1.1\cdot10^{-8}$

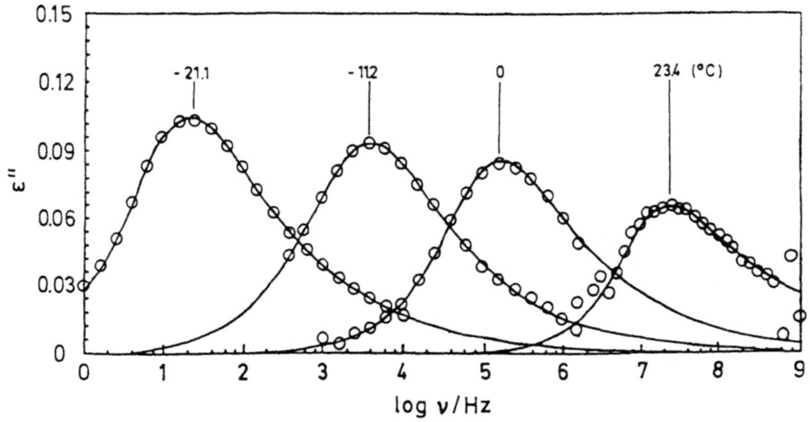

Fig. 7. Data from Fig. 6 as fitted (solid line) using the Havriliak-Negami equation [Eq. (9)]. The fit parameters are shown in Table 1

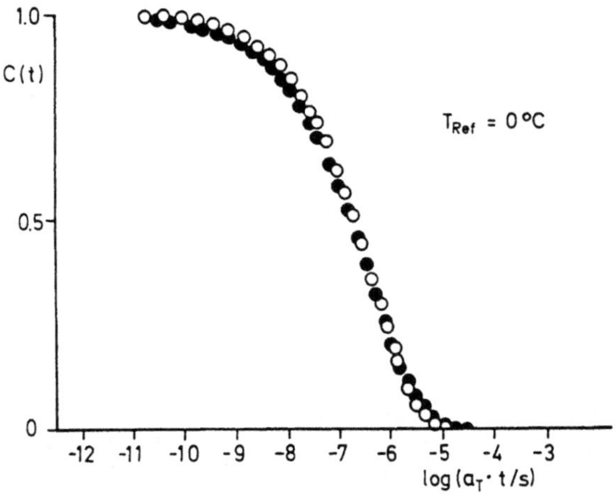

Fig. 8. Normalized correlation function calculated according to Eq. (2) using the HN parameter from Table 2. The resulting β parameter [Eq. (12)] is $\beta = 0.45$ (●). A correlation function with $\beta = 0.5$ (○), the distribution parameter from light scattering, is given for comparison for the same τ_{KWW}

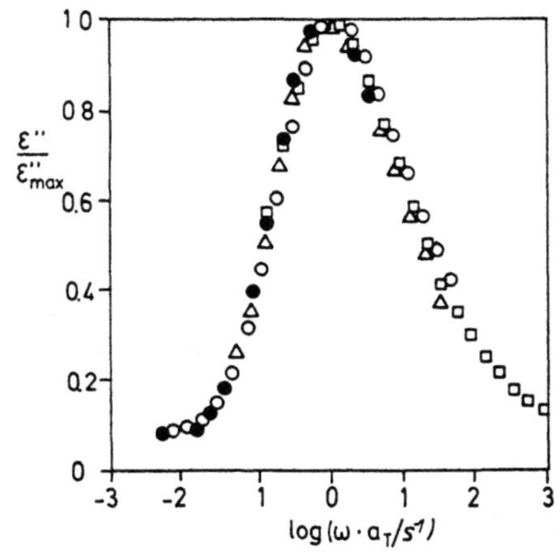

Fig. 9. Masterplot with $T_{Ref} = 0\,°C$. Plotted in normalized data $\varepsilon''(\nu)/\varepsilon_{max}$ from four different temperatures: (●): 0 °C; (△): −6 °C; (○): −17 °C; (□): −23 °C. The HN parameters of this curve are $\alpha = 0.80$ and $\gamma = 0.48$

transformation of a distribution of retardation times $L(\ln\tau)$ [26]

$$g^{(1)}(t) = \int_{-\infty}^{+\infty} \exp(-t/\tau) L(\ln\tau)\, d\ln\tau \ . \tag{16}$$

The distribution $L(\ln\tau)$ can be calculated from the experimental data using the complementary inverse Laplace transformation (ILT). From both representations mean relaxation times can be obtained via

$$\log\langle\tau\rangle = 1/2.303\cdot\ln\left(\frac{\tau_{KWW}}{\beta}\right)\Gamma(1/\beta) \tag{17}$$

Table 2. Temperature dependence of the fit parameters a, β, τ_{KWW} from the Ansatz of Kohlrausch-Williams-Watts [Eq. (15)] to describe the autocorrelation function as calculated from the measured dielectric loss at different temperatures

T/°C	a	β	τ_{KWW}/s	$\langle\tau_{KWW}\rangle/s$
−22.8	0.65	0.44	$1.6\cdot10^{-2}$	$4.2\cdot10^{-2}$
−21.1	0.64	0.44	$5.3\cdot10^{-3}$	$1.4\cdot10^{-2}$
−19.1	0.62	0.44	$1.4\cdot10^{-3}$	$3.7\cdot10^{-3}$
−17.0	0.62	0.43	$3.9\cdot10^{-4}$	$1.1\cdot10^{-3}$
−11.2	0.55	0.47	$3.3\cdot10^{-5}$	$3.4\cdot10^{-5}$
−9.7	0.54	0.47	$1.7\cdot10^{-5}$	$3.8\cdot10^{-5}$
−7.8	0.53	0.46	$7.2\cdot10^{-6}$	$1.7\cdot10^{-5}$
−6.0	0.52	0.46	$3.6\cdot10^{-6}$	$8.5\cdot10^{-6}$
−4.0	0.52	0.44	$1.6\cdot10^{-6}$	$4.9\cdot10^{-6}$
0	0.48	0.47	$6.7\cdot10^{-7}$	$1.5\cdot10^{-6}$
23.4	0.38	0.45	$3.5\cdot10^{-9}$	$8.7\cdot10^{-9}$
28.7	0.38	0.43	$1.8\cdot10^{-9}$	$5.0\cdot10^{-9}$

where Γ denotes the gamma function or

$$\langle\log\tau\rangle = 1/2.303\cdot\int_{-\infty}^{+\infty} \ln\tau L(\ln\tau)\, d\ln\tau \ . \tag{18}$$

The difference between the values from Eq. (17) and Eq. (18) for $0.4<\beta<1$ is less than 0.5 decades. A further discussion is given in the literature [27].

In Table 2 the fit-parameter according to the KWW-Ansatz are listed. The β-parameter turns out to be temperature-independent within the limits of experimental accuracy. In contrast to that the shape parameters α and γ of the Havriliak-Negami function exhibit a pronounced temperature dependence. Plotting the normalized data $\varepsilon''(\omega)/\varepsilon''_{max}$ for different temperatures leads to a master plot (Fig. 9) of the dielectric loss as one has to expect from the temperature independence of the "stretched exponential" β (Table 2).

In photon correlation spectroscopy the measured quantity is the intensity time autocorrelation function $\langle I(t)\, I(0)\rangle$ of the scattered light intensity I. This quantity is determined on the length scale of an optical wavelength. Under the assumption of homodyne detection and a Gaussian amplitude distribution of the scattered light the Siegert relation holds [28].

$$\begin{aligned}\langle I(t)\, I(0)\rangle &= \langle I\rangle^2 + A\,|\langle E^*(t)E(0)\rangle|^2 \\ &= \langle I\rangle^2(1+f|g(t)|^2) \ .\end{aligned} \tag{19}$$

It connects the autocorrelation function of the scattered electrical field $E(t)$ with the measured autocorrelation function of the light intensity, f being a constant; $g(t)$ is the normalized autocorrelation function. In the VV geometry [29] for a given q-vector it holds that

$$\langle E(t)\cdot E(0)\rangle = a_1 G_{iso}(t) + a_2 G_{VH}(t) \quad (20)$$

with constants a_1 and a_2.

$$G_{iso}(t) = + \left[\left(\frac{\partial\varepsilon}{\partial\varrho}\right)_T\right]^2 \varrho_0 N k_B T[D_e - D(t)] \quad (21)$$

$$G_{VH}(t) = a_3^2 k_B T V[J_e - J(t)] \ . \quad (22)$$

Here a_3 is a constant, D is the longitudinal compressive compliance, and J is the shear compliance; ε is the real part of the dielectric constant in the visible, ϱ_0 the average density, N is the number of scatterers, V the volume, D_e is the equilibrium bulk compliance $D_e = \lim_{t\to\infty} D(t)$ and J_e the equilibrium shear compliance $J_e = \lim_{t\to\infty} J(t)$. An experimental verification of Eq. (21) has been undertaken for the example of poly(vinylacetate) [29]. It proves that photon correlation spectroscopy as a method in the $q \to 0$ limit yields results which are consistent with those from dynamic mechanical measurements.

For the intensity autocorrelation function in VV and VH geometry – within experimental accuracy – the same time dependence was found in as different systems as polystyrene [30], orthoterphenyl [31, 32] and PMPS [15]. Thus since the time-independent prefactors in Eqs. (21) and (22) are not included in the analysis, Eq. (20) allows to deduce the time-dependence of the field autocorrelation function even in cases where the intensity in VH-geometry and hence G_{VH} is not negligible compared with G_{iso}. The latter is given by the expression

$$G_{iso}(t)_q \equiv \left[\left(\frac{\partial\varepsilon}{\partial\varrho}\right)_T\right]^2 \langle\delta\varrho(t)\delta\varrho(0)\rangle_q \quad (23)$$

containing the autocorrelation functions of the density fluctuations [28] at a constant scattering vector q. Therefore the time-dependence of the intensity autocorrelation function (in VV or VH geometry) is proportional to the time-dependence of the autocorrelation function of the density fluctuations.

The measured correlation function $|g(t)|^2$ is highly asymmetric with respect to the time scale (Fig. 10). By combining the measurements at different temperatures it is possible to construct a master curve for a reference

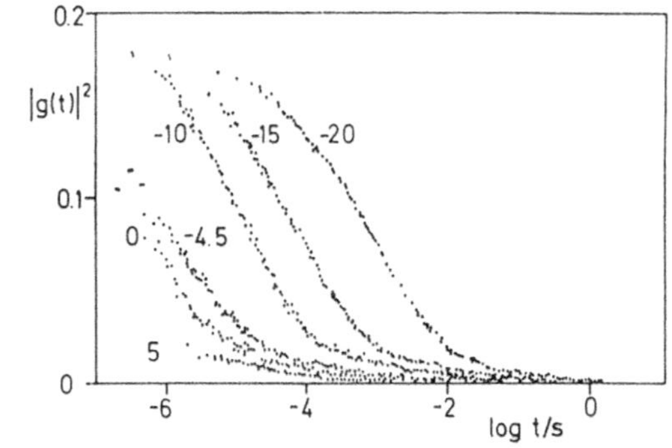

Fig. 10. Density autocorrelation function of PMPS $|g^{(1)}(t)|^2$ vs log t measured at various temperatures given in the figure in °C as indicated. Accuracy of the measurement is ±1%

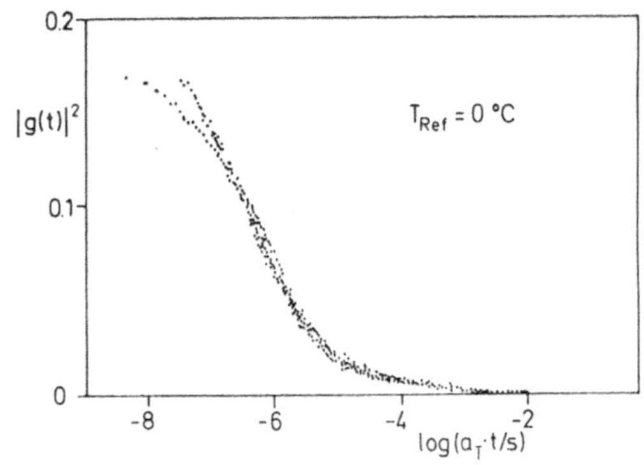

Fig. 11. Mastercurve construction of all the individually measured correlation functions shown in Fig. 10 with a reference temperature of $T_{Ref} = 0\,°C$

temperature of $T = 0\,°C$ (Fig. 11). Besides deviations at the short time side, this construction is possible, proving the applicability of the temperature-frequency superposition principle in this regime. In order to compare these results with the dielectric measurements it is indicated to construct a master plot of the normalized dielectric loss $\varepsilon''(\omega)/\varepsilon''_{max}$ using the same reference temperature $T = 0\,°C$. Neglecting deviations on the high frequency side, this construction works well as one has to expect from the temperature independence of the β-parameter listed in Table 2. To carry out the comparison between the molecular and the collective dynamics more thoroughly it is necessary to analyze the autocorrelation function measured by photon correlation spectroscopy in more detail. The attempt to fit Eq. (15) to the experimental data shows systematic deviations which are similar to the cases where primary and secondary relaxation processes are within the accessible time window of the PCS experiment [33]. In Fig. 12 the deviation from a best fit single KWW function to the correlation function obtained at $T = -20\,°C$ is plotted to show that a single KWW representation is inadequate to describe the data within the experimental accuracy. Therefore the inverse Laplace transformation (ILT), first demonstrated for polymer systems, in the case of PMMA [34] is employed to calculate the relaxation time distribution (Fig. 13). For two different temperatures it is demonstrated that two well-separated processes contribute. The physical picture connected with this additional process, which is at a given temperature slower than the α-relaxation, can be accounted for as a normal mode process, equivalent to findings from ultrasonic relaxation [4, 35] and dielectric relaxation [36–39].

The assignment of the slow maximum in Fig. 13 to a normal mode process in indicated by two experimental findings: Firstly, by changing the molecular weights M of the samples under study, the mean time

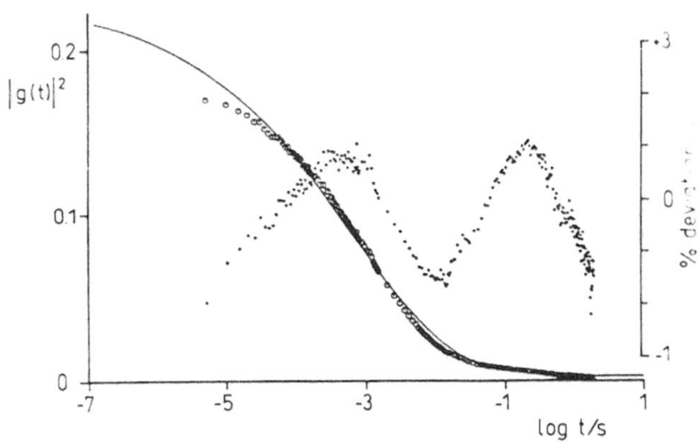

Fig. 12. The deviations in percent of a best fit single KWW function [Eq. (2)] to the correlation function of PMPS at $T = -20\,°C$

$\langle\tau\rangle$ varies with M^2, as it should according to the Rouse theory. Secondly, one finds agreement between $\langle\tau\rangle$ from quasielastic light scattering (QLS) and the prediction of $\langle\tau\rangle$ from steady flow viscosity data on the same sample of known molecular weight [40]. This mode is not observable by dielectric spectroscopy [41], since the mean square dipole moment in the polymer main chain arises from the Si-O bond [18, 19]. Due to the symmetry of the monomeric unit the dipole moment components parallel to the main axis are not dielectrically active, whereas they add up perpendicular to it. Consequently, in poly(cis-1,4-isoprene) a normal mode has been observed [36, 37, 41]. This also holds for polypropyleneglycol (PPG), where this effect has been observed for the first time [39].

From the foregoing arguments it follows that the fast process in Fig. 13 is the one which can be compared with the dielectrically observed relaxation. To analyze this process in terms of the mean relaxation time and its distribution one can use the first moment of the distribution $L(\ln\tau)$ which gives $\langle\log\tau\rangle$ and its variance is a measure of the width of the distribution. Another approach is to fit a double KWW function to the experimental data and to select the fast process for comparison with the dielectric relaxation. Both methods deliver the same results, as checked explicitly. For the distribution parameter β a mean value of $\bar{\beta} = 0.45 \pm 0.03$ is obtained. To demonstrate this small difference in the shape beween $\beta = 0.45$ and $\beta = 0.5$ in Fig. 8 for the same τ_{KWW}, both correlation functions are plotted.

The temperature dependence of the $\langle\tau\rangle$ values in Table 2 can be described by the WLF equation

$$\log a_T = \frac{C_1(T-T_0)}{C_2+T-T_0} \; . \tag{24}$$

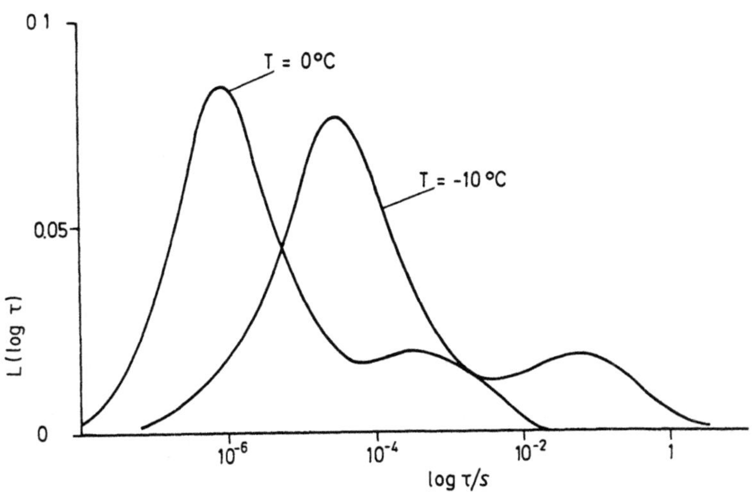

Fig. 13. Results of the inverse Laplace transformation on two measured correlation functions of PMPS at $T = 0\,°C$ and $T = -10\,°C$. From the figure the two-peak structure is evident

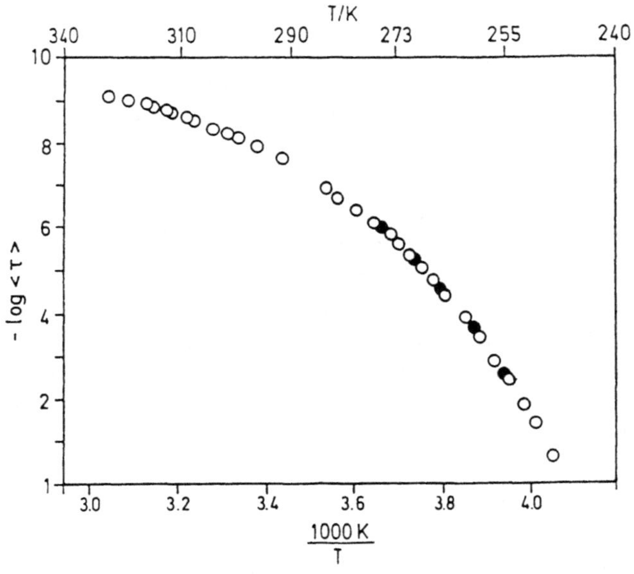

Fig. 14. Plot of the logarithm of the mean relaxation times vs the reciprocal temperature. The hollow symbols (○) denote the dielectric data, the filled symbols (●) are the results obtained by quasielastic light scattering

Here a_T is a shift factor ($a_T = \langle\tau\rangle/\langle\tau\rangle_{\text{Reference}}$). For the constants C_1 and C_2 values of $C_1 = 14.8$ and $C_2 = 55.9$ K are found using $T_0 = T_g$ as reference temperature. Determining the dielectrically measured molecular relaxation the WLF-parameter at the same reference temperature leads to $C_1 = 11.8$ and $C_2 = 67.9$ K. Thus for the activation parameter B in the Vogel-Fulcher-Taman equation $B = 2.3\ C_1 \cdot C_2$ values for the molecular and the collective relaxation are found which are in agreement within limits of a 5% error. Plotting the mean relaxation times [calculated using Eq. (17)] vs the inverse temperature (Fig. 14) shows (within the limits of experimental accuracy) a quantitative agreement, as was found already for the distribution parameter β. This indicates that the response obtained by the two different methods are caused by one and the same relaxation processes, although measured on a very different length scale. Since the WLF behavior has its origin in long-time collective motions it point towards the conclusion that the relevant relaxation times are the collective ones which govern the molecular dynamics. This is in quantitative agreement with the findings outlined in a recent publication by Ngai et al. [42].

In summary the comparison between the molecular and the collective dynamics in poly(methylphenylsiloxane) leads to the following conclusions:

1) The application of the Havriliak-Negami equation to the dielectric loss data $\varepsilon''(\omega)$ provides an adequate functional form to calculate the dipole-dipole correlation function $\phi(t)$.

2) The distribution parameters α and γ of the Havriliak-Negami equation varies significantly with temperature, whereas the β-parameter obtained from the Kohlrausch-Williams-Watts function remains constant.

3) It has been shown that the time correlation function for both dielectric relaxation and photon correlation spectroscopy respectively is found to be nearly identical.

4) For both methods the mean relaxation times were found to have the same temperature dependence and this leads to the conclusion that for poly(methylphenylsiloxane) the collective dynamics determines the molecular dynamics.

Molecular dynamics in linear and star-branched poly(cis-1,4-isoprene)

Linear and star-branched poly(cis-1,4-isoprene) was prepared by anionic polymerization in hexane at 303 K

Table 3. The fast contribution in the autocorrelation function $g(t)$ as fitted by a KWW representation [Eq. (15)] with fit parameters τ_{KWW} and β. $\langle\tau_{KWW}\rangle$ was calculated according to Eq. (17). The mean value of β is $\bar{\beta} = 0.50 \pm 0.03$

T/°C	τ_{KWW}/s	β	$\langle\tau_{KWW}\rangle/s$
−20	$2.6 \cdot 10^{-3}$	0.52	$4.8 \cdot 10^{-3}$
−15	$1.7 \cdot 10^{-4}$	0.48	$3.7 \cdot 10^{-4}$
−10	$2.4 \cdot 10^{-5}$	0.48	$5.2 \cdot 10^{-5}$
−4.5	$4.0 \cdot 10^{-6}$	0.52	$7.0 \cdot 10^{-6}$
0	$8.0 \cdot 10^{-7}$	0.46	$1.9 \cdot 10^{-6}$

Table 4. Weight-average molecular weights $\bar{M}_w$ ratio of the weight to number average molecular weight $\bar{M}_w/\bar{M}_n$ and microstructure of linear poly(cis-1,4-isoprene) samples

Code[a])	$10^{-3} \cdot M_w$	M_w/M_n	Microstructure, contents in percent of units		
			cis-1,4	trans-1,4	vinyl-3,4[b])
LPIP-05	5.1	1.03	78.8	17.9	3.3
LPIP-08	8.1	1.04	75.3	21.3	3.4
LPIP-14	13.8	1.02	79.1	15.2	5.7
LPIP-35	36.2	1.04	77.2	19.5	3.3
LPIP-60	58.0	1.07	76.9	18.6	4.5
LPIP-75	74.9	1.06	69.9	22.2	7.9
LPIP-130	130	1.05	77.4	18.1	4.5

[a]) nomenclature: LPIP-05, for example, corresponds to linear polyisoprene of molecular weight 5000;
[b]) 1-isopropenyethylene unit

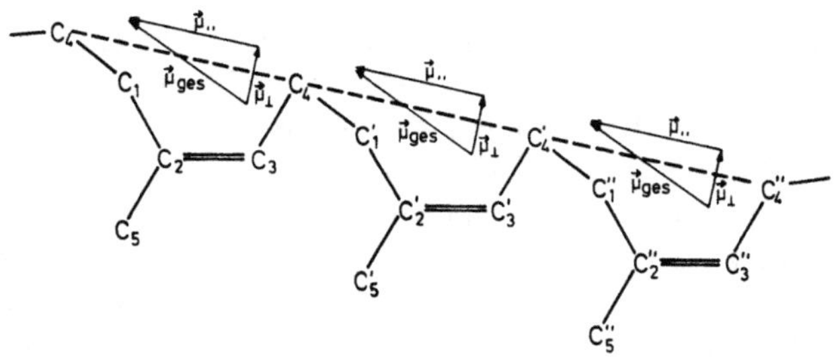

Fig. 15. Chemical structure of poly(cis-1,4-isoprene)

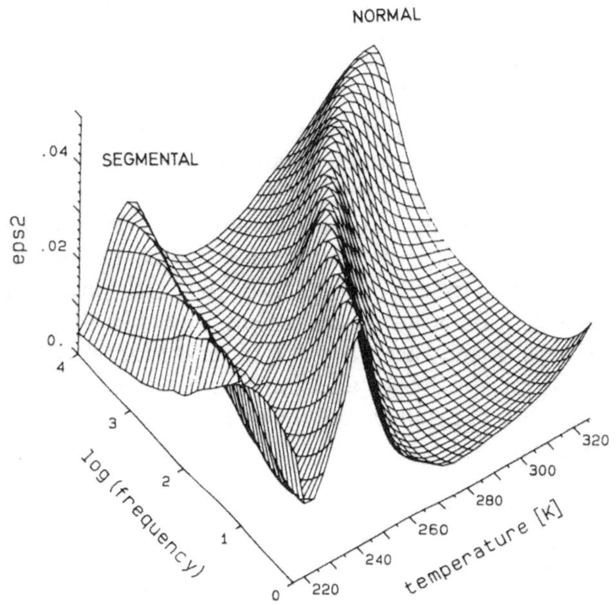

Fig. 16. Frequency and temperature dependence of the dielectric loss. Sample: linear poly(cis-1,4-isoprene), molecular weight: 13800, $\bar{M}_w/\bar{M}_n = 1.02$

Table 5. Weight-average molecular weights of the whole eight-arm, star molecule $\bar{M}_w$ (star) as well as of each arm $\bar{M}_w$ (arm) and microstructure of eight-arm poly(cis-1,4-isoprene) stars

Code[a])	$10^{-3}\cdot M_w$ *(star)*	M_w/M_n (arm)	Microstructure, contents in percent of units		
			cis-1,4	trans-1,4	vinyl-3,4[b])
SPIP-8-05	4.0	5.0	76.3	17.8	5.9
SPIP-8-08	6.0	7.5	75.4	18.5	6.1
SPIP-8-14	11.0	13.8	77.3	17.7	5.0
SPIP-8-16	12.4	15.5	80.4	15.2	4.1
SPIP-8-100	79.5	99.4	77.9	16.6	5.5

[a]) nomenclature; SPIP-8-05 e.g. corresponds to star polyisoprene with eight arms and a molecular weight of each arm of 5000; [b]) 1-isopropenylethylene unit

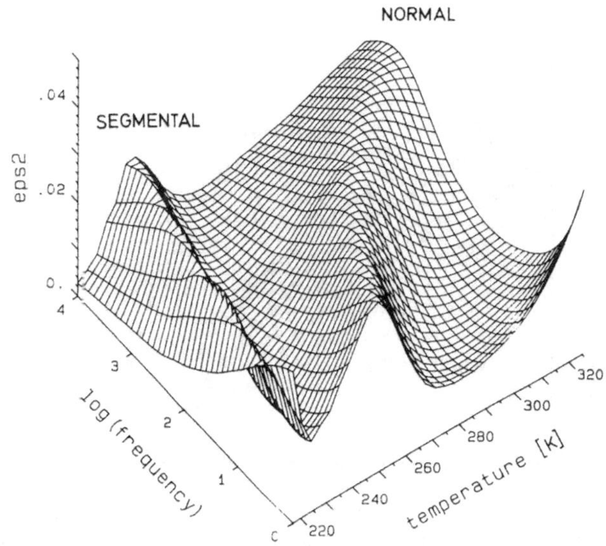

Fig. 17. Frequency and temperature dependence of the dielectric loss. Sample: Star-branched poly(cis-1,4-isoprene) with eight arms; arm molecular weight: 13800

with sec-butyllithium as initiator to produce linear polymers with narrow molecular weight distribution. Weight and number-average molecular weights ($\bar{M}_w$ and $\bar{M}_n$) of the stars and the star arms were determined by light scattering and osmometry, respectively. Size exclusion chromatography (SEC) was used to monitor the removal (via fractionation) of unlinked arms from the stars. For the linear poly(cis-1,4-isoprene) permeation chromotography (GPC) was employed to determine the weight- and number-average molecular weights. The microstructure of linear and star-shaped poly(cis-1,4-isoprene) was analyzed using ^{13}C-NMR. According to Sato et al. the signals are assigned to the cis-1,4, trans-1,4 and 3,4-vinyl units based on the resonance lines due to the methyl carbon atom [43]. The results are listed in Tables 4 and 5.

Due to the lack of symmetry in its chemical structure [Fig. 15] poly(cis-1,4-isoprene) has non-zero components of the dipole moment both perpendicular and parallel to the chain axis. Thus two dielectric relaxation processes, a segmental and a normal mode process, are present [6–10]; they are well-separated on the frequency and temperature scale (Fig. 16). The relaxation process around 220 K originates from a local segmental motion perpendicular to the main axis, while the second relaxation process at higher temperatures is assigned to motions of the dipole components parallel to the chain contour (normal mode). For star-branched poly(cis-1,4-isoprene) similar relaxation processes are found (Fig. 17). From the loss maxima, the relaxation rate of the segmental and the normal mode

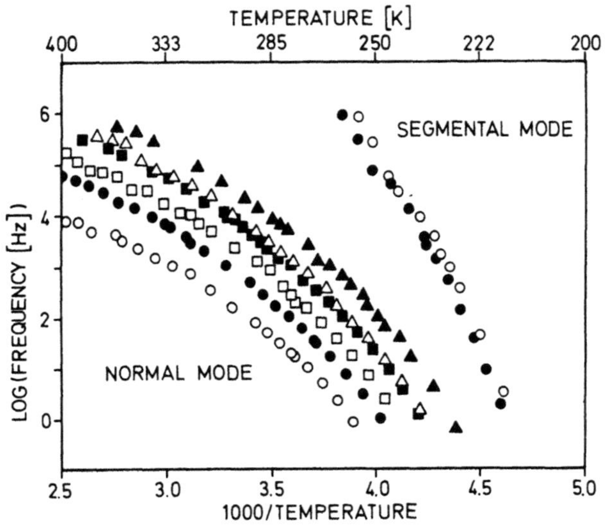

Fig. 18. WLF-plot for the segmental and the normal mode of the eight-arm star-branched and the linear poly(cis-1,4-isoprene). Because of graphical reasons, the segmental mode is only plotted for SPIP-8-14 and LPIP-14. (△): SPIP-8-05; (□): SPIP-8-08; (○): SPIP-8-14; (▲): LPIP-05; (■): LPIP-08; (●) LPIP-14

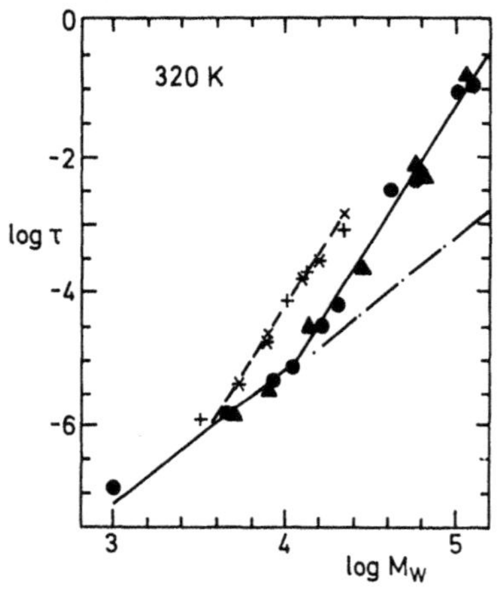

Fig. 19. Dependence of the mean relaxation time τ on the weight-average molecular weight $\bar{M}_w$, resp. arm-molecular weight for linear, resp. star-branched poly(cis-1,4-isoprene), measured at a temperature of 320 K. (▲): data reported by Adachi and Kotaka on linear poly(cis-1,4-isoprene); (●): data reported by Boese et al. on linear poly(cis-1,4-isoprene); star-branched poly(cis-1,4-isoprene) with eight-arms (*), with 12 arms (×), and with 18-arms (+)

can be determined and plotted against the reciprocal temperature. For the segmental mode, a temperature behavior is found which is typical for an α-relaxation process and can be described by the empirical WLF-equation. The normal mode processes show a strong dependence on the molecular weight or arm molecular weight, resp. (Fig. 18): i) both polymer structures exhibit nearly the same segmental mode process, ii) the normal modes have a similar temperature dependence; however, iii) the normal modes of the 8-arm star polymer are shifted to lower frequencies.

Plotting the relaxation times of the normal process vs the molecular weight or the molecular weight of the arm at a fixed temperature of 320 K (Fig. 19), the linear polymer with molecular weights above the critical molecular weight M_c ($M_c = 10000$ for poly(cis-1,4-isoprene) follows the power law with an exponent of 3.7 ± 0.2, which is characteristic for entangled macromolecules [44]. Below M_c, the dielectric relaxation time corresponds to the molecular weight dependence predicted by the Rouse theory [45]. For the eight-arm polyisoprene stars, the deduced molecular weight dependence proves that even at a molecular weight of the arm of $M_w = 5000$ the dynamics of the system is determined by entanglements. For the number of arms, *no* influence on the normal modes is found in the range of the arm molecular weight which was studied. It is very interesting that by plotting the relaxation time vs two-times the arm molecular weight, a quantitative agreement – within the limits of experimental accuracy – is observed for the normal mode of the linear and the star-branched polymer (Fig. 20). A detailed analysis of this finding is in progress [46].

Acknowledgements

The authors thank B. Momper and L. J. Fetters for synthesizing the samples.

References

1. Hedvig P (1977) Dielectric spectroscopy of polymers. Adam Hilger Ltd Bristol
2. Jonscher AK (1983) Dielectric Relaxation in solids. Chelsea Dielectric Press, London
3. McCrum NG, Read BE, Williams G (1967) Inelastic and Dielectric Effects in Polymer Solids. Wiley, New York
4. MacDonald JR (1987) Impedance Spectroscopy. John Wiley, New York
5. Ferry JD (1980) Viscoelastic Properties of Polymers, 3rd Ed. J. Wiley, New York
6. Adachi K, Kotaka T (1983) Macromolecules 16:1936
7. Adachi K, Kotaka T (1984) Macromolecules 17:120
8. Adachi K, Kotaka T (1985) Macromolecules 18:466
9. Boese D (1987) Diploma Thesis, Universität Mainz
10. Boese D, Kremer F, Fetters LJ (1988) Makromol Chem Rapid Commun 9:367
11. Pugh J, Ryan T (1979) IEE Conf on Dielectric Material Measurements and Applications 177:404
12. Elsden CT, Ley AJ (1969) Automatica 5:51
13. Rzpecka MA, Stuchly SS (1975) IEEE Transactions on Instrumentation and Measurement Vol. IM-24, No 1, 27
14. Stuchly MA, Stuchly SS (1980) IEEE Transactions on Instrumentation and Measurement Vol. IM-29, No 3, 176

15. Momper B (1986) Diploma Thesis, University of Mainz
16. Kow C, Morton M, Fetters LJ, Hadjichristidis M (1982) Rubber Chem Technol 55:245
17. Hadjichristidis N, Guyot A, Fetters LJ (1978) Macromolecules 11:668
18. Mark JE (1968) J Chem Phys 49:1393
19. Mark JE, Ko JH (1975) J Polym Sci, Polym Phys Ed 13:2221
20. Boese D, Momper B, Meier G, Kremer F, Hagenah J-U, Fischer EW (1989) accepted for Macromolecules
21. Havriliak S, Negami S (1967) Polymer 8:101
22. Williams G, Watts DC (1977) Chem Soc Ref 7:89
23. Williams G, Watts DC (1970) Trans Faraday Soc 66:80
24. Williams G, Watts DC, Dev SB, North AM (1971) Trans Faraday Soc 67:1323
25. Cook M, Watts DC, Williams GAM (1970) Trans Faraday Soc 66:2503
26. Patterson GD (1983) Adv Polym Sci 48:125
27. Hagenah J-U, Meier G, Fytas G, Fischer EW (1987) Polymer J 19:441
28. Berne BJ, Pecora R (1976) Dynamic light scattering. Wiley & Sons, New York
29. Meier G, Hagenah J-U, Wang CH, Fytas G, Fischer EW (1987) Polymer 28:1640
30. Mittag U (1983) Diploma Thesis. Universität Bielefeld
31. Fytas G, Wang CH, Lilge D, Dorfmüller T (1981) J Chem Phys 75:4247
32. Hagenah JU (1988) Thesis, Universität Mainz
33. Meier G, Fytas G, Dorfmüller T (1984) Macromolecules 17:957
34. Fytas G, Wang CH, Fischer EW, Mehler K (1986) J Polym Sci, Polym Phys Ed 24:1859
35. Alig I, Stieber F, Wartewig S, Fytas G (1988) Polymer 29:975
36. Adachi K, Okzaki H, Kotaka T (1985) Macromolecules 18:1687
37. Adachi K, Kotaka T (1987) Macromolecules 20:2018
38. Johari GP (1986) Polymer 27:866
39. Baur ME, Stockmayer WH (1965) J Chem Phys 43:4319
40. Momper B, Meier G, Fytas G, to be published
41. Boese D, Kremer F, in preparation
42. Ngai KL, Mashimo S, Fytas G (1988) Macromolecules 21:3030
43. Sato H, Ono A, Tanaka Y (1977) Polymer 18:580
44. De Gennes P-G (1979) Scaling Concepts in Polymer Physics. Cornell University Press, New York
45. Rouse PE (1953) J Chem Phys 21:1272
46. Boese D, Kremer F, Fetters LJ (1989) Submitted to Macromolecules

Received April 10, 1989;
accepted June 16, 1989

Authors' address:

Dr. F. Kremer
Max-Planck-Institut für Polymerforschung
Postfach 3148
6500 Mainz, FRG

Discussion

GOTLIB:
Do you have made some comparison between the intensities of "normal" and "segmental" motions and the squared dipoles moments for transversal and parallel (to the main chain) components? There should be definite correlations.

KREMER:
Such a comparison has not been done yet. It would require a determination of the Kirkwood-Fröhlich correlation factor which can be made only with some uncertainty. But inspite of that we will try to carry out the comparison you suggested.

GOTLIB:
It is surprising that the molecular weight dependence ($\sim M^{3.5}$) of the normal process in the branched polymer is the same or similar as for the nonbranched one (with the same $Mw = Mw$ of the arm of the item). Doe it coincide with the theoretical predictions? The mobility of the branched polymer in the melt may be different from that of nonbranched.

KREMER:
For the entanglement regime there does – to my knowledge – exist a theory which could describe our experimental results.

BURCHARD:
The properties of star-branched macromolecules can be generated from those of the linear chains by simple combinatorics. There occurs one contribution from the f arms and another from the linear chain with a weight of $f(f-1)/2$. Did you find an effect from the two-arm contribution?

KREMER:
If we plot in one diagram the relaxation rate of the normal mode of the linear chains versus the molecular weight and for the star-like molecules versus *two* times the arm molecular weight (viewgraph is presented) then we find agreement within experimental error. This is surprising especially in view of the experimental finding that the number of arms – we measured 6-, 12-, and 18-arm molecules – has nearly no effect on the relaxation rate of the normal modes.

PISSIS:
Your experimental results fit very well to the empirical Havriliak-Negami equation and the parameters of this equation can be determined. Are these parameters related to the structure of the polymer, i.e. can you draw any conclusion from the values to the parameters?

KREMER:
Not to my knowledge.

Comment SCHRADER:
There exists a recently developed approach by A. Schönhals and E. Schlosser, as shown in the paper of Prof. Hänsel which relates the two shape parameters of the relaxation curve to local and large scale interactions respectively.

Plastic deformation and mobility in glassy polymers

E. Oleynik

Institute of Chemical Physics, USSR Academy of Sciences, Moscow, USSR

Abstract: Deformation of several linear glassy polymers such as: a-PS, a-PMMA, a-PET, PC and epoxy-aromatic amine networks was investigated by deformation calorimetry, DSC, IR-Fourier spectroscopy and irreversible deformation recovery at heating. All polymers deformed in the glassy state have shown number of common features. In all cases irreversible deformation is followed by marked sample internal energy increase which reaches up to 20–50% of total mechanical input. This energy excess stored in a quite local plastic shear defects (PSD) which carry with them local irreversible deformation. The formation and growth of these PSD is the rate controlling step of macroscopic plasticity process in polymeric glasses. During of process development some PSD are converted due to defect interaction to new local structures with conformational changes of macromolecular backbone.

Due to defect's nucleation and transformation processes in glassy polymer exists steady-state concentration of plastic defects. The scheme of glassy polymer plasticity proposed which includes continuous repeating of PSD nucleation and termination. Molecular motions responsible for PSD development is discussed. It is proposed that many new kinds of motions which are not existing in unstressed polymer chains appears under sample loading. This motions are responsible for PSD behaviour in polymer glasses. Some structural features and properties of PSD are discussed.

Key words: Plastic deformation; plasticity; glassy polymers; plastic defects

1. Introduction

Plastic deformation is one of important mass transfer processes in solids and the type of species which are responsible for the process as well as the nature of molecular/atomic motions governing the shape changes of samples under mechanical stress have a fundamental importance.

The nature of crystal plasticity is well understood. It is quite clear now that the process consists of a nucleation, multiplication and growth of dislocations, thermodynamic and kinetic behaviour of which has been broadly documented in literature [1, 2].

The situation is not the same for amorphous solids. Many of them, especially glassy polymers, may show a pronounced plasticity but the mechanism of the process, the nature of plastic nuclei did not reach yet deep physical understanding [3].

Glassy polymers are intrinsically non-brittle materials [4] and this property as well as their rigidity, low density and good processability make them very attractive engineering materials. However, macroscopically many polymeric glasses show brittle behaviour due to low resistance to crack propagation. And the acceleration of the rate of plasticity in a competition with the rate of fracture events is one of important factors of polymeric materials improvement. The way to success is passing through plasticity mechanism which in that sense is one of key points in modern material science.

The role of brownian motions (α-, β- and γ-types) in glassy polymers as well as the role and rate of conformational isomerism in the chains under mechanical stresses are remained still unclear despite of great number of experimental results [5, 6] and theoretical models [7–9].

This paper deals with some new results which were mainly received in the author's laboratory at the Institute of Chemical Physics in Moscow [10–12]. These results, in our opinion make us to introduce the

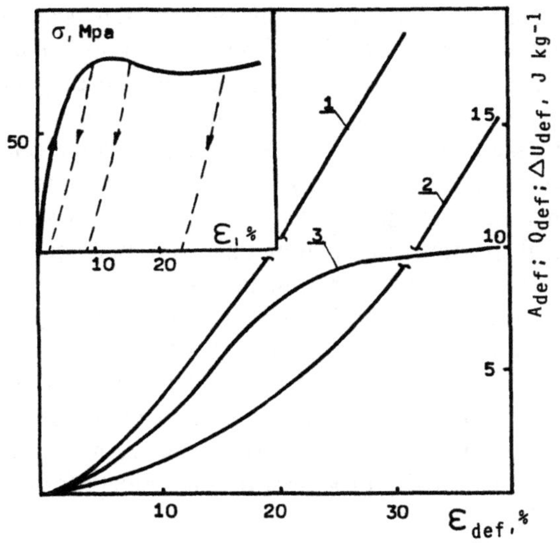

Fig. 1a. The mechanical work of deformation $A_{def} = \int_0^{\varepsilon_{def}} \sigma d\varepsilon$ (1), heat of deformation Q_{def} (2) and internal energy $\Delta U_{def} = A_{def} - Q_{def}$ (3) as a function of ε_{def} (uniaxial compression) for a-PS. All quantities were measured in loading-unloading cycles (shown in upper left part of the figure. ↑ – loading, ↓ – unloading) in a deformation calorimeter [12]. $T_{def} = 30\,°C$. $\dot{\varepsilon}_{load} = \dot{\varepsilon}_{unload} = 10^{-3}\,s^{-1}$

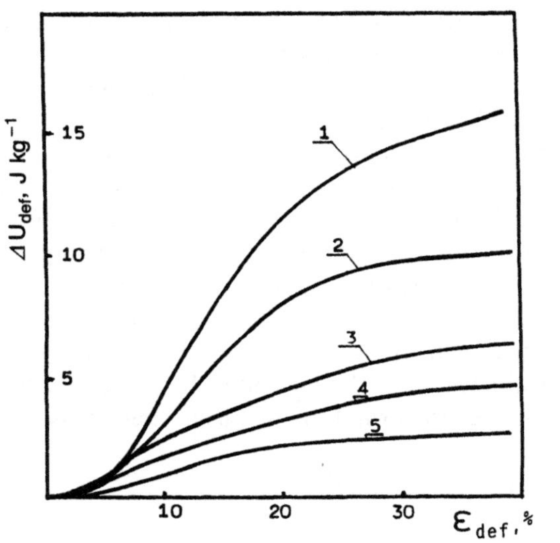

Fig. 1b. Internal energy ΔU_{def} stored in plastically deformed glassy polymers as a function of deformation. 1 – Epoxy-amine network polymer; 2 – a-PS; 3 – PC; 4 – PMMA; 5 – a-PET. Experimental conditions are the same as in Fig. 1a

new, up to some extend, picture of glassy polymers plasticity and the types of motions involved in the process.

II. New experimental results

During last years we have investigated the deformation behaviour of a number of quite typical glassy high molecular weight polymers: a-PS, PC, a-PMMA, a-PET, based on linear macromolecules and epoxy-aromatic amine networks (EAN) [10].

Here we shall summarize the most important results which bring us to the deformation mechanism.

It is quite important that all the following results were qualitatively the same for all abovementioned polymers. This situation shows that the plastic behaviour of chemically different polymeric glasses has many common features and one may discuss it in very general terms.

The main experimental results were the following:

1. We performed measurements of the heat of deformation [11, 12] Q_{def} (by deformation calorimetry) as a function of ε_{def} and ε_{irr} (see Fig. 1) in loading-unloading cycles. The comparison of Q_{def} with the total mechanical work A_{def} gives the internal energy excess ΔU_{def} stored in plastically deformed (i.e., storing ε_{irr}) samples.

The results $\Delta U_{def} = A_{def} - Q_{def}$ are shown in Fig. 1a, b. It is important to note that the exothermicity of deformation in all cases was quite large (50–85% of A_{def}) which is difficult to explain only by the dissipation of mechanical energy which is typical for glassy polymers due to internal friction. The other feature of calorimetric results is quite a high level of ΔU_{def} which is from 15% up to 50% of the total mechanical work. For metals this value is only few percents of total A_{def} [13, 14].

These results show that plastic deformation of polymeric glasses is accompanied by structural changes which create the high level of sample internal energy, i.e. a highly excited plastic state.

2. These new structures have several important features. They survive in deformed polymers as long as ε_{irr} exists in the samples. The internal energy excess ΔU_{def} and ε_{irr} may be removed from the deformed sample by heating. The shape and sizes of initial samples are completely recovered during this heating (at least in the deformation range $\varepsilon_{def} \leqslant 50\%$). In Fig. 2 are shown typical DSC traces for initial, deformed and heated after deformation samples.

The most prominent result here is the appearance of exothermic C_p anomalies for the plastically deformed samples.

The C_p-anomalies show that deformed glassy polymers release some amount of enthalpy ΔH_{def} during heating. Undoubtedly this enthalpy liberation appears due to transformation of excited deformed

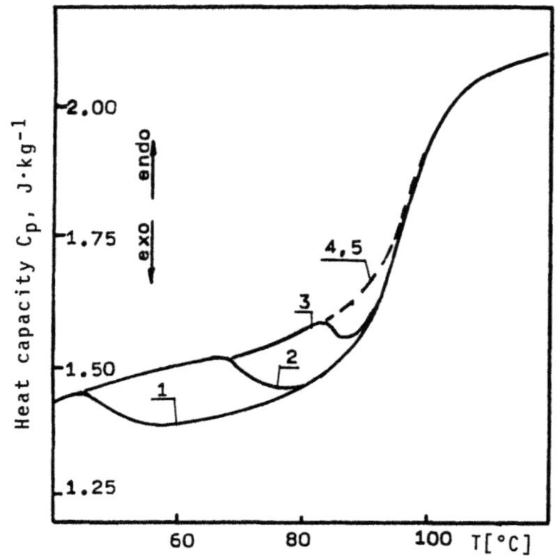

Fig. 2a. DSC curves for a-PS, deformed at different temperatures T_{def}. Heating rate 20° min^{-1}. Deformation conditions: uniaxial compression, $\dot{\varepsilon} = 10^{-3}\,s^{-1}$, $\varepsilon_{def} = 13\%$; T_{def}: *1* – 25°, *2* – 65°, *3* – 80°. 4 – second DSC runs for samples 1, 2, 3; 5 – initial undeformed sample

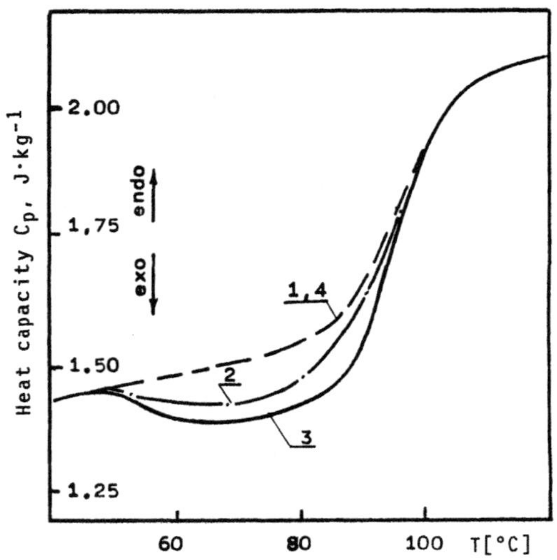

Fig. 2b. DSC-curves for a-PS samples. Heating rate 20° min^{-1}. 1. Initial undeformed sample. 2. a-PS film deformed before DSC measurements by uniaxial tension. $\varepsilon_{def} = 10\%$; $\varepsilon_{irr} = 5\%$; $T_{def} = 25°$. 3. a-PS deformed by unidirectional compression. $\varepsilon_{def} = 20\%$, $\varepsilon_{irr} = 15\%$, $T_{def} = 25°$. 4. Sample subjected before DSC measurements to hydrostatic pressure; $P = 10$ kbar at 23°.

state of the sample with increasing temperature.

The amount of ΔH_{def} is close to ΔU_{def} measured by deformation calorimetry however always smaller then ΔU_{def} [12].

The plastic state of all glassy samples completely disappear after heating. DSC curves of the samples with erased memory about the plastic state exactly correspond to those of initial samples. This means that heating of deformed samples recovers the original structure of glassy polymers or, in other words, the plastic state of glassy polymers is the metastable one.

3. Most of ΔH_{def} (ΔU_{def}) disappear at heating of deformed samples up to $T < T_g$, i.e. in the glassy state. In the glass transition region only few percents of stored ΔH_{def} escaped. This means that the structural defects which have been formed in polymer during its loading can exist only in the glassy state.

From these results it becomes clear that α-type segmental motions of macromolecules are not responsible for deformation defects structural recovery and consequently for the processes of the defects formation under stress.

4. C_p curves shown in Fig. 2b give some additional information about the plastic defects. The hydrostatic compression of the polymers up to $P = 10$ kbars (23°C) does not change C_p-curves.

The glassy polymers deformed by unidirectional compression and tension have shown the same exothermic DSC-traces at the same level of ε_{def}.

These results clearly show that the hydrostatic component of the stress tensor does not influence seriously the process of plastic state formation. This brings us to the conclusion that the process of forming plastic defects in glassy polymers is not very sensitive to free volume and hence plastic defects nucleation under load should be mainly the shear process.

Direct measurements of density changes of plastically deformed polymers [15, 16] have shown that the volume changes are very small and sample densification takes place in most cases. Heating of deformed a-PS [11] did not show of specific volume change at least $\delta V < 4 \cdot 10^{-4}\,cm^3 g^{-1}$ during complete annealing of deformation defects by heating [11]. Simple estimations of the enthalpy ΔH_{def} needed for glassy polymer to form 1 m^3 of plastic defects is

$$\Delta H_{def}/\Delta V_{def} > 10^{10}\ \mathrm{J\ m^3}$$

which is an order of magnitude higher than the typical value for vacancy (hole) formation in organic substances ($\approx 1 \div 2 \cdot 10^9\ \mathrm{J \cdot m^{-3}}$) [11]. This means that the plastic shear defects (SD) store higher amounts of local energy per unit volume than normal vacancies in organic crystals. Thus one may conclude again that the quite high level of internal energy ΔU_{def} storage is not the consequence of free volume changes.

5. Important information about the plastic state of polymeric glasses was received by measuring the recovery of the deformation ε_{irr} upon heating. In all

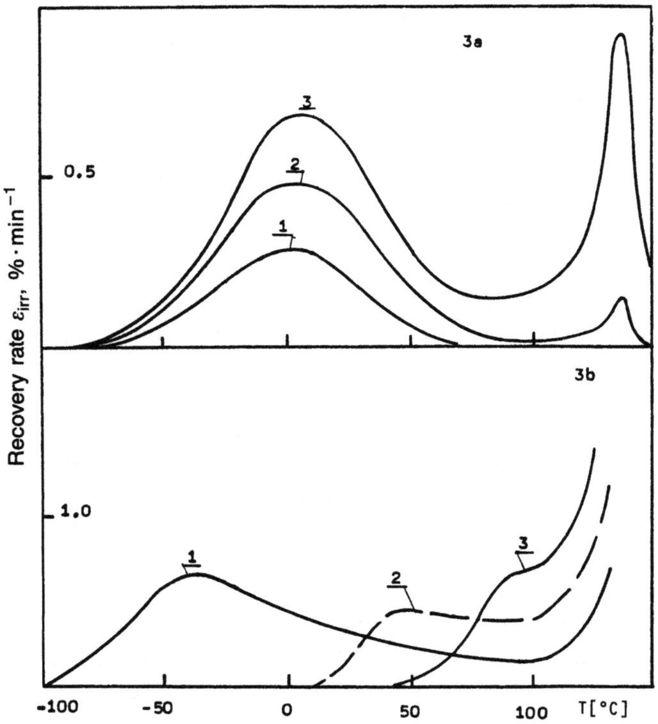

Fig. 3a. Deformation ε_{irr} recovery curves at heating (heating rate $\dot{T} = 20$ grad·min^{-1}) of deformed epoxy-amine glassy networks. 1, 2, 3 – recovery rates for polymer, deformed (compression) at $T_{def} = -85°$ up to ε_{irr}: *1* – 12%, *2* – 21%, *3* – 25%. Deformation performed under creep conditions

Fig. 3b. Recovery curves for epoxy-amine glassy networks. *1.* $T_{def} = -100°$, $\varepsilon_{irr} = 18\%$; *2.* $T_{def} = 20°$, $\varepsilon_{irr} = 9\%$; *3.* $T_{def} = 50°$; $\varepsilon_{irr} = 11\%$. Heating rate $\sim 20°$ min^{-1}. Deformation was performed at active loading by compression

our cases ε_{irr} was recovered completely. In Fig. 3a, b are given typical curves of recovery rate $\dot{\varepsilon}_{irr}$ during heating of unloaded plastically deformed samples. These curves show:

– at comparatively small ε_{irr} complete shape and sizes recovery of deformed samples is accomplished at the temperatures below T_g, i.e. in the glassy state. Deformation recovery starts at lower temperatures if T_{def} is lower (Fig. 3b). And the recovery starts usually at T_{def};

– the increase of ε_{def} and ε_{irr} make the deformation recovery curves more complicated. A second recovery peak appears at T_g. This high temperature recovery (HTR) peak becomes more and more intense with the increase of ε_{irr}. The low temperature recovery (LTR) peak remains almost constant (by the shape and area) with ε_{irr} when HTR peaks starts to grow;

– it is important to note that for all cases investigated the HTR peaks never appear without the LTR peaks if polymers were deformed in the glassy state.

The situation was completely different if the same deformation was carried out in the rubbery state (Fig. 4). A polymer deformed at $T_{def} > T_g$ and then cooled down to the glassy state at the condition $\varepsilon_{def} = $ const (keeping the crossheads unmoved) shows the deformation recovery curve only with the HTR peak.

This result immediately points out that the mechanisms of deformation of the same polymer are different in the glassy and rubbery states.

Once more we would like to emphasize that recovery curves of polymers deformed in the glassy state may show only LTR peak (at comparatively low ε_{irr}) but never only HTR peak;

– IR-analysis and $\dot{\varepsilon}_{irr}$ recovery curves for polymers deformed in the rubbery state clearly show that the HTR peak (its appearance with deformation and disappearance with heating) reflects the changes of conformational states of polymeric coils at loading and deformation recovery. However, this is not the case for LTR peaks.

Figure 5 shows combined curves of C_p and $\dot{\varepsilon}_{irr}$ changes through the heating of deformed samples.

It is clear that the exothermic C_p-anomalies and LTR peaks have the same origin. The LTR peaks disappear from the deformed plastic state of glasses

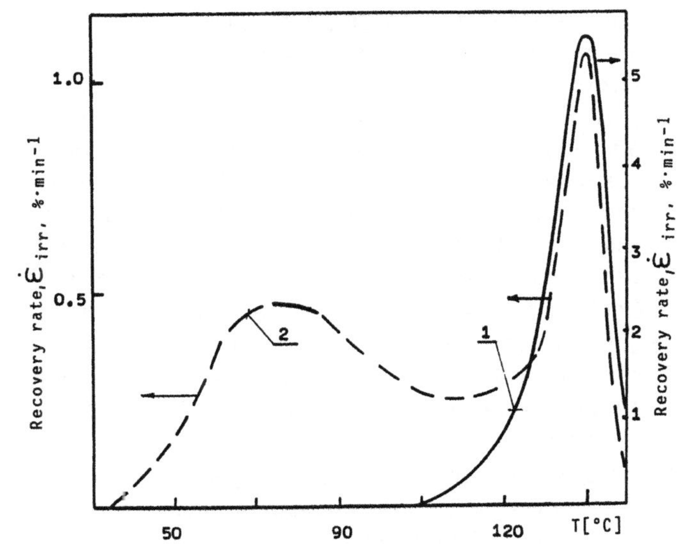

Fig. 4. Plastic deformation (ε_{irr}) recovery curves for epoxy-amine polymer ($T_g = 145°$) deformed at different T_{def} (uniaxial active compression). Heating rate $20°$·min^{-1}. *1.* $T_{def} = 160°$, $\varepsilon_{def} = 25\%$ with following cooling ($\dot{T} \approx 10°$ min^{-1}) to *RT* at the conditions $\varepsilon_{def} = $ const. *2.* The same sample deformed at $T_{def} = 30°$, $\varepsilon_{def} = 20\%$

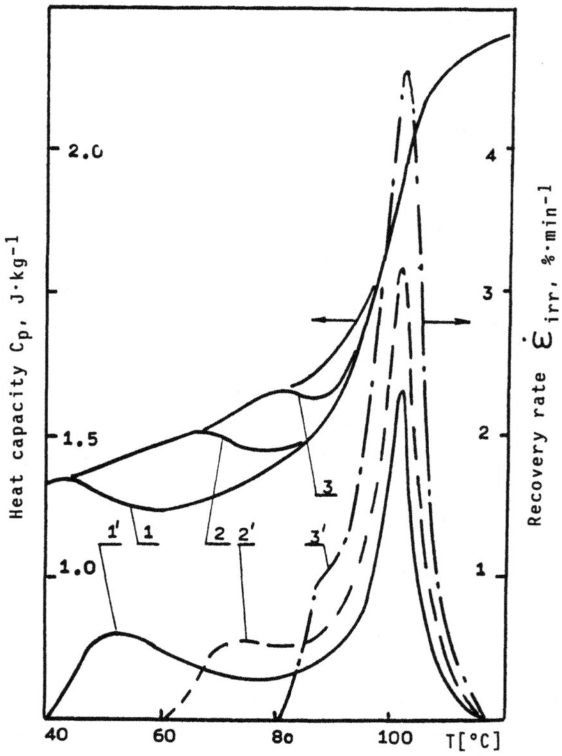

Fig. 5. Combined DSC and ε_{irr} recovery curves of a-PS deformed ($\varepsilon_{def} = 13\%$) at different T_{def} (uniaxial compression). T_{def}: 1,1'–23°, 2,2'–65°, 3,3'–80°

together with the excess of stored internal energy ΔU_{def} (ΔH_{def}). This means that LTR peaks are physically connected with plastic shear defects which appear in the glassy samples during its loading and recovery from their metastable states at heating.

IR-Fourier analysis of the polymers investigated during their deformation and temperature recovery has shown that the processes of SD formation are not connected with conformational changes in polymeric chains. In Fig. 6 are shown the results for the a-PET. Conformationally sensitive bands $\nu_G = 898\ \text{cm}^{-1}$ (rocking modes) and $\nu_T = 845\ \text{cm}^{-1}$ [17] behave in a quite different manner. At comparatively low ε_{irr} when shear defects exist in the deformed samples there are no changes of the bands intensity in vibrational spectrum (substraction spectra in Fig. 6).

Detectable conformational changes appear only at the moment when the stress-strain curve passes the macroscopic yield point (σ_y, ε_y) and HTR peak appear in $\dot{\varepsilon}_{irr}$ recovery curves.

Heating of samples being deformed in the glassy state and revealing conformational changes in the IR experiment (points A_T and A_G in fig. 6) in the temperature interval $T_{def} < T < T_g$ erases the LTR peak but does not change the intensities of the conformationally sensitive bands.

The same results have been obtained also for a-PS samples.

Exothermic anomalies of C_p for plastically deformed glassy samples also support the nonconformational nature of deformation shear defects.

Usually DSC-curves of undeformed glassy polymers show pronounced endothermic C_p-jumps at heating in T_g and T_β regions. The endotherms indicate that any new molecular motions appearing at polymer heating need some additional energy to be unfrozen.

It is not the case for deformed glassy samples. In all experiments we have found exothermic events at DSC-curves. This means that annealing events of plastic defects are performed together with the energy release and do not need unfreezing of any remarkable molecular motions.

It is evident that the disappearance of deformation defects at heating occurs with the aid of thermal activation and with quite a broad distribution of activation energies ΔE. However, there are no any pronounced endothermic effects before an exothermic effect appears. This should happen if some conformational or other type of molecular mobility control the rate of SD annealing process.

The whole set of experimental data allows to formulate the mechanism of glassy polymers plasticity and we shall try to do this in the next section.

III. Plasticity mechanism for polymeric glasses

From the above mentioned results it is clear that plastic deformation of polymeric glasses start with the formation of plastic shear defects characterized by ΔU_{def} and ε_{irr} in total.

Part of the defects should be quite local in scale because they easily appear and disappear at very low temperatures (Fig. 3, 2).

The measurements of activation volumes of the defects in the process of their annealing (Part IV) support this point of view.

Each SD is contributes some local shear deformation ε_{irr}^{loc} and local energy excess ΔU_{def}^{loc}. Macroscopic sample deformation ε_{irr} is the result of all these local events.

What is the origin of ΔU_{def}^{loc} and macroscopic ΔU_{def}? It is clear that the origin cannot belong to the conformational changes in polymeric coils. Simple estimation [11, 18] have shown that the amount of ΔU_{def}^{max} (Fig. 1) stored at deformed a-PS and epoxy-

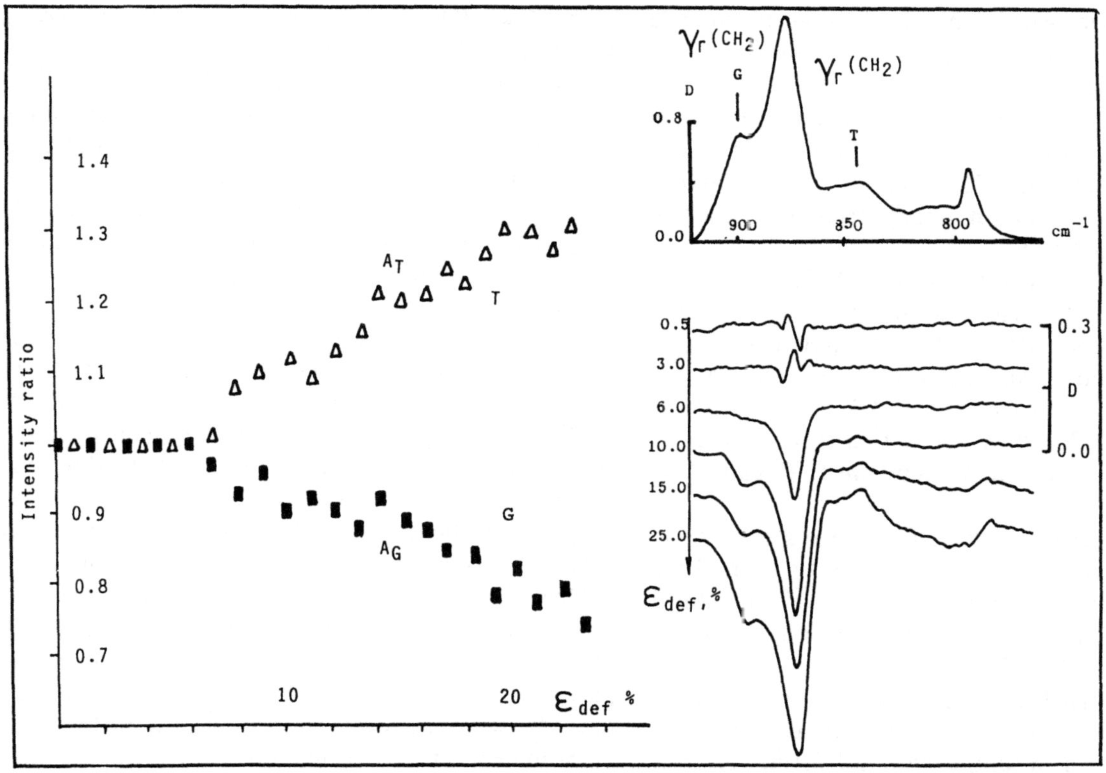

Fig. 6. Intensity D changes of trans (T) and gauche (G) sensitive bands of a-PET with deformation. Right side – subtraction spectra of a-PET for different levels of deformations (tension). $T_{def} = 23\,°$. A_T and A_G – see text

polymer may be explained quantitatively if the concentrational changes of conformations reach $[\Delta C] \geqslant 10\%$ which is easily detectable by IR-measurements. But we could not detect any changes in IR-spectra (Fig. 6) for PET and PS samples, i.e., the ΔU_{def}^{loc} and ΔU_{def} have no conformational origin.

The absence of pronounced volume changes in deformed glasses shows that creation of the new free volume cannot be the reason for the high experimental values of ΔU_{def}.

However, the reason for ΔU_{def}^{loc} still exists. The formation of each SD in a stressed solid leads to some deformation inconsistency in a local region of the sample where the SD was formed. It creates the ε_{irr}^{loc} in the core of a plastic shear defect but the deformation of this type is absent in the surrounding "matrix". This situation immediately creates elastic stresses around SD in the "matrix".

The picture is quite close to the formation of stress field around dislocations in crystals [1, 8]. We believe that this kind of stresses are responsible for ΔU_{def} appearance in the plastic state of polymeric glasses.

Evidently ΔU_{def}^{loc} and ΔU_{def} will exist in a sample during the whole time of SD existence. The curves of ΔU_{def} change with ε_{def} (Fig. 1) reflects the total picture of SD development in polymers.

At small deformation the nucleation of plastic SD is the main process. One can see that the maximum rate of ΔU_{def} rise is close to yield point of the investigated polymers.

If the general picture of SD nucleation is correct the yield tooth and yield drop may be referred to SD multiplication and growth just the same as for dislocation behaviour in crystals.

At the later stages of deformation all ΔU_{def} curves come to saturation or close to that. There are at least two possibilities of that kind of behavior. One of them is the development of a steady-state proceeding of the deformation process. In that case rates of nucleation and decomposition of plastic SD at given $\dot{\varepsilon}$ become equal at some ε_{def}. Another possibility is the real saturation. If any SD can be nucleated only in some specific regions of a polymeric glass the growth of ΔU_{def} must stop, saturate due to exhausting of this regions. In other words, this means that all possible SD's for a given sample have formed under the applied stress and saturation occurs.

In reality, with the existing experimental data it is impossible yet to come to unequivocal choice from these two possibilities. However, we think that the steady-state deformation mechanism has slightly stronger support now.

There are several reasons for that. At first the stress-strain curves in the deformation region $\varepsilon_{def} > \varepsilon_y$ show steady-state behaviour for all glassy polymers [19].

Secondly, our measurements of ε_{irr} recovery at heating in the neck of PET ($\lambda = 4.5$) (the recovery of stretched films [20]) have shown that even in the neck the curves of $\dot{\varepsilon}_{irr}$ recovery show two-LTR and HTR peaks.

This result may be interpreted as the existence of the steady-state concentration of plastic SD's in the PET neck, i.e. at quite a high deformation level. As a third point we shall return for a moment to the behaviour of LTR and HTR peaks.

Note again that HTR peak at recovery curves never appears without its precursor-LTR peak. Even more: if one take a deformed sample with a deformation level $\varepsilon_{def} > \varepsilon_y$ (two deformation recovery peaks will appear at this sample upon heating above T_g) and erase out the LTR peak and ΔU_{def} by heating of the sample in the glassy state (up to $T < T_{def}$) one will obtaine as the result a sample only with the HTR peak. The next deformation of this sample will start again by the formation of plastic SD's which may be clearly seen from the appearance of the LTR peak at heating and growth of ΔU_{def}. The HTR peak will start to grow further only when the concentration of SD (the intensity of LTR peak) almost reaches the same value observed in the originally deformed sample.

The results are shown in Fig. 7. The growth of HTR peak in a sample exposed to the second deformation cycle after LTR peak erasing starts only when LTR peak again reaches its critical value.

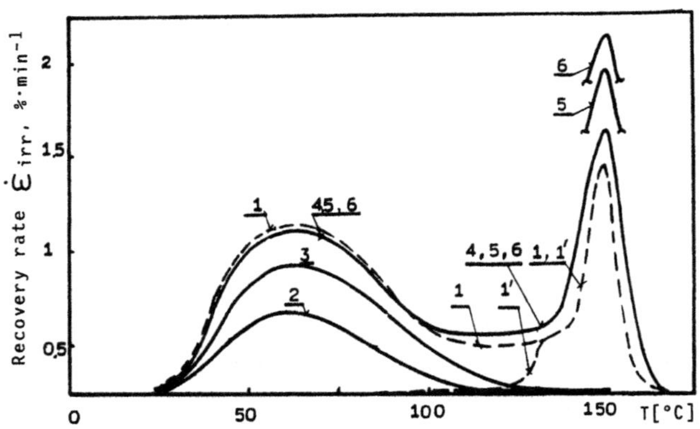

Fig. 7. Plastic deformation recovery curves for epoxy-amine glassy polymers subjected to consecutive loading (compression). $T_{def} = 23°$. *1* – $\varepsilon_{irr} = 10\%$. *1'* – sample *1* heated up to $T = 120°$. *2* – sample *1'*; deformed at $T_{def} = 23°$, ε_{def} (additional deformation) – 2%; *3,4,5,6* – sample *1'* deformed to additional ε_{def}: *3* – 5%; *4* – 10%; *5* – 14%; *6* – 18%

The last argument is the experimental values of Q_{def} [12, 18]. All the glassy polymers have shown quite a high exothermicity during their plastic deformation.

It is difficult to explain these high Q_{def} only be polymers internal friction because it is quite small outside of temperature regions of relaxation transitions.

High level of Q_{def} may be easily explained by the plastic SD decomposition during sample loading. SD decomposition should take place with enthalpy release (Fig. 2) and the decomposition may explain the high exothermicity of glassy polymers deformation.

Now the question arises: what is the process of plastic SD decomposition? This process must satisfy one very important demand: the decomposition of any plastic SD must lead to local energy release but also must conserve the local deformation ε_{irr}^{loc} in the sample volume where the SD had existed. Only in this case the sample may be deformed macroscopically up to high values of ε_{def} in the steady-state process.

Undoubtedly the plastically deformed state of the polymer is characterised by the new shape of polymeric coils and this shape change occurs due to an appearance of new backbone conformations in each coil.

This means that the process of shear defects decomposition finally lead to some conformational rearrangements.

The existing experiments do not yet give enough information about the nature of the transition "plastic SD" → "CED" (conformationally excited coil). Only one thing may be assumed now. The steady-state condition for deformation energy (Fig. 1) appear pronouncedly after the macroscopic yield point, when the concentration of plastic SD's in deformed polymer is quite high. In this case the transformations of plastic defects to new local coil conformations may take place due to defects interactions (collision or annihilation).

Of course the last proposal is only a speculation. However, it is important to say that any real steady-state plasticity mechanism for glassy polymers must include some transformations of plastic defects to new local chain conformations.

The previous discussion may be shortly summarized in the following deformation mechanism (Fig. 8).

The initial state of polymeric coil (the state "0") changes it internal energy under the applied stress in

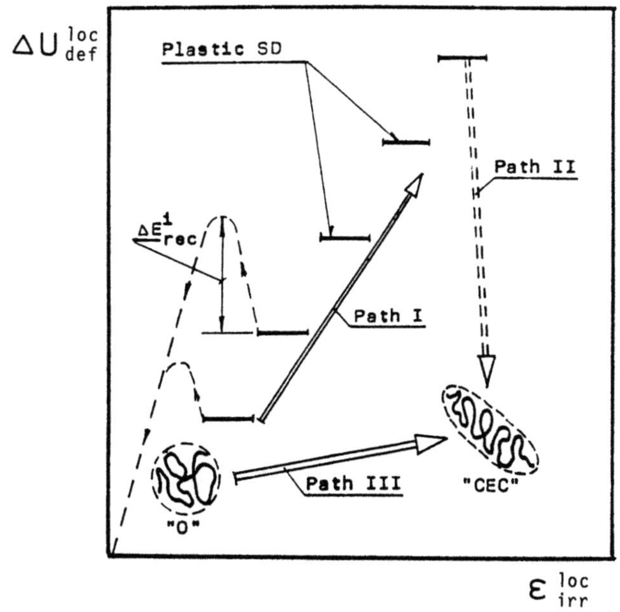

Fig. 8. The deformation mechanism. Path I – formation of plastic shear defects at loading of glassy polymers. Horizontal bars – plastic shear defects (SD's) with different internal energy and different activation energy ΔE^{i}_{rec} of their recovery to initial structure. Path II – the way of plastic SD decomposition. Path III – direct change of size and shape of polymer coil (or some fragment of a coil) with deformation due to conformational rearrangements in macromolecular backbone. Path III is the typical deformation path for the rubbery state of polymers. Glassy polymers can perform plastic deformation only through Pathes I and II. The steady-state deformation occurs when the rates of Path I and Path II become equal. "0' – initial polymeric coil (or its fragment) with the size, shape and local energy characteristics of the glassy state of polymer. "CEC" – conformationally excited coil (or its fragment) appearing in the polymer due to local plastic deformation

small local volumes of the glassy polymer. This happens through local plastic shear defects formation. The coils of the glassy polymer change their shape and size in this deformation stage but not due to conformational rearrangements.

The formation of plastic defects seriously increases the internal energy of the polymer. The energy ΔU_{def} increases to its steady-state value during sample deformation.

Some plastic defects cannot be developed further at given local stress and deformation and terminate its growth. At the same time new shear defects are created in the sample under applied stress and the rates of both processes – SD nucleation and termination equilibrate in the steady-state deformation region.

Schematically termination of the existing of plastic defects is shown as the Path II in Fig. 8.

Multiple repeating of this two processes in a polymeric glass results in its deformation.

The SD decomposition process does not bring probably any additional ε_{irr} to the sample (or bring a little) but creates a new form of polymeric coil, "CEC", the state of a conformationally excited coil. In this sense the Path II is not the deformation process itself.

It is clear that the first step (Path I) of plastic deformation process – the nucleation and growth of plastic defects – is a kinetically controlling step of the whole process.

The second stage, SD decomposition, is quite fast.

The last stage is probably sensitive to chain flexibility and this factor may be important for the levels of ΔU^{max}_{def} for different polymers.

Path III in Fig. 8 shows the typical way of rubbery polymer deformation.

It is clear that both deformation processes in the glassy and rubbery states of the same polymer have as a result the same shape of polymeric coils.

However, in the glassy state part of ε_{irr} exists in the form of plastic defects and this kind of deformation is absent in the rubbery state.

It seems that recent neutron scattering experiments [21, 22] are in agreement with our assumptions. It was found that deformed glassy samples have two different modes of plastic deformation. One is so called the affine one which manifests itself in our experiments in the form of HTR peaks. The second one is nonaffine which appears in our cases as LTR peaks.

IV. Molecular mobility and deformation of polymeric classes

Mass transport during plastic deformation must be performed by some atomic and molecular motions. The discussion of this may be organized around the two main questions:

– What is the role of characteristic brownian motions of α-, β- and γ-types well known for all glassy polymers? Many authors introduced α- and β-motions [7, 9] in a description of the yield behaviour of glassy polymers.

– Do some new modes of molecular mobility appear in polymeric glasses under external mechanical stress?

Of course, in this paper we shall try to find the answers first of all under the framework of the assumed mechanism of plasticity.

It is also evident that our discussion must be limited mainly to the process of plastic defect nucleation and annealing because the second step of plastic deformation is performed too fast in our usual experiments and we have no yet enough information to define the molecular motions responsible for this step.

Our experimental results permit us to draw some conclusions about the role of different types of molecular motions in glassy polymer plasticity.

Evidently, the segmental motions of α-type do not play an important role in the glassy polymer plasticity. This is clearly seen from DSC-curves of deformed samples.

Even the plastic SD decomposition process unlikely switches start up on the α-motions because the defect decomposition process is easily performed at very low deformation temperatures.

The more complicated question concernes the role of β-type motions. It seems to us that β-motions do not play the exclusive role in the plastic deformation process. There are several experimental findings supporting this view:

– The recovery of ε_{irr} at heating is easily performed at temperatures higher and lower then T_β.
– The plastic state of a polymer in the range $T_{def} > T_\beta$ does not show serious influence on the positions and intensity of γ- and β-motions (Fig. 9).
– It is well known that β-relaxations are well described by the Arrhenius law.

This is at variance with the following findings. We have measured the rates of isothermic relaxations for the ε_{irr} which is maintained through LTR-peaks at heating of deformed polymers. These experiments were performed by using a temperature-jump technique. After deforming the sample at a given T_{def} up to desirable ε_{def} and after unloading it was *moved* to higher T_{rel}-relaxation temperatures $T_{def} < T_{rel} < T_g$ and the kinetics of ε_{irr} recovery was measured at different T_{rel}.

It was found that the deformation recovery under these conditions was quite nonexponential and well described by fractional exponent (Kohlrausch, Williams-Watts-function)

$$\varepsilon = \varepsilon_0 \exp\left(-\frac{t}{\tau}\right)^{\beta} \quad \text{with } \beta = 0.43 \pm 0.04 \ .$$

So, if the thermal recovery of plastic deformation is supported by the β-motions it would be surprising to find this nonexponential relaxation law.

We performed the direct measurements of activation volumes for the motions which are responsible for the LTR-process in epoxy-amine networks [10]. We measured the rate of LTR process under hydrostatic

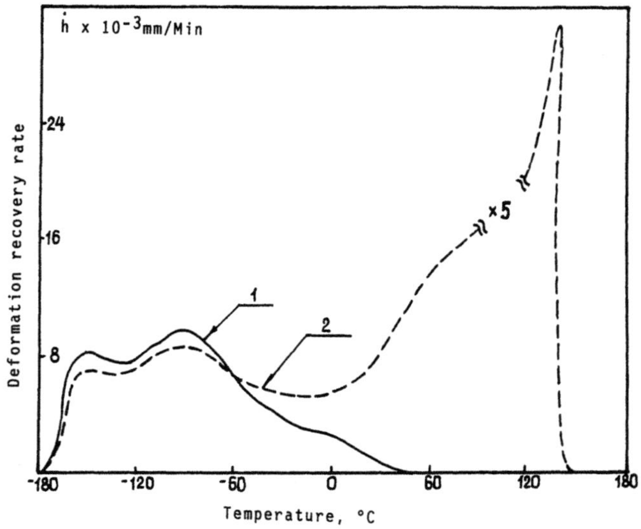

Fig. 9. Low temperature molecular mobility of undeformed (1) and plastically deformed (2) PC measured by thermo-stimulated creep experiments (bending). $\dot{h}$ – rate of deflection of bended sample at heating. Heating rate $\dot{T} = 10° \cdot \mathrm{min}^{-1}$. 1. The sample was loaded at 23° to $\varepsilon_{irr} = 20\%$ and cooled down to −165°. 2. The sample was deformed up to $\varepsilon_{irr} \approx 20\%$ at 23° and unloaded. Next procedure was the same as for 1. Two peaks at 70° and 140° temperatures are LTR and HTR peaks of PC plastic deformation recovery

pressures $P = 0.5 - 3.0$ kbar. These experiments have shown that the process of annealing of plastic shear defects is weakly sensitive to P, i.e. to free volume of polymer. It also supports the idea of the shear character of SD nucleation and annealing processes.

Measured activation volumes V_a in these experiments turned out to be $V_a = 55$ Å^3, i.e. quite small in the temperature interval 25°–85° ($T_g = 130°$).

It is well known that for local motions (including β-motions) in glassy polymers the geometrical volume of movable fragments is usually about $5 \div 6\ V_a$. This means that the volume of mobile fragment under the acting mechanical stress is $V = 250 - 300$ Å^3.

After the measurements of thermodynamics and kinetics of yield behaviour of several glassy polymers [23] Escaig with coworkers concluded that in these polymers there were two different modes of plasticity: one is the glide mode which operates at low deformation temperatures and the other is the diffusional one at higher temperatures [24].

In our experiments we did not find any change of plasticity mechanism for all the investigated polymers inside the entire region of the glassy state. We have found some change in activation parameters of the SD recovery process at heating of deformed PS about $T \approx 300$ K, which is close to the critical temperature

T_c of Escaig [25]. However we have found that the mechanism of plasticity through nucleation and growth of SD's operates above and below T_c.

The reasons for activation parameters change near T_c should be analysed more carefully in the future.

Thus our experiments have shown that motions of α-type do not play any role in glassy polymers plasticity. The β-motions and probably γ-motions also do not play any exclusive role in the deformation process. Moreover, it looks more appropriate that β- (and γ-) motions do not take part in the plastic deformation of glassy polymers completely but the last statement needs of course more experimental evidences.

However, the mass transport during deformation microevents needs some atomic motions which undoubtedly exist in the glass under load. Since plastic defects have a "sheared" structure, the motion responsible for their appearance must not be of usual "rotational type" motions which are usually observed as relaxations in individual polymeric chains due to their "axial symmetry".

New shear-type "motions" must exist in the very local region of the glass under stress and the displacement of some small portion of polymeric chain must not change the conformational state of this portion and its closest surroundings. This means that the displacement may occur only due to small changes of interatomic distances and rotational and bond angles of the chains. These displacements cannot be large.

From the high values of stored internal energy ΔU_{def} it becomes obvious that the elementary deformation events in glassy state of polymers are strongly mechanically activated.

The appearance of plastic SD's at very low T_{def} supports this point of view. One may assume that the motion responsible for the formation of SD's are also strongly mechanically activated and probably do not exist in unloaded polymeric glasses and chains.

The experimental finding of $T_{def} \approx T^0_{rec}$, where T^0_{rec} is the temperature where the C_p, ΔU_{def} and ε_{irr} recovery starts (Fig. 5) appears to be very important for understanding of glassy polymer plasticity.

The decrease of T_{def} brings the deformed and unloaded samples to the higher and higher levels of deformation energy ΔU_{def} storage (Fig. 2, 5). The result becomes clear if one accepts the assumption that T_{def} is always some critical temperature for deformed polymers.

The picture may be realized in the following terms. Under the load in the glassy state many plastic defects appear with a broad distribution of their ΔU^{loc}_{def}, ε^{loc}_{irr} and activation energies ΔE of their annealing to the unexcited structure. The unloading immediately divides all the plastic defects in two parts: with $\Delta E \lesssim KT_{def}$ and $\Delta E \gtrsim KT_{def}$. The SD's with $\Delta E \lesssim KT_{def}$ disappear from the unloaded sample quite quickly and the sample stores only SD's with comparatively high values of ΔE and ΔU^{loc}_{def}.

The defects with $\Delta E \approx KT_{def}$ represent the component of macroscopic ε_{irr} which disappears from unloaded sample at T_{def} first of all (so called fast relaxing component of ε_{irr}, well-known for all solid polymers). The situation immediately make any T_{def} a critical temperature from the point of view of SD relaxation events.

Up to which low T_{def} may this picture exist? If the lowest T_{def} where plastic SD's still may appear is T_δ, the Gillman's idea [26] about mechanical dipoles formation becomes valid.

The answer to all these and many other questions needs more experiments and new theoretical models.

V. Conclusions

The plastic deformation of the glassy state of matter including glassy polymers is a very controversial but interesting and important subject.

Modern interest of material science to glassy polymers is due to their growing practical importance.

It may be that the proposed simple scheme of glassy polymer plasticity is oversimplified and even not the real one, however it is impossible now to neglect the existence of plastic SD's in stressed polymers. The experiments have shown that the formation of SD's is the first step of plasticity and all known events like crazing, cracking, deformation softening and so on are performed in the polymer state with quite high concentration of SD's and even through those SD's.

For example, we have found SD formation in impact loaded glassy polymers. We found the interesting SD behavior in flexure and fatigue tests and the sensitivity of SD behaviour to sample prehistory and thermal treatment.

It seems to us that further investigations of kinetic and thermodynamic properties of the plastic defects may bring a new fruitful stream to the topic of glassy polymer plasticity.

Acknowledgements

The author brings sincere thanks to his coworkers at the Institute of Chemical Physics: Dr's S. Rudnev, O. Salamatina, S. Nazarenko, I. Magafurov, S. Shenogin without whose experiments this presentation could not be possible, and to doctors:: V. Bulatov, V. Topolkaraev, A. Gusev for many critical but friendly discussions of the topic, and to V. Asatryan, G. Vorobjeva for the help in manuscript preparation.

References

1. Friedel J (1964) Dislocations. Pergamon Press, Oxford
2. Krausz AS, Eyring H (1979) Deformation kinetics. Wiley-Intersc, NY
3. Argon AS (1980) In: Uhlmann DR, Kreidl NJ (eds) Glass: Science and Technology. Academic Press, NY. V 5, pp 79–132
4. Kelly A (1973) Strong Solids. Clarendon Press, Oxford
5. Bowden PB (1973) In: Howard R (ed) The Physics of Glassy Polymers. Appl Sci Publ, London, pp 279–339
6. Bauwens J-C (1982) In: Escaig B, G'Sell C (eds) Plastic Deformation of Amorphous and Semicrystalline Materials. Physique Publ., Les Ulis, pp. 175–186
7. Robertson RE (1966) J Chem Phys 44:3950
8. Bauwens-Crowet C (1973) J Mat Sci 8:968
9. Argon AS (1974) Phil Mag 28:839
10. Oleynik E (1986) Adv Polym Sci 80:49–99
11. Oleynik E (1987) Polym J 19, 1, 105–117
12. Salamatina O, Nazarenko S, Rudnev S, Oleynik E (1988) Mechanica Kompoz Mater (USSR), 6, 979–984
13. Chen HS (1976) Appl Phys Lett 29, 6, 328–332
14. Binder G, Müller FH (1961) Kolloid Z Z Polymere 157:89
15. Wang T et al (1982) Polymer 23:1407–1409
16. Kitagawa M, Yoneyama T (1988) J Polym Sci. Polym Lett 26, N 4, pp 207–212
17. Hutchison I, Ward I (1980) Polymer 21, 1, pp 55–65
18. Nazarenko S (1988) Theses of the Institute of Chemical Physics, Moscow, USSR
19. G'Sell C (1985) In: McQueen H, Bailon J, Dikson J, Akben M (eds) Strength of Metals and Alloys. Pergamon Press, N 4, 3, pp 1943–1982
20. Sheiko S, Salamatina O, Rudnev S, Oleynik E (1989) Visokomolek Soed (USSR), in press
21. Dettenmaier M, Maconnachie A, Higgins J, Kausch H, Nquyen T (1985) In: Deformation, Yield and Fracture of Polymers. Forth Int Conf Cambridge, England, pp 5.1–5.5
22. Lefebvre J, Escaig B, Picot C (1982) Polymer 23:1751
23. Escaig B, Lefebvre J (1978) Revue de Physique Applique, 13, 6, pp 285–292
24. Escaig B (1982) In: Escaig B, G'Sell C (eds) Plastic Deformation of Amorphous and Semi-crystalline Materials, Les Ulis, Les Editions de Physique, pp 187–225
25. Escaig B (1984) Polym Eng Sci 24, 10, pp 737–749
26. Gilman J (1973) J Appl Phys 44, 5, pp 2233–2236

Received June 6, 1989;
accepted June 22, 1989

Author's address:

E. Oleynik
Institute of Chemical Physics
USSR Academy of Sciences
Kosygin Str. 4
Moscow 117977, USSR

Discussion

KRÜGER:

Do you have any evidence that the shear deformation induced defects affect the quasistatic itself in its properties? Depends e.g. the transition temperature T_g on the concentration and the energy of the shear induced defects?

OLEYNIK:

If the dependence of T_g on shear defect concentration exists, it is quite small. During heating of plastically defored polymers practically all shear defects disappear below T_g and the whole polymer sample comes back to macroscopical original structure which should show and shows in experiments the T_g of the undeformed polymer.

But we had found quite a strong influence of the plastic state of polymer – the state with high concentration of shear defects – on sub-T_g transitions, γ and β for example.

In the plastic state we could not find any evidences for γ and β relaxations in proper temperature intervals.

Instead of that during heating of deformed samples and measuring of its size change the brad spectrum of molecular motion appear in temperature region from deformation temperature sometimes up to T_g. There were no any evidences for the loss peaks at T_g and T_β temperatures.

We believe this behaviour is the manifestation of new types of molecular motions which appear in polymer during its loading and corresponding shear defects formation.

Dynamic light scattering from polymer solutions

W. Burchard and W. Richtering

Institute of Macromolecular Chemistry, University of Freiburg, Freiburg, FRG

Abstract: Dynamic light scattering results from density and concentration fluctuations which by modern techniques can be recorded in a time domain from 50 ns to about 100 s. These fluctuations are correlated and can be evaluated in terms of various relaxation processes via a time correlation function which is built up by an autocorrelator. The dominating process at small scattering angles is a translational diffusion of the centers of mass of the particles. For flexible particles this translational motion is superimposed of internal modes of motion which show significant contributions at larger q-values. The conditions for a successfull determination of the first relaxation times is critically discussed, and it is demonstrated that only the first relaxation process may be separated from the translational motion on the one side and the higher relaxation modes on the other. For $qR_g > 5$ the translational diffusion no longer has influence, and the time correlation function is now determined by the internal relaxation processes alone. The strongly angular dependent correlation functions form one common curve of characteristic shape when Γt is used as a scaled delay time instead of the real delay time t. These shape functions differ for flexible chains and rod-like particles and allow to decide whether Rouse or Zimm-type relaxations with hydrodynamic interactions are effective. Experiments give clear indication for Zimm processes.

In semidilute solutions a cooperative diffusion coefficient can be measured. It is composed of a thermodynamic part, the osmotic modulus, and a frictional part. The osmotic modulus may be measured by static light scattering, and hence the frictional contribution can be studied separately. The osmotic modulus was found to scale with a parameter $X = A_2 M_w c$, but both the diffusion coefficient and the frictional coefficient do not follow such scaling laws.

At even higher concentrations a very slow diffusion motion occurs in almost all systems studied so far. Results from a low molecular weight micellar system and from a thermotropic polymer in solution are presented. The former gives evidence for clusters with a high orientation in the cluster and a fractal dimension of $D = 1$, the latter for densely packed, non-oriented clusters with $D = 2.85$.

Key words: Dynamic light scattering; translational diffusion; relaxation modes; semidilute solutions; slow motion

1. Introduction

Scattering techniques have been long recognized as one of the most important methods for determining *equilibrium* structures. Of course, only fairly large structures can be detected if visible light is used as scattering source, and this restricts the light-sacttering technique to macromolecules or colloids or to large supramolecular structures. More recently it has become possible to record the scattering intensity within time intervals as short as 50 ns. Very strong intensity fluctuations are now detected. These fluctuations are not fully independent of each other but are correlated, and the correlation results from motions within the system [1].

The motions which give rise to light-scattering fluctuations are of different origin depending on whether condensed one-component systems or solutions are considered. In one-component systems (melts, simple liquids or simple gasses) it is a *density* fluctuation which gives rise to the intensity fluctuations while in solution or two-component systems *concentration* fluctuations contribute in addition. This difference between solutions and bulk materials is of remarkable influence and will be reviewed in a separate section.

For polymer solutions, the topic of this paper, the dynamic behavior is one step more complex, even for systems which can be considered as isotropic, i.e., randomly oriented. At very dilute solutions the motion of individual molecules is observed and this is only weakly modified by the thermodynamic and hydrodynamic interaction among the macromolecules [4]. Even then macromolecules differ characteristically from common colloids, since the latter are mostly rigid, whereas the former have at least some internal flexibility. Internal motions cause likewise concentration fluctuations, and although these are locally restricted and bound to individual molecules these segment fluctuations contribute to the dynamic light-scattering function.

At concentrated solutions, in particular when considerable entanglement takes place the intermolecular interactions start to play the dominant role, and the whole system has now to be considered as a very complex many-body problem.

This contribution is divided into three parts. In the first part some basic relationships are reviewed which will demonstrate the principle differences between solutions and one-component systems. The second part deals with the dynamics of individual molecules in dilute solutions, and the third part gives initial not completely understood results from semi-dilute solutions.

2. Some principles of dynamic light scattering in solution

2.1 Osmotic modulus

Conventional or static light scattering (LS) arises from fluctuations in materials around the mean values. As already mentioned, for pure one-component liquids it is the local fluctuation of the density which contributes, but in solutions fluctuations of the concentration contribute in addition [2]. Their influence is by orders of magnitude larger than those resulting from the density fluctuations. Thus up to concentrations of about 40% the density fluctuations can be safely set equal to those in the pure solvent. These density fluctuations arise from the isothermal compressibility. This is for a polymeric melt not much different from that of a low molecular weight liquid. This allows to define a Rayleigh ratio of the scattering intensity $i(\theta)$ to the primary beam intensity I as

$$R_\theta = r^2[i(\theta) - i_{\mathrm{sol}}]/I = KcRT(\partial c/\partial \pi)_T \ . \qquad (1)$$

In this equation $i(\theta)$ is the scattering intensity of the solution at the scattering angle θ and i_{sol} that of the solvent, and r the distance of the detector from the scattering center. K is an optical contrast factor, which can be measured separately, and $RT(\partial c/\partial \pi)_T$ is the isothermal osmotic compressibility. Equation (1) is evidently equivalent to a relationship that is obtained for LS from bulk materials, i.e., melts or simple liquids, but the isothermal compressibility is replaced here by the osmotic compressibility. It is indeed a standard result of statistical thermodynamics that the mean square concentration fluctuations can be expressed by the osmotic compressibility of the system [3].

2.2 Cooperative diffusion

For rigid particles the time dependent intensity fluctuations in solution for a system at thermodynamic equilibrium result from Brownian motion, and according to Einstein's famous theory this Brownian motion is related to the translational diffusion of the particles. Macromolecules are, however, often highly flexible, and a number of internal modes of motion will be excited while the particle is pushed around in the Brownian process. These internal motions contribute to the dynamic LS function in a characteristically different manner than the translational motion as will be shown in detail in section 3.2.

Neglecting for a moment this internal mobility as an additional effect we may consider the concentration fluctuations as being solely due to motions of the centers of mass. These fluctuations are not at random but are constrained by the *interparticle* interaction. Irreversible thermodynamics prove that the mutual or cooperative diffusion coefficient at a finite concentration is given by [4]

$$D_c = [kT/f(c)][(M/RT)(\partial \pi/\partial c)] \ . \qquad (2)$$

Thus this translational diffusion coefficient contains two parts, i.e., 1) a hydrodynamic one which is determined by the *hydrodynamic* interaction among the various macromolecules, and 2) a *thermodynamic* interaction term, (the part in the second squared brackets), which for convenience will be called in the following the osmotic modulus. Its magnitude is determined by the repulsion between the particles.

It is worth having a closer look at the common static LS equation at zero scattering angle which is mostly given as

$$Kc/R_{\theta=0} = (1/RT)(\partial \pi/\partial c)_T \qquad (3\mathrm{a})$$

or if normalized to a dimensionless quantity [5]

$$MKc/R_{\theta=0} = M/M_{\mathrm{app}} = (M/RT)(\partial\pi/\partial c)_T \ . \tag{3b}$$

These relationships make evident how important it is to measure both static and dynamic scattering behavior simultaneously. A number of theories [25] are known dealing with the concentration dependence of D_c but these do not correctly describe the experimental findings as will be shown at the end of this paper. Since also theories on the osmotic modulus are available, (and this modulus can be measured independently by static light scattering), it has become possible to check whether the observed deviations from theory are based in the thermodynamic or in the hydrodynamic part.

2.3 General properties of the time correlation function

The function of interest in dynamic LS is the field time correlation function (TCF) which is defined as [1]

$$g_1(t) = |\langle E^*(0)E(t)\rangle| / |\langle E^*(0)E(0)\rangle| \ . \tag{4}$$

The function is obtained from the experimentally directly accessible scattering intensity TCF via the Siegert relationship [6]

$$\langle i(0)i(t)\rangle = G_2(t) = A + Bg_1^2(t) \tag{5}$$

where A and B are time-independent constants and t is the delay time. The Siegert relationship is strictly correct for highly dilute solutions but is usually also applied to semidilute solutions and gels where the validity may be questionable.

The denominator in Eq. (4) describes the static LS and is in general given by

$$R_\theta = KcS(q,c) \tag{6}$$

where $q = (4\pi/\lambda)\sin\theta/2$ is the magnitude of the scattering vector and $S(q,c)$ is the static structure factor of the solution. Thus Eq. (4) may be rewritten as

$$g_1(t) = S(q,c,t)/S(q,c) \tag{7}$$

or more explicitly

$$g_1(t) = \frac{\sum\sum \langle \exp[iqr_{jk}(t)]\rangle_{G(r,c,t)}}{\sum\sum \langle \exp(iqr_{jk})\rangle_{G(r,c)}} \tag{8}$$

with the abbreviations

$$r_{jk}(t) = |r_j(0) - r_k(t)| \ , \quad r_{jk} = |r_j(0) - r_k(0)| \ .$$

The sums run over all pairs of scattering elements, or segments, within the scattering volume.

The averages in the numerator and the denominator differ essentially. In the former the average has to be taken over the space and time distribution $G(r,c,t)$, whereas in the static structure factor (the denominator) the equilibrium pair distribution function $G(r,c)$ is needed. Clearly, one has

$$G(r,c) = G(r,c,t=0) \ . \tag{9}$$

A general solution of Eq. (8) appears hopeless. However, the following approximation can be made in static LS [5]

$$S(q,c) = P(q)S'(q,c) \tag{10}$$

where $P(q)$ is the particle scattering factor of the individual, spatially well separated macromolecules, and $S'(q,c)$ is often assumed to be a structure factor that arises from the arrangement of the centers of mass in the space. This interpretation is correct, however, only for non-interpenetrating particles, such as strongly repulsive interacting colloids. For penetrating architectures the physical meaning of $S'(q,c)$ is more complex and generally known only for the single contact approximation by Zimm [7]

$$S(q,c) = P(q)M_w[1 - A_2M_wcP(q)] \ , \tag{11}$$

which holds only for rather low concentrations.

Whatever the interpretation of $S'(q,c)$ may be the approximation of Eq. (10) suggests a similar one for the dynamic structure factor

$$g_1(t) = P(q,t)S'(q,c,t)/|P(q)S'(q,c)| \ ; \tag{12}$$

$P(q,t)$ describes now the change in the form of the particle as a function of time due to thermal fluctuations. This approximation facilitates calculation and discussion of the various dynamic effects.

3. Behavior of some special structures

3.1 Hard spheres

It is instructive to start the detailed consideration with monodisperse homogeneous and hard spheres which is an example for a structure of highest sym-

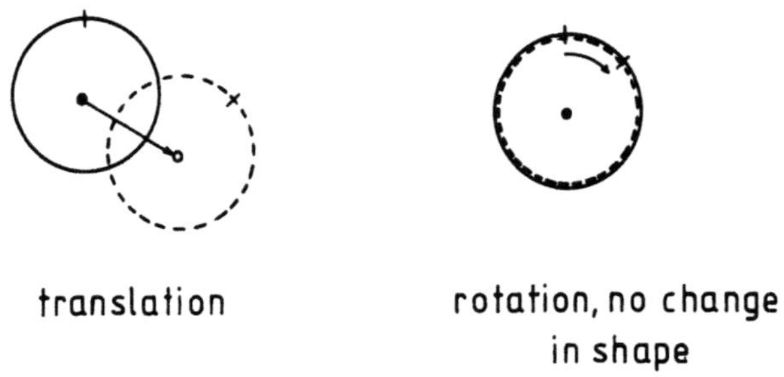

Fig. 1. Types of motion of a hard sphere in a laboratory system (left) and a coordinate system that is connected with the center of mass (right). Only translational and rotational motions occur which do not lead to a change in the shape

metry. Obviously there is no change of the shape in time; the only dynamic effect which could affect the particle scattering factor is a rotation around the center of mass, but the final result remains indistinguishable from the initial state (see Fig. 1). Thus we have $P(q,t) = P(q)$ and

$$g_1(t) = S'(q,c,t)/S'(q,c) \ . \tag{13}$$

The consideration may now be confined to sufficiently low concentrations such that no defined phase correlations occur among the various particles. Under such conditions the static structure factor is q-independent, and only the motion of the centers of mass of the individual spheres has to be taken into account. From Eq. (8) we now find [1]

$$g_1(t) = \int_0^\infty G(r,c,t) \exp(iqr) r^2 dr \tag{14}$$

which is the Fourier transform of the space-time pair correlation function. This function is known to describe a simple diffusion process

$$\partial G(r,c,t)/\partial t = D_c \Delta G(r,c,t) \tag{15}$$

where Δ represents the Laplace operator. This equation is easily solved by applying the mentioned Fourier transform which reduces the diffusion equation to a common differential equation in time. Its solution is [1]

$$g_1(t) = \exp(-D_c q^2 t) \ . \tag{16}$$

The TCF decays as a single exponential with a decay constant

$$\Gamma = D_c q^2 \ . \tag{17}$$

For a system of polydisperse spheres one has

$$g_1(q,t) = \sum_j a_j(q) \exp(-D_j q^2 t) \tag{18}$$

with

$$a_j(q) = c_j M_j P_j(q) / \sum_j c_j M_j P_j(q) \tag{19}$$

where the subscript j refers to particles with molar mass M_j and particle scattering factor $P_j(q)$ which are present in the system with a concentration c_j. Note: For a system of spheres with a size distribution the TCF becomes angular dependent if the particle scattering factors $P_j(q)$ deviate from unity.

In particular for a two-component system one has

$$g_1(q,t) = a_1(q) \exp(-D_1 q^2 t) + a_2(q) \exp(-D_2 q^2 t) \ . \tag{20}$$

This example has been written explicitly to demonstrate that each decay constant in the exponentials contains the factor q^2 and furthermore, the particle scattering factors of the individual particles no longer cancel out. Of course the uniform result is recovered if only one component is present. The physical meaning of the q^2 dependence of the decay constant is evident, because D is proportional to the mean square displacement and consequently, the diffusion process must correspond in the Fourier space to an *inverse* squared length q^2.

Equation (18) represents a Laplace transform, and inversion of this function by suitable programs (CONTIN [8], MAXENT [9]) will give the a_j distribution, which at $q = 0$ becomes the size distribution. Often the individual fraction represent particles whose dimensions are small compared to the wavelength. Then the particle scattering factors are $P_j(q) \cong 1$, and no extrapolation to zero angle is required.

Figure 2 gives an example [10] of inversion of the measured TCF $g_1(t)$ in comparison to the Schulz-Flory distribution ($M_w/M_n = 2$). Experience shows that the inversion can be performed successfully for unimodal molar mass distributions; for bimodal distributions, however, less reliable results are obtained.

3.2 Internal modes

The next question is what can be expected for flexible macromolecules. This question was answered 24

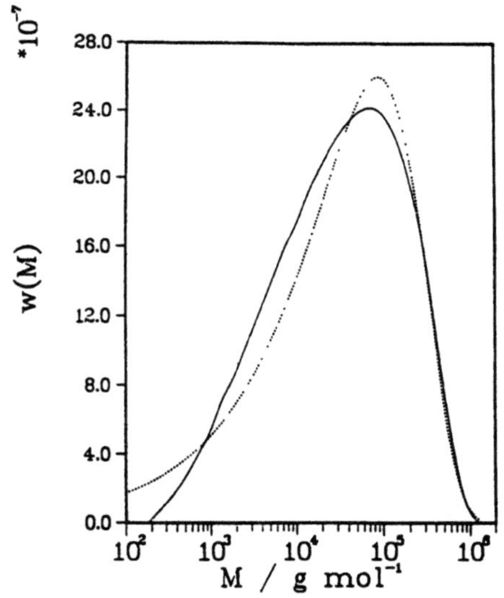

Fig. 2. Molar mass distribution (solid line) of a PTMAC (Poly-2-trimethylammonium-ethyl-methacrylate chloride) in 0.1 N NaCl aqueous solution which was obtained by Laplace inversion of a measured TCF. $M_w = 373000$. The dotted line represents the Schulz-Flory distribution [10]

years ago by Pecora [11] in a pioneering work. Figure 3 demonstrates that there will be a translational motion of the center of mass and a change of the shape with time because of internal segment fluctuations. These become clearly noticeable for an observer travelling with the center of mass. Thus, what can be done in theory is a separation of the center of mass motion from those relative to the center of mass. The center of mass motion follows the same diffusion equation as that for hard spheres. For the internal motions with respect to the center of mass a normal mode analysis has to be made. Pecora [11] started with the spring-bead model and solved the set of equations of motions for the various normal modes by means of the Wang-Uhlenbeck equation [12] for a

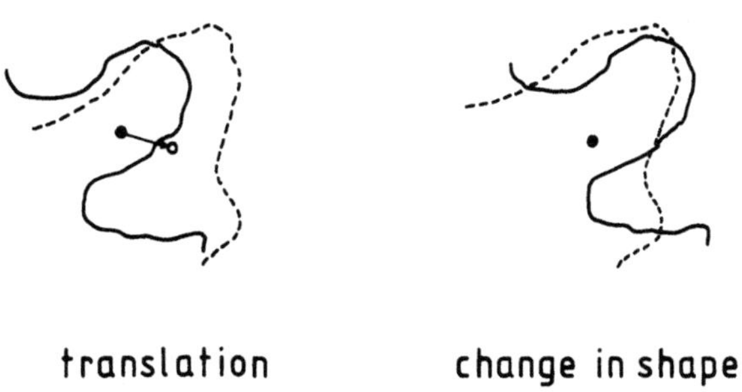

Fig. 3. Types of motion of a flexible chain molecule in a laboratory system (left) and a moving coordinate system (right). Internal modes of motion lead to a fluctuation in the shape of the molecule

Brownian particle in a harmonian (oscillating) potential which is the potential for a particle that is bound to another object via a Hookian spring. The solution for the kth individual relaxation processes is

$$g_k(q,t) = \left[\frac{3}{2\pi\langle\mu_k^2\rangle[1-\exp(2t/(\tau_k)]}\right]^{3/2} \cdot\exp\left[-\frac{3}{2\langle\mu_k^2\rangle} \cdot\frac{\{\mu_k(t)-\mu_k(0)\exp(-t/\tau_k)\}^2}{1-\exp(2t/\tau_k)}\right] \quad (21)$$

and for the total TCF

$$q_1(q,t) = \frac{\exp(-Dq^2t)}{P(q)} \times\sum_j^n\sum_j^n\prod_{k=1}^n \exp\left[-\frac{q^2\langle\mu_k^2\rangle}{6}\cdot\{Q_{jk}^2+Q_{ik}^2-2Q_{ik}Q_{jk}\exp(-t/\tau)\}\right] \quad (22)$$

where the Q_{jk} are the matrix elements in the similar transformation of the jth coordinate in the system

$$x_j = \sum_k^n Q_{jk}\mu_k$$

and μ_k are the corresponding coordinates of the normal mode system [13, 14].

Comparison with the translational motion Eq. (8) shows that these relaxation processes no longer follow a simple exponential which makes the interpretation of Eq. (22) very complex. However, if the delay time is sufficiently short the Wang-Uhlenbeck function reduces to a simple exponential [11, 12].

Pecora has used this approximation and obtained within this restricted time domain the general solution

$$g_1(t) = \exp(-D_q^2t) \times[P_0(q)+\sum_j P_j(q)\exp(-b_jt/\tau_1)] \quad (23)$$

where $\tau_1 = 2R_g/\pi^2D$ is the slowest internal relaxation time. The prefactors b_j depend on the polymer architecture and on the type of relaxation processes, i.e., whether these are coupled with hydrodynamic interaction (Zimm model) [13] or not (Rouse model) [14]. The simplest solution is obtained for a cyclically closed chain with Rouse-type relaxation processes. For this model Akcasu et al. [15] calculated explicitly the first three terms and obtained the following decay constants

$$b_1 = 2 \ ; \quad b_2 = 4 \ ; \quad b_3 = 6 \ .$$

Similar calculations have been made also for the open chain without and with hydrodynamic interaction, but because of the complexity only the first term in the sum has been evaluated explicitly [11, 15]. It is now necessary to discuss i) the decay constants in the exponentials, and ii) the prefactors $P_j(q)$.

The first exponential function corresponds, of course, to the translational diffusion and the other ones to internal modes of motion. We notice for the decay constants

$$\underset{\text{translational}}{\Gamma_0 = D_q^2} \qquad \underset{\text{internal}}{\Gamma_j = b_j/\tau_1} \ .$$

That is, the decay constants of the *internal* motions *do not* contain the angular dependent factor q^2 which is characteristic of translational motion. Indeed, the internal motion with respect to the center of mass is not connected with a translational motion, and the Fourier transform must for this reason be q^2-independent. This fact is a decisive criterium by which internal motions can be distinguished from translational diffusion.

Next, the angular dependent amplitude factors may be discussed. $P_0(q)$ decays continuously with q^2 and is closely related to the particle scattering factor. The internal mode prefactors show characteristic maxima which roughly speaking may be considered as Bragg reflexes from standing waves. The maximum positions correspond to the "wave length" of a special vibration, but this length distribution is unfortunately not well resolved as can be realized from the relationship for the first three members derived by Akcasu et al. [15]

$$g_1(t) = \exp(-D_q^2 t)[A_0 + A_1 \exp(-2t/\tau_1) + A_2 \exp(-4t/\tau_1) + A_3 \exp(-6t/\tau_1)] \tag{24}$$

with

$$A_0 = \exp[-(qR_g)^2/3]/P_{\text{ring}}(q)$$
$$A_1 = A_0 X^2$$
$$A_2 = A_0 X^4/4$$
$$A_3 = A_0 (X^3/3)(1 + X^3/9)$$
$$P_{\text{ring}}(q) = \frac{\sqrt{2}}{qR_g} \exp\left[-\frac{q^2 R_g}{2}\right] \int_0^{qR_g/\sqrt{2}} \exp(t^2)\,dt$$

where

$$X = (qR_g)^2/\pi^2 \ .$$

Figure 4 shows the qR_g dependence of the individual terms and the rest function R.

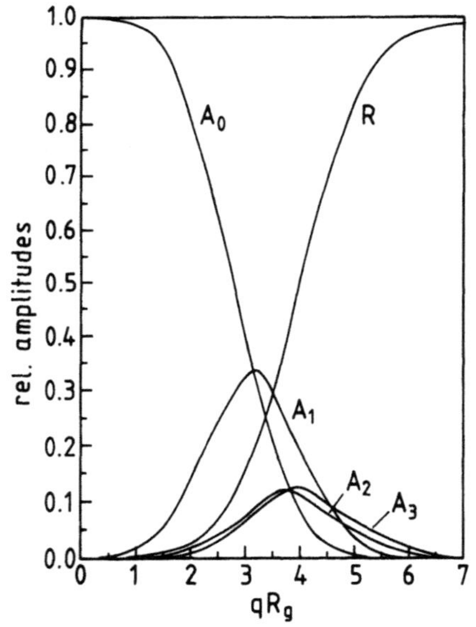

Fig. 4. Plot of the amplitude factors A_0, A_1, A_2, and A_3 for a cyclically closed flexible chain as a function of qR_g. A_0 is the prefactor for the translational diffusion; A_1, A_2, and A_3 are related to the mode length of the first three relaxation modes. R denotes the residue scattering and comprises the amplitude of all other modes larger than A_3. Note: A_3 is actually composed of two processes which are related to τ_3 and a weighted contribution of τ_2, both having a decay of $6/\tau_1$ [15]

4. Analysis and interpretation of time correlation functions

4.1 Remarks on attempts of determining the first relaxation times

The special form of the TCF as given by Eq. (23) has aroused great interest and hopes for evaluating the first relaxation times from dynamic light scattering. It has been indeed possible to evaluate the first relaxation time as a function of the molecular weight, and agreement with the Zimm model was found [16], as was expected to occur in dilute solution. The question whether a separation of the first relaxation process from the dominating translational process can be performed with sufficient accuracy by the use of Provencher's CONTIN program [8] or the maximum entropy technique MAXENT [9] remains still not fully clarified.

The evaluation of the next higher relaxation terms becomes even more involved in particular as the decay constants $\Gamma_j = b_j/\tau_1$ in the various exponentials do not simply represent the individual relaxation times, but these characteristic times result from a weighted sum of different relaxation times, and the weighting is strongly model dependent [15]. It is instructive to have a closer look at Figure 4 since it gives some hints which conditions have to be chosen to evaluate τ_1 with sufficient precision. We notice:

1) The first maximum occurs for rings at $qR_g = 3.2$ and for open chains at $qR_g = 2.12$ [11, 15]. Thus rather large macromolecules with $R_g > 100$ nm are required before the first relaxation time shows its maximum effect.
2) At this maximum the contribution from the translational diffusion is still of the same magnitude, and in addition there is a 30% contribution from the higher exponents. Because of the different decay constants one has, of course, at larger delay times [see Eq. (24)] no longer a strong influence of these faster relaxations, but their effect becomes increasingly more important at larger angles, while the contribution of the translational diffusion continuously looses its influence.

4.2 The shape function

One further point is of great relevance: If the macromolecules have dimensions larger than $qR_g = 5$ then the TCF is no longer significantly influenced by the translational diffusion but is solely determined by internal modes of motion and the specific relaxation spectrum [11, 15]. Furthermore, the Pecora approximation for the individual process is no longer valid, and the full Wang-Uhlenbech equations has to be used.

As early as 1967 de Gennes [17], Dubois-Violette and de Gennes [18] considered the extreme of very large macromolecules at sufficiently large q-values. Under these conditions they were able to derive closed expressions [17]:

$$g_1(t) = \int_0^\infty \exp\left[-u-(\Gamma t)^{1/2}\, g\left(\frac{u}{(\Gamma t)^{1/2}}\right)\right] du \tag{25a}$$

with

$$g(v) = \frac{2}{\pi}\int_0^\infty \frac{\cos(xv)\,[1-\exp(-x^2)]}{x^2}\,dx\ , \tag{25b}$$

and

$$\Gamma = \frac{kT}{12\xi_0}\cdot a^2 q^4 \tag{25c}$$

a = bond length ; ξ_0 = monomer friction coefficient

for the Rouse-model (no hydrodynamic interaction), and [18]

$$g_1(t) = \int_0^\infty \exp\left[-u-(\Gamma t)^{2/3}\, h\left(\frac{u}{(\Gamma t)^{2/3}}\right)\right] du \tag{26a}$$

with

$$h(v) = \frac{2}{\pi}\int_0^\infty \frac{\cos(xv)\,[1-\exp-(x^3/2)^{1/2}]}{x^2}\,dx \tag{26b}$$

and

$$\Gamma = (kT/6\pi\eta_0)q^3 \tag{26c}$$

for the Rouse-Zimm model (strong hydrodynamic interaction).

Thirteen years later Akcasu et al. [15] took up this problem and they found that these asymptotic equations for the TCF follow the remarkable scaling law of

$$g_1(q,t) = g_1(\Gamma t) \tag{27}$$

where Γ is the first cumulant of the TCF, i.e.,

$$\Gamma = -\partial \ln g_1(t)/\partial t|_{\text{at } t=0}\ . \tag{28}$$

One realizes from the Pecora approximation and the asymptotic relationships that Γ and even $\Gamma/q^2 \equiv D_{app}(q)$ are strongly q-dependent. Again two limits are known.

$(qR_g)^2 < 2$ [19]

$$\Gamma/q^2 = D_c(1 + Cq^2R_g^2 - \ldots) \tag{29}$$

$$D_c = D(1 + k_D c) \quad \text{and} \tag{30}$$

$(qR_g)^2 \gg 1$ [15, 19]

$$\Gamma/q^2 \to (kT/\eta_0)(a^2q^2/12) \tag{31a}$$
(Rouse-model)

$$\Gamma/q^2 \to (kT/\eta_0)(q/6\pi)\ . \tag{31b}$$
(Rouse-Zimm model)

Thus, if $D_{app}(q)$ is plotted against $q^2 + kc$ as is done in a common Zimm plot for static LS measurements then a linear initial slope should be expected which gradually bends over to a q-dependence. Such behavior is indeed observed. Figures 5a and b show results from a high molecular weight microbial polysaccharide in water [20]. The C coefficient that can be found from a combination with static LS has been shown to depend on the molecular architecture [21], but it is often difficult to measure accurately.

The asymptotic region is remarkable since now chain-length dependence no longer occurs. The reason for this effect is that large q-values correspond to short chain sections, and consequently only the internal motions of such *sections* are observed. In this region scaling behavior of $g_1(\Gamma t)$ should be valid.

We checked this prediction with very long but fairly stiff linear chains [21, 22]. The $g_1(t)$ TCF exhibited

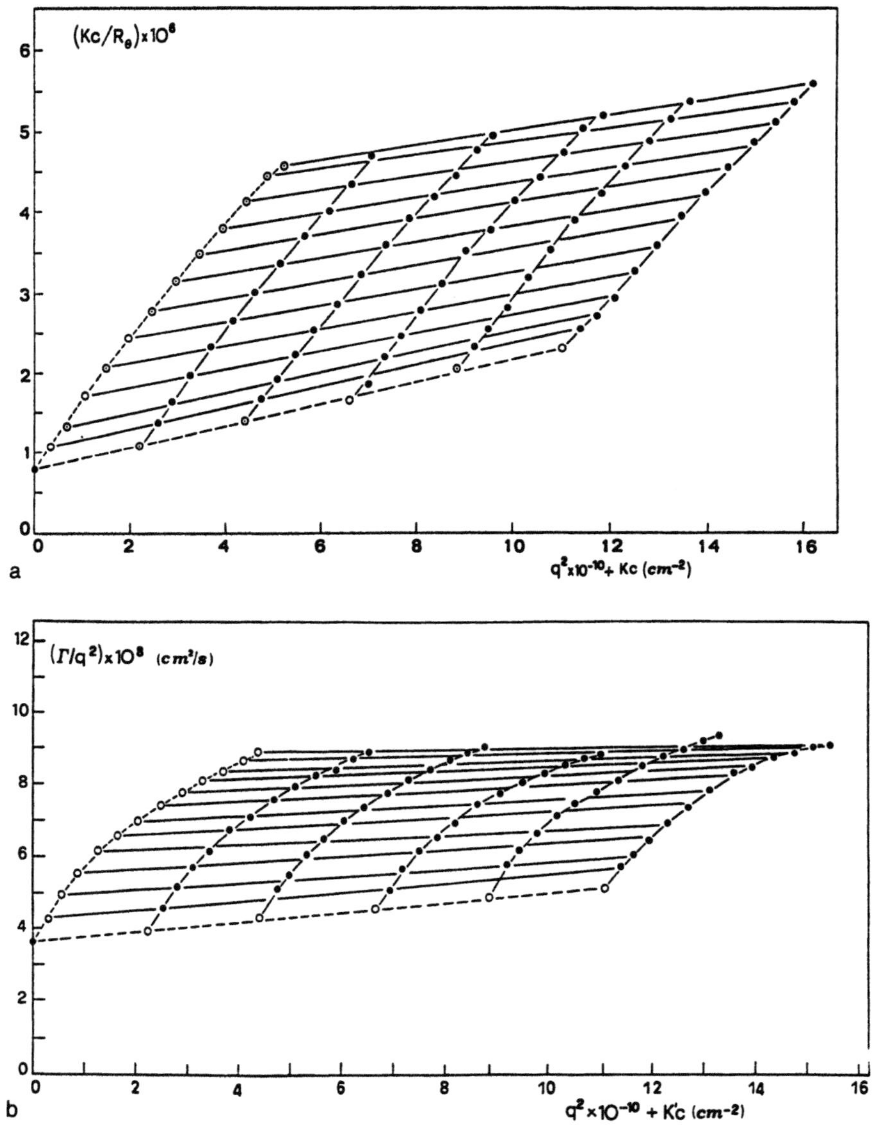

Fig. 5. (a) Static and (b) dynamic Zimm plots from a microbial polysaccharide, synthesized by Rhizobium trifolii strain TA-1 [22]. Note the strong q-dependence of the apparent diffusion coefficient $\Gamma/q_2 = D_{app}(q)$. The translational diffusion coefficient D_c at concentration c is obtained after extrapolation of Γ/q_2 against $q_2 = 0$. For further details see [22]

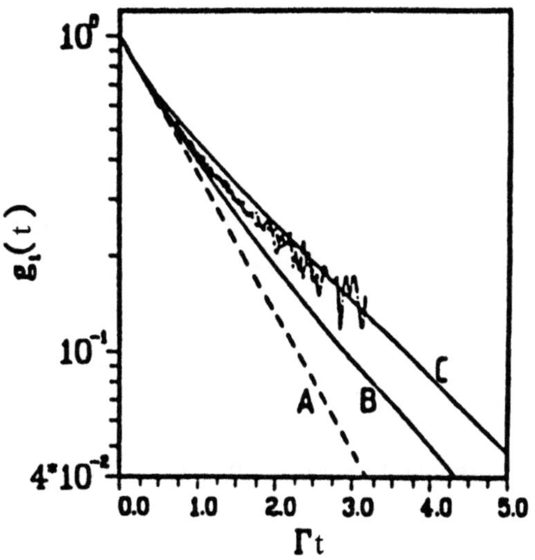

Fig. 6. Shape function, i.e., scaled TCF $g_1(\Gamma t)$, as a function of Γt for a pyruvate free xanthan (PFX) in 0.1 N NaCl solution. The strongly q-dependent curves of $q_1(\Gamma t)$ now form one common line for angles of 30°, 60°, 90°, and 120°. The curves A, B, and C denote single exponential behavior, the shape function for Zimm type relaxations [18], and the shape function of a rod molecule [23], respectively

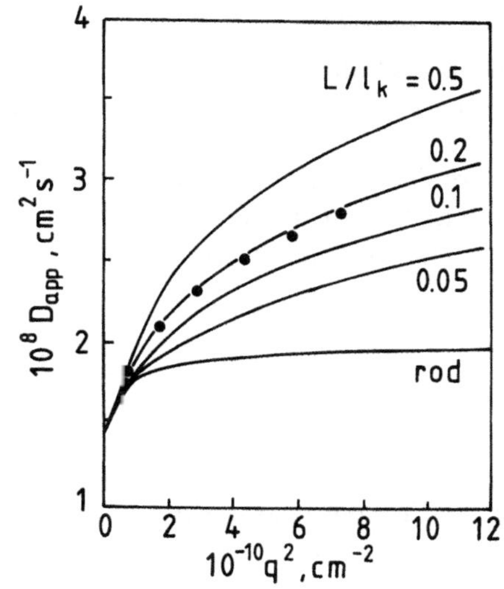

Fig. 8. q_2-dependence of $\Gamma/q_2 \equiv D_{app}$ at infinite dilution for the rod-like fd-virus [24]. Full points represent the measurements. The solid lines are theoretical curves for bending rods whose contour length L is much shorter than a Kuhn segment l_k. For further details see [24]

strong angular dependences but if instead of t the scaled time Γt is chosen all measurements made at angles between 30° and 150° now form a common curve. This behavior is shown in Figs. 6 and 7. We compared our results with the two shape functions of Eqs. (25) and (26) and with a recently derived similar relationship for stiff rods [23]. For the xanthan chain the experimental curve appears to follow first rod-like behavior but passes for large times to the curve for a flexible Zimm-type chain [22]. Such behavior appears reasonable since at short times the motion of short segments are observed, and these are close to the rod limit, whereas at large time scales the motion of larger chain sections are recorded. For other similar stiff chains (Fig. 7) we did not observe such transition but did observe Zimm behavior over the whole time domain.

Recently Fujime and Maeda [24] pointed out and proved that real rigid-rod behavior is rare with polymers, and even the stiffest objects known to us, e.g., rod-like viruses, exhibit a remarkable bending flexibility. This effect is demonstrated in Fig. 8.

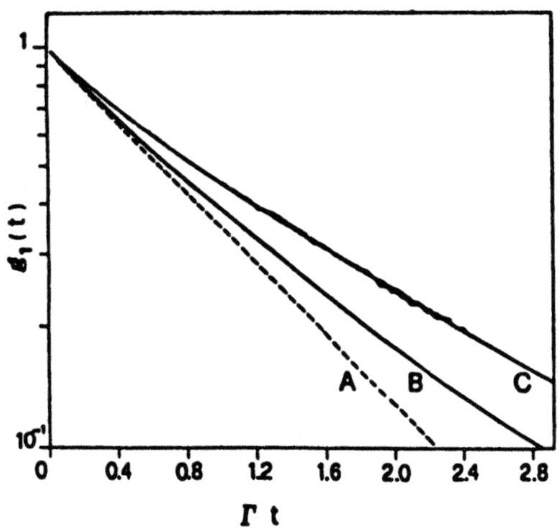

Fig. 7. The same plot as in Fig. 6 for the microbial polysaccharid TA-1

Table 1. The ϱ-parameters and C-parameters for some selected structures

Architecture	ϱ	C
Homogeneous sphere	0.778	0
Random coil, monodisperse		
θ-conditions	1.50	0.173
good solvent	1.78	–
Random coil polydisperse, $z = 1$		
θ-conditions	1.73	0.200
good solvent	2.05	–
Regular stars		
θ-conditions, $f = 4$	1.33	0.148
θ-conditions, $f \gg 1$	1.079	0.098
Rigid rod		
monodisperse	>2.0	0.042
polydisperse, $z = 1$	>2.0	0.16

4.3 The C- and ϱ-parameters

As has been mentioned already the C parameter is structure dependent. A few theoretically derived data for some models are given in Table 1. Another much stronger architecture-dependent parameter can be obtained from a combination of the dimension that are measured by static and by dynamic LS, i.e., from a parameter [19]

$$\varrho = R_g/R_h \tag{32}$$

where R_h is the hydrodynamic effective radius which can be obtained from the translational diffusion coefficient by the Stokes-Einstein relationship

$$R_h = kT/(6\pi\eta_0 D) \ . \tag{33}$$

This parameter has been calculated for different architectures and polydispersities by one of us [19], and the results are also listed in Table 1. Thus, the ϱ-parameter is a very helpful quantity for structures which are too small for the determination of the shape of the molecule from the special form of the particle scattering factor.

5. Semidilute solutions

5.1 Cooperative diffusion coefficient

The concentration dependence of D_c often remains linear in c up to rather high concentrations, i.e., where the coils of several molecules already overlap strongly. This experimental finding appears a little unexpectedly since the osmotic modulus, that is measured by static LS, already shows in this regime a strong non-linearity in c. It was pointed out in the beginning of this article that

$$D_c = [kT/f(c)]\ [(M/RT)(\partial\pi/\partial c)] \tag{2}$$

actually contains two contributions, a thermodynamic one, i.e., the reduced osmotic modulus, and a hydrodynamic one, $f(c)$, which also is concentration dependent.

We have eliminated the osmotic modulus in our experiments. This was easily done because static and dynamic LS are simultaneously recorded by our instrument, and the reduced osmotic modulus is simply the ratio of

$$M/M_{app} = MKc/R_{\theta=0} \ .$$

Figure 9 gives the result of the static measurements in which M/M_{app} is plotted against a parameter

$$X = A_2 M_w c \sim c/c^* \tag{33}$$

where A_2 is the second virial coefficient. It has been shown previously [5] that this parameter X is a good scaling parameter for the osmotic modulus, and Fig. 9 demonstrates this for three different architectures. With the exception of the stiff chains the osmotic modulus increases at large X much stronger than linearly.

Figures 10 and 11 show the corresponding results of the diffusion coefficient and the frictional coefficient which was obtained after eliminating the osmotic modulus from D_c. We notice two facts:

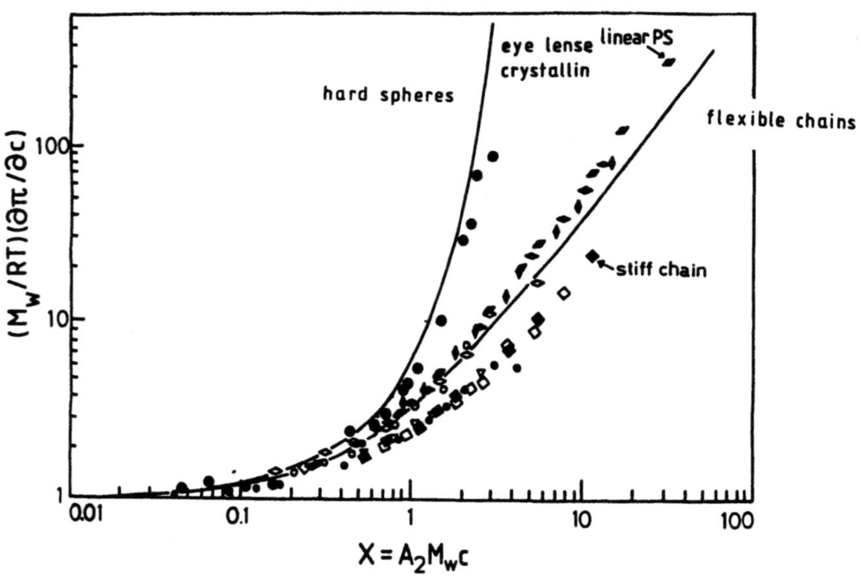

Fig. 9. The reduced osmotic modulus $M_wKc/R_{\theta=0} = (M_w/RT)(\partial\pi/\partial c) = M_w/M_{app}$ as a function of $X = A_2M_wc$ for three architectures [5]. The results from eye crystallines were obtained by Delay and Tardieu [26]. The data for the flexible chains correspond to polystyrene with molar masses between 150000 and 7.8×10^6 g/mole. Note: X can be changed by variation of (a) A_2, (b) M_w or (c) of c. Complete scaling is observed. The results of the stiff chains were obtained from microbial polysaccharides [22] and filamentous micelles of the $C_{16}G_2$ type [27], in which a maltose unit G_2 is coupled via an amide linkage to a C_{16} aliphatic chain

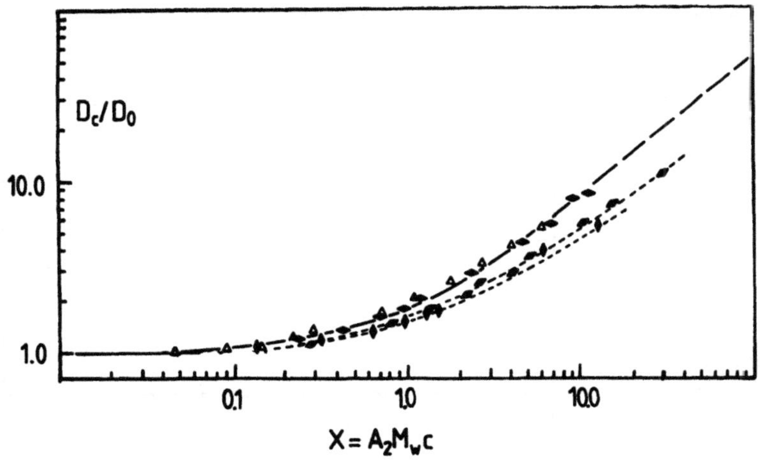

Fig. 10. Plot of $D_c/D_{c=0}$ vs $X = A_2M_wc$ for three polystyrene chains of $M_w = 150000$ g/mole (lower curve), 1.08×10^6 g/mole (middle curve), and 7.8×10^6 g/mole (upper curve), respectively. Scaling behavior is no longer observed

i) The concentration dependence of $f(c)$ is not linear in c. Thus, the linearity of D_c with c is a result of a counterbalance of the non-linear increase of the osmotic modulus and a non-linear decrease of $1/f(c)$. This pecularity suggests that theoreticians should derive separate equations for the osmotic modulus and for the hydrodynamic behavior rather than giving relationships for the diffusion coefficient alone.

ii) The hydrodynamic interaction does not scale with the parameter X. The measurements of D_c, Fig. 10, and $f(c)$, Fig. 11, show a clear molecular weight dependence which does not appear in the osmotic modulus. In fact, such scaling could not really be expected since A_2 is a parameter that describes the thermodynamic interaction among two macromolecules, and it is not obvious that this thermodynamic parameter can describe also the hydrodynamic interaction. The hydrodynamically corresponding parameter to X would be k_fc in the linear relationship $f(c) = f_0(1 + k_fc)$. Use of this parameter better brings the curves at different molecular weights together but a complete scaling is, nevertheless, not achieved.

Finally we mention the scaling prediction by de Gennes [25] according to which one should find

$$D_c \sim (c/c^*)^{3/4} \ . \tag{34}$$

In almost all experiments, however, a much lower exponent is observed. These last results indicate that our understanding of the hydrodynamics of macromolecules is still very incomplete.

5.2 Slow motion

This brief review on dynamic light scattering cannot be closed without mentioning the phenomena of slow motions. In almost all polymeric systems a very slow motion is detected if the overlap concentration c^* is significantly exceeded. A number of examples have been already presented elsewhere [5] and only two not yet fully discussed examples will be presented here.

Figure 12 shows the result from the non-ionic surfactant $C_{14}E_8$ at 20 °C and at a concentration (36%)

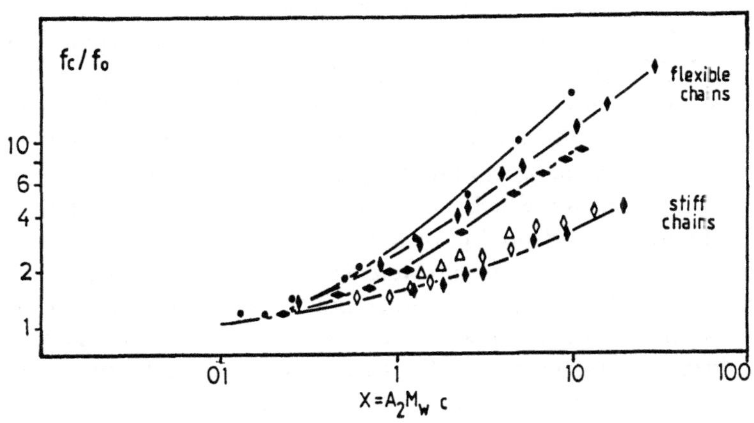

Fig. 11. The same plot as in Fig. 10 for the frictional coefficient $f(c)/f_0$. Again, X is not a good scaling parameter. The lowest M_w is represented by the upper curve, the highest is represented by the lowest curve

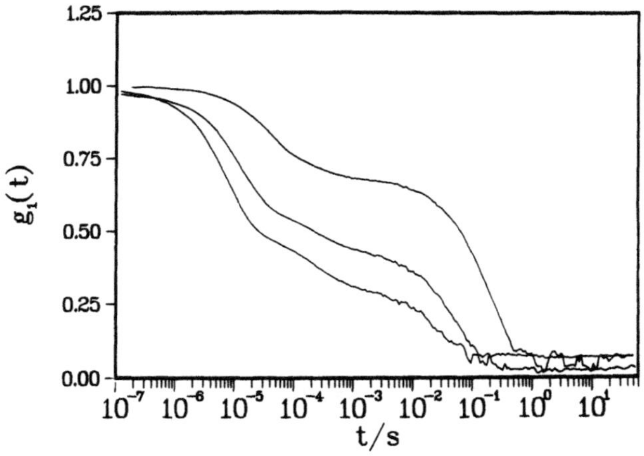

Fig. 12. Plot of the TCF for the $C_{14}E_8$ surfactant at $c = 36\%$ at three different angles (50°, 90°, 130° from above) [28]

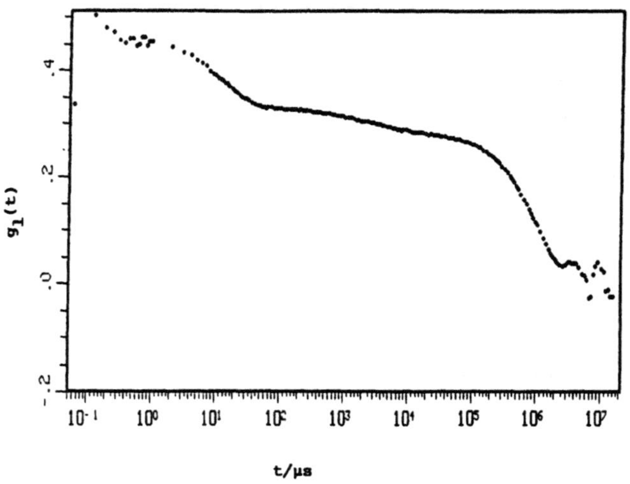

Fig. 15. TCF of the liquid crystalline polymer at $c = 36\%$ and $\theta = 70°$ [29]

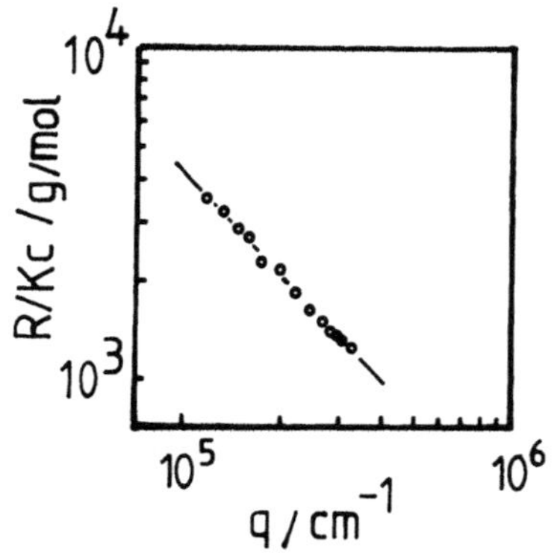

Fig. 13. Plot of the scattering intensity for $C_{14}E_8$ at $c = 36\%$

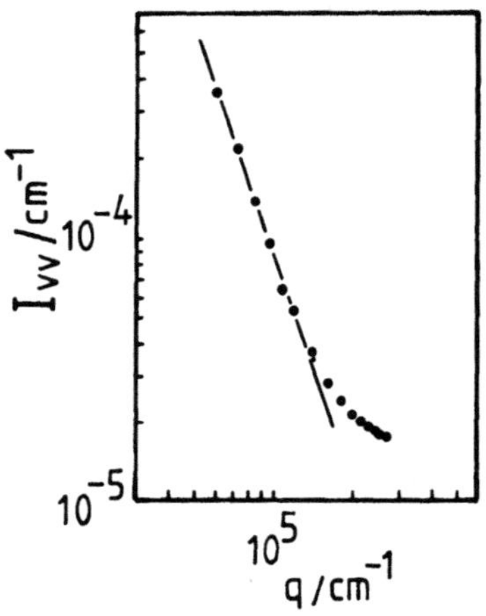

Fig. 16. Plot of the scattering intensity form the liquid crystalline polymer at $c = 36\%$

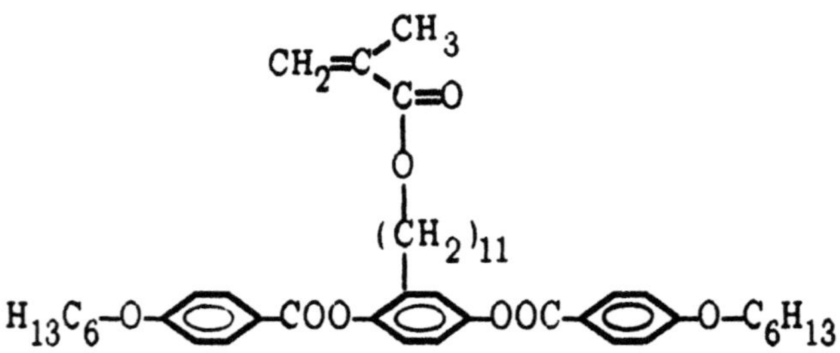

Fig. 14. Chemical formula of the monomeric unit from the thermotropic liquid crystalline polymer

close to the boundary of transition to a liquid crystalline phase [28]. Evidently at least two relaxation times are observed which are very different in the time scale (about four decades in time). These tow diffusion processes are well separated and do not disturb each other.

The interpretation of this phenomenon became rather clear in this example by the observation of a very pronounced angular dependence of the static light-scattering data, which indicates the formation of large clusters. The double logarithmic plot of these data as shown in Fig. 13 exhibits clear exponential behavior over the whole accessible q-range with a slope of $-D = 1.03$ where D, is the fractal dimension of elongated linear objects. The findings suggest the formation of liquid crystalline clusters in the isotropic phase, which means orientation within the cluster but no orientation of the clusters in solution.

Similar results were obtained, as already mentioned [5], with many other polymers. Most interesting in connection with the low molecular weight micellar system of C_{14}, E_8 is the result from a thermotropic polymer at a concentration of 36% [29]. The chemical formula is given by Fig. 14; the time correlation function at $\theta = 70°$ is shown in Fig. 15 and the double logarithmic plot of the static light scattering is given in Fig. 16. The slope of $-D = -2.85$ now indicates a densely packed cluster with no orientation within the clusters.

References

1. Berne B, Pecora R (1976) Dynamic Light Scattering. Wiley, New York
2. See monographs and textbooks on light scattering, e.g. a) Burchard W (1978) in: Applied Fibre Science. Academic Press, London, Vol 1, Chapt 10
   b) Carpenter DK (1970) in: Encyclopedia of Polymer Science and Technology. Wiley, New York, Vol 12, p 627
   c) Huglin MB (1972) Light Scattering from Polymer Solutions. Academic Press, London
3. See textbooks of statistical thermodynamics, e.g. Friedman HJ (1985) A Course in Statistical Mechanics. Prentice Hall, New Jersey
4. Yamakawa H (1971) Modern Theory of Polymer Solutions. Harper and Row, New York
5. Burchard W (1988) Makromol Chem Macromol Symp 18:1
6. Siegert AJF (1943) MIT Rad Lab Report No 465, quoted by Jakeman E (1973) in: Photon Correlation and Light Beating Spectroscopy. Plenum Press, New York, p 76
7. Zimm BH (1948) J Chem Phys 16:1093
8. Provencher SW (1976) Biophys J 16:27; also: Schulz-Dubois EO (1983) Photon Correlation Techniques in Fluid Mechanics. Springer, Berlin
9. a) Livesey AK, Licino P, Delaye M (1986) J Chem Phys 84:5102
   b) Licino P, Delaye M, Livesey AK, Leger L (1987) J Phys (Les Ulis, Fr) 48:1217
10. Schlager H (1989) Ph D Thesis, Freiburg
11. Pecora R (1968) J Chem Phys 48:4126
12. Wang MC, Uhlenbeck GE (1945) Rev Mod Phys 17:323
13. Zimm BH (1956) J Chem Phys 24:269
14. Rouse PF (1953) J Chem Phys 21:1272
15. Akcasu AZ, Benmouna M, Han CC (1980) Polymer 21:866
16. Caroline D, Jones G (1980) Light Scattering in Liquids and Macromolecular Solutions Degiorgio V, Corti M, Giglio M (eds). Plenum Press, New York, p 97
17. De Gennes P-G (1967) Physics 3:37
18. Dubois-Violette E, De Gennes P-G (1967) Physics 3:181
19. Burchard W, Schmidt M, Stockmayer WH (1980) Macromolecules 13:1265
20. Dentini M, Coviello T, Burchard W, Crescenzi V (1988) Macromolecules 21:3312
21. Burchard W (1983) Adv Polym Sci 48:1
22. Coviello T, Kajiwara K, Burchard W, Dentini M, Crescenzi V (1987) Macromolecules 20:1102
23. Hallet FR, Nickel B, Craig T (1985) Biopolymers 24:947
24. Fujime S, Maeda T (1985) Macromolecules 18:191
25. De Gennes P-G (1979) Scaling Concepts in Polymer Physics. Cornell University Press, Ithaca
26. a) Delay M, Tardieu A (1983) Nature 30L:415;
    b) Delay M, Gromiec A (1983) Biopolymers 22:1203
27. Denkinger P, Burchard W, Kunz M (1989) J Phys Chem 93:1428
28. Richtering WH, Burchard W, Jahns E, Finkelmann H (1988) J Phys Chem 92:6032
29. Richtering WH, Burchard W, Heßel F (to be published)

Received April 17, 1989;
accepted July 14, 1989

Authors' address:

W. Burchard
Institute of Macromolecular Chemistry
University of Freiburg
D-7800 Freiburg

Study of the glass transition of polybutadiene by neutron scattering

B. Frick

Institut Laue-Langevin, Grenoble Cedex, France

Abstract: The glass transition was studied on a simple backbone polymer, polybutadiene, using inelastic neutron scattering. We review our recent results and report on some backscattering measurements carried out on deuterated polybutadiene. Approaching the glass transition from high temperatures we find on the neutron spin echo spectrometer the freezing of the structural or α-relaxation. The intermediate scattering function $S(Q,t)$ exhibits a Kohlrausch-Williams-Watts relaxation behavior with an exponent $\beta = 0.45 \pm 0.02$. The experimental curves for different temperatures follow a universal curve if the time is rescaled using macroscopic viscosity data. Furthermore, we find somewhat below the glass transition a fast component, i.e., the dynamical structure factor shows an intensity enhancement between the elastic line and 2 meV. Well within the glass phase we observe harmonic oscillator behavior only.

Key words: Glass transition; neutron scattering; amorphous polymers; polybutadiene; mode-coupling theory

Introduction

Inelastic neutron scattering has recently given experimental evidence to be an adequate tool for the study of the glass transition phenomena in organic and inorganic compounds [1–11]. From the theoretical side – as far as mode coupling theories are concerned – a keyrole has even been ascribed to the neutron-scattering method for a critical test of the theoretical predictions [12–15]. The reason for this is that in the framework of the mode-coupling theories the freezing of the microscopic density fluctuations as a function of temperature or pressure lead to the observed anomalies of the macroscopic quantities at the glass transition.

For polymer glasses it is well known that they exhibit a large number of relaxation phenomena, even below the glass transition temperature. Most of them can be ascribed to local relaxations from side groups attached to the polymeric backbone. These relaxation phenomena have been extensively studied, for example, by the dielectric relaxation method and it is well known that they do not undergo a drastic change with temperature near the glass transition. However, the structural or α-relaxation, which is related to molecular transport processes, freezes at the glass transition temperature.

In the following we report on inelastic neutron-scattering experiments carried out on a backbone polymer, polybutadiene. We have chosen cis/trans/vinyl polybutadiene as a model substance as it is amorphous and thus we must not be concerned about crystallization effects. Furthermore, we concentrate on backbone relaxations due to the missing side groups. Parts of this work have been published recently [4, 8, 10, 16] and we here restrict ourselves to a short review; we will add some results from strictly elastic measurements on deuterated polybutadiene.

Experimental

With neutrons we probe molecular motions which are faster than about 10^{-8} s and which take place in the spatial range $1\,\text{Å} \leqslant r \leqslant 30\,\text{Å}$. The following measurements have been carried out at the ILL, Grenoble, France: Diffraction measurements on D1B with $\lambda = 1.28$ Å and $\lambda = 2.52$ Å). Neutron spin echo (NSE) measurements ($\lambda_i = 6.0$ and 4.55 Å), where we studied the intermediate scattering law $S(Q,t)/S(Q,O)$ in the time regime $9\cdot 10^{-12}\,\text{s} \leqslant t \leqslant 2.8\cdot 10^{-9}$ s. Neutron backscattering (BS) measurements on IN10 and on IN13 with $\lambda_i = 6.28$ Å and 2.23 Å, respectively, giving an energy resolution of $\delta E \sim 1$ μeV ($\sim 4 \cdot 10^{-9}$ s) and $\delta E \sim 8$ μeV ($\sim 3.2 \cdot 10^{-10}$ s). In the following we will often use the term "strictly" elastic scattering, for "elastic within this energy resolution". Time-of-flight (TOF) experiments on IN6 ($\lambda_i = 5.1$ Å; $\delta E \sim 90$ μeV), where we observe energy transfers up to at least 100 meV. The accessed Q-range lies roughly between 0.2 and 2 Å^{-1} on IN10 and IN6, whereas it is between 0.2 Å^{-1} and 5.3 Å^{-1} on IN13. Here we report on measurements on polybutadiene: $-(CH_2-CH=CH-CH_2-)_n-$, which is amorphous due to its irregular microstructure of cis-, trans- and vinyl-units (about 47:46:7). The samples studied by incoherent scattering on the (TOF) and BS-machines had a molecular weight of 10^5 and a DSC glass transition of $Tg = 186$ K. On the diffraction and spin echo machines we studied completely deuterated samples (C_4D_6) of molecular weight $9.3 \cdot 10^5$ and $Mw/Mn = 1.03$. The DSC glass transition temperature in this case was $Tg = 181$ K. A further deuterated sample with $Mw = 5 \cdot 10^4$ was studied on D1B and on the BS machines.

From protonated samples we observe, due to the huge incoherent scattering cross section of hydrogen, mainly the self-correlation of the protons. In the case of polybutadiene the coherent scattering cross sections amounts only to 5% of the incoherent scattering. A further advantage of inelastic incoherent scattering experiments is the sum rule for incoherent scattering, $\int S(Q,\omega)\, d\omega = 1$, which makes data interpretation easier as a loss of elastic scattering intensity must find its counterpart in the inelastic range. Also one can show that the phonon density of states may be obtained relatively easily from the inelastic incoherent scattering intensity.

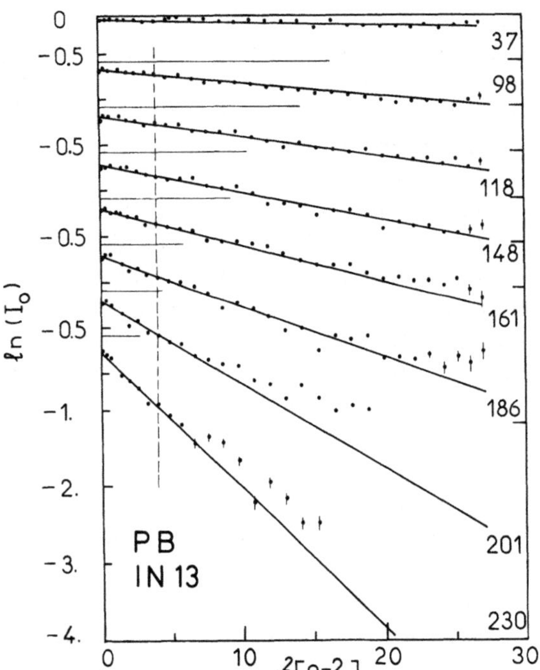

Fig. 1. Logarithm of the elastic scattering ($\delta E \sim 8$ μeV) of polybutadiene normalized to low temperatures ($T = 2$ K) and plotted Q^2. A straight line is expected for harmonic oscillations. Up to about 150 K this behavior is observerd; at higher temperatures we find deviations [4]

Experimental results in the glass

Studying protonated polybutadiene in the glass far below Tg we find from neutron backscattering and time-of-flight measurements that the dynamics appearing in the glass can be described by harmonic oscillations, i.e., i) the elastic scattering $S(Q,\omega)$ for $\hbar\omega \leqslant 8$ μeV) shows in the Q range up to 5.2 Å^{-1} the Q-dependence which is expected from the Debye-Waller factor (DWF): $I \sim I_0 \exp[-Q^2\langle r_0^2\rangle]$; ii) the inelastic phonon scattering intensity increases linearly with Q^2 and is dependent on temperature according to both Bose factors and DWF-factor; iii) the mean squared displacement $\langle r_0^2\rangle$ increases linearly with temperature. These features are summarized in Figs. 1–3. Figure 1 shows a plot of the logarithm of the normalized elastic scattering (measured on IN13 with an energy resolution of 8 μeV) vs Q^2 for several temperatures. The experimental points fall well on a straight line up to $T \sim 150$ K due to the Debye-Waller factor. From the slope of the lines, we get the mean squared displacement $\langle r_0^2\rangle$, which we may study as a function of temperature. The resulting temperature dependence of $\langle r_0^2\rangle$ is presented in Fig. 2 for the experimental data from both BS-instruments used. We note the linear temperature variation of the mean squared displacement well below the glass transition. One expects this linear behavior for a harmonic solid. The missing elastic intensity is related to the increase of the inelastic phonon scattering. As an example for the low temperature phonons of polybutadiene we show in Fig. 3 an excitation spectrum measured at $T = 80$ K. The spectra are dominated by the huge elastic scattering and on an enlarged scale the inelastic phonon scattering centered at 2 meV is observed, which shows an intensity increase with Q^2. In summary, at low temperatures the inelastic neutron scattering results on protonated polybutadiene can be explained by an harmonic oscillator behavior.

Studying deuterated samples by neutron diffraction on D1B we find that the samples are completely amorphous and have not crystallized down to the lowest temperatures (Fig. 4). The static structure factor shows two short range order peaks in the Q-range up to 3.3 Å^{-1}: a main peak at $Q_0 \sim 1.4$ Å^{-1}, which is largely due to interchain correlations [17], and a second peak at $Q_0 \sim 2.9$ Å^{-1}, ascribed mainly to intrachain correlations. The interaction between the polymer chains is of van der Waals-type, whereas within the chains the atoms are covalently bonded. Therefore it is expected that the thermal expansion will be stronger between neighboring chains than along the chain. The stronger shift of the peak position of the main peak

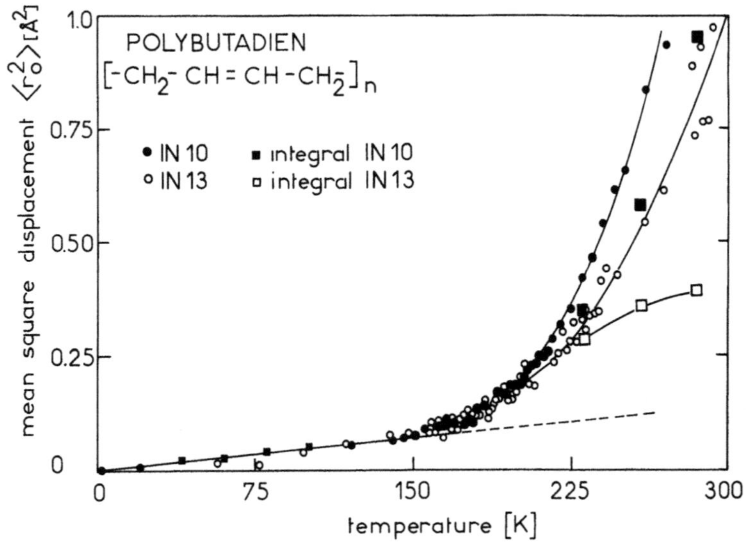

Fig. 2. Mean squared displacements $\langle r_0^2 \rangle$ calculated from the slopes of Fig. 1. Data from two BS instruments are shown. Up to $T = 150$ K, $\langle r_0^2 \rangle$ varies linearly and diverges above. Data points from integrals over inelastic spectra are included [4]

with temperature in comparison to the peak position of the second peak supports the idea that the main peak is mainly built up through correlations between monomers. In the glass the main peak position is only weakly dependent on temperature. Studying the dynamics of the system by NSE measurements at the main peak on the longest time scale accessible to neutrons, we find no decrease of $S(Q,t)$ at very low temperatures.

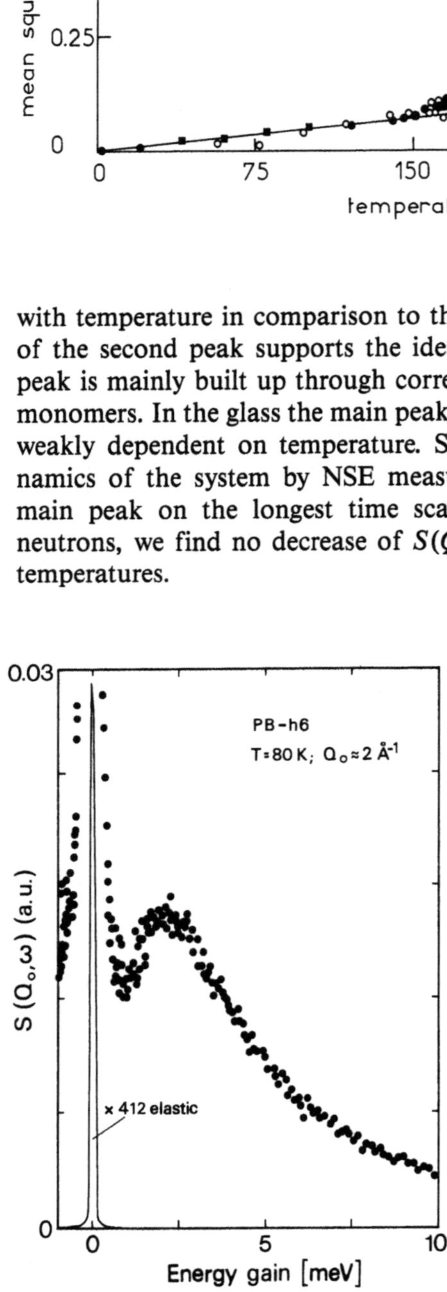

Fig. 3. Phonon spectrum from polybutadiene at $T = 80$ K for an elastic Q-value of $Q_0 = 2$ Å^{-1}. The inset shows the elastic line [20]

Experimental results near and above the glass transition

The harmonic oscillator behavior found at low temperatures changes if we approach the glass transition. Coming back to Figs. 1–3, shown above, we observed above $T \geqslant 150$ K deviations from the described Q^2-decrease of the elastic scattering intensity (Fig. 1). Furthermore, from Fig. 2 it is evident that the mean squared displacement does no longer increase linearly with temperature but diverges above $T \geqslant 150$ K from the straight line. Up to $T = 200$ K here is no dependence of $\langle r_0^2 \rangle$ on the energy resolution of the instrument used (IN10 and IN13). This indicates that the loss of elastic intensity is not due to the broadening of a quasielastic signal which arises within the dynamical window of the instrument. Quasielastic scattering from an activated process for example would be first detected on the instrument with a better energy resolution and thus the curves for this instrument would diverge first. In principle the situation is similar, if the elastic intensity is redistributed to a flat background with increasing temperature. This should be observed first with better energy resolution, as we cut out more of the constant background using worse resolution. However, depending on the ratio of elastic to flat background signal this will most probably lead to very small differences for the two resolutions ($\delta E = 1$ μeV and 8 μeV, respectively) and $\langle r_0^2 \rangle(T)$ could be the same for both spectrometer. The absence of quasielastic scattering in the energy range up to about 200 μeV has been proved by inelastic scans on the same spectrometers. A divergence of the curves for different instrumental resolutions is observed only

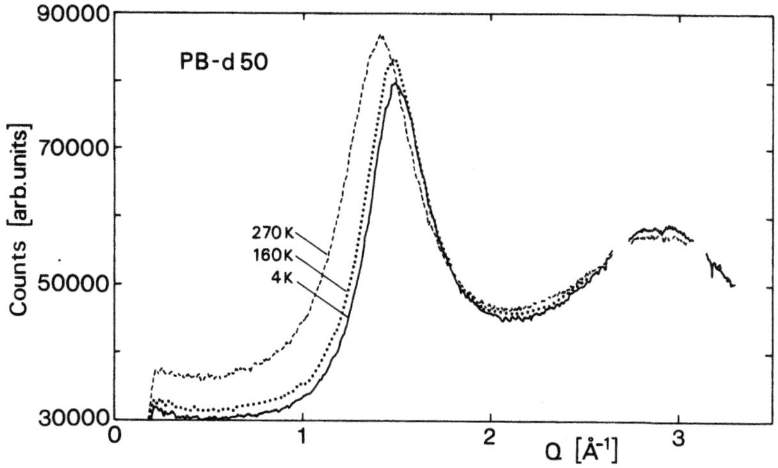

Fig. 4. Temperature dependence of the static structure factor of polybutadiene [16]

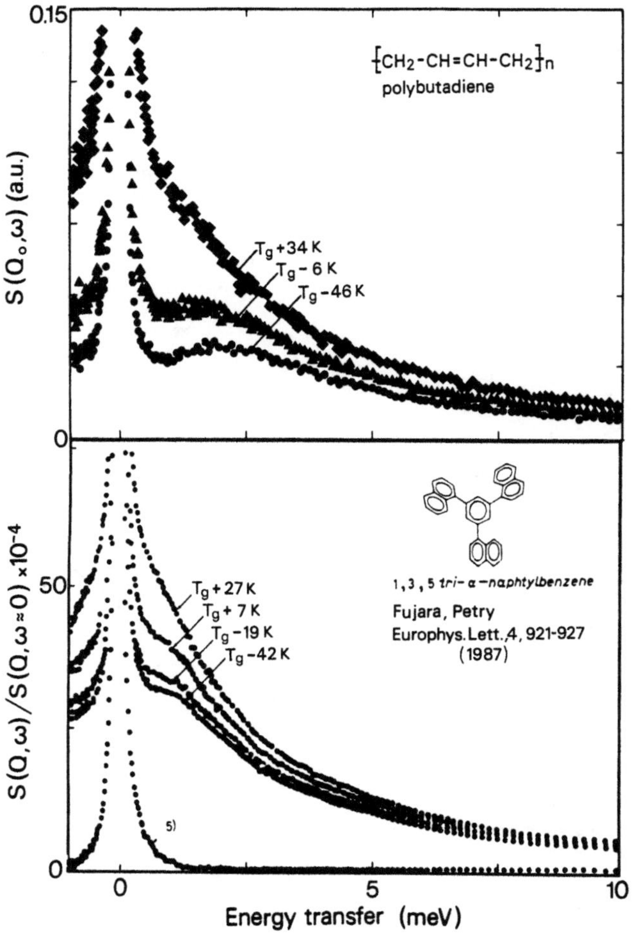

Fig. 5. Appearance of the fast component near the glass transition as observed on the TOF-instrument IN6 for a) polybutadiene and [20], b) 1,3,5-tri-α-naphthyl benzene for comparison (from [3], by permission)

above roughly 200 K; there we find quasielastic scattering as well. Due to the above mentioned sum rule we have to answer the question how the missing elastic intensity (in the temperature range 150–200 K) is redistributed in the (Q, ω)-space. Time-of-flight experiments reveal in the energy range between the elastic line and the inelastic phonon hump at 2 meV an intensity enhancement. The increase of this inelastic scattering intensity is shown in Fig. 5a for three temperatures around the glass transition temperature $T_g = 181$ K. The observation of this intensity enhancement near the glass transition seems to be general for several glasses, and has been observed by neutrons on a wide class of materials. From spin echo measurements on an anorganic glass $KCa(NO_3)$ such a fast component has been predicted first [7]. First direct observation of this fast component was made simultaneously on an organic low molecular weight glass, 1,3,5-tri-α-naphthylbenzene [3], and on polybutadiene [4]. The spectra from the last two samples are compared in Figs. 5a and 5b for a similar temperature range. Pronounced inelastic scattering intensity in the low energy range is observed for both samples. The low temperature phonon scattering is located at somewhat lower energies in the case of 1,3,5-tri-α-naphthylbenzene. This makes the decision about whether the observed additional scattering is of inelastic or quasielastic nature somewhat more difficult. At even higher temperatures a separation of the different scattering contributions becomes more difficult, because of the onset of a broadening of the elastic line due to the α-relaxation. Thus at high temperatures the quasielastic signal is observed which extends over the spectral

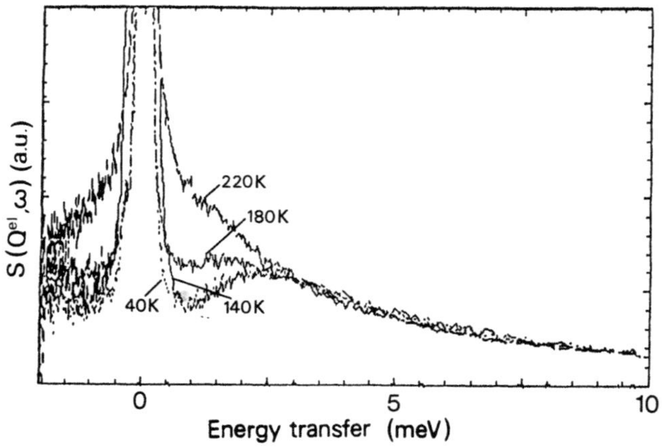

Fig. 6. Inelastic neutron excitation spectra (IN6) recalculated to the arbitrarily chosen reference temperature of $T = 100$ K. This illustrates the additional scattering arising between the phonons and the elastic line. If the phonon density of states is the same at all temperatures the phonon part should fall on a common curve

ranges of all the instruments used. An analysis of the additional scattering between 0 and 2 meV, assuming a quasielastic nature, shows that the width does neither increase with increasing Q nor with increasing temperature. The additional fast component [8] again shows up clearly if we analyze the phonon density of states and we show here a somewhat different approach. We take the low temperature phonons as a reference. Then we correct the high temperature spectra by calculating those for an arbitrary chosen reference temperature (taking DWF-factors and Bose factors into account). Figure 6 shows the spectra which are recalculated this way from the measured temperatures to the reference temperature of $T = 100$ K. From this presentation one finds that at larger energy transfers the scattering is well described by phonons and that already around *Tg* pronounced additional intensity shows up at low energies.

Neutron diffraction experiments have been carried out in steps of 10 K temperature increase between $T = 4$ K and room temperature. A selection of three temperatures has been given in Fig. 4. In Fig. 7 we summarize the main results from this diffraction work [16]. The position of the main peak as a function of temperature is shown in Fig. 7a. The observed shift of the main peak position is linear with temperature, but changes its slope around $T = 150\,\mathrm{K} \sim Tg - 30$ K, but not at *Tg* as expected from macroscopic measurements of the expansion coefficients [16]. The crossover happens at the same temperature where we find anomalies in the inelastic spectra. The linewidth Γ of the main peak is unchanged at low temperatures at a value of about $\Gamma = 0.365$ Å^{-1}, which corresponds to a correlation length of $\xi = 4\pi/\Gamma \sim 34$ Å (i.e., roughly 7–8 monomer distances). At higher temperatures the correlation length decreases (increase of the linewidth). A first decrease of the linewidth around 150 K may be due to relaxation effects because the sample was heated up in steps of 10 K with measuring times of 1 h. The equilibrium curve which is approached cuts the high temperature curve near the DSC glass transition. The appearance of different crossover temperatures for the chain distances and the correlation lengths might be an important point for the understanding of the structural changes near T_g but is not yet completely understood and affords further measurements.

In order to get more information on the dynamics in the temperature range where the structural relaxation is observed we carried out NSE measurements on deuterated samples at a Q-value corresponding to the main peak. The direct measurement in time space makes a lineshape analysis easier. In the temperature range between $T = 280$ K and $T = 200$ K we observe relaxation in the time window of the instrument (see

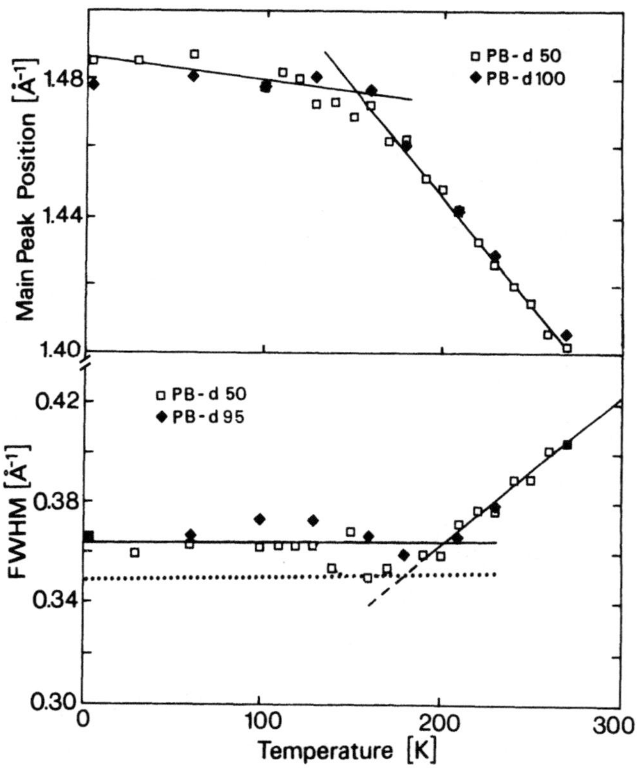

Fig. 7. Temperature dependence of the structure factor main peak position and linewidth of polybutadiene. The peak position changes its temperature behavior near $T = 150$ K simultaneously with the appearance of the fast component. The linewidth changes near *Tg*, if relaxation is taken into account [16]

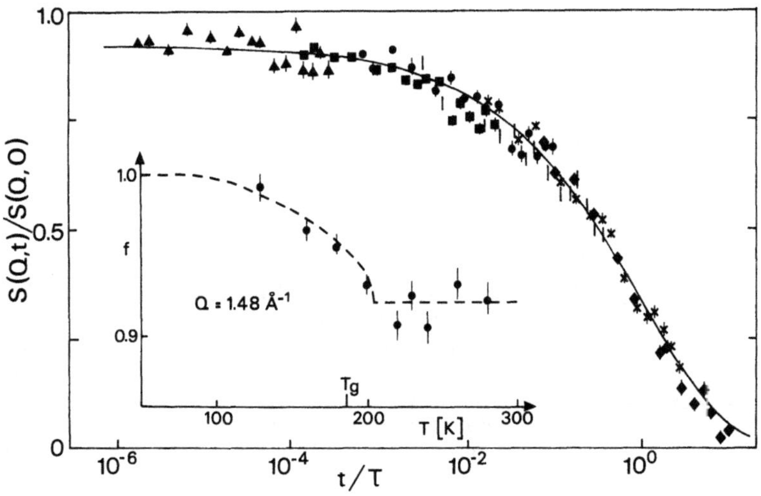

Fig. 8. Intermediate scattering law of polybutadiene vs rescaled time. The inset shows the temperature dependence of the extrapolated short time fraction ($S(Q,t)/S(Q,o)$ for $t=0$). The dashed line in the inset sketches qualitatively the mode-coupling prediction [10]

Fig. 8). The relaxation slows down with decreasing temperature and at $T = 200$ K the signal becomes elastic. A fit to the curves shows that the relaxation is not simply exponential. A rescaling of the time scale using the temperature dependence of the macroscopic shear viscosity leads to a universal curve, as shown in Fig. 8. Thereby we have used the relation between the viscosity and the monomeric friction coefficient ξ_0, which is given by the Rouse model and the parameters determined for the shear viscosity in [18]. We get then for the relevant microscopic time: $\tau = l^2 \xi_0(T)/6 k_B T$, with the Boltzmann factor k_B and the microscopic jump length l. The resulting universal curve can be described by a stretched exponential function: $S(Q,t)/S(Q,O) \sim \exp(-t/\tau)^\beta$ and $\beta = 0.45 \pm 0.02$. For the jump length we find: $l = 2.07 \pm 0.06$ Å. The β exponent is in quiet good agreement with a temperature independent value for β recently found by dielectric measurements at lower frequencies on polyisoprene [19]. It is also seen from Fig. 8 that for $t=0$ the curves do not extrapolate to 1 as expected for $S(Q,t)/S(Q,O)$. This is similar to the observation in [7] and is an indication for the presence of a fast component which we directly measured on the TOF spectrometers. NSE does not allow a decision about the nature of this fast component: fast relaxations and low energetic phonons show up in the same way. Only a careful study of the T-dependence can give more insight, and a more direct method is to use TOF measurements. Information on the Q-dependence of the dynamics may be drawn from BS measurements on deuterated samples. In Fig. 9 we again present the decrease of the logarithm of the "strictly" elastic scattering with temperature as measured on the BS instrument IN13. At low temperatures we find that the "strictly" elastic intensity decreases following in $\ln(i) \sim a - bT$ (harmonic solid) and at higher temperatures the known deviation from this behavior is found. The deviation temperatures seem to be different for different Q-values. The curve for $Q \sim 1.49$ Å near the structure factor main peak appears to be frozen up to much higher temperatures. The Q-dependence of the "strictly" elastic intensity and the static structure factor is explicitly plotted in Fig. 10 for three temperatures. We have chosen

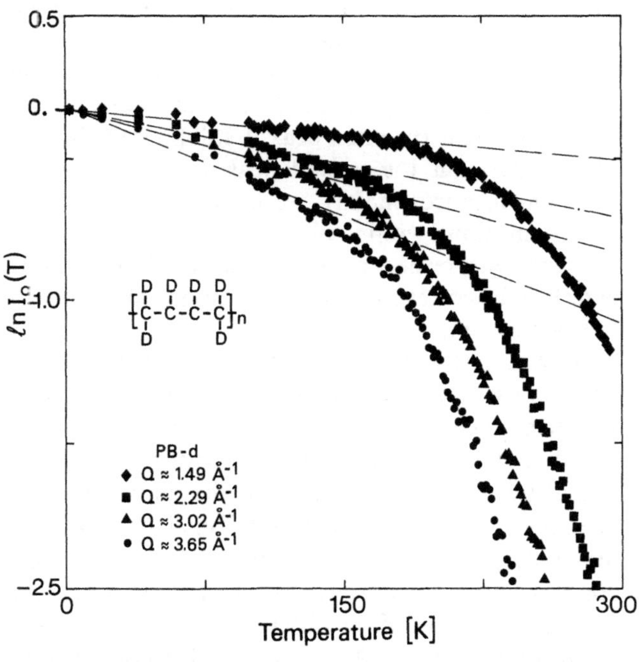

Fig. 9. Logarithm of the normalized elastic scattering ($\delta \sim 8$ μeV) vs temperature for deuterated polybutadiene ($Mw \sim 50000$) for several Q-values. The intensities were normalized to $T = 2$ K. The presented mean Q-values result from averaging over $\Delta Q \sim \pm 0.35$ Å^{-1}

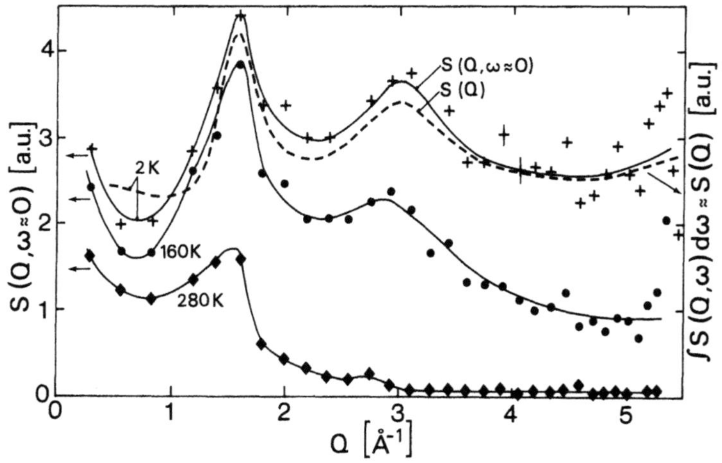

Fig. 10. Comparison of the Q-dependence of the static structure factor ($\int_{-\infty}^{+\infty} S(Q,\omega)d\omega$) and the "strictly" elastic structure factor ($S(Q,\omega < 8$ μeV) at $T = 2$ K. The temperature dependence of the "strictly" elastic structure factor shows the disappearance of the correlation peaks at higher temperatures

the same temperatures as above for the presentation of the static structure factor (see Fig. 4). The static structure factor is again presented by the dashed line ($T =$ 2 K). Within the accuracy of the experimental data from BS the agreement between the static (integral over $S(Q,\omega)$) and the "strictly" elastic structure factor ($S(Q,\omega)$ for $\omega < 8$ μeV) is good at $T = 2$ K. This indicates that the sample is frozen. With increasing temperature the "strictly" elastic structure factor data decrease strongly due to the appearance of inelastic processes. Another description is that on the observed time scale ($t \sim 3 \cdot 10^{-9}$ s) the correlation peaks of the "strictly" elastic structure factor smear out more and more with increasing temperature. At $T = 280$ K the short range order peak at 2.9 Å has nearly disappeared, whereas the main peak is still present. In view of the fact that the second peak is ascribed to intrachain distances mainly, we interpret this as being due to motions within the chain. These are already very fast with respect to the time scale of observation and thus the corresponding correlation peaks are nearly smeared out.

To summarize, we add some general remarks: We have shown by numerous inelastic and elastic neutron scattering experiments on protonated and on deuterated polybutadiene samples that neutrons can give essential information belonging to the glass transition phenomena. It is now well established from various (not only polymeric) samples that a fast component appears below or near the glass transition temperature. The interpretation of the observed scattering intensity at low energies is not yet clear. The discussion ranges from an interpretation of the fast relaxation component (β-relaxation) in the sense of the mode-coupling theories, to an interpretation in terms of fast fluctuations described as a quasi-harmonic phonon density of states. At present now we do not have experimental results that allow a decision. For comparison with the prediction of the mode-coupling theory, we refer to previous papers [2–11]; we note here only that the experimental situation is still open. Some experimental results from neutron scattering are in agreement with the predictions of the mode coupling theory, e.g., the scaling properties found in the spin echo experiments [2, 10], the temperature dependence of the fraction of the intermediate scattering function which does not freeze at the glass transition [2, 10], the presence of the fast and slow relaxation processes [2, 10] and the form factors found from incoherent inelastic scattering experiments [2, 3]. Most of these experiments, however, could not clearly fix the "critical temperature T_c", which is predicted by the mode coupling theory. Furthermore, and related to the difficulties in determining T_c, the critical exponents which are predicted by the mode coupling theory for the structural and the β-relaxation could not yet be determined reasonably.

In summary, the recent neutron-scattering results from various glass-forming substances have given new insight into the glass transition, even if they are not decisive for the mode-coupling theories that have created new interest in this field.

References

1. Galli G, Migliardo P, Bellisent R, Reichardt W (1986) Solid State Commun 57:195
2. Mezei F, Knaak W, Farago B (1987) Phys Rev Lett 58:571
3. Fujara F, Petry W (1987) Europhys Lett 4:921
4. Frick B, Richter D, Petry W, Buchenau U (1988) Z Phys B70:73
5. Richter D, Springer T (1988) (eds) Polymer Motion in

Dense Systems. Springer Proc in Physics 29, Springer, Berlin Heidelberg New York
6. Buchenau U, Monkenbusch M, Stamm M, Majkrzak CF, Nücker N in [5]
7. Mezei F, Knaak W, Farago B (1987) Physica Scripta T19:363
8. Frick B in [5], p 143
9. Fujara F, Petry W in [5], p 149
10. Richter D, Frick B, Farago B (1988) Phys Rev Lett 61:2465
11. Several contributions (1988) in: Dynamics of Disordered Materials. To be published in: Springer Proc in Physics, Workshop, Grenoble
12. Leutheusser E (1984) Phys Rev A29:2765
13. Bengtzelius U, Götze W, Sjölander A (1984) J Phys C17:5915
14. Sjögren L, Götze W (1987) J Phys C20:879
15. Götze W, Sjögren L in Ref. [11]
16. Frick B, Richter D, Ritter Cl (1989) Europhysics Lett 9:557
17. Mitchell GR, Lovell R, Windle AH (1979) Polymer 13:1273. Kilian HG, Boueke K (1962) J Polymer Sci 58:B11
18. Berry GC, Fox TG (1968) Adv Polymer Sci 5:261
19. Kremer, Boese: priv comm

Received February 20, 1989;
accepted July 7, 1989

Author's address:

B. Frick
Institut Laue-Langevin
156X
F-38042 Grenoble Cedex, France

Viscoelastic relaxation in amorphous bulk polymers

S. Wartewig, I. Alig, F. Stieber, and G. Fytas [1]

"Carl Schorlemmmer" Technical University Leuna-Merseburg, Sektion Physik, Merseburg, G.D.R. and [1]Research Center of Crete, P. O. Box 1527, Heraklion, Crete, Greece

Abstract: We report ultrasonic relaxation measurements of amorphous bulk polybutadiene ($\bar{M}_w = 13\,000$, $T_g = 222$ K) and poly(phenyl methyl siloxane) ($\bar{M} = 5\,000$, $T_g = 235$ K) in the temperature range 280 K – 360 K and frequency range 0.1 – 100 MHz. Complementary polarized photon correlation studies were performed on the same polybutadine sample.

The acoustic parameters of longitudinal and shear wave experiments were interpreted in terms of localized segmental motions of the polymer chain, coherent subchain motion ("Rouse"-like chain modes) and effects of chain entanglements. The mean relaxation times of segmental motions in polybutadiene obtained from light scattering and ultrasonic data over nine decades in time conform to a single WLF equation.

At temperature higher than T_g, photon correlation spectroscopy detects slower relaxation modes, which based on the ultrasonic results seem to be related to subchain motions and/or entanglement effects.

Key words: Polybutadiene; poly(phenyl methyl siloxane); viscoelastic function; ultrasound; photon correlation; segmental motions; subchain motions; entanglement

Introduction

Understanding chain dynamics is important in a wide variety of problems of polymer research. In recent years, there has been an increased amount of success in establishing the relationships between the complex macroscopic properties of polymers and their molecular structure [1, 2]. Nevertheless, there are many open questions.

The dynamic phenomena of polymers encompass a variety of motional modes, certain of which are very localized and others of which involve co-operative movements of large parts of the macromolecules. There are fast segmental motions and slow chain modes, e.g. "Rouse"-like and "reptation"-like motions. Because of this complexity it is necessary to apply various experimental methods for the investigation of the time- and frequency-dependent properties of polymeric systems. Comparing the results obtained by different methods, it must carefully by taken into consideration that the physical interactions involved in different experiments might also be different.

In the present paper we report ultrasonic and dynamic light scattering (also known as photon correlation spectroscopy (PCS)) studies on the viscoelastic behavior of amorphous bulk polymers near and above the glass transition temperature (T_g). The combination of these two methods offers the advantage that both probe the same mechanical properties of the system studied.

Ultrasonic investigations are based on mechanical perturbation of the system in a frequency range from a few kilohertz up to gigahertz. Accordingly, they can be considered as the direct continuation up to higher frequencies of conventional dynamic-mechanical and rheological methods. In the case of a longitudinal wave experiment the real $M'(\omega)$ and imaginary $M''(\omega)$ parts of the modulus $M^*(\omega)$ can be determined from the sound velocity and the absorption coefficient. Numerous ultrasonic relaxation measurements in polymer solutions have been reported, but there are fewer studies in polymer melts ([3 – 10] and references therein). Those studies of ultrasonic absorption in polymer melts have indicated that besides local segmental motions, coherent subchain motions ("Rouse"-like chain modes) and effects of chain entanglements also contribute to the loss component $M''(\omega)$ [6]. The longitudinal sound wave is a combination of both shear and compression. Therefore for discriminating between shear and compressional properties it is necessary to perform separately longitudinal and shear wave experiments. Such parallel measurements of both types of viscoelastic properties on the same sample and in the same frequency and temperature range provide a deeper understanding of the relaxational behavior of a polymeric system [5, 6, 9].

Photon correlation spectroscopy investigates the dynamics of the thermodynamic fluctuation in the density and optical anisotropy over the time range 10^2-10^{-6} s [11]. In the case of weakly anisotropic scatterers, such as the segments of polybutadiene chains, the dominant contribution to the light-scattering intensity arises from density fluctuations. Recently, the corresponding density time correlation function has been associated with the time-dependent longitudinal compliance $D(t)$ [12]. We were prompted to show again the correctness of this theoretical prediction by performing PCS and ultrasonic experiments on one and the same amorphous polybutadiene sample.

Experimental

The weight and number-averaged molar mass of the polybutadiene (PB) sample were determined by gel permeation chromatography to be $\bar{M}_w = 13000$ and $\bar{M}_n = 8600$ ($\bar{M}_w/\bar{M}_n = 1.5$). Its chemical microstructure determined by ^{13}C-NMR consists of 71% 1,2 (vinyl), with 62% iso and 38% syndio dyads 20% 1,4-trans and 9% 1,4-cis. The glass transition temperature T_g was measured in a DSC-2C (Perkin-Elmer) at a heating rate of 8 K/min to be $T_g = 222$ K. No crystallization effects could be detected.

For the sake of comparison of ultrasonic spectra a fractioned poly(phenyl methyl siloxane) (PPMS) sample (Petrarch, USA) with $\bar{M} = 50000$ and $T_g = 235$ K was investigated.

The ultrasonic absorption measurements were performed over a frequency range from 0.1 to 100 MHz by using the statistical reverberation method [13, 14] (0.1 – 2 MHz), the resonator method [15] (0.3 – 6 MHz) and different pulse methods [16, 17] (3 – 100 MHz) with an accuracy of between 5...10%, 3...5%, and 3...10%, respectively. Additionally, the temperature dependence of the ultrasonic absorption and velocity was measured at 1 MHz by a modified pulse overlap method [18].

The dynamic shear properties were determined at temperatures between 273 K and 353 K by means of a torsional resonator [19] (0.02 to 1.3 MHz). To construct a master curve the values were shifted by using the temperature-time superposition method [2]. The errors in the values of the shear moduli G' and G'' are estimated to be less than 10%.

The temperature dependence of the steady-state shear viscosity was investigated with a Couette-type viscosimeter (Rheotest 2, MLW Medingen).

The photon correlation spectra were obtained by using a 28-channel logarithmic-linear single clipped correlator (Malvern K 7027) with a dynamic range of 4.3 decades. The autocorrelation functions $G(t)$ of the polarized light-scattered intensity at different temperatures were recorded in homodyne detection at a fixed scattering angle of $\theta = 90°$. The light source was an argon ion laser (Spectra Physics 2020) operating at 488 nm with a power of 200 mW. The incident beam was polarized perpendicular to the scattering plane and since there was only very little depolarized intensity no polarizer was used for the scattered light.

Theory

Ultrasonic spectra

In the case of small ultrasonic absorption in an isotropic linear viscoelastic medium, the sound velocity v and the attenuation α or the absorption coefficient α/f^2 are given by the following relations:

a) longitudinal wave

$$\begin{aligned} v_l^2 &= \frac{M'}{\varrho} = \frac{K'+(4/3)\,G'}{\varrho} \\ \alpha_l/f^2 &= \frac{2\pi^2}{\varrho v_l^3}[\eta'_v+(4/3)\,\eta'_s] \\ &= \frac{\pi}{\varrho v_l^3 f}[K''+4/3)\,G''] = \frac{\pi}{\varrho v_l^3 f}M''\ ; \end{aligned} \tag{1}$$

b) transversal wave

$$\begin{aligned} v_t^2 &= \frac{G'}{\varrho} \\ \alpha_t/f^2 &= \frac{2\pi^2}{\varrho v_t^3}\eta'_s = \frac{\pi}{\varrho v_t^3 f}G'' \end{aligned} \tag{2}$$

where ϱ is the density, $\eta_s^* = \eta'_s - i\,\eta''_s$ and $\eta_v^* = \eta'_v - i\eta''_v$ are the complex shear and the volume or bulk viscosities, respectively. The relations between the components of the complex longitudinal modulus M^*, the bulk modulus K^*, the shear modulus G^* and the corresponding bulk or volume η_v^* and shear η_s^* viscosities are given by:

$$\begin{aligned} M^* &= K^*+(4/3)\,G^* \\ K^* &= K'+i\,K'' = K'+i\omega\eta'_v \\ G^* &= G'+iG'' = G'+i\omega\eta'_s\ . \end{aligned} \tag{3}$$

Taking into consideration relaxational processes it is obvious that all these moduli and viscosities are frequency dependent. In general, the frequency dependence of the ultrasonic absorption α/f^2 can be described by distributions of relaxation times for both the volume viscosity and the shear viscosity [4]. In the simplest case the absorption coefficient may be described by

$$\alpha/f^2 = \frac{A}{1+(\omega\tau_r)^2} + B \tag{4}$$

where A is the amplitude of the relaxation process, B is the classical non-relaxing absorption, τ_r is the characteristic relaxation time and $\omega = 2\pi f$. A better description can be achieved by assuming a superposition of a number of independent single processes. Such discrete relaxation processes mainly appear in the high frequency range and are connected with very localized motions, e.g., segmental modes associated with primary and/or secondary relaxation processes.

Unfortunately, the theory of ultrasonic absorption in polymeric melts is not well developed, but there are generally accepted theories for the dynamic shear properties of polymer solutions. For the molecular motion of isolated chains the Rouse model (bead-spring-model without hydrodynamic interaction) [20] yields scaling behavior as follows,

$$\eta_s' \sim f^{-1/2} \quad \text{or} \quad G', G'' \sim f^{1/2} \tag{5}$$

with a distribution of relaxation times $\tau_N < \tau < \tau_1$, where τ_N and τ_1 are the minimum and maximum relaxation time of the Rouse spectrum. Including the hydrodynamic interaction (Zimm model [21]) we obtain for the frequency dependence of the dynamic shear viscosity

$$\eta_s' \sim f^{-1/3} \quad \text{or} \quad G' \;, \quad G'' \sim f^{2/3} \;. \tag{6}$$

There are only a few theoretical investigations on longitudinal properties. Assuming a regular three-dimensional polymer network in a viscous medium where the chains between the knots behave like Gaussian chains, Gotlib et al. [22] predicted Rouse-type behavior for the frequency dependence of α/f^2. Berger and Straube [23] have treated the sound absorption of dilute polymer solutions in the framework of the Kirkwood diffusion equation and have also obtained Rouse-Type behavior. In the case of concentrated polymer solution the same authors [24] calculated the absorption within the Edwards-Grant model [25], described the dynamics by a Langevin equation, and took into account entanglements by a tube-like harmonic constraining potential. This extension of the Edwards-Grant approach leads to

$$\alpha/f^2 = \frac{2\pi^2 nkT}{\varrho v^3} \sum_{p=1}^{N} \frac{\tau_p}{1+(\omega\tau_p)^2} \tag{7}$$

with the relaxation time of the p-th mode

$$\tau_p = \tau_0 \left(\frac{N}{p}\right)^2 \left(\frac{N}{p\cdot d}\right)^2 \left\{\left[1+\left(\frac{p\cdot d}{N}\right)^2\right]^{1/2} + 1\right\}^2$$
$$\tau_0 = \frac{\zeta_k l^3}{6\pi^2 kT} \quad \text{and} \quad N = \frac{L}{l} \tag{8}$$

where n is the number of polymer molecules per unit volume, $\zeta_k \cdot l$ is the friction coefficient of the segment, L is the contour length of the chain, l is the statistical segment length and N is the number of segments. The parameter d, controlling the influence of the constraints, is given by

$$d = \sqrt{6}\,\pi \left(\frac{d_0}{l}\right)^2 \tag{9}$$

where d_0 is the tube radius. For melts and concentrated solutions it has been shown [26] that

$$\frac{d_0}{l} = 8.5 \cdot (n_s \cdot l^3)^{-1/2} \;; \tag{10}$$

n_s denotes here the number density of segments.

It follows from (7) and (8) that owing to the topological hindrance of the subchain motions, a crossover takes place from $\alpha/f^2 \sim f^{-1/2}$ to $\alpha/f^2 \sim f^{-3/4}$ at the low frequency limit of the spectrum.

For very high frequencies ($\omega\tau_1 \gg 1$) there is a decrease in the transversal mobility of the chain, which can be empirically described by a frequency dependence of the parameter d_0 as follows:

$$d_0 \sim \omega^{-\nu} \;. \tag{11}$$

As mentioned in reference [24] this leads to $\alpha/f^2 \sim f^{-n}$ with $n < 1/2$.

Because of the lack of theoretical considerations for polymer melts, we have tried to apply the forementioned theories also to melts.

Photon correlation spectra

In a photon correlation spectroscopy experiment with homodyne detection the measured autocorrelation function $G(t)$ of the scattered ligth intensity is related to the normalized density time correlation function $c(t)$ by

$$G(t) = A\ [1+b\,|c(t)|^2]\ . \tag{12}$$

A is the baseline measured at long delay times and the amplitude b is considered as a fitting parameter. Often the correlation function $c(t)$ shows a highly non-exponential shape. So far, the best known fit to the measured data is given by the Kohlrausch-Williams-Watts (KWW) fractional exponential decay function [27]

$$c(t) = \exp[-(t/\tau_0)^\beta] \quad \text{with} \quad 0<\beta\leq 1 \tag{13}$$

where b, τ_0, and B are adjustable parameters.

As shown in [12], the density correlation function $c(t)$ is related to the time-dependent longitudinal compliance $D(t)$ by

$$c(t) = [D_0-D(t)]/(D_0-D_\infty) \tag{14}$$

where D_0 and D_∞ are, respectively, the limiting compliances at low and high frequencies. The dynamic loss compliance $D''(\omega)$ can be constructed from the correlation function $c(t)$ according to

$$D''(\omega) = (D_0-D_\infty)\omega \int_0^\infty c(t)\cos\omega t\ dt\ . \tag{15}$$

If $D(t)$ is written in terms of the retardation spectrum $L(\log\tau)$ as

$$D(t) = D_\infty + \int_{-\infty}^{+\infty} L(\log\tau)\ (1-\exp\{-t/\tau\})\ d\log\tau \tag{16}$$

then $c(t)$ can also be described by using the Laplace transformation of the distribution of the retardation times

$$c(t) = 1/(D_0-D_\infty) \int_{-\infty}^{+\infty} L(\log\tau)\exp(-t/\tau)\,d\log\tau\ . \tag{17}$$

Therefore, the distribution $L(\log\tau)$ can be determined from the experimental data using the inverse Laplace transformation (ILT). We used an algorithm developed by Provencher in order to calculate $b^{1/2}\Gamma(\log\tau)/(D_0-D_\infty)$ from Eq. (17) with $b^{1/2}\ c(t)$ obtained experimentally. The average time $\langle\log\tau\rangle$ is given by the first moment of $L(\log\tau)$:

$$\langle\log\tau\rangle = \frac{\int_{-\infty}^{+\infty} \log\tau L(\log\tau)\,d\log\tau}{\int_{-\infty}^{+\infty} L(\log\tau)\,d\log\tau}\ . \tag{18}$$

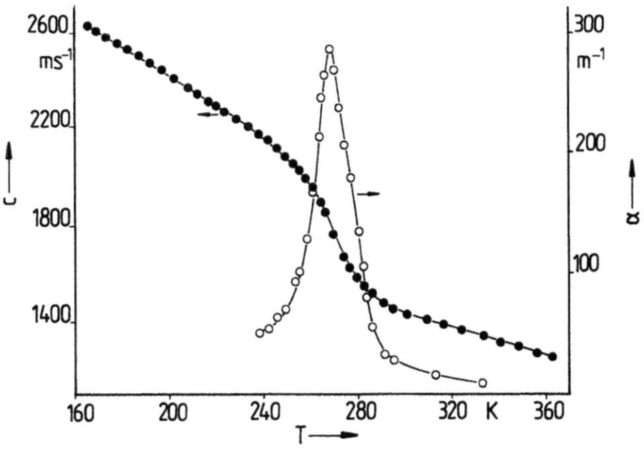

Fig. 1. Temperature dependence of the ultrasonic attenuation α_l and velocity v_l for PB at 1 MHz

These mean relaxation times $\langle\log\tau\rangle$ are to be compared with the values τ_r derived from the ultrasonic data (Eq. (4)).

Results and discussions

Segmental motion

The temperature dependence of the ultrasonic attenuation α_l was measured for PB and PPMS at different frequencies between 1 MHz and 96 MHz. The experimental results for PB at 1 MHz are shown in Fig. 1. It is evident that there is a distinct maximum of the attenuation at $T = 268$ K and a well-resolved dispersion of the sound velocity; similar behavior was observed for PPMS. These findings are related to the

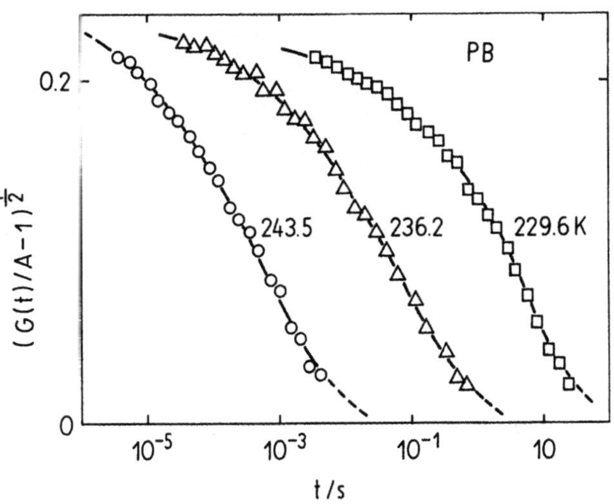

Fig. 2. Measured net correlation function for PB at three different temperatures: □ 229.6 K; △ 236.2 K; ○ 243.5 K

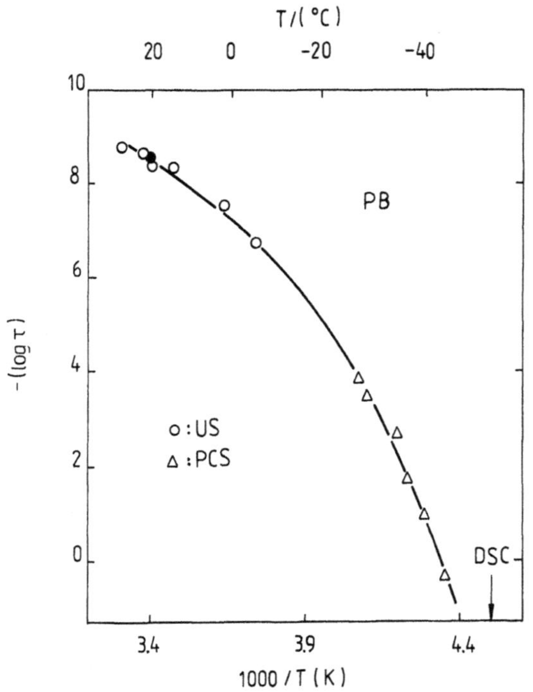

Fig. 3. Temperature dependence of the segmental relaxation time: ○ from temperature and ● from frequency-dependent ultrasonic absorption, △ from photon correlation spectroscopy. The full lines denote the fit of the VFTH equation (19) to the experimental values. The glass transition temperature determined by DSC is indicated by an arrow

dynamic glass transition. The temperature dependence of the attenuation maximum directly yields the temperature dependence of the segmental relaxation time.

The net correlation function $[G(t)/A-1]^{1/2}$ for PB is plotted vs log t in Fig. 2 for three different temperatures. The KWW function fits the experimental functions well. The value of the shape parameter β is equal to 0.36 ± 0.03. The corresponding mean relaxation times $\langle\tau\rangle$ were calculated according to Eq. (18). The results for $\langle\tau\rangle$ thus obtained from the two techniques are shown in Fig. 3. We have fitted the Vogel-Fulcher-Tamman-Hesse (VFTH) equation [2]

$$\langle\tau\rangle = \langle\tau_0\rangle \exp\,[B/(T-T_0)] \qquad (19)$$

to the experimental data. The fit yields the values $\log\,[\langle\tau\rangle/(s)] = -13.6\pm0.7$, $B=(1175\pm110)$ K and $T_0=(193\pm5)$ K. The corresponding WLF parameters with respect to T_g are $C_1=17.6$ and $C_2=29$ K.

We are certainly aware that the PCS measures longitudinal compliance under isothermal conditions while the ultrasound measures longitudinal modulus under adiabatic conditions. Nevertheless, the two sets of data which cover almost nine decades of time in the temperature range of T_g to T_g+65 K can be well represented by the VFTH equation.

The comparison of the VFTH parameters for the segmental motions with those obtained for the steady-state viscosity shows that the local segmental process has a different temperature dependence then that of the terminal one. The corresponding VFTH parameters of the steady-state viscosity are $\eta_0=3.46\cdot10^{-3}$ Pa s, $B=1325$ K and $T_0=176$ K. Based on the coupling model an explanation of the different shift factors for segmental motions and terminal relaxations is given in reference [28]. We have observed a similar behavior for PPMS.

Subchain motion

The frequency dependence of the absorption coefficient α_l/f^2 for PB at three different temperatures is shown in Fig. 4. It gives evidence of three different relaxation regions as follows:

i) The dispersion in the high frequency range (above 50 MHz) can be ascribed to local segmental motions of the polymer backbone. Because of the difficulties of spectral decomposition, the high-frequency tail of the spectrum was fitted by a single relaxation given by Eq. (4). The relaxation times τ_r thus derived also fit the VFTH equation (19) with the parameter given above.

ii) In the region 1–20 MHz the absorption is related to subchain motions generally described by "Rouse"-like motion given by $\alpha_l/f^2\sim f^{-n}$. The exponent was determined to be $n\approx0.3$.

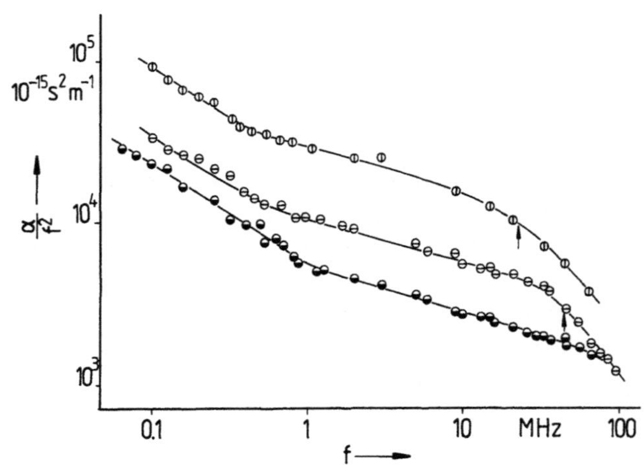

Fig. 4. Frequency dependence of the ultrasonic absorption coefficient α_l/f^2 for PB at three different temperatures: ⦶ 293 K; ⊖ 313 K; ◒ 333 K. The arrows indicate the relaxation time according to Eq. (4)

iii) In the range below 1 MHz the slope of the α_l/f^2 curve changes from $n \approx 0.3$ to $n \approx 0.65$. We attribute this low frequency part of the absorption spectrum to effects of entanglements.

The last conclusion can be supported by two experimental facts. Firstly, in the case of PPMS with $M < M_c$, entanglement effects could not be observed in the absorption spectrum (see Fig. 5). In this polymer melt only relaxation due to local segmental motions and subchain motions with $n \approx 0.3$ take place. It should be noted that at higher temperatures constant values of α_l/f^2 appear in the low-frequency range ($f < 1$ MHz). From this behavior the maximum relaxation time τ_1 of the subchain motion can be extracted.

Secondly, detailed experimental investigations on ultrasonic absorption and dynamic shear properties of the transition from dilute solution to the melt of PB have demonstrated that effects of entanglements become visible in the absorption spectra, if the volume fraction of the polymer is larger than a well defined value, i.e., $\phi > M_c/M$. (For more detail see [6].)

A surprising aspect is the finding that for both PB and PPMS the exponent of the dependence $\alpha_l/f^2 \sim f^{-n}$ within the region of the subchain motions amounts to be $n \approx 0.3$. This value indicates Zimm-behavior, e.g., subchain motions determinated by hydrodynamic interaction between polymer segments, which is usually not expected for polymeric melts.

To explain the exponent $n \approx 0.3$ we have simply divided the absorption spectrum into two parts: the single relaxation process according to Eq. (4) and the modified Rouse model represented by Eqs. (7) and (8). The shortest relaxation time τ_N of the Rouse spectrum was used as an adjustable parameter obtained from the best fit to the experimental data. The results of this approach are shown in Figs. 6 and 7. In the case of PPMS, topological constraints were of course not included and the following parameters were assumed: $N = 28$ and $\tau_N = 0.4 \cdot 10^{-9}$ s. For PB Eq. (8) was applied with the molecular parameters [29]: molar mass of the statistical segment $M_s = 240$ g/mol, $N = 55$, $l = 1.17$ nm, $n_s \cdot l^3 = 3.7$, $d_0/l = 4.5$, and $\tau_N = 2.3 \cdot 10^{-9}$ s. But using this parameter set we obtain only the simple Rouse behavior – in other words, we have to consider stronger constraints formally expressed by decreasing the parameter d_0. For $d_0/l = 2$ a fit to the experimental data was achieved (see curve 1 in Fig. 7).

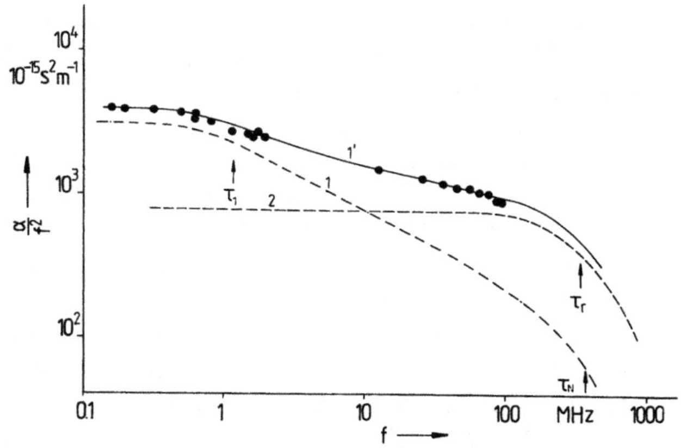

Fig. 6. Partition of the ultrasonic absorption spectrum α_l/f^2 for PPMS at $T = 328$ K into a single process (curve 2) and a Rouse spectrum (curve 1). The curve 1′ represents the sum of both. Dots are experimental data. The arrows indicate the relevant relaxation times

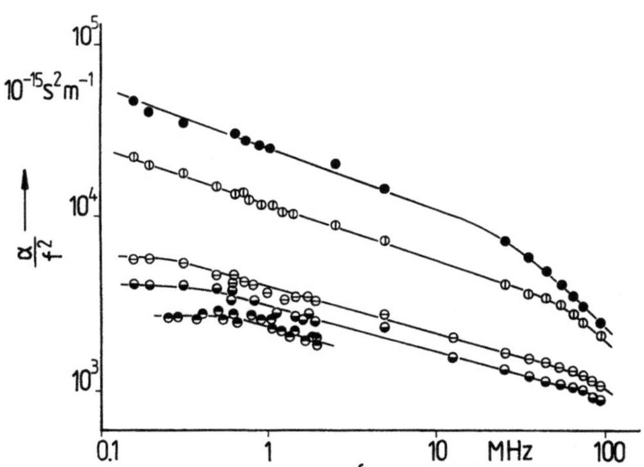

Fig. 5. Frequency dependence of the ultrasonic absorption coefficient α_l/f^2 for PPMS at different temperatures: ● 283 K; ⦶ 293 K; ⊖ 318 K; ◒ 328 K; ◓ 338 K

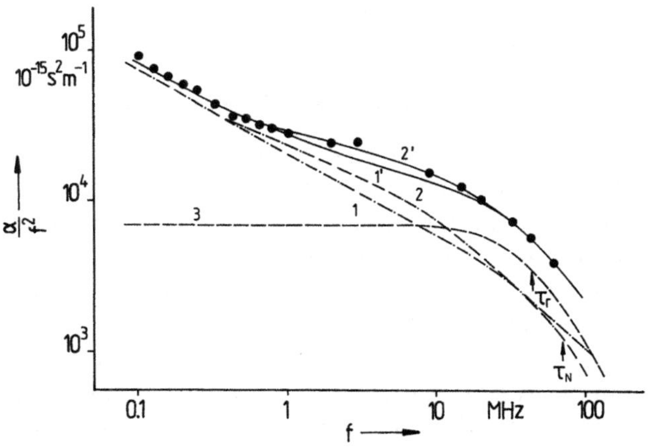

Fig. 7. Partition of the ultrasonic absorption spectrum α_l/f^2 for PB at $T = 293$ K into a single process (curve 3) and a modified Rouse spectrum (curve 1 $\nu = 0$, curve 2 $\nu = 0.5$, for parameters see text). Curves 1′ and 2′ are the sum of curve 1 and 3 and 2 and 3, respectively. Dots are experimental data. The arrows indicate the relevant relaxation times

A further improvement can be obtained by introducing the frequency dependence of the parameter d_0 according to Eq. (11). Based on the parameters $d_0(\omega) = d_0$ for $\omega < \omega_0$, $d_0(\omega) = d_0(\omega/\omega_0)^{-\nu}$ for $\omega_0 \le \omega \le \omega_\infty$ and $d_0(\omega) = d_\infty$ for $\omega > \omega_\infty$ where $\nu = 0.5$, $\omega_0 \tau_N = 5 \cdot 10^{-4}$ and $\omega_\infty = 16\,\omega_0$, the curve 2 in Fig. 7 fits the experimental data rather well.

Summarizing, we can describe the ultrasonic absorption spectrum of the PB melt as a superposition of a single relaxation process (we suppose that the segmental relaxation here described by a single process is related to the dynamic glass transition) and Rouse modes modified by effects of entanglements. In other words, the subchain motions in polymer melts are modified by the glass transition in the high frequency range of the spectrum (especially for low molecular mass polymers) and by entanglements in the low frequency range. It should be noted that the shortest relaxation time of the Rouse modes is of the same order of magnitude as the segmental relaxation. It is not appropriate that several parameters of the models assumed are freely adjustable. Another possible description of the absorption spectrum is the formal extension of the Rouse-Zimm theory to the melt, as discussed in [6].

The unambiguous detection of the subchain motion by ultrasound suggests we should observe in the time window of the correlator light scattering from PB at 20 °C. Figure 8 shows a weak correlation from about 10^{-6} to 10^{-2} s, which cannot be confused with the tail of the segmental relaxation clearly observed at lower temperatures; at 20 °C it would be less that 10^{-7} for the KWW equation. From the viewpoint of the ultrasonic spectra, the slow motions of Figure 8 seem to be contributed to subchain motions and entanglement effects.

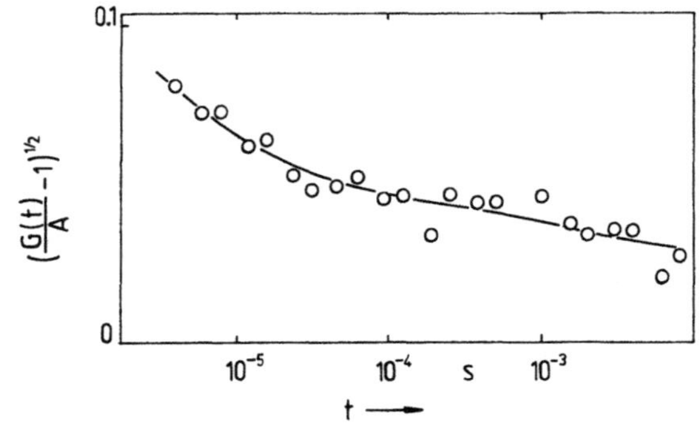

Fig. 8. Measured net correlation function for PB at 293 K

Dynamic modulus and dynamic compliance

As expressed by the relation (1), the loss component $M''(\omega)$ of the longitudinal modulus $M^*(\omega)$ can be estimated from the ultrasonic absorption spectrum. Because it is difficult to carry out frequency-dependent measurement of the velocity we have used the velocity value at 1 MHz for calculating $M''(\omega)$ at all frequencies. That means we have neglected the dispersion of v_l or M'. This approximation would only lead to small deviations at the high frequency limit of the range studied.

The dynamic moduli M' and M'' and the dynamic loss compliance D'' are related by

$$D'' = \frac{M''}{(M')^2 + (M'')^2} \,. \tag{20}$$

Since M' is greater than M'' by at least one order of magnitude (for PB at 20 °C $M' = 1.91$ GPa and $M'' = 0.23$ GPa), we can approximate

$$D''_{us} \approx \frac{M''}{(M')^2} = \frac{\alpha_l/f}{\pi \varrho v_l} \,. \tag{21}$$

The subscript, us, indicates ultrasonic.

Alternatively, D''_l can be constructed from the autocorrelation function $c(t)$ according to Eq. (15). As shown in a previous paper [30] the normalized curve for $D''(\omega)/(D_0 - D_\infty)$ characterized by $\beta = 0.36$ is broader than that for $D''_{us}(\omega)$. The experimental data D''_{us} were obtained at a higher temperature than D''_l, so that it is reasonable to assume a narrower distribution for the former. Detailed numerical calculations [18] have shown that for PB the high frequency behavior of $M''(\omega)$ is fitted by the KWW function with $\beta = 0.8 \ldots 0.9$. Hence, the combination of PCS and ultrasound allows us to investigate the temperature dependence of the distribution parameter.

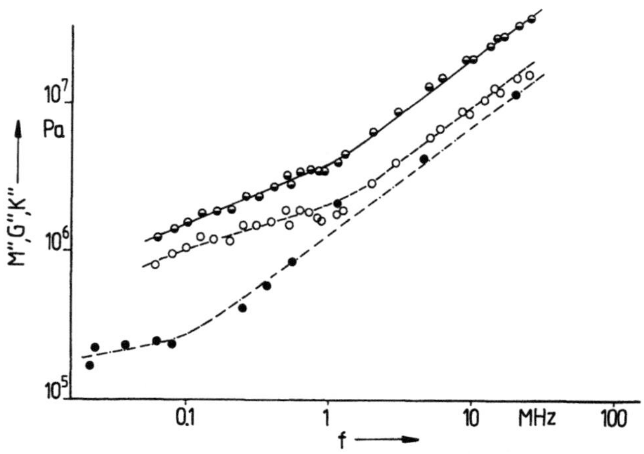

Fig. 9. Frequency dependence of $M''(\omega)$ (⊖), $K''(\omega)$ (○) and $G''(\omega)$ (●) for PB at 333 K

Because it was possible to measure the dynamic shear modulus $G''(\omega)$ separately [6], we could also determine the dynamic bulk modulus $K''(\omega)$. In the case of PB, the results for $M''(\omega)$, $G''(\omega)$ and $K''(\omega)$ are shown in Fig. 9. For frequencies higher than 3 MHz the three curves are parallel to each other and the ratio K''/G'' is constant and equal to 1.4 (compare [31, 32]). In the lower frequency range there is evidence that entanglements influence both shear and bulk properties. Furthermore, there are hints that for polymer melts shear and bulk properties in the entanglement region show different shift factors. This seems to be in contrast to the findings for solutions [31, 32]. For a better understanding of this behavior more precise measurements should be done.

Conclusions

By the combined application of ultrasound spectroscopy and photon correlation spectroscopy we have been able to provide a detailed description of the viscoelastic behavior of amorphous bulk polybutadiene. It was shown that the segmental relaxation times obtained from the two techniques over nine decades of time conform to a single WLF equation. The segmental relaxation process can be described by the Kohlrausch-Williams-Watts function. The ultrasonic absorption spectrum of the PB melt can be described as a superposition of a single relaxation process and a Rouse spectrum which is modified by effects of entanglements. The subchain motions could also be detected as weak correlation by dynamic light scattering. Separate shear wave investigations have shown that the subchain motions and entanglement effects influence both shear and bulk properties.

Acknowledgements

This work has been facilitated by the bilateral scientific agreement between the GDR and Greece.

I. A. thanks Dr. Yu. S. Manučarov and Dr. A. D. Bakhramov, Leningrad State University, for making available their equipment and for carrying out the shear measurements.

G. F. would like to acknowledge the financial support of the Research Center of Crete.

References

1. Bailey R, North AE, Pethrick RA (1981) Molecular Motion in High Polymers. Int. Series of Monographs on Chemistry. Clarendon Press, Oxford
2. Ferry JD (1980) Viscoelastic Properties of Polymers, 3rd Ed. John Wiley & Sons, New York
3. Pethrick RA (1983) Prog Polym Sci 9:197
4. Grigor'ev SB, Mikhailov IG (1981) Vysokomol Soed 22:1907
5. Alig I, Bakhramov AD, Manučarov YuS, Solovyev VA, Stieber F, Wartewig S (1987) Polymer 28:1543
6. Ahlig I, Stieber F, Manučarov YuS, Bakhramov AD (1989) Polymer 30, in press
7. Bell W, North AE, Pethrick RA, Teik PB (1979) J Chem Soc, Faraday Trans II 15:1115
8. Abubakar AM, Pethrick RA (1986) Polym Comm 27:194
9. Alig I, Grigor'ev SB, Manučarova YuS, Manučarov SA (1986) Acta Polymerica 37:699 and 733
10. Onabajo A, Dorfmüller Th, Fytas G (1987) J Polym Sci, Polym Phys Edn 25:749
11. Patterson GC (1983) Adv Polym Sci 48:125; and Fytas G, Wang CH, Fischer EW, Mehler KJ (1986) J Polym Sci, Polym Phys Edn 24:1859
12. Wang CH, Fischer EW (1985) J Chem Phys 82:632; and Meier G, Hagenah JU, Wang CH, Fytas G, Fischer EW (1987) Polymer 28:1640
13. Manučarov YuS, Mikhailov IG (1974) Akust Z 20:288
14. Hauptmann P, Alig I, Säuberlich R (1982) Acustica 50:189
15. Alig I, Hempel G, Lebek W, Stieber F (1989) Experimentelle Technik der Physik 38, in press
16. Borodin VN, Grigor'ev SB, Kulesov B (1974) Akust Z 20:360
17. Hauptmann P, Säuberlich R (1981) Exp Tech Phys 30:21
18. Stieber F (1988) Dissertation A, Merseburg
19. Grigor'ev SB, Mikhailov IG, Moiseev AI (1979) Vestnik LGU Fiz Khim 10:42
20. Rouse PE (1953) J Chem Phys 21:1272
21. Zimm BH (1956) J Chem Phys 24:269
22. Gotlib JJ, Salichov KM (1963) Akust Z 9:301; and Gotlib JJ, Darinskij AA (1973) Akust Z 19:67
23. Berger HR, Straube E (1985) Acta Polymerica 36:652
24. Berger HR, Heinrich G, Straube E (1986) Acta Polymerica 37:226
25. Edwards SF, Grant JWV (1973) J Physics A 6:1169 and 1186
26. Heinrich G, Straube E (1984) Acta Polymerica 35:115
27. Kohlrausch R (1847) Pogg Ann Phys 12:393 and Pogg Ann Phys IV-91:56, 79
28. Ngai KL, Rendell RW, Rajogopal AK, Teitler S (1986) in: Angell CA, Goldstein M (eds) Annals of the New York Academy of Science, New York, vol 484:150–180
29. Aharoni SM (1983) Macromolecules 16:1722
30. Alig I, Stieber F, Wartewig S, Fytas G (1988) Polymer 29:975
31. Solovyev VA, Manucharov YuS, Alig I (1989) Acta Polymerica 40, in press
32. Alig I, Stieber F, Bakhramov AD, Manucharov YuS, Solovyev VA (1989 submitted for publication in Polymer)

Received January 30, 1989;
accepted July 6, 1989

Authors' address:

Prof. Dr. S. Wartewig
Technische Hochschule "Carl Schorlemmer"
Leuna-Merseburg, Sektion Physik
Otto-Nuschke-Str. 42
DDR-4200 Merseburg

Progress in Colloid & Polymer Science Progr Colloid & Polymer Sci 80:180–186 (1989)

Small-angle neutron scattering investigation of phase separation kinetics of a diblock copolymer

J. G. Connell[1]) and R. W. Richards[2])

[1]) Department of Pure and Applied Chemistry, University of Strathclyde
[2]) Department of Chemistry, Science Laboratories, University of Durham, UK

Abstract: The application of radiation scattering methods to investigate the phase separation kinetics of block copolymers is discussed from the view point of the random phase approximation. Experimental data for a series of equilibrium and time-resolved experiments have been obtained using small-angle neutron scattering (SANS). Values of the interaction parameter and the Onsager coefficient (characterising copolymer interdiffusion processes) have been obtained by non-linear least squares fitting to data. The values of the Onsager coefficient are compared with available theoretical estimates.

Key words: Block copolymers; random phase approximation

Introduction

There has been considerable interest in polymer blends particularly with modelling of phase separation mechanisms, kinetics of phase separation processes [1, 2], polymer interdiffusion and relaxations *within* homogeneous one phase regions [3]. Less interest has been shown in block copolymers due to an absence of detailed theory and a scarcity of experimental data. We present here an outline of our recent work on the application of the random phase approximation to phase separation processes of block copolymers. Small-angle neutron scattering (SANS) has been used to provide the experimental data and a summary of the salient theory is included. More detailed discussions of the work will be presented elsewhere.

Background theory

We consider a general, two-component polymer system, poly A and poly B, that could be mixtures of homopolymers or a block copolymer [4–6]. The state of the system is described by a local monomer density, $\varrho_i(\boldsymbol{r})$, where $\varrho_i(\boldsymbol{r})$ is defined as the number of monomer units of type i at vector position $\boldsymbol{r}$ divided by the total number of monomers in the system. In the homogeneous single phase state

$$\varrho_A(\boldsymbol{r})+\varrho_B(\boldsymbol{r})=1 \ , \tag{1}$$

and the average composition, x, is given by

$$x=\langle\varrho_A(\boldsymbol{r})\rangle \ . \tag{2}$$

Although a homogeneous phase exists there will be local thermally driven fluctuations in $\varrho_i(\boldsymbol{r})$ which ultimately cause phase separation when the fluctuation amplitude exceeds a critical value. The phase of the polymer system (i.e. homogeneous, one-phase or phase separated, two-phase) is described by an order parameter, $\psi(\boldsymbol{r})$, that is

$$\psi(\boldsymbol{r})=\langle\langle(1-x)\varrho_A(\boldsymbol{r})-x\varrho_B(\boldsymbol{r})\rangle\rangle \tag{3}$$

$$=\langle\varrho_A(\boldsymbol{r})-x\rangle \ . \tag{4}$$

For a single-phase system $\psi(\boldsymbol{r})=0$. The intensity of radiation scattering from the polymer system is determined by the square of the Fourier transform of the order parameter,

$$I(Q)\alpha\,|\psi(Q)|^2 \tag{5}$$

where Q is the scattering vector given by $Q=(4\pi/\lambda)\sin\theta$ for radiation of wavelength λ in the polymer system and scattering angle 2θ. Application of the random phase approximation to this situation by de Gennes [4] has produced detailed expressions for $I(Q)$ involving the Flory-Huggins polymer-polymer interaction parameter, χ.

$$I(Q)=[(S(Q)/W(Q))-2\chi]^{-1} \tag{6}$$

where $S(Q)$ is the sum of all possible monomer-monomer correlation functions and $W(Q)$ is the determinant of those correlation functions. Detailed expressions for homopolymer mixtures have been provided by de Gennes [4]; the important parameter of the system is identified as the products χN, where N is the degree of polymerisation of poly A *and* poly B. For a homopolymer mixture with $x = 1/2$, the critical value of χN for spinodal decomposition is $(\chi N)_c = 2$, and at this point the scattered intensity at $Q = 0$ is infinite.

For linear diblock copolymers, the detailed solution of Eq. (6) was provided by Leibler [5]:

$$I(Q) = N/(F(u) - 2\chi N) \tag{7}$$

$$N = N_A + N_B$$

$$F(u) = g_D(1, u)/\{g_D(x, u)\, g_D((1-x), u) - \tfrac{1}{4}[g_D(1,u) - g_D(x,u) - g_D((1-x),u]^2\}$$

$$g_D(x, u) = (2/u^2)(xu + \exp(-xu) - 1)$$

$$u = Q^2 R_{gi}^2$$

R_{gi} = radius of gyration of component i .

For a diblock copolymer with $x = \frac{1}{2}$, the critical value of $(\chi N)_c = 10.495$ [7]. Expressions for other types of copolymers and mixtures of copolymers and homopolymers have been provided by Morii [6]. At the critical value of χN the scattering is again infinite. However, this infinite value of the scattered intensity is at a *finite* value of Q by contrast with that of homopolymers which is at $Q = 0$. Figure 1 shows the calculated scattering profile for a linear diblock copolymer in the homogeneous region. As the value of χN increases (equivalent to temperature *decreasing* for a system with UCST behavior), the maximum increases in amplitude but does *not* move from the initial value of $(Q_{\max} R_g)$.

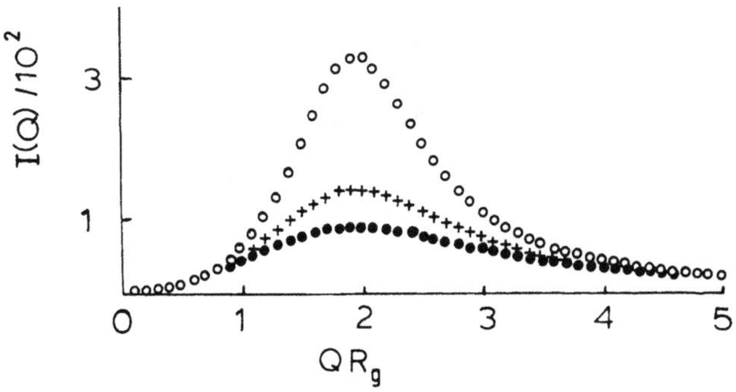

Fig. 1. Scattering profile calculated from the random phase expression for a diblock copolymer with $x = \frac{1}{2}$; $\chi N = 9$; $+ \chi N = 7$; $\chi N = 5$

Kinetics of phase transition

Phase transition of homopolymer blends are generally described using variations on linear Cahn-Hilliard spinodal decomposition theory [8]; there are many discussions of this available in the literature [9].

For diblock copolymers, Hashimoto [10] has proposed a phenomenological time-dependent Ginzburg-Landau model. The Q-Fourier component of the order parameter is given by the diffusion equation

$$\delta\psi(Q,t)/\delta t = L_o \nabla^2 (\delta \Delta F/\delta\psi)_Q \tag{8}$$

with L_o = *Onsager coefficient*

ΔF = coarse-grained free energy function $= \frac{1}{2} \sum S^{-1}(Q)_{|\psi Q|^2}$,

where $S^{-1}(Q)$ is the Fourier transform of the monomer density correlation function which is $I(Q)$ for the equilibrium disordered state of the block copolymer and random thermal fluctuations have been neglected. Eventually one obtains

$$\psi(Q,t) = \psi(Q,O) \exp(L_o Q^2(-S^{-1}(Q))t) \tag{9}$$

whence the evolution of the scattered radiation intensity following a sudden 'quench' of the system from the homogeneous to a phase separated state is:

$$I(Q,t) = \langle|\psi(Q,t)|^2\rangle = I(Q,O)\exp(2R(Q)t)$$

with

$$R(Q) = L_o Q^2(-S^{-1}(Q)) \tag{10}$$

= first order relaxation rate constant.

Similar to the scattered intensity, $R(Q)$ also has a maximum at a finite value of Q, however, unlike the intensity $R(Q)$ does *not* become discontinuous when $(\chi N)_c$ is exceeded. Figure 2 shows calculated values of $R(Q)$ for a diblock copolymer with $x = \frac{1}{2}$ as a function of QR_g.

Unlike a homopolymer mixture, block copolymers have an *organised* phase separated state with well defined long-range order. Indeed, Leibler [5] points out that on cooling block copolymers they will

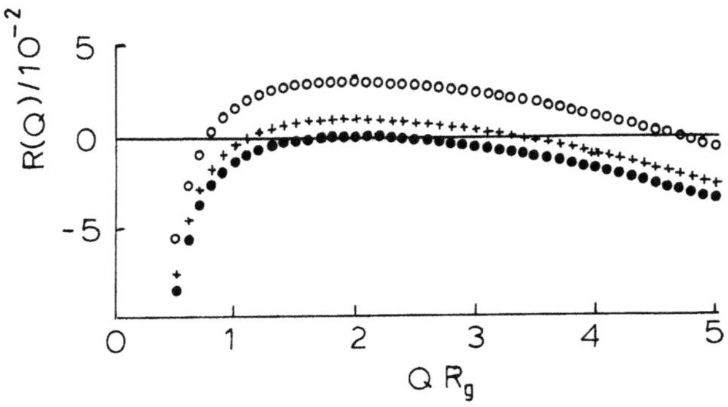

Fig. 2. Calculated variation of $R(Q)$ with QR_g for a diblock copolymer with $x=\frac{1}{2}$; $\chi N = 25$; $+\chi N = 15$; $\chi N = 10$

transform to this ordered state *before* the critical value of χN is reached.

Experimental

Small-angle neutron scattering (SANS) experiments were performed on two types of styrene-isoprene (SI) diblock copolymers, the molecular characteristics of which are given in Table 1. For the SANS experiments the styrene blocks were prepared from perdeutero-styrene and measurements were made on concentrated solutions of these copolymers in cyclohexane with $40 \leqslant C\% \leqslant 77$. Concentrated solutions were chosen for three reasons:

i) reduction of the order-disorder temperature to a more amenable value;
ii) as an aid to heat transfer in quenching experiments;
iii) by mixing H and D isomers of cyclohexane together, only the scattering from the polydeuterostyrene component was observed.

Two types of SANS experiments were performed;

i) At a series of fixed temperatures to evaluate the order-disorder temperature and the value of χ.
ii) Time-resolved experiments where the change in scattered intensity as a function of time was followed after a sudden quench of the system to a lower temperature.

All SANS experiments were performed at the Institut Laue-Langevin, Grenoble, France, using the D11 neutron diffractometer.

Table 1. Molecular characteristics of linear diblock copolymers

Copolymer	$M_w/10^3$	M_w/M_n	w_s*
S	85.63	1.04	0.19
C	84.78	1.03	0.38

* Weight fraction of deuterostyrene

Results and discussion

Equilibrium experiments

Figure 3 shows the typical scattering profiles for copolymer C at two temperatures. At the lower temperature a series of reasonably well resolved maxima are evident which are compatible with a hexagonally close packed array of cylinders [11]. At higher temperatures the higher order maxima disappear and eventually only a broad featureless maximum is observed which decreased in amplitude as the temperature is increased still further. From Eq. (7) we have

$$\tilde{I}(Q)^{-1} = (F(u)/N) - 2\chi \tag{11}$$

where $\tilde{I}(Q)$ is the experimentally measured scattering intensity and $\tilde{I}(Q) = K\,I(Q)$ with K = contrast factor, and generally,

$$\chi = A + B/T \tag{12}$$

where A and B are dependent on the polymer system, hence

$$\tilde{I}(Q)^{-1} = (F(u)/N) - 2(A + B/T)\ . \tag{13}$$

Thus within the homogeneous disordered phase at a fixed value of Q, $I(Q)^{-1}$ is linear in $1/T$. Figure 4 shows $\tilde{I}(Q_{max})^{-1}$ plotted in this manner for solutions of copolymer C, two branches are clearly evident in each set of data. At high values of T the data is strongly dependent on $1/T$ and these temperatures are within the homogeneous region. At lower temperatures, the dependence of $\tilde{I}(Q_{max})^{-1}$ on $(1/T)$ is much reduced as expected in the two phase region.

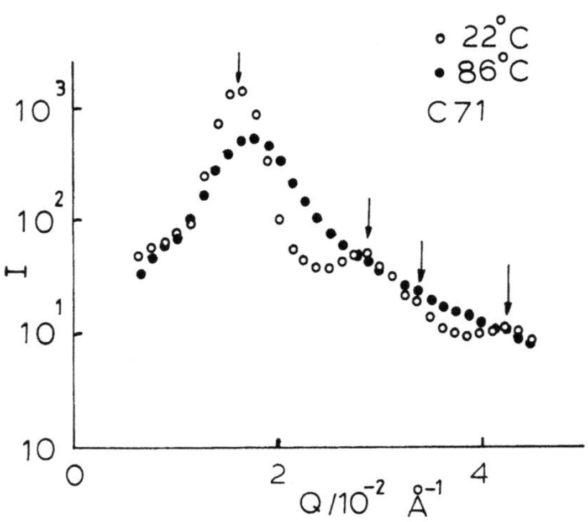

Fig. 3. SANS profile for copolymer C at a concentration of 71% (w/v). The arrows indicate the position of maxima expected for a hexagonally close-packed array of cylinders

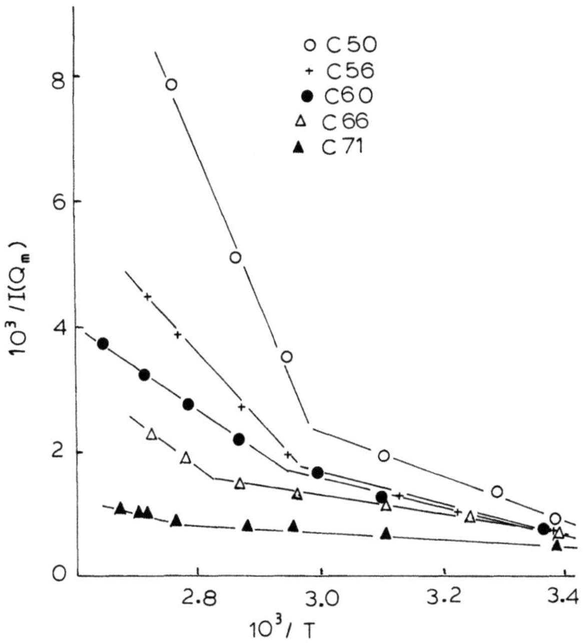

Fig. 4. Reciprocal maximum intensity plotted as a function of $1/T$ for solutions of copolymer C

For the evaluation of χ (and R_g) Eq. (7) was fitted to the scattering profiles by a non-linear least squares procedure with χ and R_g as the adjustable fitting parameters; Fig. 5 shows a typical result of this fitting procedure. A cautionary note is appropriate here: it is possible to obtain reasonable fits of the theoretical expression to the data even in the phase separated region (Eq. (7) is only strictly valid in the homogeneous phase). However, the values of χ obtained, and more particularly the dependence on temperature, clearly show the onset of the two-phase region. Due to the presence of solvent the χ value obtained is an effective

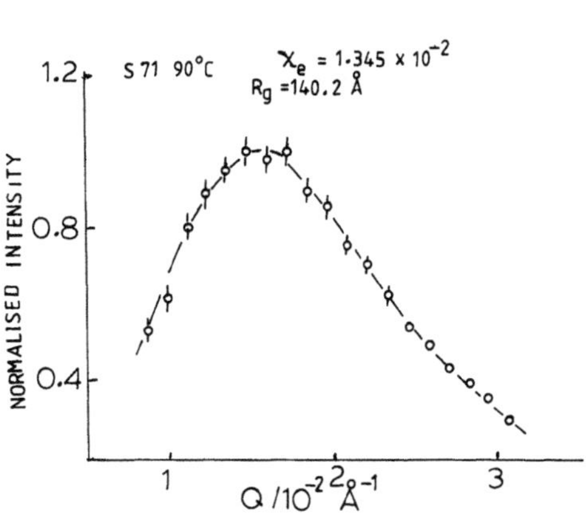

Fig. 5. Fit of random phase approximation equations (solid line) to SANS data for copolymer S at 71% concentration and 363 K

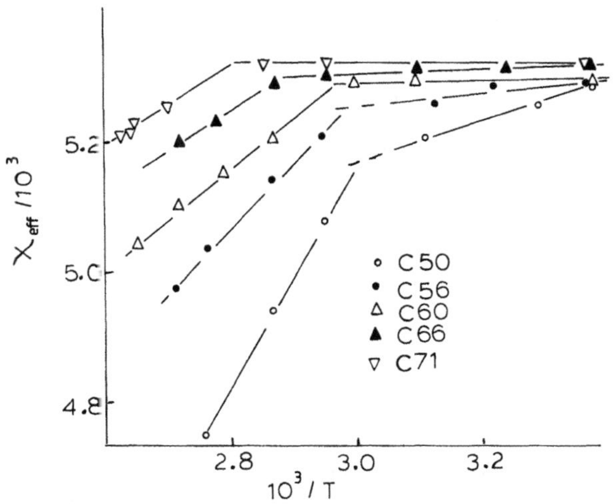

Fig. 6. χ_{eff} as a function of $1/T$ for copolymer C solutions. Cnn refers to the concentration of nn% w/v

χ value (χ_{eff}), nonetheless as Fig. 6 shows, the homogeneous region is clearly evident as the strongly temperature dependent part of the plot.

An expression involving the χ parameter for the bulk polymers (χ_B) and the volume fraction of polymer (Φ_P) can be writen for χ_{eff} as

$$\chi_{eff} = \chi_B - c\chi_B + c\chi_B\Phi_P$$

$$= \chi_B - c\chi_B(1-\Phi_P) \tag{14}$$

where c is a constant for a particular value of Φ_P and temperature.

By extrapolation of χ_{eff} as a function of Φ_P for all temperatures in the *disordered* phase of the copolymer systems, the relations below for the two copolymers are obtained.
Copolymer S

$$\chi_B = -0.073 + (40.49/T)\ ; \tag{15}$$

copolymer C

$$\chi_B = 0.32 \times 10^{-3} + (4.71/T) \tag{16}$$

with T in Kelvin.

Time-resolved experiments

A series of scattering intensity profiles as a function of Q for a range of times after a sudden quench to a lower temperature is shown in Fig. 7. The intensity increases as time increases, as expected from the model for the phase separation kinetics discussed above. For

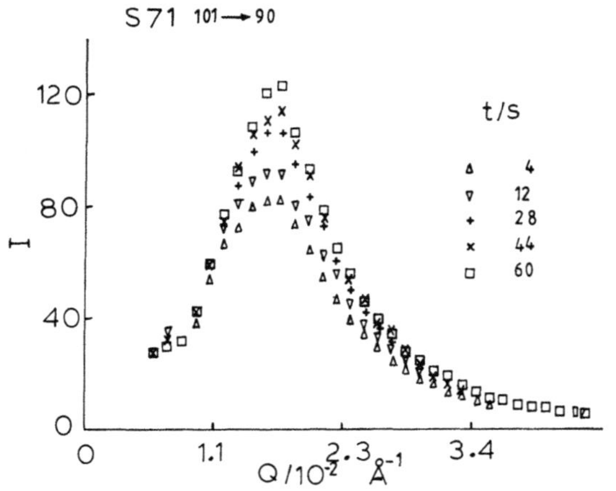

Fig. 7. Evolution of SANS intensity for copolymer solution S71 after a quench from 374.6 K to 363 K for the times (in seconds) indicated. The quench is within the disordered homogeneous phase

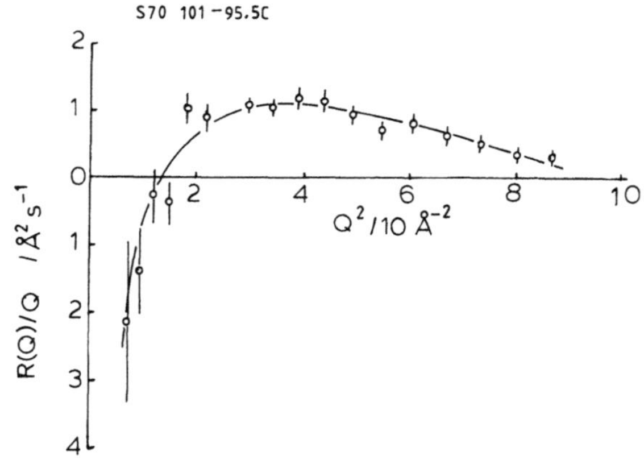

Fig. 9. $R(Q)/Q^2$ as a function of Q^2 for a quench of copolymer solution S71 from 374.6 K to 368.8 K. Solid line is a fit of theoretical equation to the data

a fixed value of Q, Eq. (9) predicts that ln $I(Q,t)$ should be linear in t. Figure 8 confirms this prediction and values of $R(Q)$ in Eq. (9) were extracted from the slopes of these lines. The values of $R(Q)$ obtained in this manner were used to obtain values of the Onsager coefficient by non-linear least squares fitting of Eq. (10). For this purpose $R(Q)/Q^2$ was plotted as a function of Q^2 and $S^{-1}(Q)$ was replaced by the Leibler random phase expression, Eq. (6); the adjustable parameters of the fit were L_o and x. An example of the quality of the fit is shown in Fig. 9, with values of L_o obtained for one solution given in Table 2. A notable feature of Fig. 9 is the occurrence of *negative* values of $R(Q)/Q^2$ at low Q values, i.e., the thermal fluctuations at these wave vectors are decaying in amplitude and there is no resultant driving force for phase separation on length scales given (approximately) by the reciprocal of these wave vectors. Notwithstanding this agreement between experimental data and theoretical predictions at low scattering vector values, there is disagreement at high values of Q.

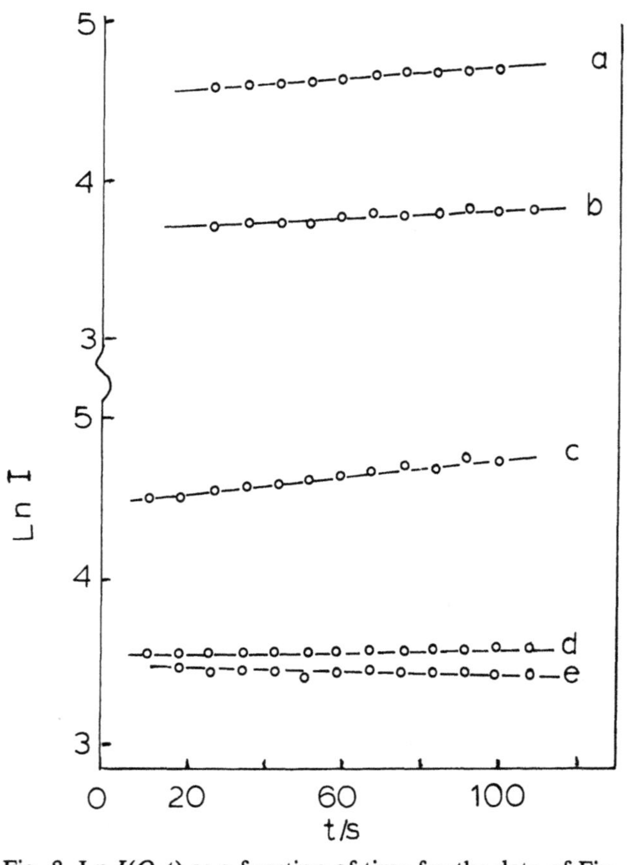

Fig. 8. Ln $I(Q,t)$ as a function of time for the data of Fig. 7 at selected Q values. a) $Q = 0.019$ A^{-1}; b) $Q = 0.026$ A^{-1}; c) $Q = 0.014$ A^{-1}; d) $Q = 0.009$ A^{-1}; e) $Q = 0.005$ A^{-1}

Table 2. Onsager coefficients (L_o) obtained by fitting RPA equations to $R(Q)/Q^2$ vs Q^2 data

Quench temperature/K	$L_o/(A^2\,s^{-1})$
368.7	107 ± 3
363.1	223 ± 1
359.2	872 ± 1
356.4	1324 ± 6
353	1507 ± 5
350.7	2482 ± 22
343.6	3123 ± 8
333.2	3893 ± 7
323.0	4329 ± 7
316.0	3300 ± 3

Data for copolymer S at a concentration of 71% (w/v) quenched from 374 K.

The experimental data does not have the negative values of $R(Q)$, that is, an upper critical wavenumber limit to phase growth, predicted by the Landau-Ginzburg model. This could be due to the neglect of either any Q-dependence of the Onsager coefficient, L_o, or the neglect of random thermal fluctuations in the system as discussed by Cook [12]. However, the latter have been shown to be negligible compared to the background *incoherent* scattering in SANS experiments [13] and this latter scattering has been subtracted from the data presented here.

Binder [9] has given an expression for a Q-dependent Onsager coefficient in homopolymer blends:

$$L_o(Q) = (1-\Phi_A)\sigma^2 W_{AB} S_A(Q)/N_A \tag{17}$$

where σ is the length of statistical step and W_{AB} is the rate of motion of a subunit in a polymer molecule. For homopolymer blends it is estimated that $10^9 \leqslant W_{AB}(/s^{-1}) \leqslant 10^{11}$ and hence $10^9 \sim L_0(/A^2 s^{-1}) \sim < 10^{11}$. Our values here are some seven orders of magnitude smaller. This reduction may possibly be associated with the very strong long-range ordering in these copolymer systems. Such long-range ordering requires significant mass transport over long distances through the system, so that like components can eventually aggregate in an ordered way. This long-range periodicity becomes very evident in the Q-dependence of the growth rates of the Q Fourier components of the thermal fluctuations, when systems are deeply quenched (Fig. 10).

Conclusions

We have clearly demonstrated that the random phase approximation first expounded by Leibler is a good description for the equilibrium homogeneous disordered state of linear diblock copolymers. Values of the effective interaction parameter, χ_{eff}, can be evaluated with accuracy from the intensity variation with Q of scattered radiation. By making measurements as a function of copolymer concentration and temperature, the bulk interaction parameter and its temperature dependence can be established.

The initial stages of the relaxation of the block copolymer system on quenching from a one-phase system to a temperature where eventually a two-phase system will be obtained is reasonably well described by a time-dependent Ginzburg-Landau model. The Onsager coefficients obtained for the block copolymer interdiffusion process are much smaller than values estimated for homopolymer blends. It is speculated that this may be due to long-range mass transfer needed to eventually establish the para-crystalline of the final microphase separated solid state.

All of the theory is applicable to small-angle x-ray scattering and for many systems the appropriate data could be obtained by this technique. However, many polymers have very similar electron densities, making the signal from small-angle x-ray scattering weak. On the other hand, the prospect of selectively deuterium-labelling one component of a block copolymer makes SANS much more widely applicable for the examination of random phase approximation theories.

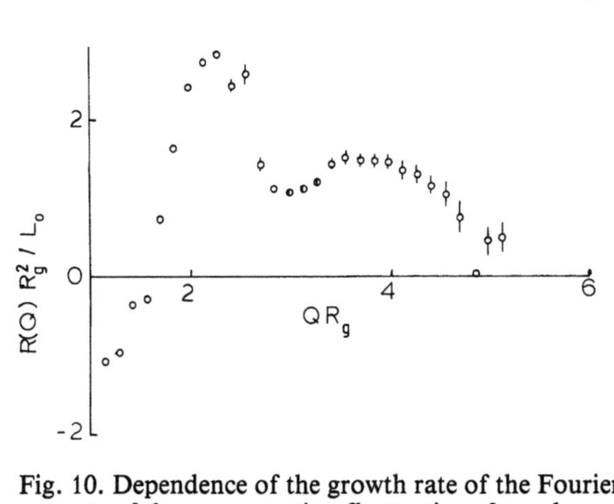

Fig. 10. Dependence of the growth rate of the Fourier components of the concentration fluctuations for a deep quench of copolymer solution S71 from 374 K to 316 K

Acknowledgements

The Science and Engineering Research Council of the UK is thanked for financial support of this research. Both authors are particularly grateful to Dr. A. R. Rennie of the Institut Laue-Langevin for his assistance in the time-resolved experiments.

References

1. Binder K (1987) Coll Polym Sci 265:273
2. Higgins JS (1987) In: Polymer Motion in Dense System. Springer, Berlin Heidelberg New York
3. Strobl G (1987) In: Polymer Motion in Dense Systems. Springer, Berlin Heidelberg New York
4. de Gennes PG (1979) Scaling Concepts in Polymer Physics. Cornell University Press, London
5. Leibler L (1980) Macromolecules 13:1602
6. Mori K, Tanaka H, Hashimoto T (1987) Macromolecules 20:381
7. The model used by Leibler is that pertaining to weak segregation of the components in the block copolymer, i.e. $\psi(r)$ changes smoothly over the wavelength of a concentation fluctuation. Models for the strong segregation limit more appropriate to microphase separated copolymers have been provided by Ohta and Kawasaki (Macromolecules 19:2621 (1986)) and Fredrickson and Helfand (J Chem Phys 87:697 (1987)). By this model the latter obtain $(\chi N)_c = 10.495 + 41.02\,N^{-1/3}$

8. Cahn JW, Hillard JE (1958) J Chem Phys 28:258
9. Binder K (1983) J Chem Phys 79:6387
10. Hashimoto T (1987) Macromolecules 20:465
11. Richards RW, Thomason JL (1983) Macromolecules 16:982
12. Cook HE (1970) Acta Metall 18:297
13. Koberstein J (1982) J Polym Sci Polym Phys 20:593

Received January 27, 1989;
accepted May 16, 1989

Authors' addresses:

J. G. Connell
Department of Pure and Applied Chemistry
University of Strathclude
Glasgow G1 1XL, UK

R. W. Richards
Department of Chemistry
Science Laboratories
University of Durham
South Road
Durham DH1 3LE, UK

The role of diffusion of free volume at volume relaxation of amorphous polymers

B. A. Rozenberg, V. I. Irzhak and L. M. Bogdanova

Institute of Chemical Physics, Academy of Sciences of the USSR, Moscow Region, USSR

Abstract: Scale effect of density, elasticity modulus and annealing rate of amorphous polymers in glassy state were discovered. The indicated phenomena were due to the action of diffusion mechanism in volume relaxation process. They are described by formulated mathematical model. Superposition of diffusion and local volume relaxation mechanisms results in the appearance of property gradient in the surface layer of polymer.

Key words: Volume relaxation; diffusion of free volume; density fluctuation; surface layers; scale effect of properties

Introduction

It is assumed that the process of volume relaxation in amorphous polymers proceeds locally. This implies that the correlation between the rise and disappearance of free volume excess in space is absent. However, Frenkel [2] considered that the motion of free volume in amorphous bodies can proceed by the same mechanism as in crystalline materials.

In crystalline materials the main carrier of free volume is a hole (i.e., vacancy). The mechanism of its formation has been suggested by Frenkel [2]. The hole arises by the mechanism of "inner evaporation of atoms", i.e., by the migration of atoms into interstitial sites that can freely diffuse through the volume of a body until it interacts with an atom from the interstitial site (i.e., the opposite process to the formation of a vacancy) or until it is absorbed by free surface. The latter process is actually volume relaxation (if the system is supersaturated with holes).

The hole formation in crystal requires high energy for the deformation of a lattice. That is why rather large holes (pores) may arise only in defect areas, e.g., in a joint of crystal grain. The change of pore dimensions also proceeds by the vacancy mechanism, i.e., it is due to the dispersion of pores into vacancies or coalescence of vacancies into a pore.

The diffusion mechanism of volume relaxation in crystalline materials causes the scale effect: apparent viscosity (or characteristic relaxation time) varies as the square of the distances between the places of rise and disappearance of the vacancy. The characteristic relaxation time is proportional to the square of sample linear dimension when the places of rise and disappearance of the vacancies is sample surface. In crystalline materials the lifetime of the vacancies is long due to the rate of their absorption with the sample surface. Therefore the lifetime of a hole in amorphous material is extremely short [4]. This is the reason to consider that the diffusion mechanism of volume relaxation (the relaxation due to the diffusion of a hole to the sample surface) does not take place in amorphous polymers. Today this concept is generally accepted.

However, our study [5–8] on the dependence of density and annealing rate upon the film thickness of amorphous polymers has shown that the scale effect is also observed in this case.

Experimental

The films of cured epoxy resins were obtained by the condensation of epoxy oligomers with *m*-phenylenediamine (Ia), *p*-phenylenediamine (Ib), or phenol-formaldehyde resin of resol type (II). The polymers Ia and Ib were cured to full conversion at 70 °C for 5 h and at 150 °C for 3 h, respectively. Amine was dissolved in liquid resin, and air bubbles were evacuated. The mixture was then filled between the glass plates treated with dimethyldichlorosilane to avoid adhesion. Thin films were formed by casting the initial oligomer mixtures on aluminum foil or aluminum. The film thickness was measured at five points chosen arbitrarily. The foil plates were mechanically removed from the films. The aluminum plates were stripped with 5% alkali solution.

The polymer II films of different thicknesses were obtained by applying the solution of epoxydiane oligomer (85 wt%, mw $4 \cdot 10^3$) and phenol-formaldehyde oligomer of resol type (15 wt%, mw 600 – 700) in ethyl-cellosolve to the base. Phosphoric acid was used as catalyst (0.35 wt%). The solvent was removed in a vacuum to the level weight of the sample. The samples of polymer II were cured at 250 °C for 10 min.

Polymethylmethacrylate films were obtained by applying the polymer solution of different concentration in dichloroethane to a horizontal glass surface treated with antiadhesive. The solvent was then evaporated.

The films thickness was measured on the Microtest instrument. The density was determined by the method of hydrostatic weighing.

The Instron-1122 and 1231 V-10 instruments were used to determine the tensile strength of the films at the deformation rate of $3.3 \cdot 10^{-5}$ and $0.8 \cdot 10^{-5}\,m \cdot s^{-1}$, respectively. The relative error at the determination of elasticity modulus E for the polymer II samples to 20 µm in thickness was 2.6%, and 10% – at the thickness less than 5 µm.

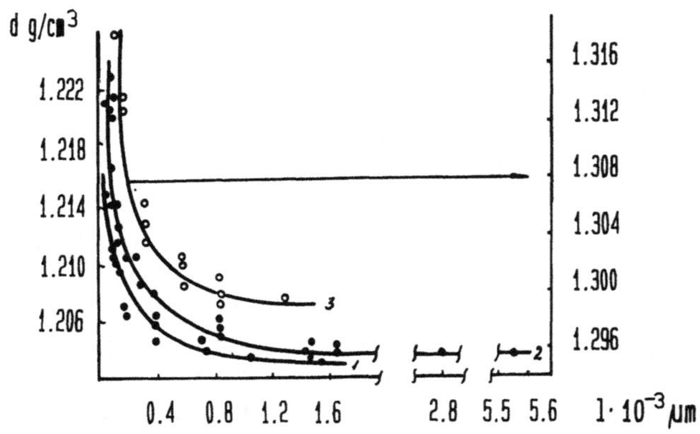

Fig. 1. Dependence of density (d) on polymer film thickness. 1) diepoxide of bisphenol $A + m$-phenylenediamine on aluminum foil; 2) the same on glass treated with dimethyldichlorosilane; 3) diglycidyl ether of resorcinol + m-phenylenediamine on aluminum plate. (Data taken from [7])

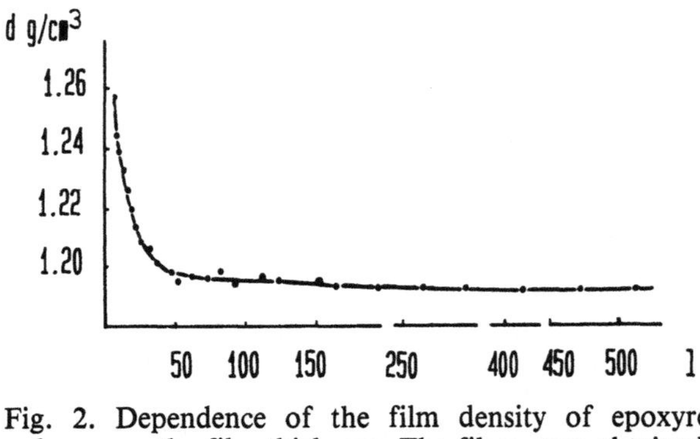

Fig. 2. Dependence of the film density of epoxyresol polymer on the film thickness. The films were obtained by the curing of resin on aluminum foil [7]

The polymers I and II were annealed, respectively, at 155 ° and 110 °C before slowly cooling them at a rate not exceeding 20 °C/h.

Vitrification temperature was determined by thermomechanical method.

The film structure of the polymers Ia and Ib was investigated by the method of angle dependence on birefringence.

Scale effect of amorphous polymer density

Figures 1 – 4 demonstrate the dependence of density of amorphous polymer films (obtained by different methods) on their thickness. As can be seen (recognizing that there is no dependence on the production methods: polycondensation or solvent evaporation

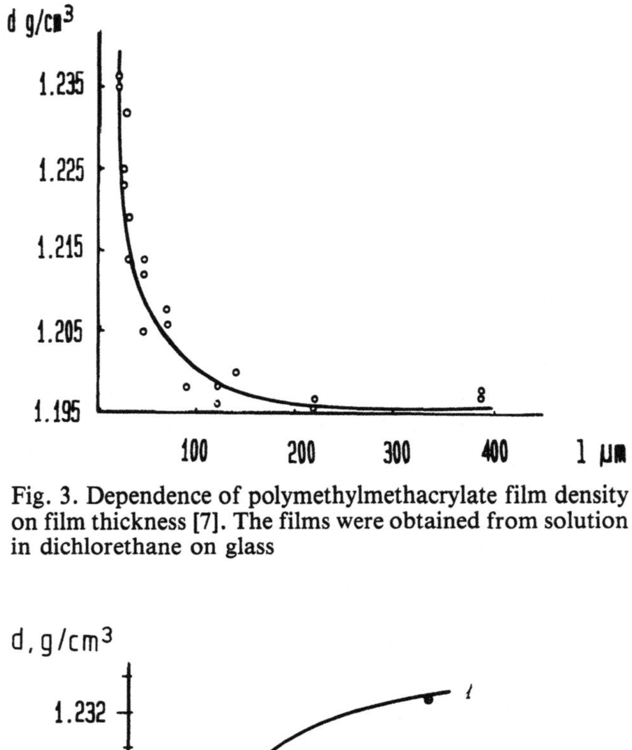

Fig. 3. Dependence of polymethylmethacrylate film density on film thickness [7]. The films were obtained from solution in dichlorethane on glass

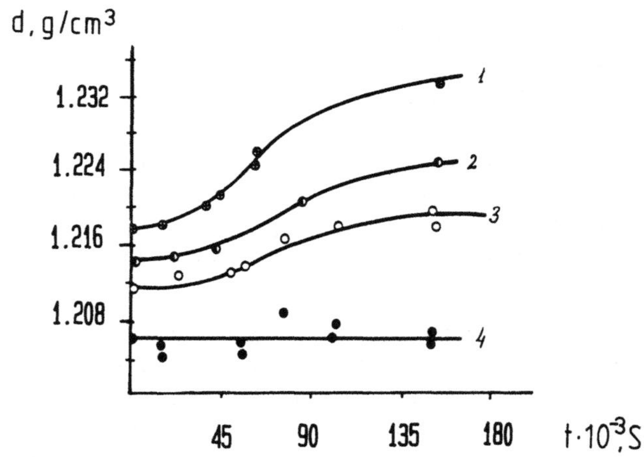

Fig. 4. Dependence of density of the films obtained on the glass treated with antiadhesive upon the annealing time at temperature 5 °C above T_g. The system is diepoxide of bisphenol $A + m$-phenylenediamine. The film thickness is (µm):1) 70; 2) 200; 3) 470; 4) 770

from polymer solution) thinner films have higher density, i.e., a distinct scale effect is observed. The films have been obtained on various surfaces: aluminum, glass treated with antiadhesive substances, and Vood alloy. Therefore, it is safe to say that the scale effect here is not due to the interaction of polymer with the base, but is inherent in the polymer system itself.

This conclusion for amorphous materials can be confirmed by the comparison of the distances of phenomenon scale with the characteristic distances of adsorption forces. As follows from the figures, the film density variations are observed up to a thickness of 600 μm. Usually the adsorption surface interactions in liquids and amorphous solids take place within several molecular layers [8]. For polymers this distance is identical to macromolecule radius [9]. Therefore these distances must not exceed a part of micrometer. The annealing of the films separated from the base experimentally proves that the observed scale effect is inherent in the polymer system itself.

As is obvious from Figs. 4 and 5, the thinner the films the higher the rate of film density increase at isothermal annealing. In this case the scale effect cannot be considered otherwise than the property of polymer body itself. All the kinetic curves can be described by the unique dependence in the case where the diffusion coordinates are used (the dependence of polymer density on the parameter of t/l^2, where l is film thickness).

Thus, it is quite natural to suppose that the increase of density of amorphous polymer films (in the process of their production and annealing) with decreasing thickness is caused by the diffusion of free volume excess from the film, i.e., by the phenomenon usually characteristic of crystalline bodies. This conclusion has been made by the authors [5–8, 10] on the basis of the above-mentioned experimental data.

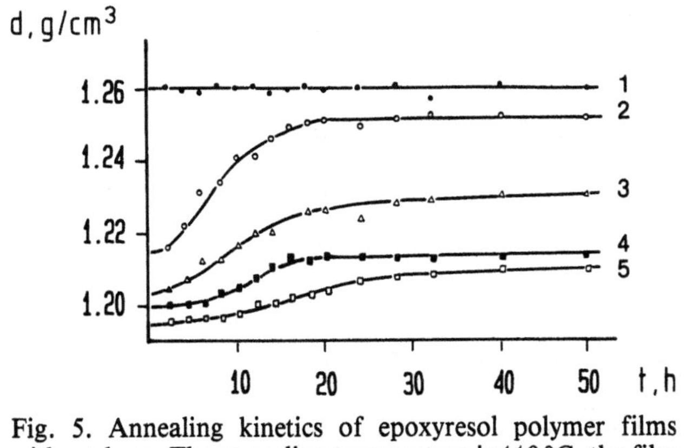

Fig. 5. Annealing kinetics of epoxyresol polymer films without base. The annealing temperature is 110 °C, the film thickness is (μm): 1) 4; 2) 20; 3) 35; 4) 90; 5) 340. (Data taken from [7])

It ought to be noted that there are a lot of indirect "hints" in the literature (see, e.g., [9]) pointing to the existence of this phenomenon, although it has been explained by the interaction of polymer system with the base or filler [9].

It has been stated [10], that the films of thermostable polymers have a higher packing factor (0.691) compared to the bulk polymer (0.685).

Analogous effect has also been found for powder polymers. Thus, in the work [11], devoted to comparative investigation of specific volume, coefficient of thermal expansion, and glass transition temperature of polystyrene in bulk and powder states, Kovacs et al. considered that there were no differences in their properties. However, it can be concluded from the data presented in their work that in some cases the specific volume of bulk polymer is higher than of powder one. For octane (polymers were studied in immersion liquids), say, the difference reaches 0.0022 cm^3/g.

As has been shown elsewhere [12, 13], in powder epoxy polymers the relaxation rate of stresses, accumulated during milling at temperature of liquid nitrogen, decreases with increasing particle dimensions. Since inner stresses are directly connected with particle deformation, specifically with expansion or compression on all sides, their relaxation should correlate with that of the volume. In that way the scale effect of volume relaxation is corroborated.

The experiments indicated above show that the scale effect of properties – first of all, density – is also restricted to amorphous polymers. This implies that volume relaxation in amorphous polymers is connected to some extent with the realization of diffusion mechanism as it takes place in crystalline bodies.

Diffusion model of volume relaxation of amorphous polymers

The volume relaxation in crystalline bodies is of diffusion nature because the free volume carrier is a long-living vacancy. Such long-living formations are absent in amorphous bodies.

How is the diffusion of the excess free volume realized in amorphous body? What is the carrier of free volume here? The works [14–16] on computer simulation of dence amorphous systems answer these questions.

According to these works, the defects resulting from local density deviation from average value ought to be characterized by local (on a given atom) tensor of stresses (its diagonal terms) or by respective tensor of deformation. In terms of quoted works the excess quantity of free volume around a given atom corresponds to n-defect, i.e., the concept of n-defect corresponds completely to that of a hole in Frenkel terms. Analysis shows that a single n-defect is an extremely instable formation, with a short lifetime.

A much more stable formation in an amorphous body is a cluster of n-defects including 10–20 particles. Suppose that the cluster of n-defects (CND) is a carrier of free (excess) volume, then its lifetime and the characteristic of wandering in space should determine the behavior of amorphous body volume relaxation by diffusion mechanism.

Density fluctuations with high volume value are in fact the CND [1]. Therefore, their lifetime should be calculated on the basis of the theory [1], i.e., from stochastic equation of fluctuations wandering in the space of their densities (volumes). It is obvious that sooner or later the CND must increase its density, i.e., disappear and transform into a cluster of p-defects*). This process could be precisely described, however the uncertainty of the concept of n- and p-defects prevents us from doing so. Apparently, some constant values of reversible quasi-chemical reactions of the CND appearance and disappearance can be assumed with a sufficient degree of accuracy. The rate constant of the CND disappearance determines the CND lifetime. Let us assume that in the first approximation a value of the CND lifetime is the function of temperature and pressure.

If a vacancy movement in crystalline body is connected with a jump of a particle into this vacancy, then this process seems to be more complicated for the CND, because the latter can be considered as a macroscopic formation in comparison with Frenkel's hole. Some possible mechanisms of the movement of macroscopic formation including pores have been analyzed in a monograph [17].

The movement of the CND is apparently realized due to the transition of kinetic units from one "wall" to another through the cluster space. The notion "wall" (being applied to fluctuation) is very conventional, however it is obvious that material redistribution inside the CND without considerable change of its mean density can be perceived as movement in space.

Such movement mechanism suggests that the rate of the CND motion is independent of the medium and only depends on the parameters of the CND itself, i.e., on its local density.

*) According to the works [14–16], p-defect corresponds to fluctuation with higher density as compared with average density

This implies that the movement of the CND in space does not depend on the total fraction of free volume as in the case of atom or molecule diffusion in amorphous body.

In the suggested model the CND with finite lifetime and constant diffusion coefficient is the carrier of excess free volume. Having reached the sample surface the CND can only diminish its excess volume and thus decrease the total volume of the system.

Proceeding from research [18] let us consider a simple one-dimensional scheme of the CND wandering within the section -1, $+1$. Taking into account the assumptions and considerations stated above it can be described by the following scheme:

$$\frac{\partial c}{\partial t} = D\frac{\partial^2 c}{\partial x^2} - \frac{\partial N}{\partial t}$$
$$\frac{\partial N}{\partial t} = k_1 c - k_2 N \tag{1}$$

with boundary

$$c(l,t) = c(-l,t) = 0 \; ; \quad \frac{\partial c}{\partial x}(0,t) = 0$$

and initial conditions: $c = c_0$ at $t = 0$, $x \neq \pm l$.

Here C and N are fluctuation concentrations, respectively, with low (CND) and high local densities. The latter practically do not move in space, because the wandering of the kinetic units inside them proceeds rather slowly due to high local density. k_1 and k_2 are effective constants of the CND appearance and disappearance.

As has been mentioned above, the approximation to the process of fluctuation wandering in local density space is rather rough, but it is of no importance for the quality analysis carried out below.

The solution of Eq. (1) leads to the following relationship:

$$c = \frac{4c_0}{\pi}\sum_{n=0}^{\infty}\frac{(-1)^n\cos(k_n x)}{(2n+1)\gamma(\bar{\bar{P}}_{1n} - P_{2n})}\Big\{p_{2n}(p_{1n}\tau - \gamma) \times e^{-p_{1n}t} - p_{1n}(p_{2n}\tau - \gamma)e^{-p_{2n}t}\Big\} \tag{2}$$

where $\tau = k_2^{-1}$ is the lifetime of the CND, $\gamma = c_0/N_0$, and $\bar{\bar{P}}_{1n}, \bar{\bar{P}}_{2n}$ are the roots of the equation

$$Dk_n^2 = p_n[1 - (p_n\tau - \gamma)^{-1}]$$
$$k_n = \frac{2n+1}{2}\frac{\pi}{l} \; .$$

If the CND quantity is small, i.e., $\gamma \ll 1$, then

$$p_{1n} \approx \frac{1}{\tau}(1+k_n^2 D\tau) \qquad p_{2n} \approx \frac{k_n^2 D}{1+k_n^2 D\tau}\gamma \tag{3}$$

provided $p_{1n} \gg p_{2n}$.

Subject to Eq. (3), Eq. (2) has the form:

$$c = \frac{4c_0}{\pi}\sum_{n=0}^{\infty}\frac{(-1)^n \cos(k_n x)}{2n+1} \times \frac{D\tau k_n^2 \exp\{-(1+k_n^2 D\tau)(t/\tau)\} + \exp\{-[(k_n^2 D\tau)/(1+k_n^2 D\tau)]\gamma(t/\tau)\}}{1+k_n^2 D\tau} . \tag{4}$$

The total excess free volume v is

$$v = \frac{1}{2lc_0}\int_{-l}^{l} c\,dx = \frac{8}{\pi^2}\sum_{n=0}^{\infty}\frac{D\tau k_n^2 \exp\{-(1+k_n^2 D\tau)\} + \exp\{-[(k_n^2 D\tau)/(1+k_n^2 D\tau)]\gamma(t/\tau)\}}{(2n+1)^2(1+k_n^2 D\tau)} . \tag{5}$$

As can be seen, the dependence of free volume on time is determined by two exponents, the index of the first being appreciably higher in comparison with the second. The first index may increase infinitely with increasing n, whereas the second does not exceed the value of γ.

It is clear that at rather low values of t the second exponent does not very much differ from 1 and

$$v = \frac{8}{\pi^2}\sum_{n=0}^{\infty}\frac{1 + Dk_n\tau \exp\{-(1+Dk_n^2\tau)(t/\tau)\}}{(2n+1)^2(1+k_n^2 D\tau)} . \tag{6}$$

At fairly high values of t the first exponent becomes zero, so

$$v \simeq \frac{8}{\pi^2}\sum_{n=0}^{\infty}\frac{\exp\{-[(k_n^2 D\tau)/(1+k_n^2 D\tau)]\gamma(t/\tau)}{(2n+1)^2(1+k_n^2 D\tau)} . \tag{7}$$

If the parameter of $D\tau$ is sufficiently high (high diffusion coefficient or long CND lifetime), then Eq. (5) changes into a simple diffusion one:

$$v = \frac{8}{\pi^2}\sum_{n=0}^{\infty}\frac{\exp\{-k_n^2 D\tau\}}{(2n+1)^2} .$$

Figure 6 presents the kinetic curves of volume relaxation of films calculated from Eq. (5). There are two regions of the process course respectively of long and short times. They are divided by the time interval, which is longer the lower the γ value. The interval is so great that it can be easily taken for the limit, because practically no changes take place in the system for a long time.

The limit of the v_∞ value is determined by the parameter value of $D\tau/l^2$, which is indicated in Fig. 7. Its sense is absolutely clear: a surface layer, whose thickness is of the order of magnitude of $\sqrt{D\tau}$ (free volume being in deficiency) is formed during the times commensurable with τ. This time is too short for the excess free volume to wander from the inside to polymer surface.

As can be seen from Fig. 7, at fairly low values of $D\tau/l^2$ ($D\tau/l^2 < 0.1$) we obtain

$$1 - v_\infty \simeq \left(\frac{D\tau}{l^2}\right)^{1/2} . \tag{8}$$

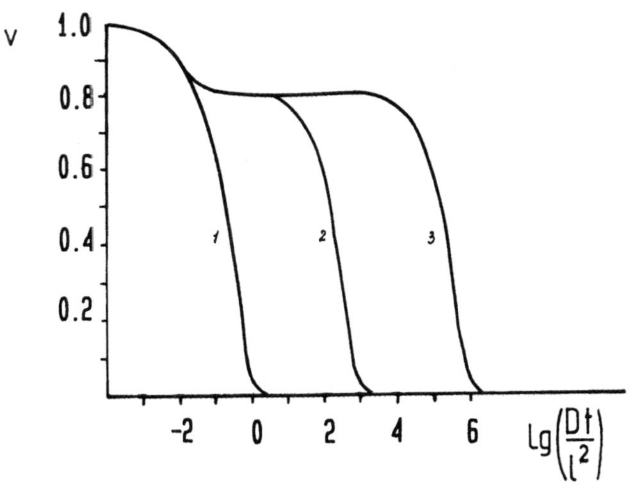

Fig. 6. Kinetic curves of the relaxation of the excess free volume v at $D\tau/l^2 = 4 \cdot 10^{-2}$. Curves 1–3 correspond to the values $\gamma \gg 1$, $\gamma = 10^{-3}$, $\gamma = 10^{-6}$, respectively.

At high values of $D\tau/l^2$, $(D\tau/l^2>1)$, as follows from Eq. (5)

$$v_\infty = \frac{1}{3}\frac{l^2}{D\tau} .$$

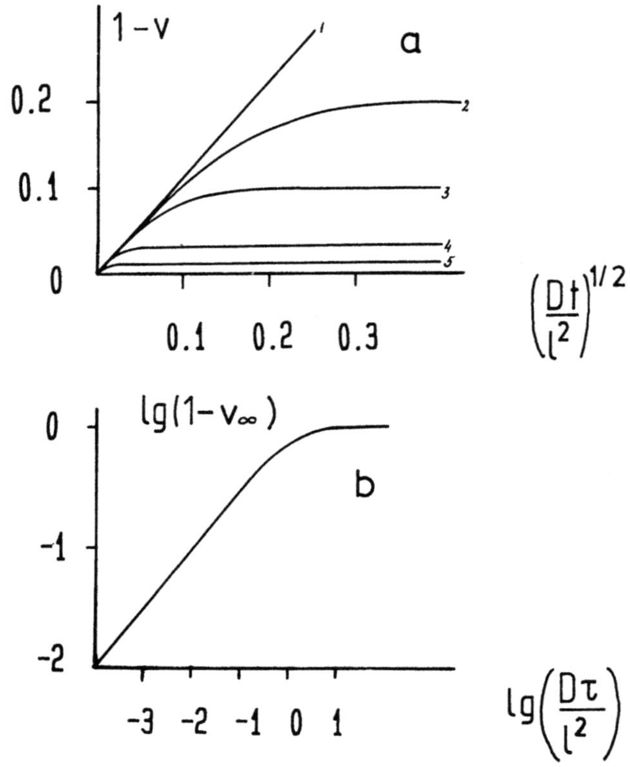

Fig. 7. Kinetic curves of the excess free volume relaxation at $D\tau/l_2 \gg 1$; $4\cdot10^{-2}$; 10^{-2}; 10^{-3}; 10^{-4} (curves 1–5, respectively); $\gamma = 10^{-6}$ (a); dependence of the limit excess free volume on the value of $D\tau/l^2$; $\gamma = 10^{-6}$ (b)

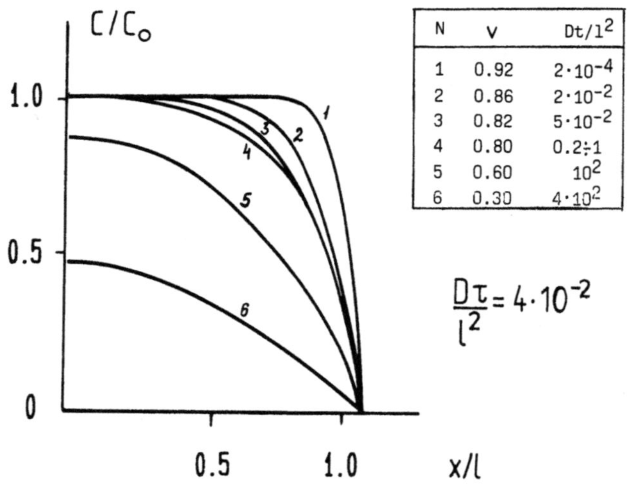

N	v	Dt/l^2
1	0.92	$2\cdot10^{-4}$
2	0.86	$2\cdot10^{-2}$
3	0.82	$5\cdot10^{-2}$
4	0.80	0.2÷1
5	0.60	10^2
6	0.30	$4\cdot10^2$

Fig. 8. Time variation of the concentration profile of the excess free volume. Curves 1–6 correspond to the time, at which the excess free volume has the values of: 0.92; 0.85; 0.82; 0.60; 0.30, respectively. $D\tau/l^2 = 4\cdot10^{-2}$; $\gamma = 10^{-6}$

The concentration profiles of free volume and their time variations are given in Fig. 8. Within small times the free volume concentration only changes in the region close to the sample boundary. Central regions of the sample are only changed at considerably longer times, which corresponds to the fall of the kinetic curves after plateau in Fig. 6.

Thus, the diffusion of the CND with limited lifetime results in the formation of dense surface layer of macroscopic thickness determined both by the CND lifetime and by the value of its diffusion coefficient.

Analysis shows that at the conversions, which are not very great $(v \geqslant 0.7)$, Eq. (5) approximates in the following way:

$$v \simeq 0.9\left(\frac{D\tau}{l^2}\right)^{1/2} . \tag{9}$$

Figure 9 presents the kinetic curves of films annealing (data from Figs. 4 and 5) on respective coordinates. As can be seen, the experimental data are de-

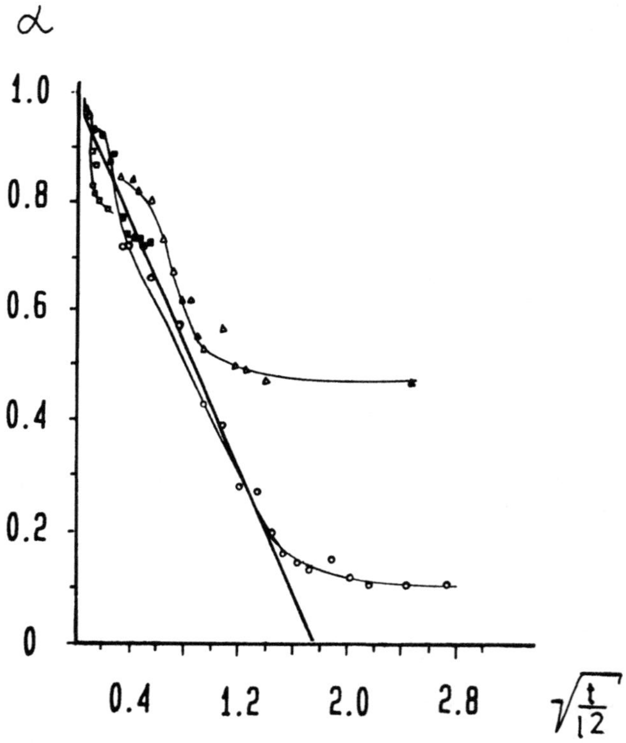

Fig. 9. Dependence of epoxy film density on the film thickness and time in diffusion coordinates. (Data and symbols from Fig. 5)

$$\alpha = \frac{d_\infty - d}{d_\infty - d_0} .$$

The thick line is the tangent of the annealing curves

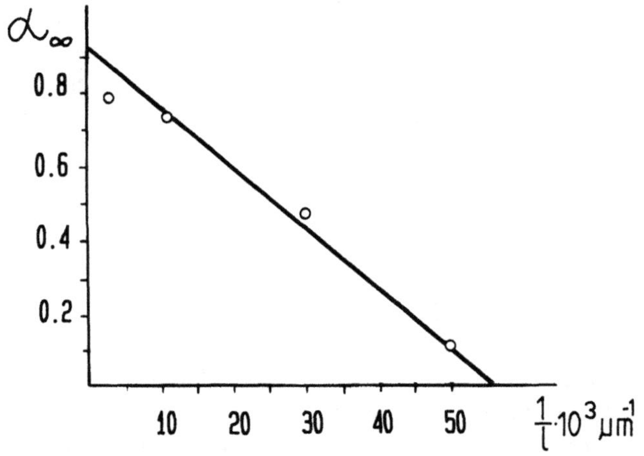

Fig. 10. Rectification of the limits of annealing kinetic curves (data taken from Fig. 5) in Eq. (8) coordinates

scribed by Eq. (9), the diffusion coefficient being equal to $10^{-14}\,m^2/s$.

The data on limits from Fig. 10 (experimental results taken from Fig. 5) are well described by Eq. (8); the CND lifetime is 10^3 s.

It is obvious that the free volume fraction, which in principle is unable to relax, cannot exceed $0.10 \div 0.13$ (the so-called Simha-Boyer constant K_1). The error in determination of body density is $0.001 \div 0.0005$ g/cm^3 (which is practically the same in terms of fraction, 0.001). Therefore, the limit value of the relaxed free volume cannot be less than 0.01. It is impossible to measure lesser values by the existing methods. So, the diffusion mechanism of volume relaxation may only be observed in the case, when

$$D\tau/l^2 \geqslant 10^4 \ . \tag{10}$$

Relationship (10) imposes a limit on the realization of the diffusion mechanism of free volume relaxation: at predetermined values of D and τ the geometrical dimensions of a sample cannot exceed some definite values. Proceeding from the values $D = 10^{-14}\,m^2/s$ and $\tau = 10^3$ s we obtain that the film thickness, where diffusion mechanism is realized, must not exceed $l = 1$ mm. The order-of-magnitude estimates correspond to the experimental results obtained.

Since the contribution of the diffusion mechanism into the total relaxation process is determined by the product of D and τ, the temperature conditions of its realization may be rather complicated. Thus, in rather low-viscous amorphous bodies the CND lifetime τ may be extremely short, so that the product of D and τ will be small inspite of the great value of D. The $D\tau$ value may also be small in glassy state, because of small D, although τ may be comparatively great. Temperature intervals for the realization of the diffusion mechanism of volume relaxation may be different. Unfortunately, the experimental data for well-founded conclusion are in deficiency.

However, in any case (at any values of $D\tau$) the diffusion mechanism will be realized when corresponding characteristic times are either commensurate or smaller than those of the local mechanism of relaxation. Since these times drop with increasing temperature, the samples must be thin to give the possibility of realization to the diffusion mechanism. Thus, the film where the diffusion mechanism of free volume relaxation may be detected, must be thinner, the higher the temperature of its annealing.

Note that the diffusion mechanism of free volume relaxation in amorphous polymers has been suggested earlier in the works [19, 20] to explain the observed features of amorphous polymers' volume relaxation [21–23].

All the basic features of volume relaxation such as non-linearity of isothermal relaxation, asymmetry of relaxation, "memory"-effects, and hysteresis of temperature changes of body volume are explained in these works. The authors managed to do so, allowing for the existence of non-linear dependence of the diffusion coefficient on the value of free volume.

This assumption is correct when the diffusion of material particles into the medium is considered. However, in the case of free volume diffusion in amorphous solids it has no physical meaning, because the diffusion coefficient is constant, due to the specific mechanism of diffusion. That is why all the known features of the volume relaxation in amorphous polymers cannot be explained by the diffusion mechanism.

It will be noted that all the features of amorphous polymers volume relaxation can be explained naturally by local relaxation mechanism in the framework of lately developed theory [1]. The diffusion mechanism is responsible for scale effect in thin films. However, in solids only the properties of surface layers are changed due to this mechanism. This problem will be considered below.

Surface layers in amorphous polymers

As has been shown, amorphous polymers in solid state must be characterized by the existence of surface layer with increased density as compared with the mean one, the size of these layers being of macroscopic values. The higher density of polymer may be realized only owing to the higher degree of order in the surface layers in comparison with the inner ones, because high packing factor is typical of many polymers [10, 24].

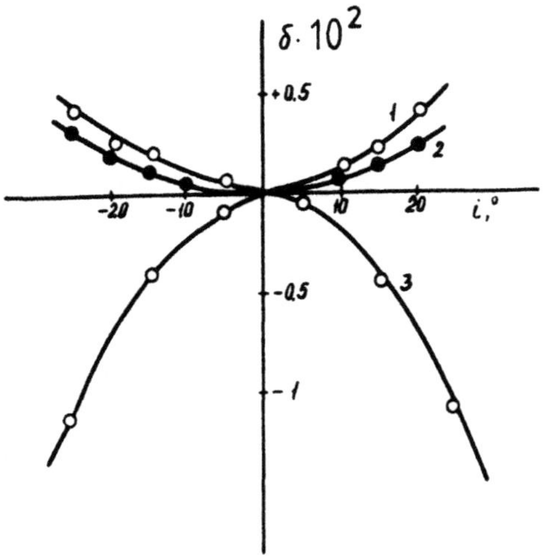

Fig. 11. Dependence of the phase difference value of δ upon angle of beam incidence on polymer film surface. Curves 1 and 2 represent the system of diepoxy of bisphenol $A+m$-phenylenediamine, the thicknesses of the films are 190 and 370 μm, respectively, curve 3 – diepoxy of bisphenol $A+p$-phenylenediamine, the thickness of film is 820 μm. (Data from [26])

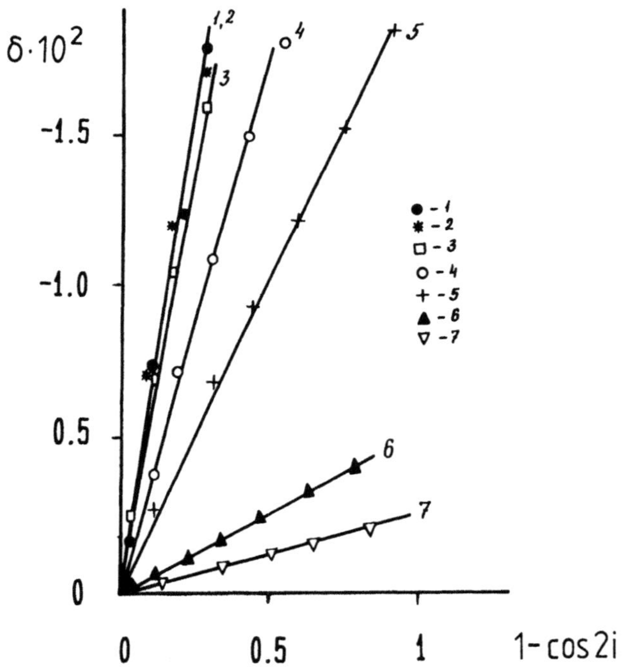

Fig. 12. Dependence of δ on $(1-2\cos 2i)$ for the films of polymer obtained on the base of diepoxy of bisphenol $A+m$-phenylenediamine. The thickness of film is (μm): 1) 2880; 2) 1650; 3) 1250; 4) 820; 5) 380; 6) 190; 7) 80. (Data from [26])

The investigation of birefringence under different angles [25, 26] shows that macromolecules are oriented in the surface layers of various polymers, the chains being arranged parallel or perpendicular to the surface.

Theory [25] gives the following expression for the dependence of birefringence δ on the angle of beam incidence (i) upon the film surface:

$$\delta = B(1-\cos 2i) \tag{11}$$

where B is proportionality factor related both to optical properties of macromolecules and to their orientation degree of order. Factor B is proportional to the thickness of oriented film.

Figure 11 presents the dependence of birefringence on beam incidence for some epoxy polymers. Figure 12 gives the same dependence in Eq. (11) coordinates. Obtained from these data proportionality factor B as the function of film thickness of polymer is given in Fig. 13.

The specific view of curves in Fig. 13 implies that at low film thicknesses the surface layer with oriented chains is greater, the greater the film thickness. At high film thickness the factor of B stops changing. This limit value corresponds to the thickness of modified boundary layer of polymer.

The thicknesses of the surface layers for various polymers [25, 26] are given in Table 1.

As follows from the Table, the thickness of oriented surface layer varies over a wide range. However, it remains rather great in all cases and agrees with the data of measurements of polymer films density on the order of magnitude.

The variation of curing agent nature, e.g., metha- by paraphenylenediamine changes the sign of bire-

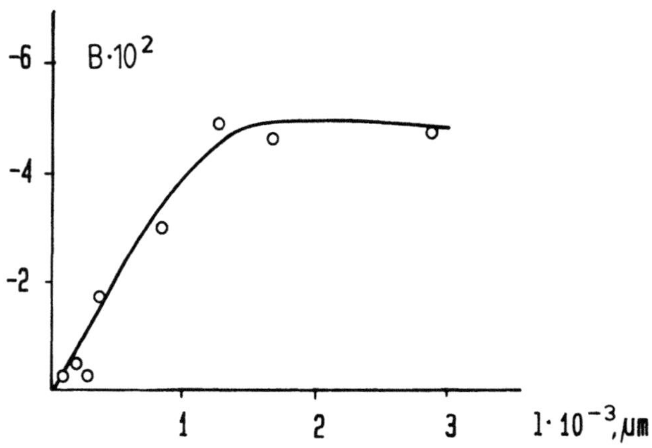

Fig. 13. Dependence of coefficient B in Eq. (11) upon the thickness of polymer films obtained on the base of diepoxy of bisphenol $A+m$-phenylenediamine. (Data from [26])

fringence, as is clear from the data in Fig. 11. It is not clear today what factors cause parallel and perpendicular orientations to the surface layer.

As has been revealed elsewhere [27, 28] by IR-spectroscopy, polystyrene surface layers consist of chains with increased trans-conformation content ($540\,cm^{-1}$), as compared with bulk polymer. Due to this fact, glass transition temperature, probably grows with decreasing film thickness.

The increase of trans-isomers fraction in the chains of polystyrene surface layers is actually the orientation of polymer chains (registrated by birefringence method), i.e., the same phenomenon, which has been mentioned above.

Table 1. Thickness of surface layers in polymers [25, 26]

No.	Polymer	H_0, µm
1	Polystyrene	12
2	Polymethylmethacrylate	9
3	Cyclolinear polyphenylsilseskvioxane	200
4	Poly-α-methylstyrene	44
5	Ethylcellulose	70
6	Monophenylacetate of cellulose	170
7	Diphenylacetate of cellulose	90
8	Phenylcarbonate of cellulose	120
9	Silica glass	0
10	Polyimide	200
11	Epoxy polymers on the base of diepoxy of bisphenol A and methaphenylenediamine	600
12	Same as 11, plus paraphenylenediamine	600

The increase of glass-transition temperature with decreasing film thickness is also characteristic of epoxy polymers (Fig. 14). Our attention is attracted by non-monotonic character of this dependence. At film thickness 20 µm the T_g drops instead of the expected further growth.

Analogous features are observed for the scale dependence of elasticity modulus of the polymers obtained on aluminum base (Fig. 15).

The data on the mechanodestruction under the action of inner stresses at network polymer synthesis [12, 13, 29] help to understand the nature of the extreme dependence. These data clearly demonstrate that if stress gradient appears on phase boundary during network structure formation, then corresponding property gradient (decreased glass-transition temperature) takes place in polymer phase.

The decrease of inner stresses by some means or other results in the monotonic, not extreme dependence of T_g and E on film thickness (Fig. 15, curve 4).

Conclusion

It has been shown that volume relaxation in amorphous body may proceed both by local and by diffusion (generally characteristic of crystalline bodies) mechanisms, realizing simultaneously. That is why the total rate of volume relaxation W is the sum of the rates of each process

$$W = W_{loc} + W_{dif}$$

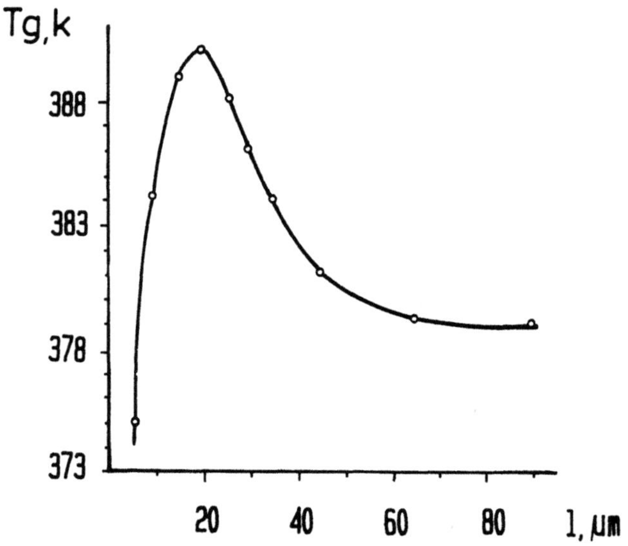

Fig. 14. Dependence of T_g on epoxyresol film thickness

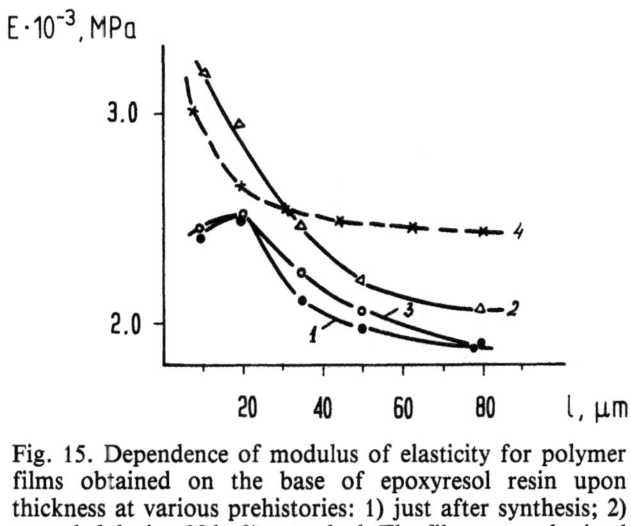

Fig. 15. Dependence of modulus of elasticity for polymer films obtained on the base of epoxyresol resin upon thickness at various prehistories: 1) just after synthesis; 2) annealed during 30 h; 3) quenched. The films were obtained on aluminum base, except film 4, which was obtained on Vood alloy. (Data from [7])

where W_{loc} is the strong function of the free volume value of v. In a first approximation one can consider that

$$W_{loc} \sim v \exp\{B/v\} .$$

At the same time the relationship of W_{dif} to v is much weaker, manifesting itself only by the parameter of γ. Therefore, at high values of v (high values of T_g, i.e., in the high-elasticity state) the local mechanism of free volume relaxation is the most important.

If $v \to 0, W_{loc} \to 0$ then the diffusion component will prevail.

On the other hand, W_{dif} greatly depends on sample dimensions, $W_{dif} \sim 1/l^2$. Thus, the decrease in sample size results in the increased contribution of the diffusion mechanism to the total process of relaxation.

Nevertheless, the diffusion mechanism in large samples does not cause any appreciable change in the body volume. However, it results in property gradient of surface layers. The value of $(D\tau)^{1/2}$ determines the scale, at which the difference in properties is substantial.

The diffusion mechanism features at volume relaxation of amorphous polymers, described in this paper, may obviously be used in property regulation of polymer films, coatings, adhesives, and composite materials. The work in this direction is just beginning.

References

1. Rostiashvili VG, Nekhoda AR, Irzhak VI, Rozenberg BA (1984) J Polym Sci Polym Phys Ed 22:1041
2. Frenkel YaI (1959) Sobranie nauchnykh trudov, Kineticheskaya teoriya Zhydkosti, M. – L. izdanie Akad Nauk SSSR, vol 3, 460 pp
3. Geguzin YaE (1979) Fizika spekaniya, Moscow, Nauka, 344 pp
4. Bennet CH, Chaudhari P, Moruzzi V, Steinhardt P (1979) P Phil Mag, A40 No 4:485
5. Bogdanova LM, Irzhak VI, Rozenberg BA, Enikolopyan NS (1983) Dokl Akad Nauk SSSR 268:1139
6. Bogdanova LM, Ponomareva TI, Irzhak VI, Rozenberg BA (1984) Vysokomol Soedin A26:1400
7. Aleksandrov VN, Bogdanova LM, Naidovskaya VI, Tarasov AI, Irzhak VI, Rozenberg BA (1985) Vysokomol Soedin A27:914
8. Bogdanova LM, Irzhak VI, Rozenberg BA (1984) Mekhanika Kompositnykh Materialov 3:554
9. Lipatov YuS (1980) Mezhfaznye yavleniya v polimerakh, Moscow Khimiya
10. Askadskii AA (1981) Struktura i svoistva teplostoikikh polimerov, Moscow, Khimiya, 320 pp
11. Braun G, Kovacs A (1963) J Phys Chem Glasses 4 No 4:152
12. Glotova YuK, Ponomareva TI, Shteinberg VG, Zakirov IN, Derinovskii VS, Irzhak VI, Rozenberg BA (1982) Vysokomol Soedin B24:339
13. Glotova YuK, Irzhak VI, Rozenberg BA (1981) In: Polimer-polimernye kompozitskii v narodnom khozyaistve, Moscow, Znanie, 104
14. Egami T, Maeda K, Vitek V (1980) Phil Mag, A41:833
15. Srolovitz D, Maeda K, Egami T, Vitek V (1981) Phil Mag A44:847
16. Srolovitz D, Egami T, Vitek V (1981) Phys Rev B24:6936
17. Geguzin YaE, Krivoglaz MA (1971) Dvizhenie makroskopischeskikh vklyuchenii v tverdykh telakh, Moscow, Metallurgiya, 344 pp
18. Bogdanova LM, Irzhak VI, Rozenberg BA (1985) Rol diffuzionnogo mekhanizma v relaksatsii ob'ema amorfnykh polimerov, preprint, Chernogolovka, 19 pp
19. Gurro JG, Lagasse RR, Simha R (1982) Macromol 15 No 6:1621
20. Oleinik EF, Bulatov VV, Gusev AA (1984) Makromol Chem Suppl 6:305
21. Kovacs A (1982) J Macromol Sci Phys B21:513
22. Kovacs JA, Aklonis JJ, Hutchinson JM, Ramos AR (1979) J Polym Sci, Polym Phys Ed 17:1097
23. Sidorovich AV, Kuvshinskii EV (1972) In: Relaksatsionnie yavleniya v polimerakh, red Bartenev GM, Zelenev YuV, Leningrad, Khimiya, 373
24. Irzhak VI, Rozenberg BA, Enikolopyan NS (1979) Setchatye polimery, Moscow, Nauka
25. Cherkasov AN, Vitovskaya MG, Bushin SV (1976) Vysokomol Soedin A18:1628
26. Grishchenko AE, Ruchin AE, Koroleva SG, Skazka VS, Bogdanova LM, Irzhak VI, Rozenberg BA, Enikolopyan NS (1983) Dokl Akad Nauk SSSR, 269:1384
27. Zhbankov RG, Tretinnikov ON, Tretinnikova GK (1984) Vysokomol Soedin B26:104
28. Zhbankov RG, Tretinnikov ON (1984) Vysokomol Soedin B26:146
29. Glotova YuK, Ponomareva TI, Shteinberg VG, Irzhak VI, Rozenberg BA (1983) Vysokomol Soedin B25:407

Received October 5, 1988; accepted July 7, 1989

Authors' address:

B. A. Rozenberg
Institute of Chemical Physics
Academy of Sciences of the USSR
142432 Chernogolovka, Moscow Region, USSR

Discussion

CHUDACEK:

Because many of the material constants of our polymeric thin films are functions of thickness, do you accept that your systems have fractal structure with self symmetry?

ROZENBERG:

We did not investigate this problem from the point of view of fractal structures, but I think it will be interesting to do so in the future.

HOLZMÜLLER:

The fluctuation of density leads to a fluctuation of thermal expansion coefficient, and therefore heating leads to internal stresses. These stresses create holes in the structure and are the main reason for glass transition. Have you measured fluctuations of the thermal expansion coefficient?

ROZENBERG:

I fully agree with your suggestion. The density gradient in amorphous polymers must lead to the gradient of thermal expansion coefficient. Unfortunately, we have not yet measured the scale effect of thermal expansion coefficient in our samples, but I believe that this effect must, in reality, exist.

DUSEK:

Can you explain what parameters control the size of clusters of holes, what is their physical meaning, and what is the possibility to obtain their values from experiments? Why has the system with relatively large holes lower free energy than that in which the holes are distributed randomly?

ROZENBERG:

In my opinion the sizes of free volume fluctuations are controlled by the rate of cooling through the glass transition range. The size distribution function of free volume fluctuations can be determined by the fluctuation theory of polymer vitrification recently developed by V. G. Rostiashvili, V. I. Irzhak, and B. A. Rozenberg ("Vitrification of polymers" CHIMIA L 1987 [in Russian]). Experimental determination of size distribution of density fluctuations is possible principally from x-ray scattering measurements, which we did not perform. Concerning the last point you made, I think there is some misunderstanding. The presence in glassy polymer of the large free volume fluctuations and excess of total free volume fraction are the main driving forces for macroscopic volume relaxation; as a result of that the randomization and the fluctuation of sizes appears in a relaxing sample. As I mentioned earlier, I think that the concentration of large-scale density fluctuations depends on the rate of sample cooling through glass transition region.

DUSEK:

Have the time dependencies of glass properties been recorded after annealing of free films above T_g?

ROZENBERG:

Yes, eventually some properties of the glassy state depend on the temperature and time of annealing the sample above T_g, because some chemical reactions can run at high temperatures. But if the chemistry of the polymer remains unchanged normally one does not see any changes of properties in the glassy state.

Dynamics of density fluctuations in glass forming liquids and polymers as measured by light scattering

E. W. Fischer, Ch. Becker, J.-U. Hagenah, and G. Meier

Max-Planck-Institut für Polymerforschung, Mainz, FRG

Abstract: In the static light-scattering patterns of glass-forming polymers a strong q-dependent ($q = (4\pi/\lambda)\sin\theta/2$) "excess" scattering is generally observed also at temperatures above T_g. Consequently, the Landau-Placzek ratio (LPR) in the Rayleigh-Brillouin spectra is much higher than theoretically predicted for viscoelastic liquids. It is shown that these features are also observed in low molecular weight glass-forming liquids, namely o-terphenyl and chlophen, which were investigated by Brillouin spectroscopy over a wide temperature inverval ranging from shortly below the glass transition temperature to very deep into the melt region. The experimentally determined Brillouin shifts and line widths are consistently describable using updated theories of light scattering from viscoelastic liquids in the hypersonic regime. However, the LPR is much higher than expected from theory, therefore simple liquids behave in the neighborhood of T_g very much like polymers. To explain this behavior we introduce an order parameter in the melt whose mean square fluctuation gives rise to "excess" scattering. It is postulated that in the supercooled liquid above T_g "pre-freezing" phenomena occur that can be qualitatively discussed in the framework of recent free volume glass transition theories. This concept is supported by data from photon correlation spectroscopy and angular dependent elastic light-scattering measurements.

Key words: Viscoelastic liquids; Rayleigh-Brillouin spectroscopy; Landau-Placzek ratio; glass transition; order parameter

Introduction

It is a well known fact that the physical properties of amorphous polymers are subject to drastic changes in a small temperature interval usually known as the glass transition temperature. These changes are accompanied with a decrease of molecular mobility which can be studied by a number of different experimental techniques probing the molecular dynamics, such as by dielectric relaxation [1], nuclear magnetic resonance [2] or quasielastic neutron scattering [3]. These techniques are sensitive towards the reorientation of single dipoles, bond vectors or single segment reorientation. However, the relevant relaxation times characterizing the glass transition process are in the order of seconds, thus collective phenomena must play a role. In a recent comparison [4] between dielectric relaxation and quasielastic light scattering of poly(methylphenylsiloxane) we were able to show that the relaxation times characterizing the α-process for this model system are identical when studied with either method, which clearly indicated that the collective dynamics govern the molecular one. Hence collective phenomena are most important for studying the dynamics of the glass transition. Here of course the scattering of light is of great advantage as it has a small transfer of momentum during the scattering process and is thus sensitive towards cooperative modes. Usually the time-resolved quasielastic scattering (photon correlation spectroscopy) is performed when dealing with the long time end of the relaxation time spectrum, that means $\tau > 10^{-6}$ s. The accessible quantity here is the density autocorrelation function of the scattered light [5]. Its Fourier transform is connected via the Wiener-Khintchine theorem with the spectrum measured in the range of $10^{-8} - 10^{-11}$ s, by performing a polarized Rayleigh-Brillouin spectrum using Fabry-Perot interferometry [6].

In this paper we will mainly deal with the analysis of Rayleigh-Brillouin spectrum of glass forming non-

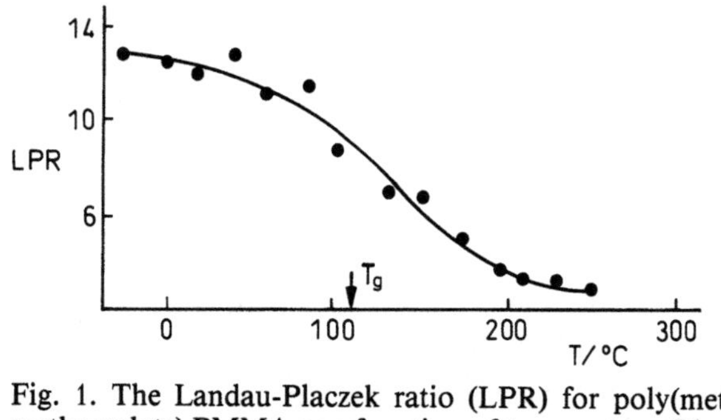

Fig. 1. The Landau-Placzek ratio (LPR) for poly(methylmethacrylate) PMMA as a function of temperature. The line is a guide for the eye. Data taken from [7]

polymeric liquids. Here we are interested to learn more about the question of the spectral distribution of the scattered intensity. One quantity describing this feature is the Landau-Placzek ratio (LPR), which is the ratio of the non-frequency shifted intensity in a spectrum over twice the photon intensity. This ratio in the high temperature limit is equal to $\gamma-1$ where $\gamma = C_p/C_v$. A remarkable effect for polymers and non-polymeric glass formers is a strong increase of LPR with decreasing temperature above T_g, although γ is roughly a constant (see [7, 8]). To demonstrate the remarkable effect for polymers in Fig. 1 the data for poly(methylmethacrylate) PMMA is shown (from [7]), which clearly indicates that at almost 150 K above T_g ($T_g = 105\,°C$) an additional intensity contribution shows up. The LPR show a gradual increase over a very broad temperature range in the melt when approaching T_g. This extra increase is much greater than predicted by any theory of Rayleigh-Brillouin spectra of viscoelastic liquids [9–13]. All these theories give more or less similar formulas for the LPR as a function of temperature [9–13] and show that a value of roughly LPR = 2 in the vicinity of T_g is what can be expected, whereas one observes values of LPR $\simeq$ 12 [7, 8].

Since a liquid above T_g is usually assumed to be in thermodynamic equilibrium this effect is even more striking. We would have assumed this effect to be more pronounced in the glassy state where additional scattering into the central line has been reported due to density fluctuations that are in equilibrium with a fictive temperature such that it controls the structural relaxation in the glass since no further structural rearrangements are possible by further lowering of the temperature. This theoretical concept [14] accounts for the high Landau-Placzek ratio in the glassy state, but not in the melt as there the fictive temperature is equal to the actual temperature and the equation for the LPR [14] reduces to those for normal viscoelastic theories [9–13].

For our study we have chosen non-polymeric glass-forming liquids *o*-terphenyl (OTP) and chlophen. We are interested to know whether the mentioned behavior is a property typical for polymer systems and further, of course, how this intensity-temperature behavior relates to the glass transition phenomenon. To do so we perfom temperature-dependent Rayleigh-Brillouin spectra over a wide temperature range shortly below T_g and at roughly 180 K above T_g. We have analyzed the spectra using up-to-date theories for the relaxation time behavior of viscoelastic liquids in the fast relaxation time limit and put the dynamic results into the common interpretation pattern with photon correlation data on the same material.

Theory of the Rayleigh Brillouin spectra of viscoelastic liquids

In contrast to simple liquids the Rayleigh-Brillouin spectrum shows a four-peak structure. In a simplified manner these are first the Rayleigh peak, which is centered at the frequency of the incident light, then two peaks shifted in frequency (Brillouin) by an amount $\Delta\omega$, and a fourth, the Mountain peak, also centered at the incident frequency. These experimental findings are predicted by theories describing the scattering from a viscoelastic medium [9–13]. The common statements of all these theories concerning the physical nature of the four peaks are easily summarized. Accordingly, the measured spectra can be theoretically explained by assuming a frequency dependent bulk viscosity $\eta_V(\omega)$ and adiabatic compressibility $K_s(\omega)$. Following the description by Pinnow [12] we show in Fig. 2a a theoretical spectrum as a function of temperature. For rather high temperatures any viscoelastic liquids behaves like an ordinary simple liquid such that its Rayleigh-Brillouin spectrum can be understood in terms of a three-peak structure with the intensity of the Brillouin line $I_B \propto K_{s,0}$, the adiabatic compressibility being at zero frequency. The central line is connected with the thermal diffusivity and hence need not be considered further. By lowering the temperature, a coupling between structurally relaxing variables and the translational degrees of freedom are generally assumed and that led to the appearance of a Mountain peak with an intensity $I_M \propto K_{s,0} - K_s(\omega_p)$ with $K_s(\omega_p)$ being the compressibility at the phonon frequency at that temperature. In that temperature range the intensity of the Mountain line, which is centered around ω_o, takes a considerable part of the frequency region between the Rayleigh line and the Brillouin doublets. For still lower temperatures then

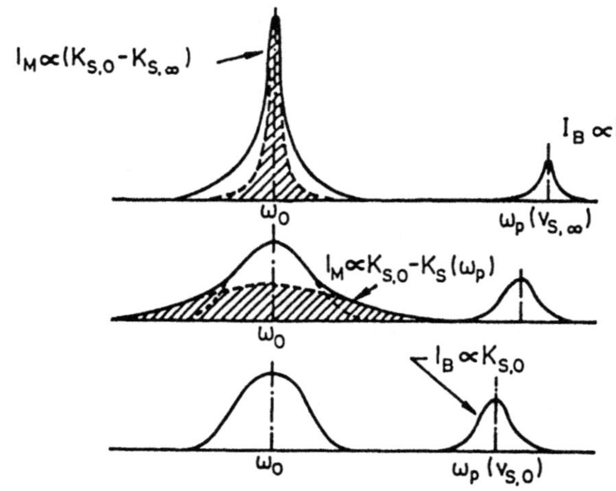

Fig. 2a. Changes of the polarized Rayleigh-Brillouin spectrum from a viscoelastic liquid for various temperatures. Upper part at low temperatures, intermediate temperature in the middle, the high temperature shown at the bottom. The symbols are explained in the text. Scheme adapted from [14]

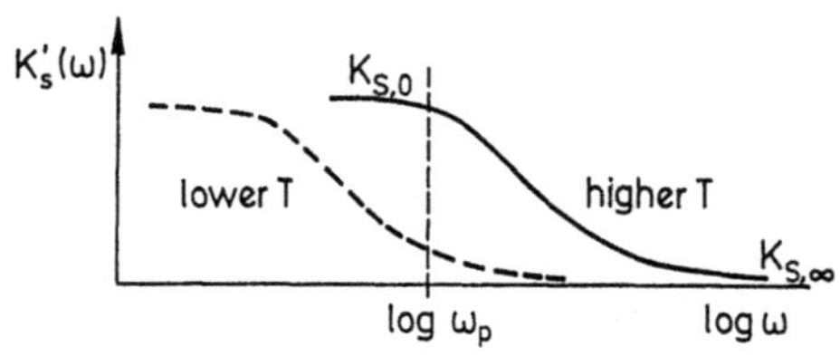

Fig. 2b. The real part of the adiabatic compressibility vs the logarithmic frequency (schematically). The symbols are explained in the text

the Mountain relaxation has become very slow such that it is hidden completely in the central peak with an intensity $I_M \propto K_{s,0} - K_{s,\infty}$, where $K_{s,\infty}$ is the adiabatic compressibility at inifinite frequency. This qualitative description of the change of spectrum with temperatures does not take into account the shear modes. The change of K_s is visualized by Fig. 2b where the real part of the compressibility $K_s'(\omega)$ is plotted vs the logarithmic frequency. The full and broken lines are possible dependencies of $K_s'(\omega)$ with varying temperatures between the limiting values $K_{s,0}$ and $K_{s,\infty}$. Then, at a given phonon frequency $\log \omega_p$, the curve for $K'(\omega)$ is shifted along the $\log \omega$ axis while changing temperature. Hence one notices that if $I_M \propto K_{s,0} - K_s(\omega_p)$ the linewidth of the Brillouin peak Γ_B will show a maximum at the phonon frequency ω_p, such that $\omega_p \tau_s = 1$ where τ_s is the structural relaxation time. It is exactly this coupling which makes the structural relaxation accessible by measuring the frequency shift f_B and, more marked, the Brillouin linewidth Γ_B.

To describe the spectra consistently we use the theory of Lin and Wang [13] which seems to be the most accurate way to describe all experimental features. The theory reformulates Mountain's approach in terms of the Mori-Zwanzig formalism and applies it to viscous molecular liquid consisting of polymeric molecules. As one main assumption a single exponential decay for the structural relaxation time τ_s is assumed. The theory gives for the Brillouin shift f_B and the Brillouin linewidth Γ_B the following expressions

$$f_B = \pm \omega_s \left[1 + \left(\frac{\alpha_s}{\omega_s^2} \right) \frac{\omega_s^2 \tau_s^2}{1 + \omega_s^2 \tau_s^2} \right] , \tag{1}$$

and

$$\Gamma_B = \frac{\alpha_s \tau_s}{1 + \omega_s^2 \tau_s^2} + \left(1 - \frac{1}{\gamma} \right) D_T q^2 . \tag{2}$$

The ω_s is the adiabatic sound frequency; $\gamma = C_p/C_v$, the ratio of the heat capacities, and D_T is the thermal diffusivity. The quantity α_s is related to the relaxation strength $R = \omega_\infty^2/\omega_s^2$ given by

$$R - 1 = \frac{K_{s,0} - K_{s,\infty}}{K_{s,\infty}} \tag{3}$$

via

$$\alpha_s = (R-1)\,\omega_s^2 . \tag{4}$$

Thus the precise determination of the limiting value of ω_∞ is of great importance to properly determine R.

We are further concerned with the evaluation of the Landau-Placzek ratio (LPR); for simple liquids $\mathrm{LPR} = \gamma - 1$. For relaxing liquid this value is reached at high temperatures. Moutain's and Rytov's theories give for LPR the equation

$$\mathrm{LPR} = (\gamma - 1) + \gamma \left(\frac{v_\infty^2 - v_0^2}{v_0^2} \right) , \tag{5}$$

where v_0 and v_∞ are the limiting sound velocities given generally by $\omega = v \cdot q$ with q being the scattering vector. The Lin-Wang theory does not provide a simple algebraic expression for the LPR. Oh et al. [15] have proposed a method of evaluating the LPR from the measured spectra to compare them with theory. It turns out that near the glass transition the values of LPR on the basis of the Lin-Wang theory are almost the same as those calculated from Eq. (5).

Experimental

The polarized Rayleigh-Brillouin spectra were taken with an optical system described in detail elsewhere [16]. The excitation light with about 100 mW power was an etalon selected single frequency line at 514.5 nm from a spectra Physics 165 Argon Ion laser. The scattered light collected at a fixed angle of $\theta = 90°$ was spectrally analyzed with a Burleigh DAS 10 stabilized Fabry-Perot interferometer operating in triple pass. The finesse of the complete optical interferometer was about 45. We choose a free spectral range of 24.5 GHz for the *o*-terphenyl and 22.5 GHz for the chlophen.

To obtain dust-free samples so that accurate intensities could be measured, the samples were filtered with a hot syringe through 0.22 μm Millipore Teflon filters into 1/2-inch outer diameter dust-free scattering cells which were then flame-sealed under vacuum. The purity of the samples was manifested by a low Landau-Placzek ratio of 0.5 at the highest temperature of measurement.

The spectra data were taken over two Fabry-Perot orders by means of a multichannel analyzer with a Rayleigh-peak separation of at least 400 channels. The instrumental function was numerically measured and deconvoluted either with one central Lorentzian plus two shifted sidebands or with a central Lorentzian plus an additional Mountain line and two shifted side bands, depending on whether the structural relaxation is dominant or not.

For the evaluation of the sound velocity the knowledge of the temperature dependent index of refraction n is essential. Thus $n(T)$ was measured for *o*-terphenyl (OTP) with an Abbe refractometer in the temperature range from 30° up to 120 °C. The data interpolated to the laser line are well-represented by $n(T) = 1.6537 - 5.13 \cdot 10^{-4}$ T/°C. For chlophen we measured only at room temperature ($n = 1.6471$) assuming then the same temperature dependence as for OTP. Values outside that range were determined by linear extrapolation.

The materials used were spectral grade OTP from Merck, Darmstadt, FRG, and chlophen, a mixture of polychlorated biphenyls of mean chlorine no. 5 purchased from Bayer AG, Leverkusen, FRG, (tradename Chlophen A 50).

The glass transition temperatures of the materials were determined with DSC and amount to $T_g = -29$ °C for *o*-terphenyl and $T_g = -20$ °C for chlophen.

Results and discussion

We have performed polarized Rayleigh-Brillouin spectra of *o*-terphenyl in the temperature range from −40° to 155 °C and for chlophen from −70° to 165 °C. Thus in both cases we have started from a temperature range shortly below and around T_g to very deep into the melt region. We show in Fig. 3 a spectrum of *o*-terphenyl at a temperature of 267 K = T_g + 24 K which is characterized by a strong central line such that the phonon contribution at $\simeq \pm 10.5$ GHz is a bit difficult to see. As usually found, the spectra do change considerably with increasing temperature. In Fig. 4a and b spectra for OTP at 379 K = T_g + 154 K and chlophen at 417 K = T_g + 164 K are shown. Here we note the additional process now contributing considerably to the spectral range between the phonons and the central line. This effect is known to be due to the Mountain peak [8, 9, 13, 15, 17]. As was already stated in the section on theory we usually determine the Brillouin shift $f_B = v_s/q$ at each temperature together with the damping of the phonon determined by the linewidth of the Brillouin-line Γ_B. The parameters are accessible by fitting the measured spectra (after deconvolution) with an expression assuming a superposition of four Lorentz-curves being appropriate to describe the spectra. In Fig. 4 the results of such a fit are shown as the Mountain part and a central part together with the phonon contribution plus a common background. From such a fit f_B

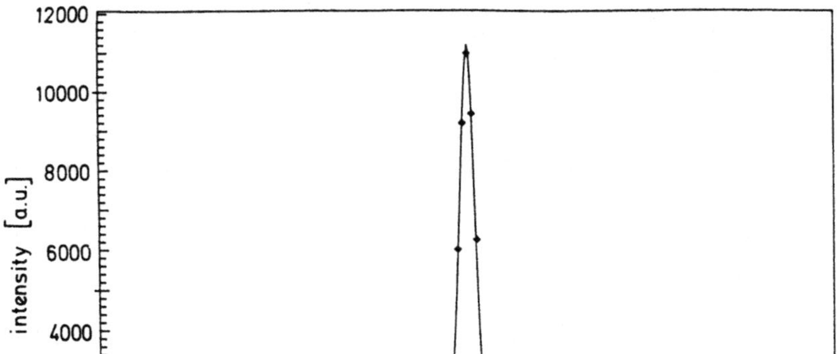

Fig. 3. Polarized Rayleigh-Brillouin spectrum of *o*-terphenyl at 267 K = T_g + 24 K

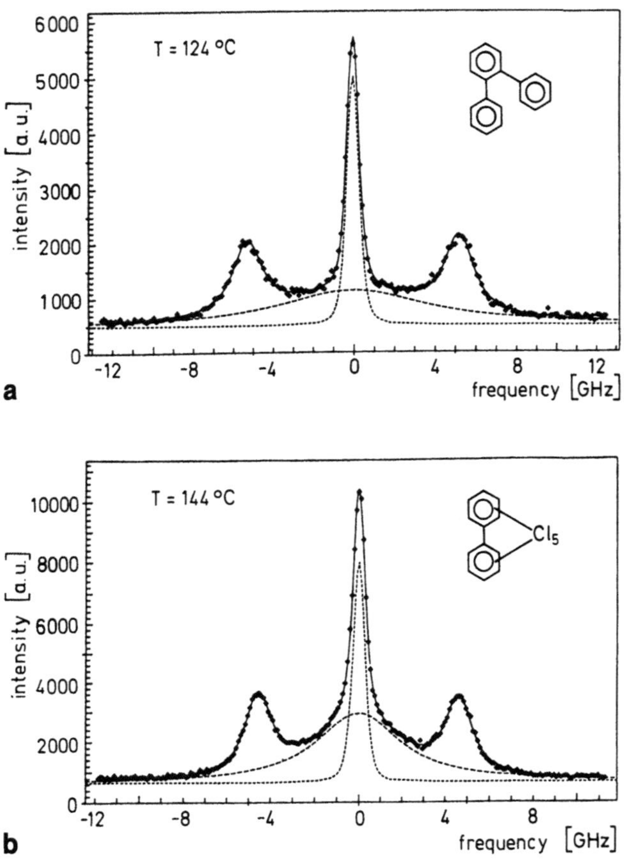

Fig. 4. Polarized Rayleigh-Brillouin spectra of *o*-terphenyl at $379\,K = T_g + 154\,K$ and chlophen at $417\,K = T_g + 164\,K$. The dashed lines indicate the Mountain line, the dotted lines represent the central component plus background. The phonon contribution is added to obtain the measured spectra

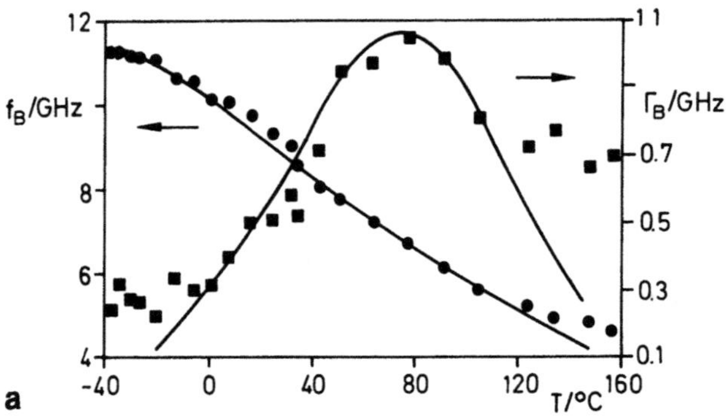

Fig. 5a. Brillouin shift f_B (●) and linewidth Γ_B (■) for *o*-terphenyl as a function of temperature. The solid lines are results of a fit of the Lin-Wang theory [13] to the data

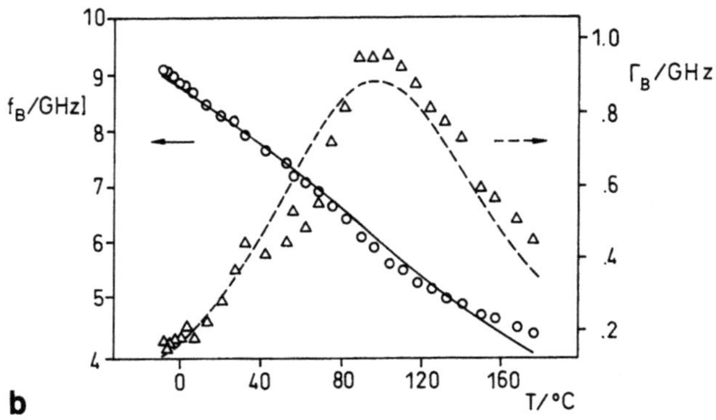

Fig. 5b. Brillouin shift f_B (○) and linewidth Γ_B (△) for chlophen as a function of temperature. The solid and dashed lines are results of a fit of the Lin-Wang theory [13] to the data

and Γ_B is immediately calculated. The results of the temperature dependent measurements of f_B and Γ_B for OTP are given in Fig. 5a and for chlophen in Fig. 5b, together with the results of the fit of Lin-Wang theory [13] to the data. From the figures several aspects are clearly readable. First the phonon velocity displays a dispersion marked by the *s*-type shape. This is important to note because for the proper determination of ω_∞ the high frequency (or high temperature) limiting value of ω_s is needed. This value enters in Eq. (4) and thus influences α_s via the relaxation strength. In Fig. 6 we show thus the f_B vs T data of OTP plus ultrasonic data measured at 30 MHz [18] which are clearly on the high frequency side of the dispersion in f_B. From Fig. 6 it is evident that the extrapolation of the ultrasonic data towards the Brillouin data at high temperatures lead to the same values; thus proving the proper determination of ω_∞. Since the glass transition of OTP and chlophen are similar and further in Fig. 5 we note great similarities between both materials, we assume the same argument in determining ω_∞ for chlophen to be valid. However, there is no ultrasonic data available for chlophen.

Secondly, the phonon damping displays a clear maximum at about 75 °C for OTP and 85 °C for chlophen, which is roughly 100 K above the thermodynamic glass transition temperature. As already extensively discussed by Patterson [19], this is a common feature for viscoelastic systems reflecting the fact of a frequency-dependent glass transition. The solid lines in Fig. 5 are results of a best fit according to the Lin-Wang theory to the data. Only qualitative agreement is reached, but proper changes in the parameters led to good fits only for one of the two experimental quantities. Further, we note a clear deviation from the prediction of Γ_B of OTP at higher temperatures. This behavior is similar to that reported by Fytas et al. [17] for poly(phenylmethylsiloxane) where a residual

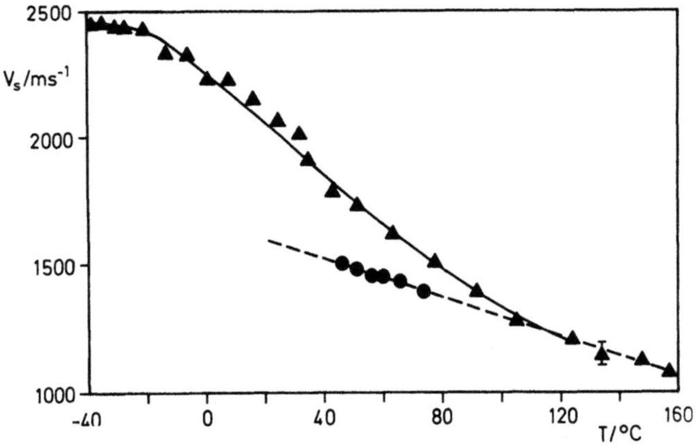

Fig. 6. Phonon velocity v_s for *o*-terphenyl as a function of temperature. Data points from Brillouin experiment (▲) at frequencies 9–11 GHz. Ultrasonic data from [18] (●) at 30 MHz

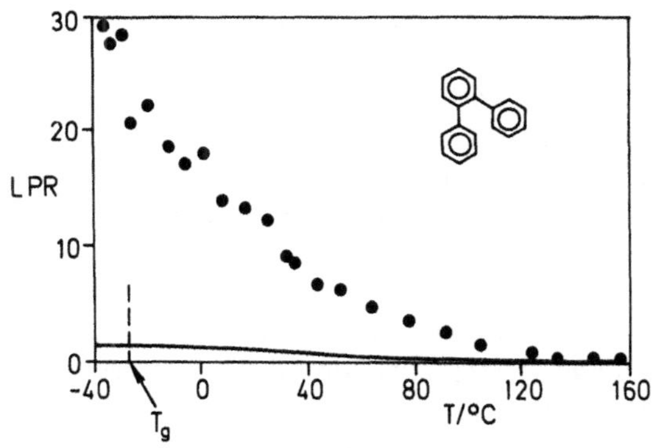

Fig. 8. The Landau-Placzek ratio (LPR) of *o*-terphenyl as a function of temperature (●). The solid line is a result of a calculation of the LPR according to the Lin-Wang theory [13]

linewidth at high temperatures has also been found. There this effect was attributed to the coupling of vibrational modes of the phenyl groups to translation of the molecules.

These ideas are supported by findings from substituted benzenes where also such energy tranfers were reported, [20] hence indicating that this effect may hold for the OTP-system. The discrepancy may also be due to the fact that in the Lin-Wang theory only a single relaxation time τ_s is assumed, whereas it is

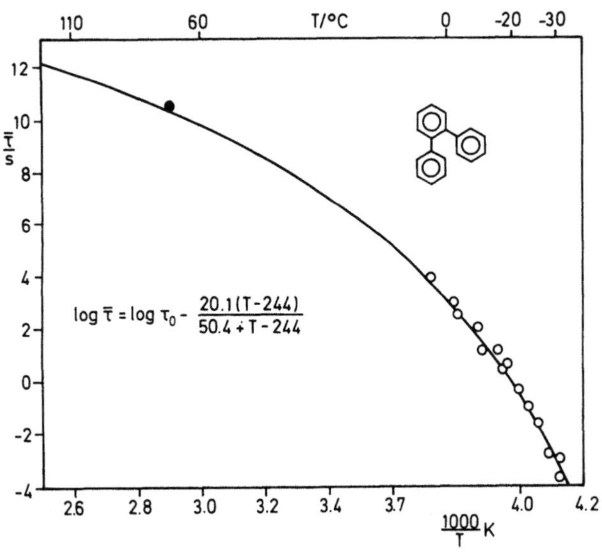

Fig. 7. Activation plot showing $\log\bar{\tau}$ vs T^{-1} for *o*-terphenyl as measured by photon correlation spectroscopy (○). The solid line is a fit of the Williams-Landel-Ferry (WLF) equation to the photon correlation data. The data point from Brillouin scattering (●) is calculated using the $\omega_B^{\max}\cdot\tau = 1$ condition. The result of the WLF fit is also shown in the figure, giving $T_0 = 244$ K, $C_1 = 20.1$ and $C_2 = 50.4$ K

well known form PCS measurements [21] that the α-relaxation exhibits a broad assymetric relaxation time distribution. The simplification of a single exponential decay function is discussed in literature [19]. A recent publication [22] states that a stretched exponential decay function, a so-called Kohlrausch-Williams-Watts function (KWW), is applicable over the whole accessible time range for light-scattering experiments, thus including also the photon correlation range (10^3 s $< \tau < 10^{-6}$ s), together with results from dielectric relaxation. The discussion is not conclusive [23, 24], but despite the different viewpoints one can estimate a characteristic relaxation time τ from the maximum in Γ_B via $\omega_{\max}^B\cdot\tau = 1$ neglecting possible distributions of relaxation times for the moment. As a matter of fact, the error in replacing the first moment of a distribution in τ by the maximum value is not crucial, especially bearing in mind the experimental uncertainties in measuring Γ_B.

It is commonly believed [25] that the relaxation described by the Mountain peak in the hypersonic regime is the same physical quantity as the measurement of the structural relaxation with a photon-correlation experiment. For OTP, Fytas et al. [26] have measured the temperature dependence of the structural relaxation. Hagenah [21, 25] has extended the accessible time range and thus temperature range of the measurements. The temperature dependence of the mean relaxation times was fitted with a Williams-Landel-Ferry (WLF) equation. The result of the fit is given in Fig. 7 as a solid line. From the figure it is further evident that the data point from Brillouin scattering according to the maximum condition meets the WLF line and by that supports the foregoing statements. Without going further into detail it should be noted that even more sophisticated methods of data evaluation give agreement with the WLF prediction within at least the same order of magnitude [25, 27].

The preceding considerations have dealt mainly with dynamic aspects. As already stated in the introduction of this paper we have closely investigated the temperature dependence of the intensity ratio, the Landau-Placzek ratio (LPR). Using the procedure described by Oh et al. [15] we have determined the LPR of OTP (and chlophen) over the entire measured temperature range. The results for OTP are depicted in Fig. 8. Here the value for LPR ($T = 155\,°C$) equals 0.5, thus approaching $\gamma - 1$ for $\tau \to \infty$. Further, we note a strong increase of LPR with decreasing temperature. At the glass-transition temperature, indicated by an arrow on the figure, a step by nearly 10 units in the LPR can be noticed; we have found a similar qualitative behavior for chlophen – but due to the limited number of data points below T_g a more precise statement is not possible. Thus we concentrate only on the results above the glass transition temperature.

The only other available set of data of LPR of OTP was published by Wang et al. [8] who state that the LPR raises to 8 for $T = -20\,°C$. Since there the LPR from the isotropic spectrum has been plotted, we have to modify our data via the depolarization ratio $\varrho = 0.4$ [26], according to $I_{VV} = I_{ISO} + (4/3) I_{VH}$; this is then identical with the Wang result within the experimental error. Furthermore, our values for f_B and Γ_B for OTP (see Fig. 5a) agree well with the results by Wang et al. [8].

We are now concerned with the interpretation of the temperature dependence of the Landau-Placzek ratio LPR. In Fig. 8 the expected behavior is drawn as a solid curve; it was calculated by computing the spectrum $I(q, \omega)$ according to the Lin-Wang theory [13] by using the fit parameters (cf. Fig. 5a) to describe f_B and Γ_B simultaneously. The spectrum was then analyzed by the procedure described by Oh et al. [15]. At this stage we recall Eq. (5) for the LPR according to Rytov's or Mountain's theories. Using $1/\gamma = \beta_s/\beta_T$ the ratio of adiabatic and isothermal compressibility and reading the correct sound velocities from Fig. 5a we compute LPR ($T = -20\,°C$) = 1.4 which is almost identical with the value calculated from the Lin-Wang theory. Therefore we can conclude that the different theories yield approximately the same temperature dependence, but none of them is appropriate to describe the experimentally observed behavior which deviates significantly from the calculated curve, as shown in Fig. 8. The effects of shear waves can be neglected because of their small intensities [8].

It may be mentioned that the theoretical curve in Fig. 8 is in clear contradiction to the calculated curve given by Wang et al. [8], although their and our experimental data agree within the experimental error.

The anomalous increase of the LPR can be due either to an anomalous behavior of the phonon contribution in the spectrum or it is due to an extra increase of the central non-shifted intensity. To decide for either possibility we calculated the Brillouin intensity. It is given by [6]

$$2I_B = \frac{\pi^2}{\lambda^4}\left[\varrho_0\left(\frac{\partial\varepsilon}{\partial\varrho}\right)_s\right]^2 \beta_s k_B T\,, \tag{6}$$

where β_s is the adiabatic compressibility being $\beta_s = 1/\varrho_0 v_s^2$ with v_s being the phonon velocity. The other symbols in Eq. (6) have the usual meanings. If we set $\frac{\partial\varepsilon}{\partial\varrho} = \frac{\partial\varepsilon}{\partial T}\cdot\frac{\partial T}{\partial\varrho}$ we can calculate I_B as a function of temperature by taking into account $n = n(t)$ (measured with an Abbe refractometer, as given in the ex-

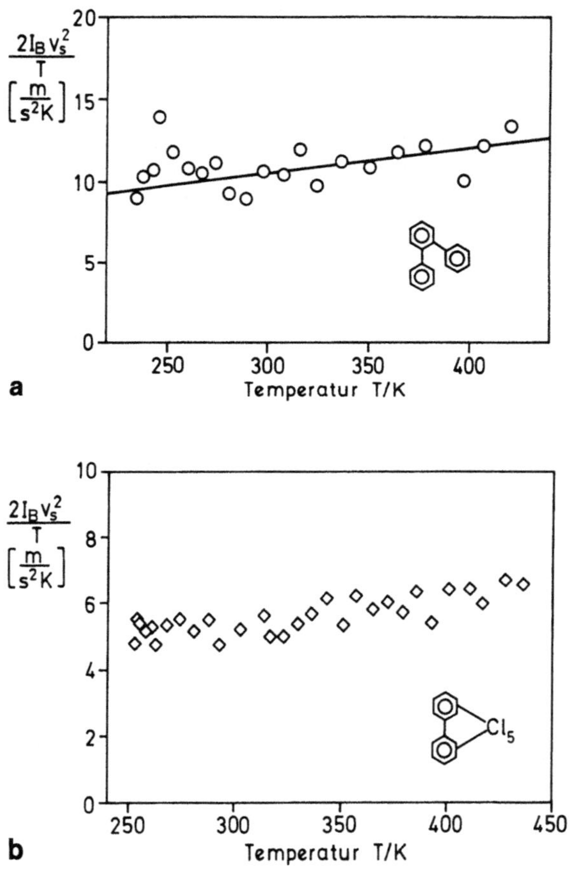

Fig. 9. The quantity $2I_B v_s^2/T$ vs temperature of o-terphenyl (○). The solid line is calculated using $n = n(T)$ and $\varrho = \varrho(T)$ data according to Eq. (6). For chlophen (◇) only experimental data is shown, since not all quantities entering in Eq. (6) are known. Nevertheless, a behavior qualitatively similar to that of o-terphenyl is expected for chlophen

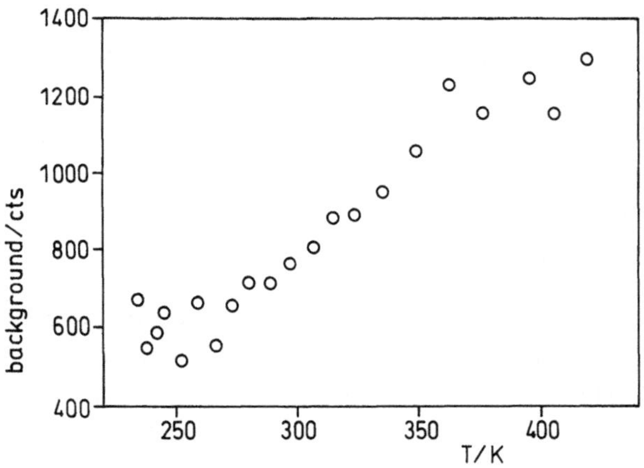

Fig. 10. The background of the spectra of *o*-terphenyl vs temperature. At the highest temperatures the background intensity is roughly 10% of the total intensity

perimental part) and $\varrho = \varrho(T)$ from specific volume data [28]. In Fig. 9a we show the result of our calculations together with the data for OTP, which clearly indicates that the phonon intensity contribution behaves regularly (and it does for chlophen), thus the extra increase of LPR with decreasing temperature must be due to additional scattering into the central line.

This contribution will be called "excess scattering" in the following; it is well known and appears often in polymeric systems [29–31]. More precisely, it was found for many polymers that not only in the glassy state but also above T_g the total scattered intensity I_{total} is much higher than expected from the well known thermodynamic relation [6]

$$\lim_{q \to 0} I_{\text{total}}(q) = kTV\varrho_0^2\beta_T \tag{7}$$

where β_T is the isothermal compressibility. In closely examining Fig. 3 and Fig. 4a for OTP again we notice a much higher scattering intensity for the lower temperatures, which is a clear contradiction to Eq. (7).

This discrepancy is partly extenuated if one takes into account the considerable background in the high temperature spectra of OTP (and also for chlophen). In Fig. 10 we have plotted the background in counts for the temperatures of measurement for OTP clearly indicating an increase with increasing temperature. Thus we conclude that the apparent excess of the intensity of the central line is due to changing or switching of the total intensity from the background to the central line with decreasing temperature. Before discussing the problem of the "excess scattering" further we summarize the main features of the background scattering: First, it shows no frequency dependence in the Fabry-Perot range meaning that its relaxation time τ_c is faster than $\cong 10^{-11}$ s. Second, its intensity increases strongly with rising temperature, and third, its intensity is shifted to the central line when approaching T_g in way similar to the Mountain intensity of the α-process.

Since the Fabry-Perot interferometer used in our study has to be set to a constant free spectral range, there is naturally a cut-off frequency involved and we know little about the intensity which scatters outside the free spectral range setting. Using other optical devices with enhanced frequency ranges we encounter the problem of collision-induced scattering [32] and that has to be treated carefully. Appropriate measurements on OTP and related glass-forming systems by means of optical gratings and interferometers with large free spectral range are in progress in our laboratory [33].

The existence of very fast relaxation processes in OTP is supported by quasielastic incoherent neutron experiments where from the linewidth of the spectrum $I_{\text{inc}}(q, \omega)$ a time of $\tau_c = 6 \cdot 10^{-12}$ s at $T = 40\,°\text{C}$ can be calculated [34]. A similar result was obtained for a related non-polymeric glass-forming system, tri-α naphthylbenzene, where τ_c was found to be $4 \cdot 10^{-12}$ s [35]. These times resemble the characteristic time scale for a fast localized relaxation process taking place on a spacial distance of 0.2 to 5 Å^{-1} as given by the q-range of the neutron spectrometer used in the studies. Although in those experiments a single particle propagator is measured, the corresponding light-scattering experiment should also reflect this property (cf. the comparison between the dielectric relaxation and light scattering in [4]).

The question that is not yet answered is whether and how this process is related to the glass-transition phenomena. The neutron results are put into the framework of recent theories of glass transition [36] where the fast processes are understood to act as precursors of a freezing temperature, usually to be the glass-transition temperature. A similar behavior is proposed within phenomenological free-volume concepts [37] where in the frozen state the particles are allowed to perform fast motions within their local cells (free volume). The onset of the structural relaxation takes place only at higher temperatures by exchange of local cells.

As we have shown above, the ultrahigh frequency background scattering in the Brillouin spectra is related to the excess scattering of the central line in such a way that the former decreases if the latter increases with lowering temperature. In order to discuss the origin of the excess scattering I_{exc} we have to take into account several observations which will be described in detail in subsequent papers:

i) The appearance of I_{exc} is connected with a strong q-dependence of the static total intensity $I_{tot}(q)$. Correlation lengths ξ in the range of 300–1000 Å are found [38].

ii) The time correlation function $g^{(2)}(t)$ measured by photon-correlation spectroscopy [39] shows a long time tail characterized by a relaxation time τ_z in the range of $10^{-1}-10^2$ s for samples that exhibit a q-dependent $I_{tot}(q)$. Those times are about 10^3 to 10^5 slower than the ordinary relaxation times of the density fluctuations.

iii) All three features, I_{exc}, ξ, and τ_z depend on the thermal history of the samples. At rather high temperature, say $T \simeq T_g + 150\,°C$, the viscoelastic liquids behave as expected. Lowering the temperature very slowly, one finds high I_{exc}, large ξ, and long τ_z. Fast quenching of the samples leads to small deviations from the "normal" behavior of a liquid in thermodynamic equilibrium. The latter observation excludes the possibility of strain fields as the origin for I_{exc}.

These observations indicate that a super-cooled viscoelastic liquid *above* T_g is generally not in a thermodynamic equilibrium state as it is mostly assumed. In order to describe the state of the liquid we have to introduce an additional "order parameter" $z(p, T)$, which depends also on thermal history. The use of such order parameters is well known [40] from the thermodynamic treatment of polymeric and other glasses *below* T_g.

Without speculating about the physical nature of z, we are able to describe our observations phenomenologically by a set of equations derived in analogy to concentration fluctuations. The excess intensity is given by

$$I_{exc} \propto \left(\frac{\partial \varepsilon}{\partial z}\right)^2 \cdot \langle \delta z^2 \rangle \,, \tag{8}$$

where $\langle \delta z^2 \rangle$ is the mean square fluctuation of the order parameter z. This quantity can be described [41] by

$$\langle \delta z^2 \rangle = k_B T \left(\frac{\partial \mu}{\partial z}\right)^{-1}_{p,T,A = \text{const}} \,, \tag{9}$$

where μ is the chemical potential and A is the affinity that is the thermodynamically conjugated variable to z and describes the deviation from equilibrium. Regarding the temperature dependence of I_{exc} or the Landau-Placzek ratio LPR, respectively, we may assume that there exists some "critical temperature" T_0, where $\left(\frac{\partial \mu}{\partial z}\right)_{p,T,A} = 0$ and that

$$\left(\frac{\partial \mu}{\partial z}\right)_{p,T,A} = C(T-T_0)^\gamma \,. \tag{10}$$

Here γ is a critical exponent. In order to test Eq. (10) in Fig. 11a, b LPR^{-1} was plotted vs temperature T for OTP and chlophen. In each case we find by extrapolation to LPR→∞ a value of $T_0 = 193$ K for

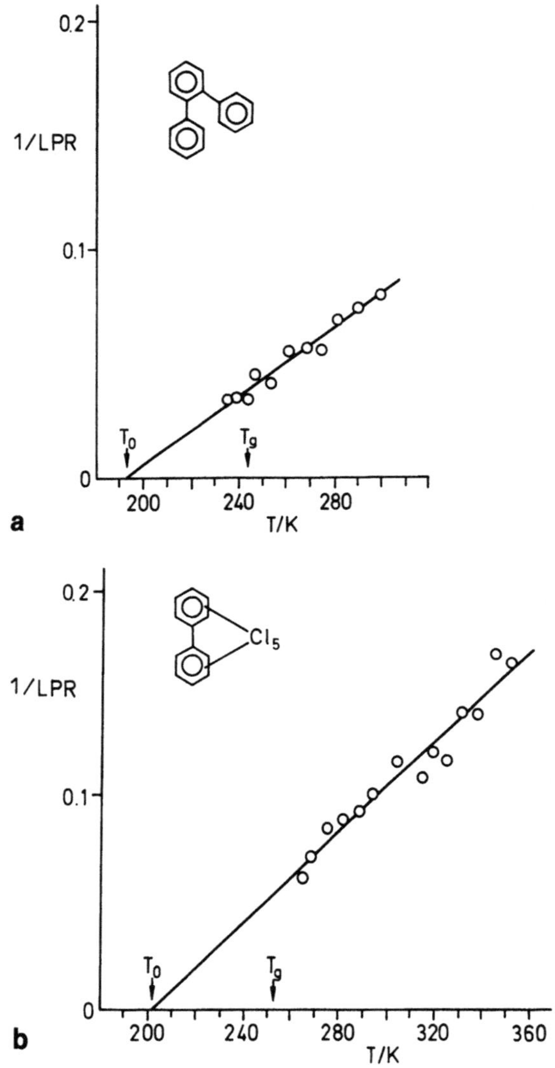

Fig. 11 a, b. The inverse of the Landau-Placzek ratio vs temperature for *o*-terphenyl and chlophen. The glass transition temperatures are indicated by an arrow. The solid lines are results of a least-squares-fit to the data yielding a value of $T_0 = T_g - C_2$ for both systems when $\text{LPR}^{-1} \to 0$

Fig. 12. Schematic representation of the "pre-freezing" effect for various temperatures

OTP and $T_0 = 202$ K for chlophen. Those temperatures are roughly 50 K below the glass transition temperatures and can thus be interpreted to be the ideal glass transition temperature T_0 from the Vogel-Fulcher-Tamman representation of a temperature dependence of the viscosity. Indeed for OTP the Williams-Landel-Ferry C_1 and C_2 parameters ($C_2 = T_g - T_0$) can be calculated from fitting the log τ vs T^{-1} curve of Fig. 7. They amount to be $C_1 = 20$ and $C_2 = 50.4$ K, thus supporting our speculative hypothesis surprisingly well. The exponent γ is found to be $\gamma = 1$ in a good approximation. Additional experiments have to be undertaken for the further clarification of our observations. The same behavior as in Fig. 11 was already found [42] for the case of bis-cresol-cyclohexan-dimethylether (BKDE) which is a glass-forming liquid of similar behavior.

Although the generality of the results described above has to be proven in the future, we may speculate about the physical nature of the observed effects. We believe that in the liquid above T_g "pre-frozen clusters" exist and give rise to the scattering phenomenon and to the long relaxation times τ_z (about 10^3 to 10^5 times slower than the density fluctuations). Such "pre-freezing" was postulated by Ubbelohde [43] long ago, but for quite other reasons that we believe not to be true. The pre-frozen clusters may also be related to the "crystal embryos" discussed by Frenkel [44]. There exists also some resemblance to the free volume theory of Cohen and Grest [37], where it is assumed that one can differentiate above T_g between "liquid-like" and "solid-like" cells. The distribution of clusters of solid-like cells may be visualized, as drawn in Fig. 12 for different temperatures. The glass transition temperature T_g then is defined as the temperature where the pre-frozen clusters percolate.

Acknowledgements

G.M. thanks the Sonderforschungsbereich 262 for financial support. The work was supported in part by the Hoechst AG.

References

1. Nozaki R, Mashimo S (1987) J Chem Phys 87:2271
2. Rössler E, Sillescu H, Spiess HW (1985) Polymer 26:203
3. Fujara F, Wefing S, Spiess HW (1980) J Chem Phys 84:4579
4. Boese D, Momper B, Meier G, Kremer F, Hagenah J-U, Fischer EW (1989) Macromolecules, accepted
5. Patterson GD (1988) Adv Polym Sci 48:175
6. Berne BJ, Pecora R (1976) Dynamic light scattering. Wiley, New York
7. Li BY, Jiang OZ, Fytas G, Wang CH (1986) Macromolecules 19:778
8. Wang CH, Zhu XR, Shen JC (1987) Mol Phys 62:749
9. Mountain RD (1966) J Res Nat Bur Standards 70A:207
10. Montrose CJ, Solovjev VA, Litovitz TA (1967) J Acoust Soc Am 43:117
11. Rytov SM (1970) Soviet Physics JETP 31:1163
12. Pinnow DA, Candou SJ, LaMacchia JT, Litovitz TA (1967) J Acoust Soc Am 43:131
13. Lin YH, Wang CH (1979) J Chem Phys 70:687
14. Schroeder J (1977) In: Tomozava M, Doremus RH (eds) Treatise on Material Science and Technology. Academic Press, New York
15. Oh TG, Fischer EW, Hellmann GP, Russell TP, Wang CH (1980) Polymer 27:261
16. Oh TG (1983) Ph D Thesis, Mainz
17. Fytas G, Lin YH, Chu B (1981) J Chem Phys 74:3131
18. D'Arrigio G (1975) J Chem Phys 63:61
19. Patterson GD (1983) Ann Rev Mater Sci 13:219
20. Mersch W, Fytas G, Dorfmüller T (1977) Ber Bunsenges Phys Chem 81:419
21. Hagenah J-U (1988) Ph D Thesis, Mainz
22. Börgesson L, Stevens JR, Torrell LM (1987) Polymer 28:1803
23. Ngai KL, Rendell RW, Wang CH (1989) Polymer 30:369
24. Börgesson L, Torell LM, Stevens JR (1989) Polymer 30:370
25. Meier G, Fischer EW (1988) In: Richter D, Springer T (eds) Polymer Motion in Dense Systems. Springer, Berlin
26. Fytas G, Wang CH, Lilge D, Dorfmüller T (1981) J Chem Phys 75:4247
27. Meier G, Fujara F, Petry W (1989) submitted to Macromolecules
28. Greet RJ, Turnbull D (1967) J Chem Phys 47:2185
29. Debye P, Bueche AM (1949) J Appl Phys 20:518
30. Dettenmaier M, Fischer EW (1973) Kolloid Z u Z Polym 251:922
31. Koike Y, Tanio N, Ohtsuka Y (1989) Macromolecules 22:1367
32. Dardy HD, Volterra V, Litovitz TA (1973) J Chem Phys 59:4491
33. Steffen W, Meier G, Fischer EW (1989) unpublished results
34. Bartsch E, Kiebel M, Fujara F (1989) to be published
35. Fujara F, Petry W (1987) Europhys Lett 4:921
36. i.e. Bengtzelius U, Götze W, Sjölander A (1984) J Phys C 17:5915
37. Cohen MH, Grest GS (1979) Phys Rev B 20:1077
38. Thönnes W, Meier G, Fischer EW (1989) to be published
39. Krug H (1989) Ph D Thesis, Mainz

40. Haward RN (1973) The Physics of Glassy Polymers. Applied Sci Publ, London
41. Wendorff JH, Fischer EW (1973) Kolloid Z u Z Polym 251:876
42. Rabenau T, Meier G, Fischer EW (1989) to be published
43. Ubbelohde AR (1978) The Molten State of Matter. Wiley, New York
44. Frenkel JE (1957) Kinetische Theorien der Flüssigkeiten. DVW

Received April 10, 1989;
accepted August 15, 1989

Authors' address:

Prof. Dr. E. W. Fischer
Max-Planck-Institut für Polymerforschung
Postfach 3148
D-6500 Mainz, FRG

Discussion

OLEYNIK:

Could you please say something about physical nature of high frequency motions which form the background scattering. It is not strong that high frequency motions may freeze at glass transition interval during sample cooling? Or may be some other explanation of strong dependence of the background intensity on temperature?

FISCHER:

As I pointed out already, the origin ofknown. This is true for both cases, the Brillouin background scattering and the inelastic neutron scattering. Very active research is going on in this field and there exist several attempts (see [35]) to correlate these effects to modern theories of glass transition, but a final conclusion could not be reached.

KRÜGER:

According to the constancy of the β-parameter I agree completely with your results. We found several years ago the same result for arylate-oligomers.

But I would like to ask you about the contribution of the shear phonons to the dynamical structure factor. I think, we agree, that at least at the quasistatic T_g the transverse phonons should become underdamped. We have found for the above mentioned arylates that the shear phonons were underdamped 40 K above T_g but they should transform to a Rayleigh wing at higher temperatures. Finally, is the T_0 you have obtained from the diverging Landau-Placzek ratio in accordance with dielectric results?

FISCHER:

With regard to the β-parameter, I am glad to hear about this agreement. According to the results from a larger number of systems which we studied and which are not mentioned here, the constancy of β with regard to temperature seems to be a general rule if no other process with a different temperature dependence merges.

We did not find a contribution of the shear phonons to the structure factor. Regarding the last part of your question: The T_0 defined by Fig. 11 is in agreement with the results from DK relaxation (see [21]).

WARTEWIG:

According to our experimental results of ultrasonic and PCS studies on polybutadiene melt the β parameter of the KWW-function is temperature dependent. But, β is constant in the temperature range where the photon correlation spectra are taken.

FISCHER:

It may be that the constancy of the β parameter is restricted to some temperature range above T_g. The point is that the temperature dependence of β is especially important in the neighborhood of T_g.

RICHTER:

I'd like to address your fast "background" component in the Brillouin spectra. In inelastic neutron scattering experiments the observation of fast relaxation or extra low frequency vibrational components in the dynamic structure factor almost appear to be a signature of the glass transition. This phenomenon has been found in ionic as well as van der Waals and polymeric glasses. In general it is observed that this fast motion reaches well into the glassy regime. Do you have any evidence for such a behavior in the light scattering data?

FISCHER:

I agree with your opinion that the fast background is an important feature of the glass transition. Its behavior below T_g is not known to us, since no measurements in this regime were performed for the cases of OTP and chlophen.

Progress in Colloid & Polymer Science Progr Colloid & Polymer Sci 80:209–220 (1989)

Relaxation in permanent networks

H. Ambacher, H.F. Enderle, H.G. Kilian, and A. Sauter

Abteilung Experimentelle Physik, Universität Ulm, FRG

Abstract: By linking together the theory of finite elasticity and thermodynamics of irreversible processes one is led to a very general description of deformation. The clue to the simplicity of describing relaxation in permanent networks is the existence of the van der Waals network model which allows a unique description of the nonlinear equilibrium states of reference. Having then a set of relaxation processes coupled to the network ("global level") one arrives at the prediction that only shear relaxation is important. This is proven to hold true by describing relaxation during stress-strain experiments at different temperatures in different deformation modes including the regime of large strains.

Key words: Finite elasticity; irreversible thermodynamics; van der Waals-networks; relaxation; large deformations

Introduction

Static stress-strain behavior of permanent networks is fully interpreted in terms of a van der Waals theory [1]. It is the strain-induced reduction of conformational entropy of chains of finite lengths modified by "weak interactions" that are important. Strain-energy is fully determined de facto by "global phenomena" [2, 3]. On the "local level", liquid-like anisotropic clusters become oriented and their intrinsic properties are left nearly strain-invariant [4].

In the rubbery state chain-segments move rapidly within the time scale of macroscopic deformation [5, 6]. When a rubber is deformed slow processes on the global level are behind deviations from equilibrium. It is due to the presence of fast "Rouse-motions" that these deviations from equilibrium remain small. For this reason, relaxation in networks should be tractable in terms of the classical thermodynamics of irreversible processes [7–10].

According to general postulates of thermodynamics of irreversible processes the strain-energy function of the van der Waals model may be extended by introducing hidden variables. These variables are functions of time [8–15]. It is then required to define how the underlying processes are connected with the "nonlinear" van der Waals strain energy function.

The description of various experiments – many of them taken from literature – serves our concern to typify relaxation in networks. We focus our attention to the central question whether the time dependence of processes may in any case be interpreted in terms of linear physical processes.

Theory

General relations will be derived by combining the theory of finite-elasticity and irreversible thermodynamics. The treatment is then specified by introducing the van der Waals model.

The hidden variables

The equilibrium strain energy is defined as a function of T and C where the variable T is the temperature field. C is the strain field defined by the Cauchy-Green tensor [16–18]. Deviations from equilibrium are described by a set of hidden tensor variables ζ_k (local extensive variables [8, 11–15]). During relaxation each of the ζ_k varies with time and strain. These phenomenological variables may be considered to account for internal constraints, the molecular interpretation of which is mostly difficult to achieve.

According to thermodynamics of irreversible processes the dissipation function δ is written as

$$\delta = \sum_{k=1}^{n} (A_k : \dot{\zeta}_k) , \tag{1}$$

where the affinity A_k is the generalized force tensor with which the tensor variable ζ_k is associated. By writing $(A_k : \xi_k)$, a scalar quantity results that is given by the trace of the tensor product. In equilibrium A_k disappears; hence, A_k is defined to characterize the momentary "distance to equilibrium". $\dot{\zeta}_k$ gives the rate with which the variable $\zeta_k(t)$ runs toward its equilibrium values, $\zeta_{k0}(T,C)$. δ describes the rate with which strain-energy is transferred into heat.

In the system of coordinates (T, C, ζ_k) the free energy-density change $d\psi/dt$ reads [19]

$$\dot{\psi} = (T : \dot{C}) - \eta \dot{T} - \delta . \tag{2}$$

η is the entropy density, T is the second Piola-Kirchhoff tensor related with the Cauchy stress tensor σ according to [20]

$$\sigma = (\varrho/\varrho_R) F T F^T . \tag{3}$$

F is the deformation gradient tensor, F^T is its transposed version. ϱ/ϱ_R is the ratio of densities

$$\varrho/\varrho_R = 1/|J| , \tag{4}$$

both being related by the Jacobi determinant $|J|$. This determinant defines how the volume increment dV_R (in the coordinate frame of a "particle") is transformed into the volume increment dV fixed in space

$$dV = J dV_R . \tag{5}$$

The Onsager-Ansatz

"In the neighborhood of thermodynamic equilibrium", the generalized flows $\dot{\zeta}_k$ are set to be proportional to the "momentary distance from equilibrium" [21–24]

$$\dot{\zeta}_k = \Sigma \alpha_{ki} A_i . \tag{6}$$

where the hidden variables are defined as a tensor of second rank

$$\zeta_k = (\zeta_{nm}^{(k)}) \; ; \quad n, m = 1, 2, 3 . \tag{7}$$

A_i is the affinity defined as symmetrical tensor of Onsager coefficients $(\alpha_{ki} = \alpha_{ik})$. Nonzero affinities lead to nonzero fluxes. Together with Eq. (1) we obtain

$$\delta = \sum_k \sum_i \beta_{ki} \dot{\zeta}_k \dot{\zeta}_i \geq 0 , \tag{8}$$

where β_{ki} is the inverse of α_{ki}. The rate of production of entropy is the sum of products of each flux with its associated affinity. δ is always positive and it disappears in equilibrium.

The free-energy density

The simplest formulation of the free-energy density $\psi(T, C, \zeta_\nu)$ is obtained by defining the homogeneous quadratic form

$$\psi(T, C, \zeta_\nu) = \Phi_0(T, C) + \Sigma \mathrm{tr}(\Phi_1^{(\nu)}(T, C)\zeta_\nu) + (1/2)\Sigma\Sigma \mathrm{tr}(\zeta_\mu \Phi_2^{(\mu\nu)} \zeta_\nu) . \tag{9}$$

The term $\Phi_0(T, C)$ represents the equilibrium strain energy of the network stored on the global level. The second term on the righthand side of Eq. (9) is the connecting link between the set of relaxation mechanism ζ_ν and the global level. The energy coefficients $\Phi_2^{(\nu\mu)}$ are taken not to depend on strain (independence of C). The coefficients $\Phi_1^{(\nu)}(T, C)$, on the other hand, are defined so as to describe how local and anisotropic processes ζ_ν and the potential $\Phi_0(T, C)$ are interconnected during deformation. Hence, normal-modes embracing both global and local processes typify the time-dependent behavior of the system.

The equations of state

The set of "generalized equations of state" (including description of time-dependent processes) is straightforwardly deduced.

Mechanical equation of state

$$T = 2(\partial\psi/\partial C)_{T, \zeta_\nu} . \tag{10}$$

Caloric equation of state

$$-\eta = (\partial\psi/\partial T)_{C, \zeta_\nu} . \tag{11}$$

Internal equation of state
("chemical equation of state")

$$-A = (\partial\psi/\partial\zeta_\nu)_{T, C} . \tag{12}$$

Differential-equations of the hidden variables

From Eqs. (7), (10), and (12) we are led to

$$T/2 = \partial\Phi_0(T, C)/\partial C + \Sigma \partial \mathrm{tr}(\Phi_1^{(\nu)}(T, C)\zeta_\nu)/\partial C \tag{14}$$

$$-A_\nu = \Phi_1^{(\nu)}(T,C) + \Sigma \Phi_2^{(\mu\nu)} \zeta_\mu \; . \tag{15}$$

From Eqs. (6) and (15) we deduce differential equations for each of the hidden variables expressed by

$$\dot{\zeta}_\nu/\alpha + \Phi_1^{(\nu)}(T,C) + \Sigma \Phi_2^{(\mu\nu)}(T)\zeta_\mu = 0 \; ; \quad \nu = 1,\ldots,n \; . \tag{16}$$

In the normal-mode representation with orthogonal ζ_ν, these equations take the form

$$\dot{\zeta}_\nu/\alpha + \Phi_1^{(\nu)}(T,C) + \Phi_2^{(\nu\nu)}(T)\zeta_\nu = 0 \; . \tag{17}$$

How ζ_ν, the ν-th hidden variable of n orthogonal relaxation mechanism, depends on deformation is characterized by the strain tensor $\boldsymbol{C}$. This is the consequence of having the processes linked to the global level. The set of $\Phi_1^{(\nu)}(T,C)$ determines how the relaxation mechanism and the global level are interconnected.

Isotropic systems

For isotropic systems the free-energy coefficient matrix degenerates: each of the modes is now characterized by two coefficients $\varrho_i^{(\nu)} (\Phi_{2ijkl}^{(\nu\nu)} = (1/3)\varrho_1^{(\nu)}\delta_{ij}\delta_{kl} + \varrho_2^{(\nu)}\delta_{ij}\delta_{kl})$ [25]. We arrive at

$$\dot{\zeta}_\nu + \alpha\varrho_2^{(\nu)}\zeta_\nu + (\alpha/3)\varrho_1^{(\nu)}\mathrm{tr}(\zeta_\nu)\mathbf{1} = -\alpha\Phi_1^{(\nu)}(T,C) \; ; \quad \nu = 1,\ldots,n \; . \tag{18}$$

By splitting ζ_ν into isotropic and deviatoric components

$$\zeta_\nu = \zeta_{\nu v}\mathbf{1} + \zeta_{\nu s} \; , \tag{19}$$

with

$$\mathrm{tr}\,\zeta_{\nu s} = 0 \; ; \quad \zeta_{\nu v} = (1/3)\,\mathrm{tr}\,\zeta_\nu \; , \tag{20}$$

and by writing the matrix coefficients accordingly

$$\Phi_1^{(\nu)} = -(1/3)\Phi_{1v}^{(\nu)}\mathbf{1} + \Phi_{1s}^{(\nu)} \; . \tag{21}$$

Equation (18) is represented by two independent differential equations

$$\dot{\zeta}_{\nu v} + \alpha(\varrho_1^{(\nu)} + \varrho_2^{(\nu)})\zeta_{\nu v} = (\alpha/3)\Phi_{1v}^{(\nu)}(T,C), \text{ and} \tag{22}$$

$$\dot{\zeta}_{\nu s} + \alpha\varrho_2^{(\nu)}\zeta_{\nu s} = -\alpha\Phi_{1s}^{(\nu)}(T,C) \; . \tag{23}$$

These differential equations, each with constant coefficients can be solved

$$\zeta_{\nu v}(t) = \zeta_{\nu v0}\exp\{-t/\tau_{\nu v}\} + (\alpha/3) \times \int_0^t \exp\{-(t-t')/\tau_{\nu v}\}\Phi_{1v}^{(\nu)}(t')dt' \tag{24}$$

$$\zeta_{\nu s}(t) = \zeta_{\nu v0}\exp\{-t/\tau_{\nu v}\} - \alpha \times \int_0^t \exp\{-(t-t')/\tau_{\nu s}\}\Phi_{1s}^{(\nu)}(t')dt' \; , \tag{25}$$

where the "volume-relaxation time" is written as

$$\tau_{\nu v} = 1/\alpha\varrho_2^{(\nu)} \; , \tag{26}$$

while the "shear-relaxation time" is set equal to

$$\tau_{\nu s} = 1/[\alpha(\varrho_2^{(\nu)} + \varrho_1^{(\nu)}] \; . \tag{27}$$

From the above equations we are led to the general relationship

$$\begin{aligned} T/2 &= \partial\Phi_0(T,C)/\partial C + (\alpha/3)\Sigma(\partial\Phi_{1v}^{(\nu)}(T,C)/\partial C) \\ &\times \int_0^t \exp\{-(t-t')/\tau_{\nu v}\}\Phi_{1v}^{(\nu)}(t')dt' \\ &- \alpha\Sigma(\partial\Phi_{1sij}^{(\omega)}(T,C)/\partial C) \\ &\times \int_0^t \exp\{-(t-t')/\tau_{\nu s}\}\Phi_{1sij}^{(\nu)}(t')dt' \; . \end{aligned} \tag{28}$$

For thermodynamic stability an isothermal volume relaxation time must always be smaller in comparison with the associated isothermal shear relaxation time [8, 10]

$$\tau_{\nu s} > \tau_{\nu v} \; . \tag{29}$$

The relaxation-time distribution

We introduce the strain-invariant relaxation-time spectrum [16] by writing the coefficients in the free energy as

$$\Phi_{1v}^{(\nu)} = \Phi_{1v}(C)h_{\nu v} \tag{30}$$

$$\Phi_{1sij} = \Phi_{1sij}(C)h_{\nu s} \; . \tag{31}$$

Equation (28) is then equal to

$$T/2 = \partial \Phi_0(T,C)/\partial C + \int_0^t \Sigma(\alpha/3) h_{vv}^2 \times \exp\{-(t-t')/\tau_{vv}\} \Phi_{1v}(t')^2 (\partial \Phi_{1v}(t)/\partial C) dt' - \int_0^t \Sigma \alpha h_{vs}^2 exp\{-(t-t')/\tau_{vs}\} \Phi_{1sij}(t') \times [\partial \Phi_{1sij}(t)/\partial C] dt' \ . \quad (32)$$

Φ_{1sij} is the sum of the tensor coefficients expressed by the Einstein convention.

Incompressible systems

With the foregoing considerations to guide us, volume relaxation is neglected. Rubbers are assumed to behave like incompressible elastic systems. To enforce incompressibility the hydrostatic pressure p is introduced. We are then led to the relation [25]

$$T/2 + pC^{-1} = \partial \Phi_0(T,C)/\partial C - [\partial \Phi_{1sij}(t)/\partial C] \times \int_0^t m_s(t-t') \Phi_{1sij}(t') dt' \ . \quad (33)$$

$m_s(t-t') = \Sigma \alpha h_{vs}^2 \exp\{-(t-t')/\tau_{vs}\}$ is the memory function characterizing shear relaxation [16]. As a fundamental feature it is predicted that relaxation in networks should be controlled by shear processes only. That volume effects [4, 14] are negligible must be proven by experiment.

Isotropic and scalar coupling

It is reasonable to assume that relaxation modes are in the average isotropically coupled to the global level constituted by the network. The matrix elements of coupling may then be formulated as

$$\Phi_{1sij} = \varphi(I_1, I_2)\delta_{ij} \ . \quad (34)$$

I_1 and I_2 are the strain-invariants of interest ($I_3 = 1$ for incompressible continua); δ_{ij} is the Kronecker symbol.

Isotropic coupling is suggested because of having the relevant Platzwechsel running within sufficiently small chain-segments clusters. The order within these clusters is liquid-like. Due to strain invariance of liquid-like properties in networks [4] molecular modes of motion are supposed to also be approximately strain invariant. The utility of this suggestion can be checked by describing relaxation in highly strained networks with the aid of the relations derived.

It turns out that the coupling can be brought into the scalar version [14]

$$\varphi(I_1, I_2) = k(\Phi_0(I_1, I_2))^{1/2} \ . \quad (35)$$

Each of the a priori orthogonal relaxation mechanisms (v) is now "linearly and in an isotropic manner" coupled with the network, the elasticity of which is characterized by the potential $\Phi_0(I_1, I_2)$. We are thus led to

$$\sigma + p\mathbf{1} = 2F(\partial \Phi_0/\partial C)F^{-1}(1 - G_0(t)) \ , \quad (36)$$

where

$$G_0(t) = (k^2/2) \int_0^t m_s(t-t') \times \{\Phi_0(t')/\Phi_0(t)\}^{1/2} dt' \ . \quad (37)$$

Later we shall explicitly formulate the first term on the righthand side of Eq. (36).

Equations of state of permanent networks

Because of having the strain-energy function of an isotropic network written as [26, 27]

$$W = W(I_1, I_2, I_3) = \psi_0(T, \varrho_R, I_1, I_2, I_3) \ , \quad (38)$$

the Cauchy stress tensor σ reads [28, 29]

$$\sigma = 2(\varrho/\varrho_R) F(\partial W/\partial C) F^T$$

$$\sigma = 2(\varrho/\varrho_R)\{\partial W/\partial I_1 + I_1(\partial W/\partial I_2)B - (\partial W/\partial I_2)B^2 + I_3(\partial W/\partial I_3)\mathbf{1}\} \ , \quad (39)$$

where $B = FF^T$ is the right Cauchy-Green deformation tensor. For incompressible systems ($I_3 = 1$) this becomes equal to

$$\sigma + p\mathbf{1} = 2\{(\partial W/\partial I_1 + I_1 \partial W/\partial I_2)B - (\partial W/\partial I_2)B^2 \ . \quad (40)$$

In homogeneous networks the deformation gradient is defined by

$$F = \text{diag}\ (\lambda_1, \lambda_2, \lambda_3) = \begin{bmatrix} \lambda_1 & & \\ & \lambda_2 & \\ & & \lambda_3 \end{bmatrix} \ , \quad (41)$$

with λ_i as macroscopic strain parameters in the direction of the axis i. The well known relations result [30]

$$B = \text{diag}\,(\lambda_1^2, \lambda_2^2, \lambda_3^2) = C \tag{42}$$

$$I_1 = \text{tr}(C) = \lambda_1^2 + \lambda_2^2 + \lambda_3^2 \tag{43}$$

$$I_2 = ((\text{tr}(C)^2 - \text{tr}(C^2))/2 = (\lambda_1\lambda_2)^2 + (\lambda_2\lambda_3)^2 + (\lambda_3\lambda_1)^2 \tag{44}$$

$$I_3 = \det(C)(\lambda_1\lambda_2\lambda_3)^2 = 1{:}\ \lambda_3 = 1/\lambda_1\lambda_2 \ . \tag{45}$$

The principal nominal forces in the biaxial deformation modes are then explicitly written as

$$f_i = \sigma_i/\lambda_i = 2(\lambda_i - 1/(\lambda_i^3\lambda_j^2)) \times(\partial W/\partial I_1 + \lambda_j^2 \partial W/\partial I_2) \ ; \quad i = 1,2 \ . \tag{46}$$

The van der Waals strain-energy function

A van der Waals version of Kuhn's network theory (Gaussian network [2]) was given recently [1]. One of the strain-independent van der Waals parameters is the maximum strain λ_m. This parameter accounts for the maximum chain extensibility in networks with finite chain lengths. The second van der Waals parameter a is supposed to describe global interaction between network chains [31–33].

The shear-modulus of a permanent network is then written as [31, 33, 34]

$$G_\infty = \varrho R T/M_c = \varrho R T/\lambda_m^2 M_0 \tag{47}$$

where ϱ is the density; T is the absolute temperature. The mean molecular weight of the network chains M_c is then equal to

$$M_c = y M_0 = \lambda_m^2 M_0 \ . \tag{48}$$

y is the number of stretching-invariant units of molecular weight M_0. M_0 was found to be in the magnitude of order of a monomer unit [34]. The last term on the righthand of Eq. (47) implies the Gaussian formulation of the maximum strain of a single phantom chain

$$\lambda_m^2 = y \ , \tag{49}$$

which is reasonable as an approximation for not too short chains only. λ_m is therefore a material constant related to the structure of the network.

To come to an adequate formulation of the van der Waals potential we define the maximum modulus attained at highest frequencies

$$G = G_\infty + \Gamma \ , \tag{50}$$

where Γ is the relaxation strength. In the present formulation the van der Waals formulation of the potential Φ_0 is then equal to

$$\Phi_0 = (G/G_\infty)\, W(I_1, I_2, y, a) \ , \tag{51}$$

with

$$W(I_1, I_2, y, a) = -G_\infty\{(y-3)\ln(1-\Theta) + \Theta\} - (2a/3)[(I-3)/2]^{3/2} \ . \tag{52}$$

Θ is given by

$$\Theta = \{(I-3)/(y-3)\}^{1/2} \ . \tag{53}$$

The strain-invariant I is assumed to have the generalized form [35]

$$I = \beta I_1 + (1-\beta) I_2 \ , \tag{54}$$

where β is an adjusting parameter.

The nominal force in simple extension is equal to

$$f = \partial W/\partial\lambda = G_\infty(\lambda - 1/\lambda^2)\{1/(1-\Theta) - a(\Phi)^{1/2}\} \ , \tag{55}$$

where

$$\Phi = (I_1 - 3)/2 = (\lambda^2 + 2/\lambda - 3)/2 \ . \tag{56}$$

Quasi-static experiments

The quality of fitting calculations to measurements carried out in the modes of equibiaxial and simple extension [36, 37] is evident from Figs. 1 and 2. Uniaxial compression and simple extension can both be described if the parameter β is assigned to values which are smaller than one. The molecular reason behind these effects is still unknown.

Biaxial deformation of polyisoprene has been carried out by Kawabata et al. [38]. First, the network parameters were adjusted by fitting the principal forces in the mode of pure shear (see Fig. 3). Leaving these network-parameters invariant, it is then possible to describe the principal forces in the mode of biaxial deformation with the aid of Eq. (46) (see Figs. 4, 5 and 6). In contrast to what has been discussed by

Kawabata [38], the interpretation presented is possible without introducing the second strain invariant ($\beta = 1$).

Heat exchange and volume effects as observed during simple extension can be computed with the aid of the thermo-elastic equations of state of a van der

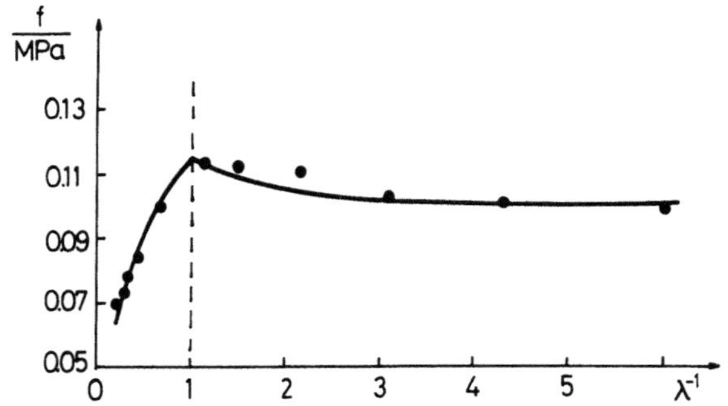

Fig. 1. Mooney-plot of poly-dimethylene-siloxane (PDMS) covering the range of simple extension and uniaxial compression according to Pak and Flory [36]. Solid line calculated with Eq. (56) with the use of the parameters $G = 0.115$ MPa, $y = 180$, $a = 0.29$, $\beta = 0.93$

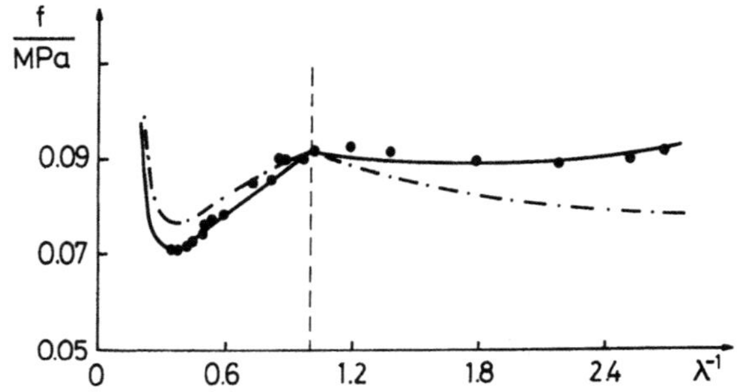

Fig. 2. Mooney-plot of PDMS covering simple extension and uniaxial compression according to Macosco et al. [37]. Calculation with the aid of Eq. (56) setting the parameters to $G = 0.0915$ MPa, $y = 64.8$, $a = 0.35$; solid line $\beta = 0.89$; dotted and dashed line shows the calculation with $\beta = 1$

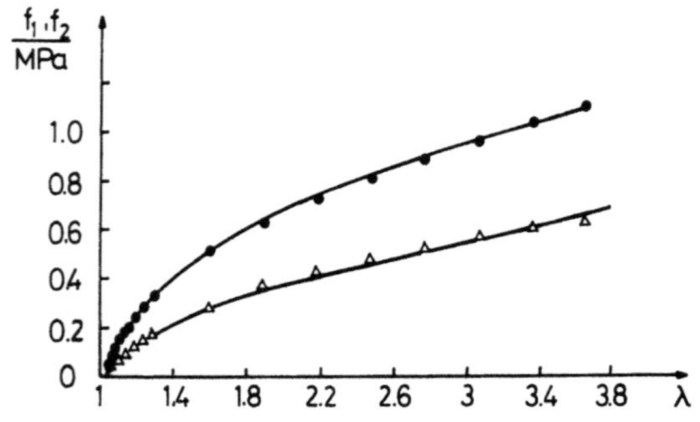

Fig. 3. Nominal principal forces of polyisoprene in pure shear ($\lambda_2 = 1$) according to [38]. Calculation with the aid of Eqs. (46) and (52) by introducing the network parameters $G = 0.38$ MPa, $\lambda_m = 18$ ($y = 324$), $a = 0.22$, $\beta = 1$ (● principal force f_1; △ principal force f_2)

Waals network. Representative examples are shown in Figs. 7 and 8 [39–42]. It is in fact sufficient to introduce the coefficients of thermal expansion and of isothermal compressibility, both measured in the unstrained network. This result supports the hypothesis that liquid-like properties in networks are de facto strain-invariant (except for the small increase in volume which must occur for thermodynamic reasons [2, 3, 4, 43]).

Hence, the van der Waals model allows a full description of quasi-static deformation in networks including thermoelasticity [4, 40, 42]. Because of knowing the van der Waals stress-strain energy, we have a solid basis for predicting where relaxation will take the network.

In a linear response theory the momentary driving force is per definition proportional to the distance from equilibrium. But, because networks behave "nonlinearly" in the mode of simple extension (see Figs. 1–6) stationary dissipation of strain energy is not achievable de facto: visco-elastic behavior of rubbers is in general nonlinear.

Relaxation in permanent networks

We discuss some representative examples.

1) Simple extension data of styrene-butadiene rubber (SBR) at different constant strain rates were described by Chang et al. with the aid of a special empirical approach [44]. These data are now fitted with the aid of the equation deduced from the Eqs. (37) and (51),

$$\sigma(t) = W'(t)\left\{1+(\Gamma/G_\infty)\right.$$
$$\left.\times\left[1-2/\Gamma\int_0^t m_s(t-t')(W(t')/W(t))^{1/2}dt'\right]\right\}, \tag{57}$$

which describes simple extension at the constant strain rate $\dot{\varepsilon} = \text{const}$. The macroscopic strain is then defined accordingly

$$\lambda(t) = \begin{bmatrix} 1 & t \le 0 \\ 1+\dot{\varepsilon}t & t > 0\,. \end{bmatrix} \tag{58}$$

The first derivative of the van der Waals potential formulated in Eq. (52) is equal to

$$W' = G_\infty D\{1-\Theta)-a[(I_1-3)/2]^{1/2}\}$$
$$D = \lambda - 1/\lambda^2\,. \tag{59}$$

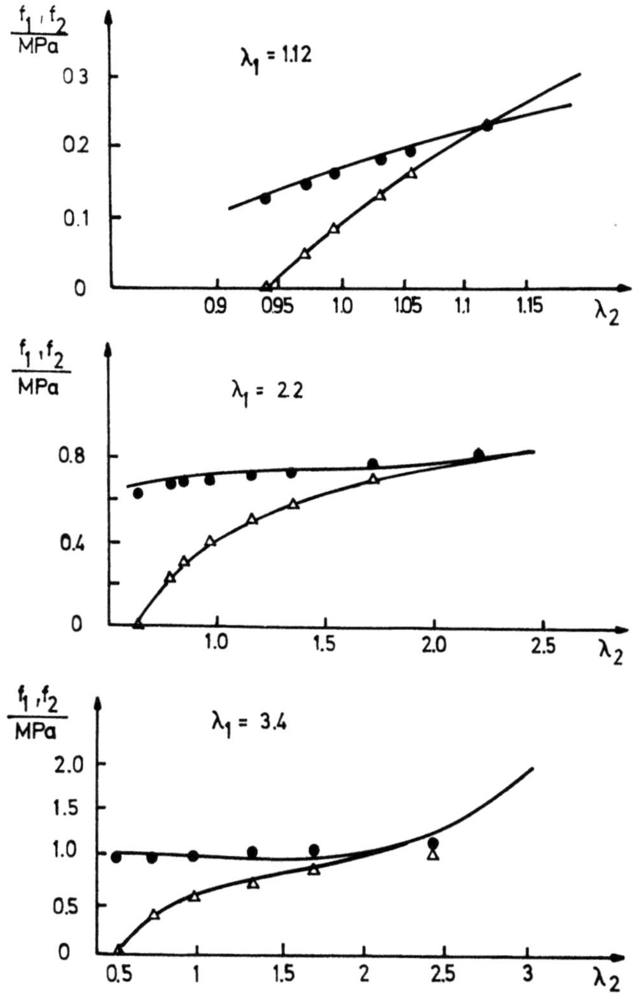

Figs. 4–6. Nominal principal forces of polyisoprene in biaxial extension with constant λ_1 according to Kawabata et al. [38]. Calculation with the aid of the Eqs. (46) and (52) by using the same parameters as listed in the caption of Fig. 3 (● principal force f_1; △ principal force f_2)

The shear-relaxation modulus $G(t-t')$ is derived from $m_s(t-t') = -dG(t-t')/d(t-t')$ to be given by

$$G(t-t') = \Sigma\alpha(h_{\nu s})^2\tau_{\nu s}\exp\{-(t-t')/\tau_{\nu s}\} . \qquad (60)$$

An empirical representation of the shear-relaxation modulus was given by Ferry et al. [45] (Fig. 9)

$$G(t) = \Gamma/\{1+(t/\tau)^k\}+G_\infty . \qquad (61)$$

The quality of the fit is evident from Fig. 10.

The non-linear visco-elastic behavior of rubber in the mode of simple extension is apparently understandable in terms of the linear response-theory presented. Relaxation seems to be relatable to the strain-independent "shear-relaxation memory function" of the system.

2) We want to prove the reliability of our approach by describing relaxation in a rubber stretched to $\lambda_0 > 1$ being submitted then to periodic extensions with small amplitudes. When the amplitudes of the periodic extension are kept sufficiently small the components of the complex modulus have been shown to be equal to [25]

$$E'(\lambda_0,\omega) = W''(\lambda_0)+(\Gamma/G_\infty)(W'^2(\lambda_0)/4\,W(\lambda_0)) \times\{1-(1/\Gamma)\Sigma\Delta g_\nu/(1+\omega^2\tau_{\nu s}^2)\} \qquad (62)$$

$$E''(\lambda_0,\omega) = (1/G_\infty)(W'^2(\lambda_0)/4\,W(\lambda_0)) \times\omega\Sigma\Delta g_\nu\tau_{\nu s}/(1+\omega^2\tau_{\nu s}^2) . \qquad (63)$$

The second derivative of the van der Waals potential – taken at $\lambda = \lambda_0 = \text{const}$ – is given by

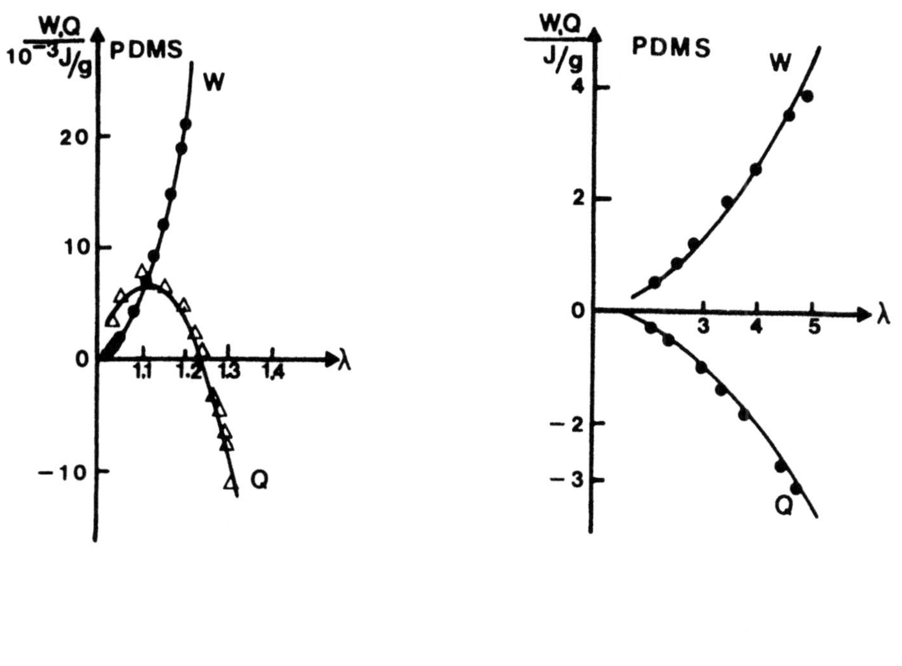

Fig. 7. The strain-energy $W(\lambda)$ and the heat exchange $Q(\lambda)$ for PDMS in the mode of simple extension under quasi-isothermal conditions, according to Godovsky [39]. The solid lines were computed [40] (parameters $G = 0.36$ MPa, $\lambda_m = 10$ ($\nu = 100$), $a = 0.2$; linear thermal expansion coefficient: $\beta_{th} = 2.2\cdot10^{-4}\,K^{-1}$; temperature coefficient of the rotational isomers $\mu = 9.6\cdot10^{-4}\,K^{-1}$)

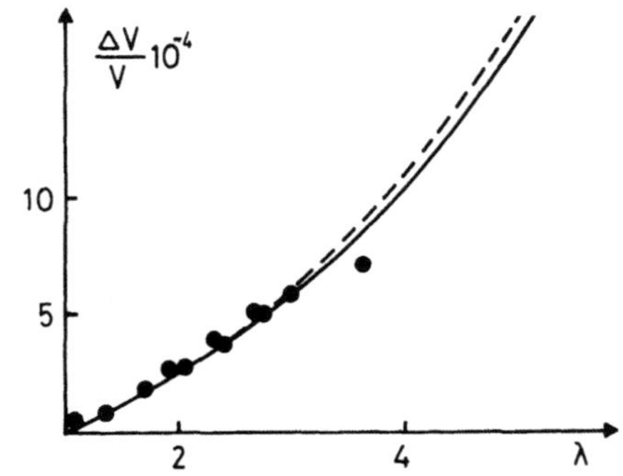

Fig. 8. The relative volume increase of NR in simple extension measured by Göritz [41]. The solid line was computed with the aid of the van der Waals model [40] setting the parameters to $G = 0.36$ MPa, $\lambda_m = 10$, $a = 0.18$, $\beta = 1$; isothermal compressibility $\kappa = 2.5 \cdot 10^{-8}\,\mathrm{m^2\,kg^{-1}}$. The dotted line was computed with the aid of the Gaussian network model

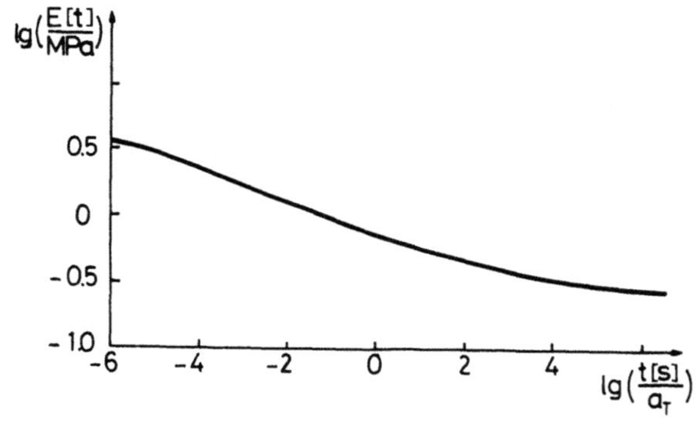

Fig. 9. Relaxation function for styrene-butadiene rubber (SBR) computed with Eq. (61) at $T = 296$ K

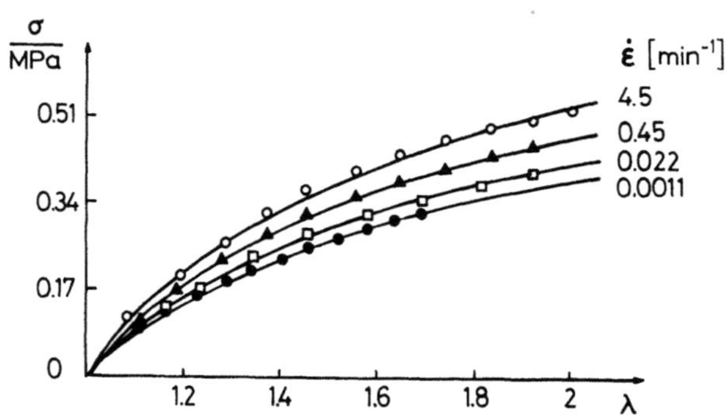

Fig. 10. Uniaxial extension experiments carried out by Chang et al. [44]. Solid lines were calculated with the aid of Eq. (57) by using the relaxation function as depicted in Fig. 9

$$W''(\lambda_0) = (D'/D)\,W'(\lambda_0) + G_\infty\,(D^2/(I_1-3)^{1/2}) \times \{(1/(y-3)^{1/3})/(1-\Theta)^2 - a/\sqrt{2}\}\,, \tag{64}$$

where

$$D' = \partial D/\partial\lambda = 1 + 2/\lambda^3\,. \tag{65}$$

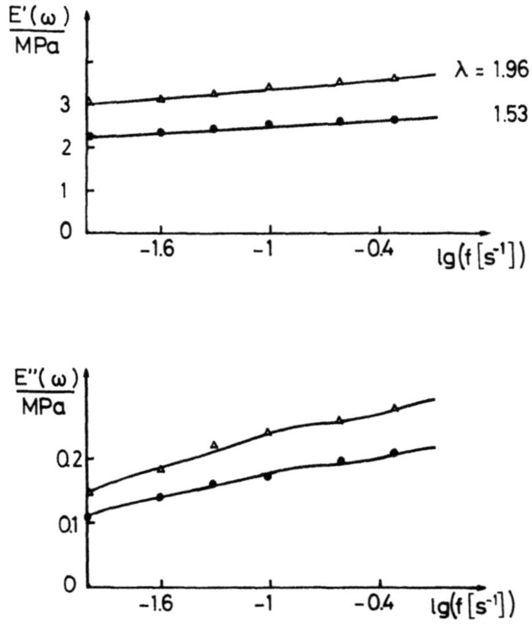

Fig. 11. Elastic moduli of styrene-butadiene rubber (SBR) at $T = 273$ K at various prestrains. The solid lines were calculated using Eqs. (62) and (64). Parameters of the relaxation spectrum are given in Table 1. The network parameters are $G = 0.355$ MPa, $\lambda_m = 10$, $a = 0.25$, $\beta = 1$

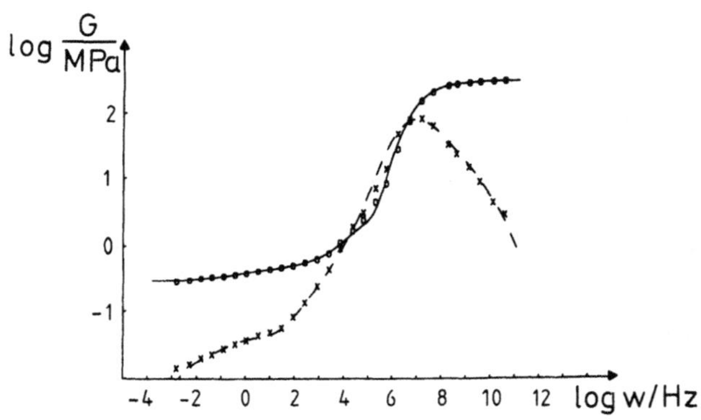

Fig. 12. Master curve for natural rubber measured in simple shear at $T = 275$ K. The lines are calculated with the aid of Eqs. (68) and (69), using the spectrum shown in Fig. 14 for proving consistency of the procedures applied

The relaxation strength has defined in Eq. (50) now takes the special form

$$\Gamma = \Sigma g_\nu\,. \tag{66}$$

The quality of the fit to the experimental data of Goldberg and Lianis [46, 47] is to be seen in Fig. 11. Parameters used are listed in Table 1; the agreement is satisfactory. This result supports the hypothesis that relaxation in networks should always be related to the same shear-relaxation spectrum irrespective of the

Table 1. The relaxation time spectrum and the network parameters used in the calculations depicted in Fig. 11

τ_i/s	850	520	100	50	18	8	4	0.91
Δg_i/MPa	0.145	0.073	0.110	0.290	0.070	0.447	0.116	0.841

mode of deformation applied. Until now, this was shown to hold true in the small-strain regime [44, 48]. The response depends, of course, on the static strain-energy function that is uniquely related to the invariant network structure parameters (M_0, λ_m, a) for each mode of deformation. The result is so simple because of having the elementary processes of relaxation ζ_v all coupled to the adequate strain energy function according to Eq. (35).

3) Stress-strain cycles carried out for natural rubber under constant strain rates mostly display a typical hysteresis. As long as crystallization does not occur the time-dependent phenomena should be related to the strain-invariant shear-relaxation spectrum of the network (see Eq. (57)). Cyclic deformation delivers a convenient possibility for studying relaxation at large strains at various temperatures. The response depends

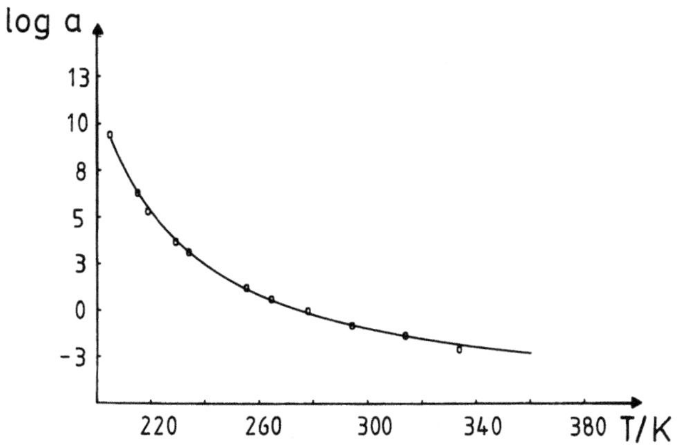

Fig. 13. WLF-fit for the master curve in Fig. 12. The solid line was calculated with Eq. (67) setting the parameters to $c_1 = 5.3$, $c_2 = 111$ K, $T_0 = 275$ K

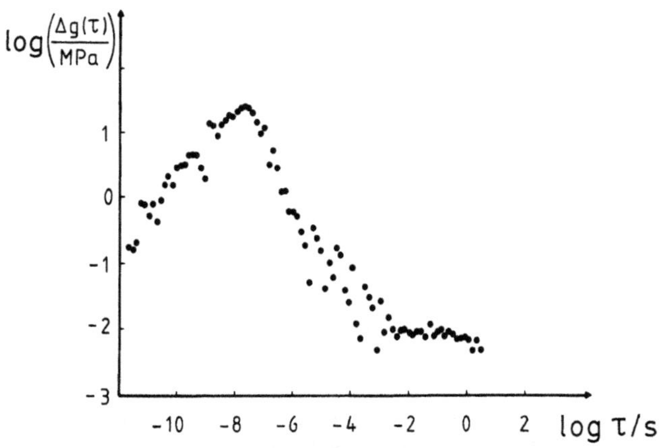

Fig. 14. Relaxation spectrum for NR calculated from the data points of the master curve depicted in Fig. 12. T_0 is set to 275 K

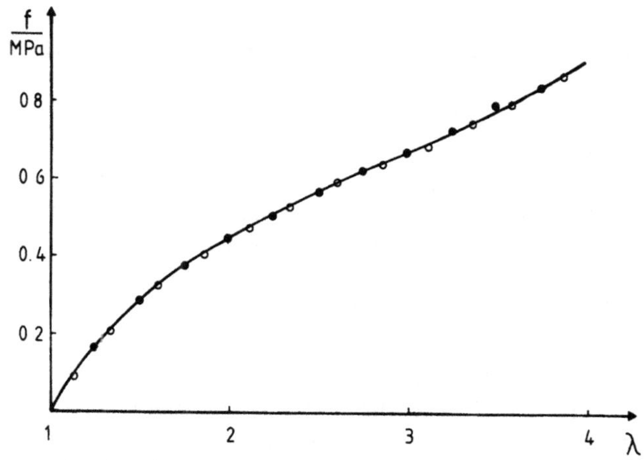

Fig. 15. Quasi-static stress strain curve for natural rubber at $T = 353$ K (● extension data; ○ retraction data). The solid line is calculated with the aid of Eq. (55) setting the parameters to $G = 0.3$ MPa, $a = 0.3$, $\lambda_m = 10.5$

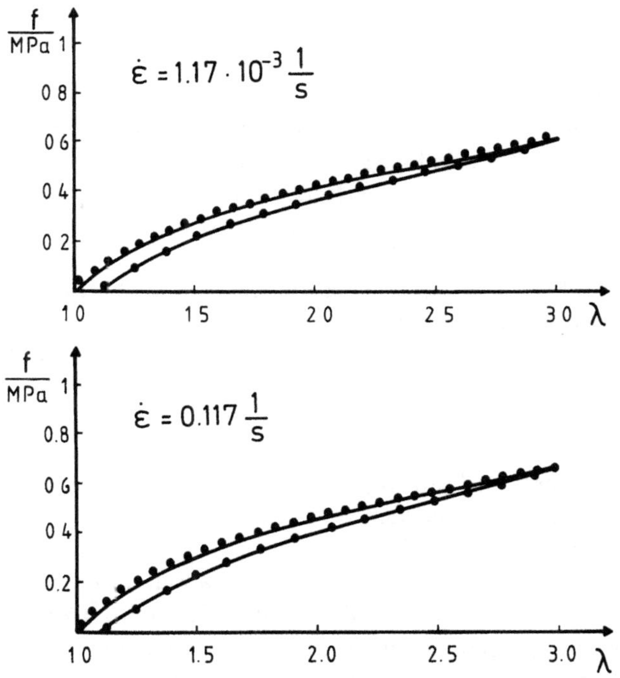

Figs. 16, 17. Stress strain cycles of natural rubber measured at different deformation rates at $T = 293$ K. The solid lines were calculated using Eqs. (52), (57), and (59) with that network parameters as listed in the capture of Fig. 15 and with the relaxation spectrum in Fig. 14. This spectrum was shifted to room temperature by using Eq. (67) whereby the parameters are listed in the capture of Fig. 13

in a very peculiar manner on the shape of the relaxation time spectrum.

Measurements on natural rubber (NR) carried out at different temperatures yield to master-curves which are shown in Fig. 12. The shift-factors obey the WLF-equation

$$\log(a_T) = -c_1(T-T_0)/(c_2+T-T_0) \ . \tag{67}$$

The fit is shown in Fig. 13 with the parameters depicted in the legend. The shear-relaxation spectrum is then computed by applying a modified Stavermann-Schwarzl method [49, 50] (Fig. 14). The dominance of the short-time region is in evidence. This region includes relatively fast processes due to the occurrence of which it is impossible to enforce "large deviations from equilibrium".

First, we deduce the shear-relaxation time spectrum from shear-deformation measurements at small strains covering a frequency range of $10^{-3} \leq \nu \leq 150$ Hz [51]. The components of the complex shear modulus, measured at small strains, are then given by [15]

$$G'(\omega) = G_\infty + \Sigma \Delta g_i(\omega\tau_i)^2/\{1+(\omega\tau_i)^2\} \tag{68}$$

$$G''(\omega) = \Sigma \Delta g_i(\omega\tau_i)/\{1+(\omega\tau_i)^2\} \ . \tag{69}$$

The network parameters M_0, λ_m, and a were now adjusted by fitting stress-strain cycles carried out at 80 °C: Strain-induced crystallization is suppressed here [52]. Irreversible effects are exceedingly small so that a hysteresis cannot be identified within the limits of experimental accuracy (see Fig. 15).

When at room-temperature the maximum strain is kept below a certain limit that depends on the degree of crosslinking, strain-induced crystallization does not occur. With the aid of the WLF-parameters the spectrum is shifted accordingly. Stress-strain cycles can then be fitted straightforwardly without changing parameters at all (invariant network structure (M_0, λ_m, a) and strain-independence of the relaxation time spectrum) (Figs. 16, 17).

At temperatures in the range of about 240–260 K crystallization is difficult to avoid [53–55]. However, when the experiment is carried out at temperatures just above the glass-dispersion region, viscosity grows so as to impede crystallization within the time scale of experiment, and a very pronounced hysteresis is observed (Figs. 18–20). By applying the WLF-shift procedure stress-strain cycles carried out under sufficiently low-strain rates are also fitted without parameter adjustments.

The successful representation verifies that the time-dependent, often macroscopically "non-linear" response can in fact be understood even at large strains in terms of the Onsager approach. Even at temperatures at the root of the glass dispersion region relaxation in NR is dominated by linear shear processes.

In the dispersion region itself new processes come into play which can no longer be simply related with the shear-relaxation spectrum (Figs. 20 and 21). Hence, stress-strain behavior in the glass-transition display additional effects, the interpretation of which will be discussed elsewhere.

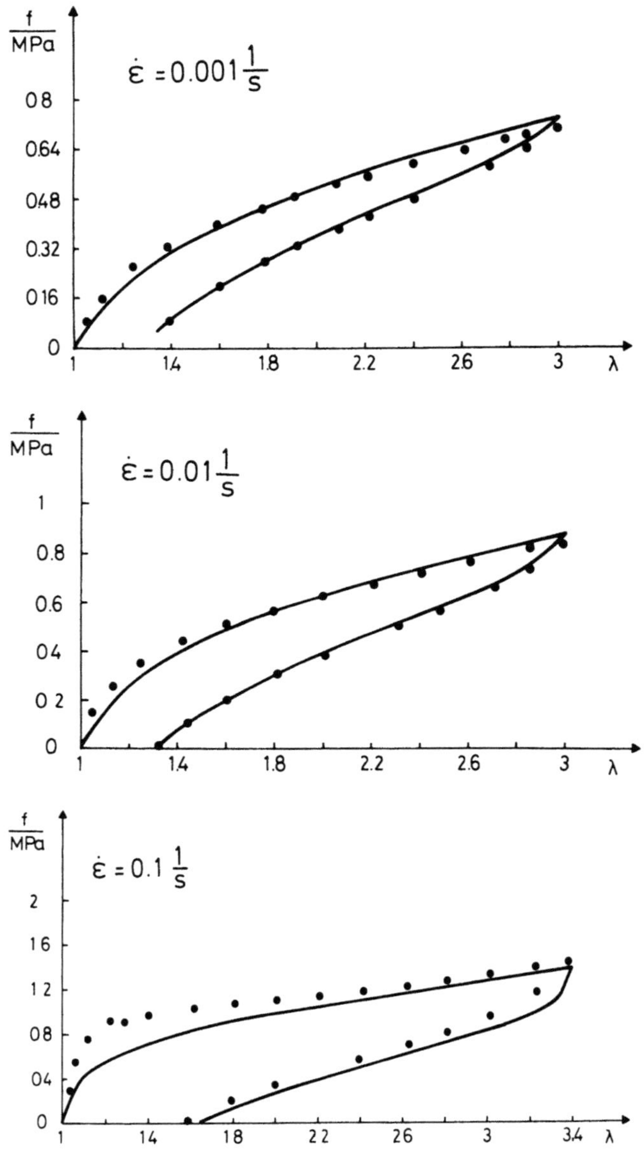

Figs. 18–20. Stress strain cycles of natural rubber for $T = 223$ K and different deformation rates. The solid lines were calculated with the aid of the Eqs. (52), (57), and (59) with the parameters as depicted in the legends of Figs. 16 and 17

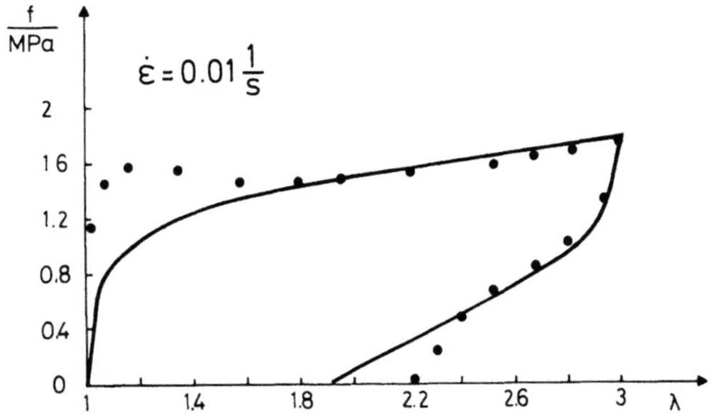

Fig. 21. Stress strain cycle of natural rubber for $T = 213$ K. The solid line was calculated with the aid of the Eqs. (52), (57), and (59) with the parameters as listed in the legends of Figs. 16 and 17

Conclusions

Thermostatics of van der Waals networks and the theory of finite elasticity linked together provide a basis that permits us to develop a tensorial description of static deformation in different deformation modes. The extension of thermodynamics which has reference to the rates of relaxation processes is the theory of irreversible thermodynamics. Put together with the above theories one arrives at very general formulations: it is predicted that relaxation in strained networks should be controlled by normal modes which are constituted by coupling a set of a priori orthogonal hidden variables with the network level ("global level"). Linear relaxation in van der Waals networks is thus related to nonlinear equilibrium states to which the system is inevitably taken to.

Most significant is the unique understanding of relaxation in strained networks by applying irreversible thermodynamics. Relaxation even at large strains is described in different deformation modes at different temperatures by using the same shear-relaxation spectrum. Because of knowing the van der Waals strain-energy it is possible to verify that nonlinear viscoelasticity in networks is simply the consequence of having a nonlinear elastic equilibrium state of reference in accord with proposals as discussed by Smith [56] and Scholtens [57, 58].

From the results presented it is evident that permanent networks cannot be brought into "large distance to equilibrium". This and the theoretical conception developed in this paper might be of some interest also for discussing relaxation in transient networks such as polymer melts [59].

Acknowledgement:

The authors wish to thank the Deutsche Forschungsgemeinschaft and the Degussa AG for financial support.

References

1. Kilian HG (1981) Polymer 22:209
2. Kuhn W (1936) Kolloid Z 76:256
3. Treloar LRG (1958) The physics of rubber elasticity. Clarendon Press, Oxford
4. Kilian HG (1982) Coll Polym Sci 260:895
5. Helfand E, Pearson DS (1984) Macromol 18:888
6. Lodge AS, Armstrong RC, Wagner MH, Winter HH (1982) Pure Appl Chem 54:1359
7. de Groot S, Mazur P (1962) Non-Equilibrium Thermodynamics. North-Holland, Amsterdam
8. Meixner J, Reik HG (1959) Thermodynamik der irreversiblen Prozesse. Handbuch der Physik, Vol III/2. Springer, Berlin Göttingen Heidelberg
9. Hasse R (1963) Thermodynamik der irreversiblen Prozesse. Fortschritte der Physikalischen Chemie, Vol 8. Steinkopff, Darmstadt
10. Baur H (1984) Einführung in die Thermodynamik der irreversiblen Prozesse. Wiss Buchgesellschaft, Darmstadt
11. Schlögl R (1964) Stofftransport durch Membranen. Steinkopff, Darmstadt
12. Keller U (1977) Thermodynamik der irreversiblen Prozesse, Thermostatik und Grundbegriffe. De Gruyter, Berlin New York
13. Valanis KC (1971) Irreversible Thermodynamics of Continuous Media. CISM-Course 77. Springer, Wien
14. Kilian HG, Vilgis Th (1984) Coll Polym Sci 262:691
15. Enderle HF, Kilian HG, Vilgis T (1984) Coll Polym Sci 262:696
16. Green AE, Atkins JE (1970) Large Elastic Deformation. Claredon Press, Oxford
17. Truesdell C (1984) Rational Thermodynamics, 2nd ed. Springer, New York
18. Haupt P (1977) Viskoelastizität und Plastizität (Ingenieurwiss Bibl). Springer, Berlin
19. Coleman B, Nole N (1961) Rev Mod Phys 33:239
20. Betten J (1985) Elastizitäts- und Plastizitätslehre. Vieweg, Braunschweig
21. Onsager L (1931) Phys Rev 37:405
22. Onsager L (1931) Phys Rev 38:2265
23. Meixner J (1949) Z Naturforsch 4a:594
24. Meixner J (1953) Koll-Z 234:3
25. Enderle HF (1988) Thesis. University of Ulm
26. Alts T (1976) Arch Rat Mech Anal 51:253
27. Eringen C (1960) Phys Rev 38:2265
28. Rivlin RS (1949) Phil Trans Royal Soc A242:173
29. Spencer AJM (1980) Continuum mechanics. Longman, London
30. Rivlin RS, Saunders DN (1950) Phil Trans Royal Soc A242:256
31. Kilian HG (1985) Coll Polym Sci 263:30
32. Kilian HG (1980) Polym Bull 3:351
33. Eisele U, Heise B, Kilian HG, Pietralla M (1981) Angew Makromol Chem 100:67
34. Rivlin RS, Saunders DN (1950) Phil Trans Royal Soc A242:256

35. Enderle HF, Kilian HG (1987) Prog Coll Polym Sci 75:55
36. Pak H, Flory PJ (1979) J Polym Sci. Phys Ed 17:1845
37. Macosko W, Benjamin GS (1981) Pure Appl Chem 53:1505
38. Kawabata S, Matsuoka M, Tei K, Kawai H (1981) Macromol 14:154
39. Godovsky YK (1977) Vysokomol Soed A19:2359
40. Kilian HG (1982) Coll Polym Sci 260:895
41. Göritz H (1982) Coll Polym Sci 260:193
42. Kilian HG (1987) Prog Coll Polym Sci 75:213
43. Flory PJ (1953) Principles of Polymer Chemistry. Cornell University Press, London
44. Chang WV, Bloch R, Tschoegl NW (1976) Proc Natl Acad Sci USA 73:4:983
45. Mancke RG, Ferry JD (1968) Trans Soc Rheol 12:335
46. Goldberg W, Lianis G (1968) Trans ASME, J Appl Mech 35:433
47. Goldberg W, Bernstein B, Lianis G (1969) Int J Nonlin Mech 4:277
48. Arenz RJ, Landel RF (1982) J Poly Sci: P Phys Ed 20:363
49. Schwarzl FR, Staverman AJ (1953) Appl Sci Res A4:127
50. In: Stuart HA (1956) Die Physik der Hochpolymeren, Vol IV. Springer, Linear deformation behavior in high polymers
51. Sauter A (in preparation) Thesis. University of Ulm
52. Holl B, Kilian HG, Schenk H (in press) Coll Polym Sci
53. Katz JR (1925) Naturwissenschaften 13:410
54. Treloar LG (1941) Trans Farad Soc 37:84
55. Gottlieb M, Macosco CW, Lepsch TC (1981) J Polym Sci, Phys Ed 19:1603
56. Smith TL (1962) Trans Soc Rheol 6:61
57. Scholtens BJR, Broij HC, Leblans PJR (1986) J Rheol 30:301
58. Scholtens BJR, Leblans PJR (1986) J Rheol 30:313
59. Demarels A, Meissner J (1986) Coll Polym Sci 264:829

Received January 30, 1989;
accepted July 19, 1989

Authors' address:

H. Ambacher
Universität Ulm
Abt. für Experimentelle Physik
Albert-Einstein-Allee 11
7900 Ulm

Progress in Colloid & Polymer Science Progr Colloid & Polymer Sci 80:221–235 (1989)

Relaxation processes in flexible mesomorphic polymers

V. S. Papkov and Yu. P. Kvachev

Institute of Synthetic Polymeric Materials, USSR Academy of Sciences, Moscow, USSR

Abstract: Some data on molecular mobilities in flexibles polymers capable of forming thermotropic mesophases are discussed. The main attention is paid to polymers with inorganic backbones, viz. poly(diethylsiloxane) and some polyphosphazenes.

An approach to mesophases in these polymers as to strongly disordered crystals is the basis for consideration. It is supported by the analysis of data on gradual changes in molecular mobilities accompanying phase transitions in poly(diethylsiloxane) from the low-temperature polymorph via the high-temperature crystal to the mesophase. Relaxation phenomena in polyphosphazenes are also treated from the same point of view. Specific features of thermal and mechanical relaxation accompanying mesophase formation in the course of extension of crosslinked poly(diethylsiloxane) films are considered and it is shown that in many aspects this process is similar to crystallization during extension of polymer networks.

An attempt is also made to correlate phase transition in crystalline phases and mesophases of the polymers with relaxation transitions in their amorphous phases.

Key words: Mesomorphic state; molecular mobilities; poly(diethylsiloxane); polyphosphazenes; relaxation

Introduction

Recent extensive studies of LC-polymers have distinguished a special set of polymers which, despite the absence in their chemical structure of any mesogenes, are able to form thermotropic equilibrium-ordered phases possessing the structure and properties intermediate between the crystalline phase and the amorphous one. In the most striking form the tendency to form such phases is displayed by polymers with flexible inorganic backbone, namely poly(diethylsiloxane) (PDES), poly(dipropylsiloxane), and various polyphosphazenes [1]. In principle, the high temperature polymorph of polytetrafluoroethylene, the high temperature hexagonal modification of 1,4-trans-polybutadiene, and the high-pressure phase of PE can be also attributed to similar phase [2]. Having analyzed the structure and molecular mobilities in these polymers, Schneider et al. [2] suggested to consider these phases as disordered crystals. Wunderlich and Grebowisz [3] have further developed this concept. Based upon the classification of small molecular mesophases by Smith [4] they proposed that all these mesophases may be considered a new specific type of mesophase, i.e., a conformationally disordered crystal (condis crystal) "in which macromolecules can undergo relatively hindered rotation to change conformation without losing positional or orientational order".

A high level of molecular mobilities in disordered crystalline phases should evidently affect mechanical and rheological behavior of the polymers and gives rise to specific relaxation phenomena. In this connection the consideration of molecular mobilities and their correlations with macroscopic properties of flexible mesomorphic polymers is of great importance, not only in understanding the origin of the mesophase but also for potential technological applications of these polymers.

Some aspects of this problem are discussed in the present lecture. In principle, it is a review of corresponding published or submitted for publication data on phase and relaxation transitions, molecular mobilities in PDES and a number of polyphosphazenes, and partially on their mechanical properties. For the most part these data are results of our own measurements.

PDES is chosen as the main subject under consideration, not only because of our extensive experience with, but also for two additional reasons:

First, PDES can exist in three temperature polymorphs (including mesophase) and gives thus an opportunity to follow step-by-step development of molecular mobilities in the crystalline phase until it converts into the isotropic melt.

Second, PDES as an elastomer affords a unique possibility to study the influence of formation of the mesophase on rubber-like properties and relaxation behavior of polymer net-works. Both questions will be considered in this paper.

Finally, we shall touch upon phase transitions and molecular mobilities in some polyphosphazenes with rather long fluorinated alkoxy side groups. These data give additional examples of step-by-step disordering crystalline phases and their influence on mechanical properties of the polymers. In addition, they show how long side substituents can act in formation of disordered phases.

Experimental

High molecular weight samples of PDES obtained by ionic polymerization of hexaethylcyclotrisiloxane in the presence of KOH as the initiator were investigated. All the samples were diactivated. The sample of intrinsic viscosity $[\eta] = 0.4$ dL/g (in toluene at 298 K) was the main subject in pulse nuclear magnetic resonance (NMR) studies. An uncrystallizable oligomeric PDES sample of $M_W = 2000$ with triethylsiloxy end groups was also used.

Crosslinked PDES films were prepared by peroxide vulcanization of the sample with $[\eta] = 0.5$ dL/g. Vulcanization was carried out at 423 K for 30 min using 0.5 and 1.0 wt.% of dicumylperoxide as the crosslinking agent.

Poly(fluoroalkoxyphosphazenes) studies were kindly provided by Dr. D. R. Tur of the Institute of Organoelement Compounds, USSR Academy of Sciences. They are

$[P(OCH_2CF_3)_2 = N]_n$ ($[\eta] = 4.45$ dL/g in THF at 298 K);
$[P(OCH_2CF_2CF_2H)_2 = N]_n$ ($[\eta] = 7.38$ dL/g);
$[P(OCH_2C_3F_6CF_2H)_2 = N]_n$ ($[\eta] = 6.4$ dL/g; and
$[P(OCH_2C_5F_{10}CF_2H)_2 = N]_n$ ($[\eta] = 9.6$ dL/g).

All these polymers were synthesized by replacement of chlorine in unbranched poly(dichlorophosphazene) by corresponding fluoroalkoxy groups using the methods developed earlier [5, 6].

Pulse NMR, dynamic mechanical, and dielectric methods were used for the investigation of molecular mobilities in the polymers. Proton spin relaxation times were measured on a Bruker SXP spectrometer at 88 MHz using different pulse sequences.

Temperature-dependent data on the mechanical modulus and the loss factor were recordered for polymer films cast from solutions with a torsion pendulum operating at roughly 1 Hz. Dielectric losses were measured at the frequency of 1 KHz. Stress-strain curves and heat evolution during stretching polymer films were monitored on specially constructed devices. More detailed information about samples studied and experimental procedures is given in pertinent references cited in the following.

Discussion

Poly(diethylsiloxane) Phase transitions

The consideration of changes in molecular mobilities with temperature is instructive to start with a general description of phase transitions in PDES. These transitions were investigated in detail [7 – 10]. DSC traces of PDES-samples of different thermal

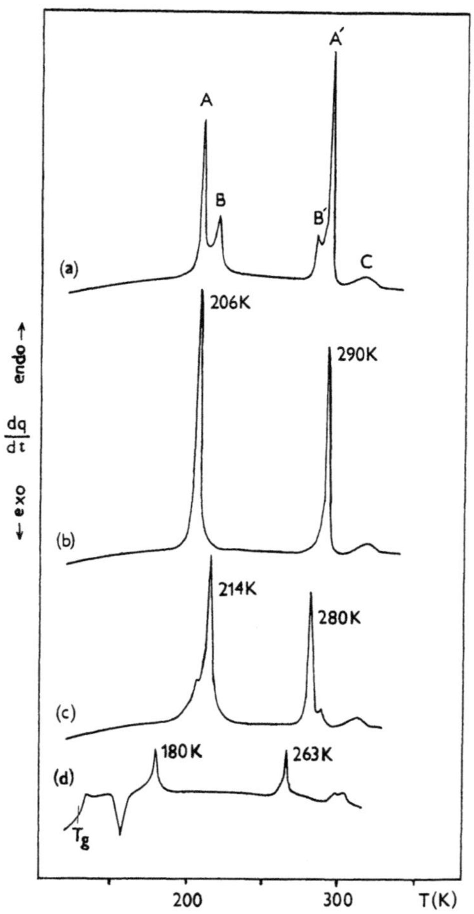

Fig. 1. DSC traces of different PDES samples: a) containing the α and β crystalline forms, b) crystallized into the β form, c) crystallized into the α form, d) amorphized by quenching into liquid nitrogen. The peaks indicated correspond to: A) $\beta_1 - \beta_2$ transition, B) $\alpha_1 - \alpha_2$ transition, A') β_2-form melting, B') α_2-form melting, and C) mesophase-isotropic melt transition. Heating rate is 10 K/min [9]

history are presented in Fig. 1. The samples crystallized from mesomorphic phases possess high crystallinity (>0.9) and glass transition is not distinguishable on their curves. Intensive endothermal peaks correspond to the solid-solid transitions: monoclinic α_1 crystal → monoclinic α_2 crystal; tetragonal β_1 crystal → tetragonal β_2 crystal, and α_2 and β_2 polymorphs → mesophase. The transition of mesophase into amorphous state (isotropization) is manifested as a small hump on DSC curves. Equatorial diffractograms of oriented PDES sample and its x-ray fiber pattern illustrating the structure changes with temperature are shown in Fig. 2. All these data can be summarized as a general diagram of phase transitions (see Fig. 3). It is worth noting that

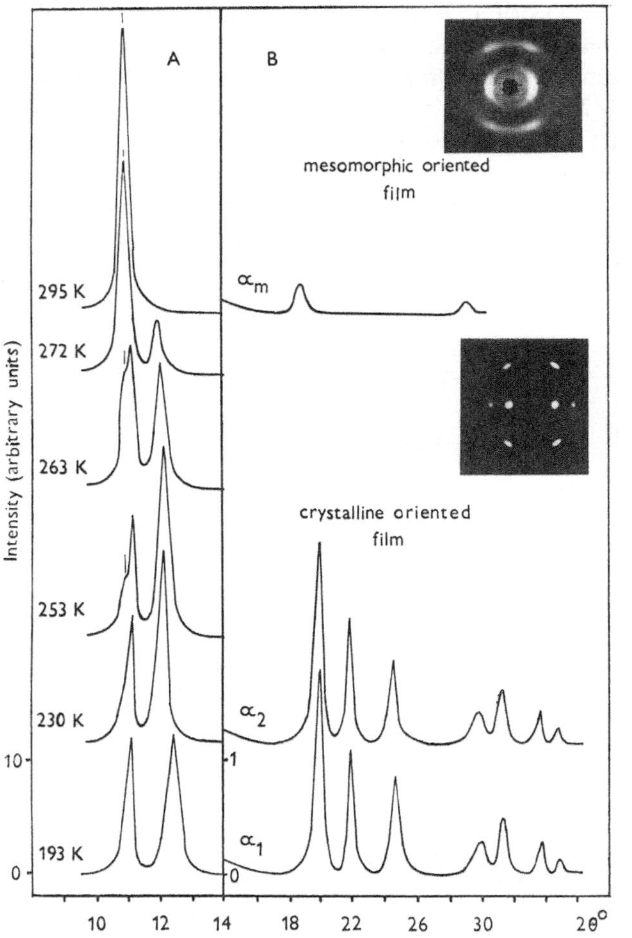

Fig. 2. WAXS equatorial diffractograms of oriented PDES film at different temperatures with its corresponding x-ray fiber patterns (top righthand inserts). The intensity range in the region of $2\theta = 8-14°$ (A) is 10 times as high as that in the region $2\theta > 14°$ (B). The position of the reflections in the $2\theta > 14°$ region at temperatures between 253 and 272 K are the same as at 230 K and are not shown [10]

all these transition are marked by two facts. First, the low temperature crystal-crystal transition is not accompanied by the change of crystalline lattice type and the parameters of the unit cell increase only jumpwise. Second, the mesophase structure is different from a pseudohexagonal packing of chains but bears rather close resemblance to the monoclinic α_2 crystalline phase. Jumpwise changes of unit cell parameters imply that due to an increase in the thermal vibrational energy the intermolecular interaction is loosened and a new level of molecular mobilities is gained.

Molecular mobilities

Pulse NMR, dynamic mechanical, and dielectric data support this suggestion. For the first time nuclear spin relaxation of PDES was studied by Froix et al. [11]. More detailed information was obtained by Litvinov et al. [12] who took account of the thermal

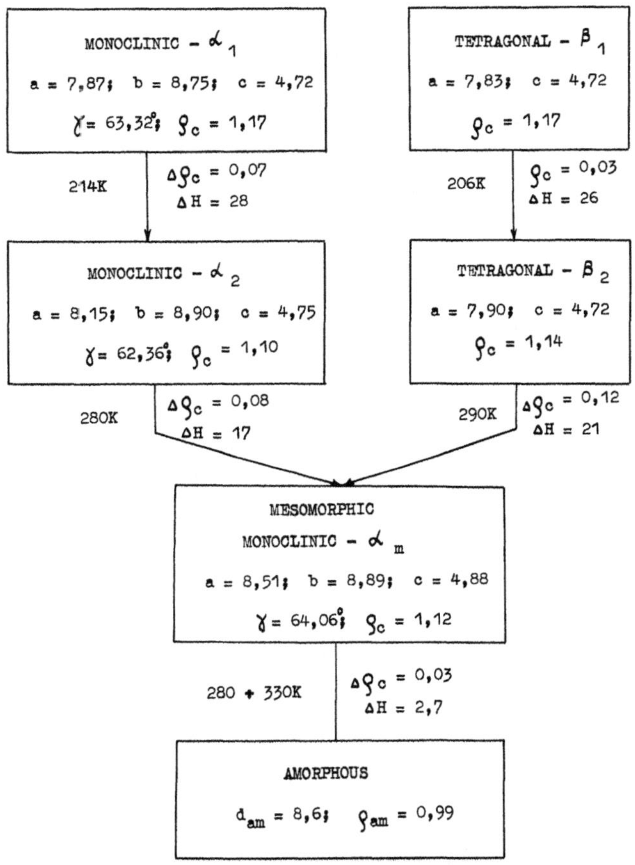

Fig. 3. General diagram of the structural changes accompanying the phase transitions in PDES. Crystallographic density ϱ_C and its changes at phase transitions $\Delta\varrho_C$ are in g/cm³; the cell unit parameters in Å; the temperatures in the absolute scale, and ΔH in J/g [10]

history on the phase state of PDES samples. In Fig. 4 spin-lattice relaxation time T_1, and spin-spin relaxation time T_2 in local fields of a sample of PDES crystallized from mesomorphic state and possessing correspondingly high crystallinity are presented as a function of temperature. According to the free induction decay (FID) with long spin-spin relaxation time T_2, the content of amorphous phase in this sample does not exceed 0.05 (see Fig. 4). There is a deep minimum at 138 K and two rather sharp breaks (or shallow minima) on the T_1 curve. Burnett and Durrett [13] using their semi-empirical calculations have attributed the T_1 minimum (75 ms) at 138 K to rotational reorientation of methyl groups. A similar T_1 minimum is also observed in the case of amorphized samples of PDES (see below) and this fact confirms the above assignment of relaxation mechanism. The frequency of C-rotation of methyl groups at the T_1 minimum is equal to about 10^8 Hz. Different slopes of the low and high temperature branches of T_1 minimum point out an asymmetrical distribution of relaxation times. The activation energy calculated for Cole-Davidson distribution was found to be as large as 10 kj/mol and the preexponent factor is equal to $10^{13}\,s^{-1}$. The phase transitions are accompanied by the jumpwise increase in spin-spin relaxation time (T_2 increases from 17 up to 25 ms at α_1; $\beta_1 \rightarrow \alpha_2$; β_2 transitions). Due to the low content of amorphous phase the increase in T_2, corresponding to glass transition is not recognizable in this case. However, it is clearly seen in amorphized samples of PDES. Temperature dependences of both relaxation times T_1 and T_2 for such a sample are presented in Fig. 5. Amorphization of this sample was achieved by quenching of isotropic melt preheated to 353 K into liquid nitrogen. Additionally, this figure furnishes corresponding data for uncrystallizable PDES oligomer ($M_W = 2000$) at a temperature range above T_g where high molecular samples of PDES crystallize. In the vicinity of glass transition, T_2 sharply increases and at the temperature of half decrease in the transverse relaxation rate $1/T_2$ (at 149 K) the correlation frequency is equal to $1.1 \cdot 10^4$ Hz. The temperature dependence of spin-lattice relaxation time T_1 obtained for the crystalline sample of PDES in low temperature region enable to separate contribution of methyl groups reorientation from the overall spin-lattice relaxation in the amorphous sample; it also allows to reveal an extra T_1 min-

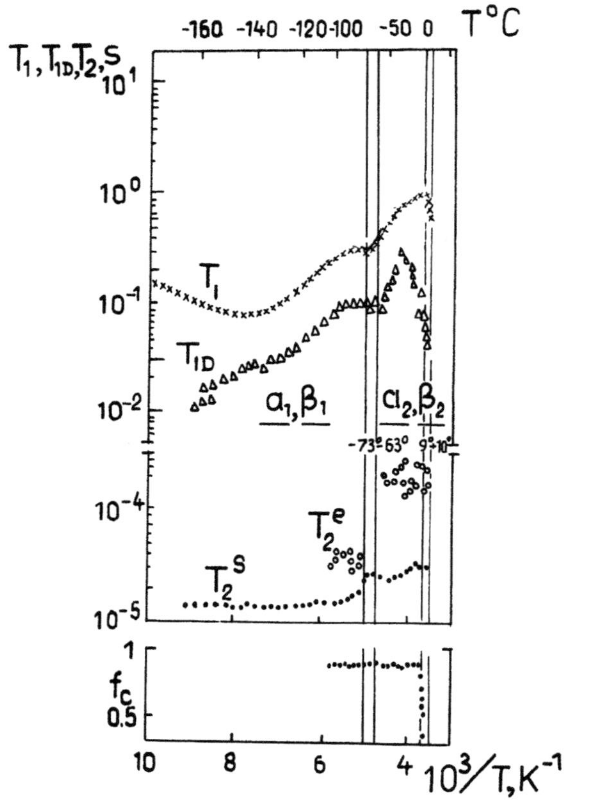

Fig. 4. Spin-spin (T_2), spin-lattice (T_1, T_{1D}) relaxation times, and the short T_2^S component (f_C) characterizing the content of crystalline phase as a function of temperature for PDES sample with high crystallinity containing α and β polymorphs. Vertical lines indicate the temperature regions of phase transitions [12].

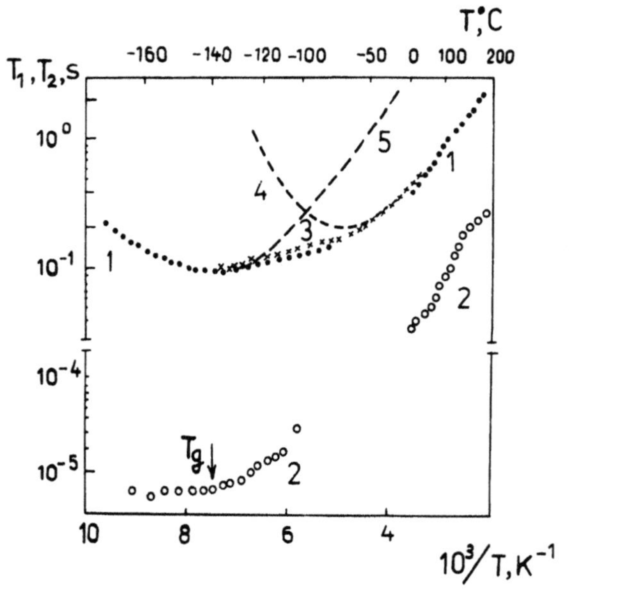

Fig. 5. Spin-lattice T_1 (1, 3, 4, 5) and spin-spin T_2 (2) relaxation times as a function of temperature (1) for high molecular and (3) oligomeric amorphous samples of PDES. Broken lines correspond to the extracted contributions in T_1 from: (5) rotation of methyl groups, (4) reorientation motion of ethyl groups around Si-C bond and segmental mobilities [12]

imum in the temperature region from 190 to 200 K which may be associated with glass transition at frequencies of about 10^8 Hz (correlation frequency is equal to $1.4 \cdot 10^8$ Hz). Some information concerning relaxation mechanisms of the transitions in PDES can be gained by comparing theoretical and experimental values of the second moment of line shape M_2 and its changes at these transitions. Although no theoretical calculation of M_2 for PDES was carried out, one can assess in the first approximation its values, corresponding to different proposed motions of ethyl groups on the basis of the known theoretical data for such a molecule as $(C_2H_5)_3P{=}O$. This molecule and monomer unit of PDES $[-SI(C_2H_5)_2-O-]$ contains only protons of ethyl groups and the difference in the intramolecular contribution is not expected to be significant. All the possible modes of ethyl group motions, the variation of experimental values of M_2 with temperature, and theoretical values of M_2 for corresponding motions of ethyl groups in $(C_2H_5)_3P{=}O$ [14] are shown in Fig. 6. For rigid ethyl groups the magnitude of M_2 was calculated to be 29.6 G^2. When the rotation of the methyl group around the threefold axis freezes out, M_2 would drop down to a value of about 14 G^2. Reorientation of ethyl group around C'-axis must cause a further decrease in M_2 to the values of 5.5 G^2. In the case of reorientation of ethyl groups around C''-axis due to rotation of Si-O bonds

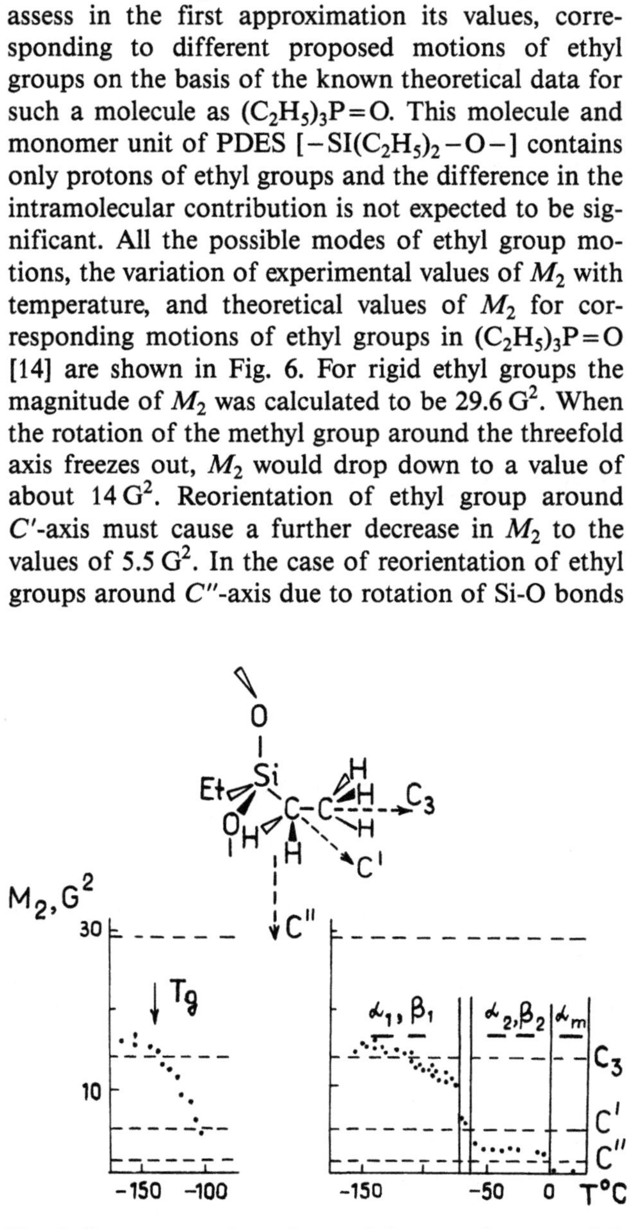

Fig. 6. Temperature dependence of the second moment M_2 for crystalline PDES sample (points). Vertical solid lines indicate the temperature boundaries of the existence of crystalline (α_1, β_1 and α_2, β_2) and mesomorphic (α_m) polymorphs. Horizontal broken lines correspond to theoretical values of M_2 expected for reorientation of ethyl group around axes C_3, C' and C'' [12]

one can expect a value of M_2 equal approximately to 1.5 G^2. Comparison of the theoretical and experimental values of M_2 leads to the following conclusions:

In the vicinity of the glass transition temperature the second moment decreases progressively and strives to 0, indicating the development of all the possible motion modes of ethyl groups. The most interesting fact is that M_2 drops down sharply to a value of about $2-3$ G^2 at the temperature region of the low temperature crystal-crystal transition, i.e., to an intermediate value between those corresponding to reorientation of ethyl groups around C' and C'' axes. This suggests that in α_2 and β_2 polymorphs not only rotation of ethyl groups but also intrachain rotation of Si-O bonds are experienced. It should be noted that the high temperature crystal-mesophase transition is accompanied by the additional decrease in M_2 to some tenths of G^2. This fact reflects a very high level of intrachain mobilities in the mesophases but meanwhile it indicates that conformational disorders in the high temperature polymorph and in the mesophase are apparently significantly similar. The latter is also verified by results of Raman spectroscopy investigations published by Friedrich and Rabolt [15]. These authors found a significant shift in the frequency of symmetric stretching vibration of Si-O bond (ν_S) from 490 cm^{-1} to 477 cm^{-} at the temperature region corresponding to the low temperature crystal-crystal transition. According to the authors' opinion, the shift in ν_S (Si-O) definitely indicates a change in chain conformation. The concomitant change in bandwidth reflects a sizeable increase in molecular motion of the backbone, most likely attributable to the introduction of disorder. The further increase in bandwidth of ν_S at the crystal-mesophase transition was interpreted as a further introduction of backbone motion. However, changes in this case are less than during the former solid-solid transition. Our preliminary IR spectroscopy data [16] presented in Fig. 7 are consistent in principle with the abovementioned results, although there are some difficulties for unequivocal assignments of the observed bands.

The transition of the high temperature α_2 and β_2 polymorphs into the mesophase is accompanied by an increase in the spin-spin relaxation time T_2 by a factor of about one order. Correspondingly, the second moment M_2 decreases to 0.5 G^2. Such values of the parameters suggest a very high level of conformational disordering. The observed further increase in bandwidth of the 477 cm^{-1} band is in line with this conclusion. Unfortunately, the mesophase in PDES exists in the very narrow temperature interval and in some degrees after its appearance it begins to convert

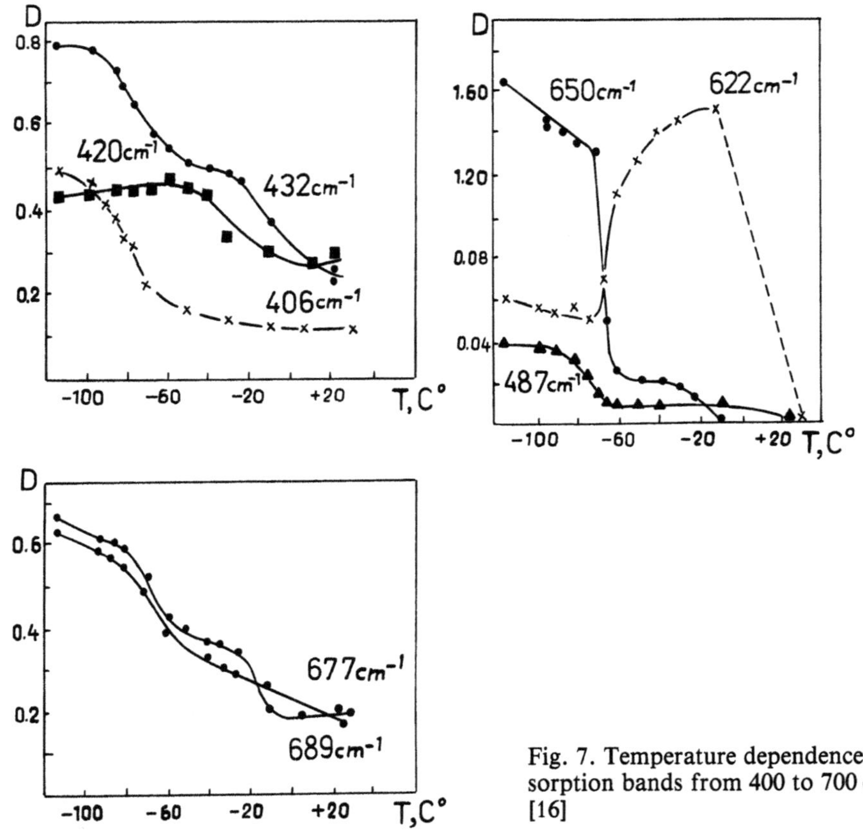

Fig. 7. Temperature dependence of the optical density at the various absorption bands from 400 to 700 cm^{-1} observed in IR spectrum of PDES [16]

gradually into isotropic melt. The mesophase and the amorphous phase coexist at the transition region. However, the mesophase can be supercooled by about 10 K and this enables more detailed measurements of its relaxation parameters. FID of mesomorphic samples of PDES consists of two components: a component with short relaxation times $T_2^S = 1.1 - 1.3 \cdot 10^{-4}$ s that can be attributed to the relaxation in the mesophase, and a long component with relaxation times $T_2^l = 3.5 - 5 \cdot 10^{-3}$ s characterizing relaxation in amorphous regions [17]. The fraction of the short T_2 component increases with decreasing temperature and just before the onset of crystallization it attains a value as high as 0.95 (see Fig. 8). This fact indicates that PDES can be almost completely converted into the mesomorphic state. In a plot of T_2^S vs temperature one can observe a rather small but reproducible jump near 281 K (the correlation frequency is equal to about 10^3 Hz). This increase in T_2^S points out a further introduction of molecular motion and reflects dynamics of the mesophase-isotropic melt transition. In this connection it is also interesting to consider the change of birefringence of mesomorphic regions. The melting of a crystalline-oriented film of PDES at 273 K leads to an approximately five-fold decrease in negative birefringence [18]. At further heating its absolute value continued to decrease and achieved zero near 285 K, and then it becomes positive (see Fig. 9). Since the anisotropy of polarizability of monomer unit in polydimethylsiloxane is positive [19] the anisotropy of polarizability of siloxane chain backbone must be also positive. Consequently, the negative optical sign of the stretched film below 275 K would arise from negative polarizability of ethyl group or, more specifically, from the corresponding difference between the transverse and longitudinal polarizability of C-C bond. The above-mentioned changes in birefringence with temperature may, in this case, be treated as a result of the transition from hindered rotation of ethyl groups around Si-C bond to a free one.

According to our data [17], the longitudinal magnetization recovery is of complex nonexponentional character caused apparently by spin-spin diffusion in a two-phase system. The dependence of spin-lattice relaxation times measured by zero-point method on temperature exhibits very shallow minimum located near 293 K; its origin is not entirely clear. This minimum may reflect a change in T_1 due to isotropization, but it may be also associated with high

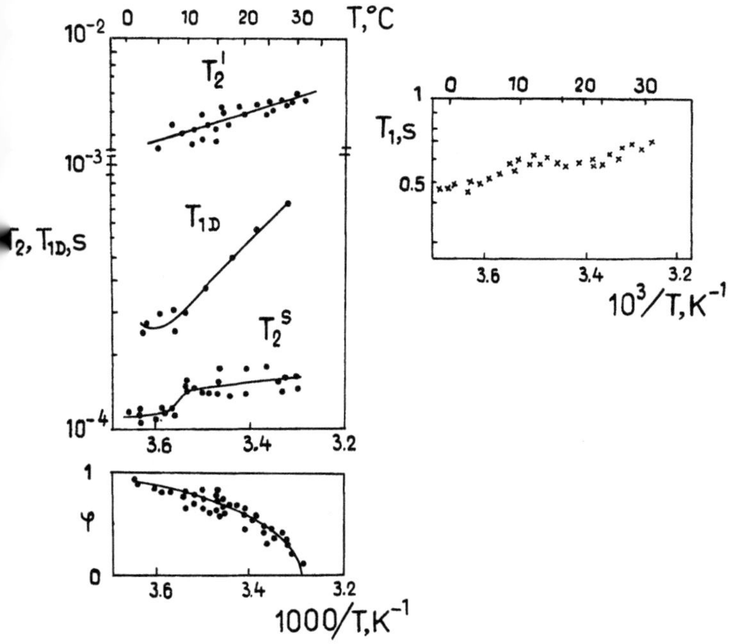

Fig. 8. Spin-spin (T_2), spin-lattice (T_1, T_{1D}) relaxation times, and the content of mesophase (f_C) as a function of temperature for mesomorphic PDES sample, f_C is assumed to be equal to the short T_2^S component of FID [17]

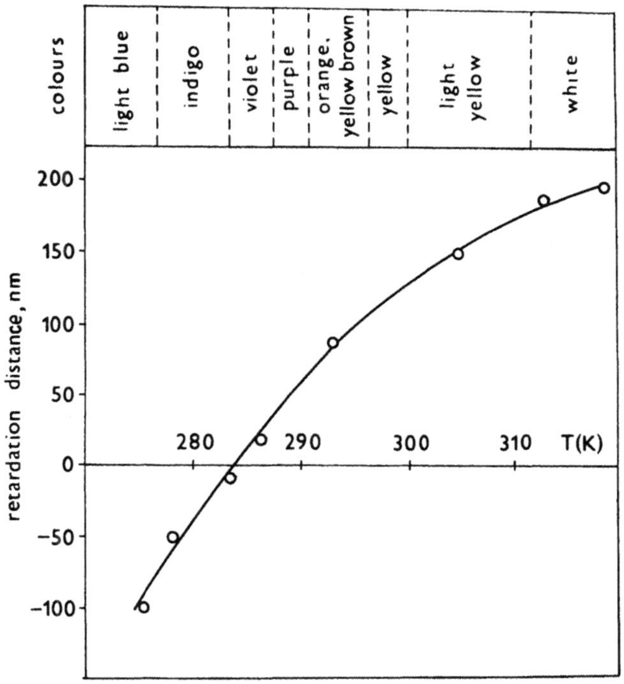

Fig. 9. Retardation distance in stretched PDES film as a function of temperature. The colors of the film observed between crossed polars are shown above. Broken lines mark the approximate boundaries of color transition. The tickness of the film is 330 μm [18]

frequency motions responsible for the above-mentioned relaxation process in the mesophase. Also, in this temperature region one can expect superposition of an extra high-frequency relaxation mechanism arising from additional transition in amorphous phase near the onset of melting of the high temperature polymorphs. An introduction of further motion in amorphous phase near 250 K is evidenced by Froix et al. data on spin-spin relaxation times [11]. This is also seen readily in Fig. 10 where our data for a partially crystalline sample of PDES are presented. This process is displayed clearly as a peak located at 235 K in

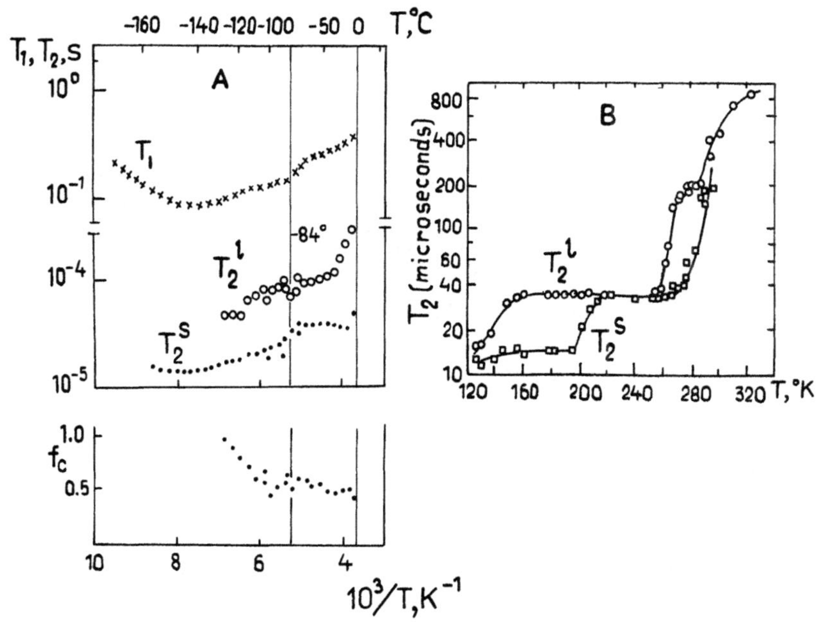

Fig. 10. Spin-spin and spin-lattice relaxation times, and the short T_2^S component (f_C) as a function of temperature for semicrystalline PDES samples: A) data from [12], b) data from [11]

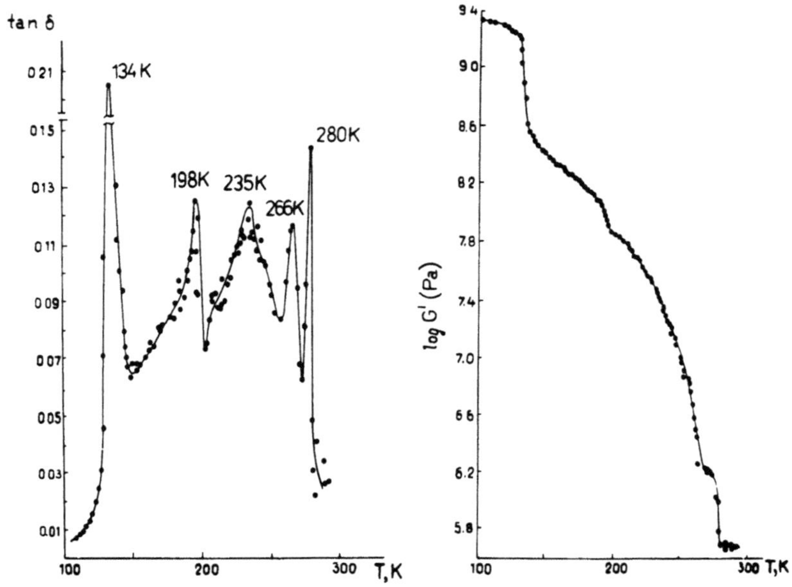

Fig. 11. Shear storage modulus G' and mechanical loss factor as a function of temperature for crosslinked PDES film. Data are obtained by using torsion pendulum operating at about 1 Hz [20] (see also Fig. 12)

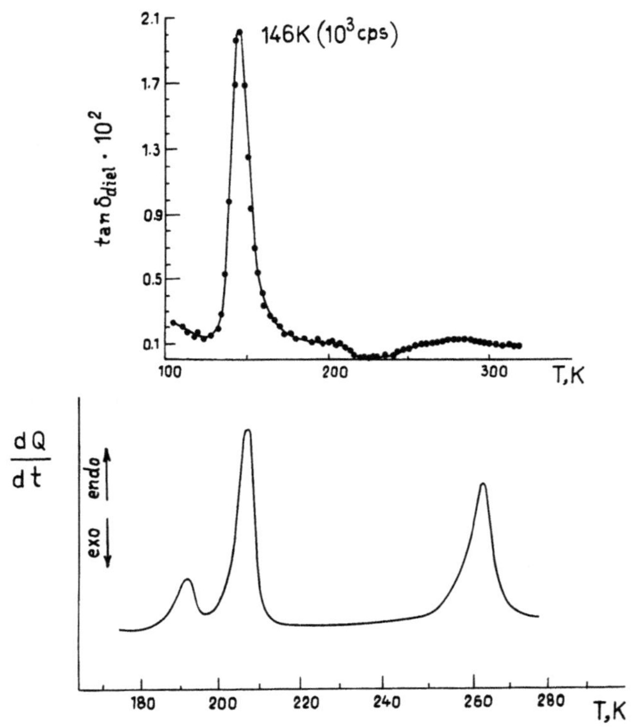

Fig. 12. A) Temperature dependence of tangent of dielectric losses and B) DSC trace at heating rate 8 K/min for crosslinked PDES film [20] (see also Fig. 11)

a plot of dynamic mechanical losses vs temperature for a cross-linked sample of PDES (see Fig. 11) [20]. On this curve there exist also the loss peaks corresponding to the glass transition, crystal-crystal transition, melting, and isotropization. However, dielectric measurements reveal clearly only the glass transition as is seen in Fig. 12. The lower graph in this figure is a DSC trace for the same crosslinked sample of PDES. Note that there is no thermal event on the thermogram connected with melting of crystalline phases near 235 K, i.e., at the temperature region of the location of the above-mentioned mechanical loss peak. Based on this temperature and the temperature of a half-decrease of the transverse magnetic relaxation rate $1/T_2^a$ ($T \cong 270$ K, $\nu_c = 4 \cdot 10^3$ Hz, see Fig. 10) one might expect an additional minimum on T_1 curve to arise at the temperature region between 290–300 K pertaining to the second transition in amorphous phase of PDES at the frequency of about 10^8 Hz.

The shift of glass transition temperature with increasing frequency is shown in Fig. 13 which presents the corresponding correlation times evaluated from the above NMR, dynamic mechanical, and dielectric data as a function of reciprocal temperature.

It is of interest that the existence of additional transition in amorphous phase above glass transition temperature is seemingly no unique feature of PDES. Apparently, a similar situation is observed in the case of another polymer characterized by the existence of low and high temperature polymorphs, namely PTFE. As known, the high temperature polymorph of this polymer is also regarded as mesophase (condis-crystal) [3]. The assumption about two transitions in amorphous phase of PTFE is prompted by the consideration of data on thermal expansion of its amorphous phase. Such data obtained by using x-ray techniques were published by Zubov et al. [21] for high temperature region and Tsvankin et al. [22] for low temperature region.

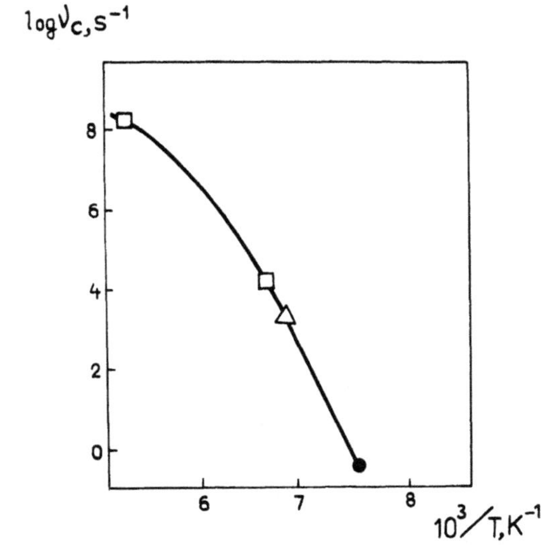

Fig. 13. Correlation frequency as a function of reciprocal temperature for PDES. Correlation times are evaluted from (□) NMR relaxation times, (●) dynamic mechanical and (△) dielectric data (see Figs. 5, 11 and 12)

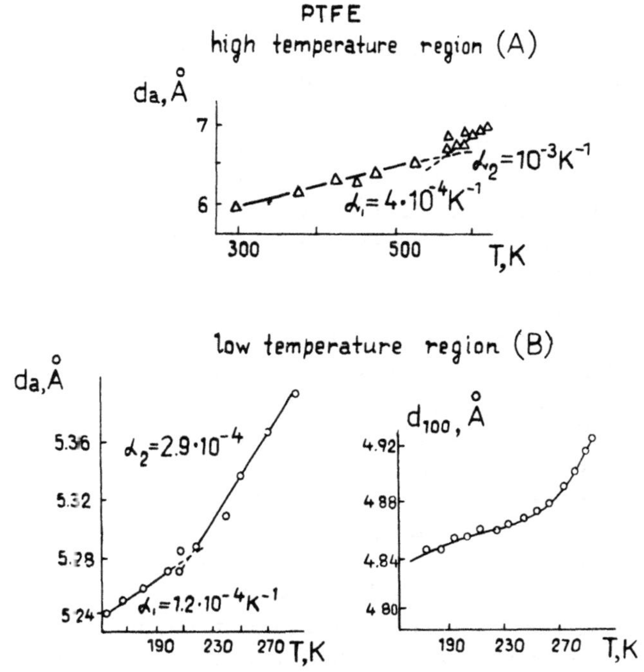

Fig. 14. *d*-Spacing of amorphous (d_a) and crystalline (d_{100}) phases of PTFE as a function of temperature in A) the high temperature [21], and B) low temperature [22] regions. α_1 and α_2 correspond to thermal expansion coefficients below and above transitions in amorphous phase, respectively

In Fig. 14 a break on a plot of the intermolecular spacing vs temperature at the 213 K is clearly seen. The slope increases at this point by a factor of about 2.5. The calculated values of linear thermal expansion coefficients are correspondingly equal to $1.21 \cdot 10^{-4}$ and $2.9 \cdot 10^{-4}\,K^{-1}$ below and above this temperature. Both values of the coefficient are close to those usually observed for most polymers in glass and rubbery states; consequently, in this respect this transition might be treated as glass transition. Note that simultaneous increase in intermolecular crystalline spacing occurs as well and these changes precede the well known solid-solid transition at 294 – 303 K. Near the melting of the high temperature crystalline phase a similar picture can be observed: expansion coefficient undergoes the jumpwise increase from $4 \cdot 10^{-4}\,K^{-1}$ to $10^{-3}\,K^{-1}$ at 563 K and this phenomenon is accompanied by a corresponding increase in thermal expansion coefficient of the crystalline phase. As known, in the same temperature region both a decrease in storage modulus and an increase in internal friction coefficient were found [23, 24]. In this connection the above transition may also be regarded as an additional transition in amorphous phase. It is obvious that the existence of several temperature polymorphs should arise from the presence of some types of inter- and intramolecular interaction. With increasing temperature they are released subsequently and this should lead to solid-solid transitions or melting. Meanwhile, glass transition temperature is also determined mainly by intermolecular interaction and segmental motion as experienced after the "scission" of some types of intermolecular bonds. Thus, a question arises of whether a correlation exists between a number of polymorphs and a number of transitions in amorphous phase or if the above-mentioned transitions are merely a result of enhanced molecular mobilities of amorphous macromolecules due to premelting and pretransitions in crystalline phases.

To summarize, the above data reveal very high molecular mobilities in the mesophase of PDES close to molecular mobilities in amorphous phase but, nevertheless, are appreciably different from them. The preceding high temperature α_2 and β_2 polymorphs are also characterized by the high level of conformational disorder. All these phases appear to match the definition of a condis-crystal and the high temperature crystalline phases are probably closer to the condis crystals of organic polymers than the mesophase.

Mechanical properties

Various levels of molecular mobilities that are determined by various levels of intermolecular interaction must obviously affect macroscopic mechanical properties of PDES in different phase states. An analysis of stress-strain curves of crosslinked PDES film at different temperatures confirms this suggestion. Since, according to our previous data [18], crystalline phases form from mesophase and vice versa without any considerable morphology changes, the difference in the stress-strain curves would reflect rather strictly the difference in the ability of the phases to plastic flow. As seen in Fig. 15, the transition of the low temperature polymorph into the high temperature one leads to a decrease in yield stress by approximately one order, whereas the yield stresses in the cases of the mesophase and the high temperature polymorphs differ only by a factor of about two. This fact supports the above conclusion about similarity of both phases in the sense of disordering. However, it is most interesting to compare stress-strain curves of amorphous and mesophase PDES networks.

As expected, the presence of mesomorphic phase and its formation in the course of stretching significantly influence rubber elasticity of PDES. To a greater extent the difference in stress-strain characteristics of mesomorphic and amorphous networks is revealed in the case of slightly crosslinked films of PDES [26]. On stretching such a mesomorphic film behaves, in principle, similar to polymers with low crystallinity that elongate without well-defined necking (see Fig. 16). Stress-strain curves for amorphous crosslinked films consist of three sections: the first at low extension ratios λ (less than 2.5) which looks like a deformation curve of noncrystallizable polymer networks; the second is an *S*-shaped part, and the third makes a rather steep upward turn. A comparison of these curves shows that orientation of chains in mesomorphic domains and their reorganization require additional work. The initial section of the curves is insensitive to strain rate changed by two

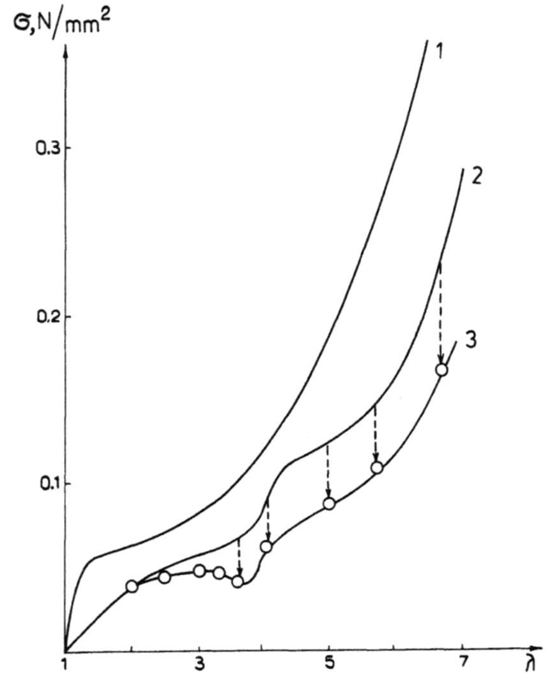

Fig. 16. Stress as a function of extension ratio for crosslinked PDES film: 1) mesomorphic film, 2) amorphized film, 3) after 6 min relaxation. Strain rate 0.0065 s^{-1} [26]

orders. At this extension ratio region stress relaxation is not observed. These facts indicate equilibrium behavior of PDES network at such strains. The first section can be rather well fitted by Mooney-Rivlin equation. Sharp increase in stress at λ higher than 3

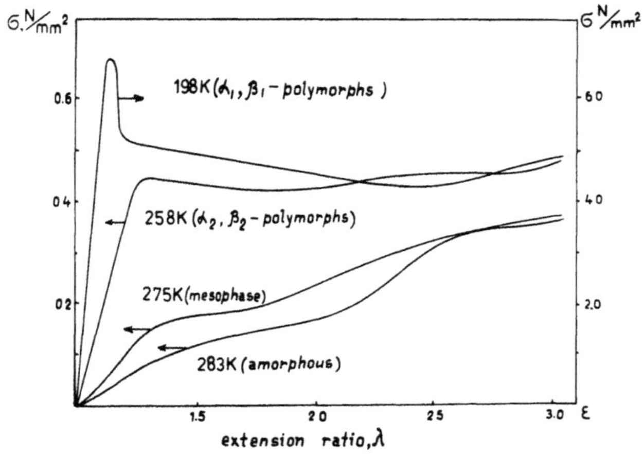

Fig. 15. Stress-strain curves for crosslinked PDES film in different phase states. Strain rate 0.0065 s^{-1} [25]

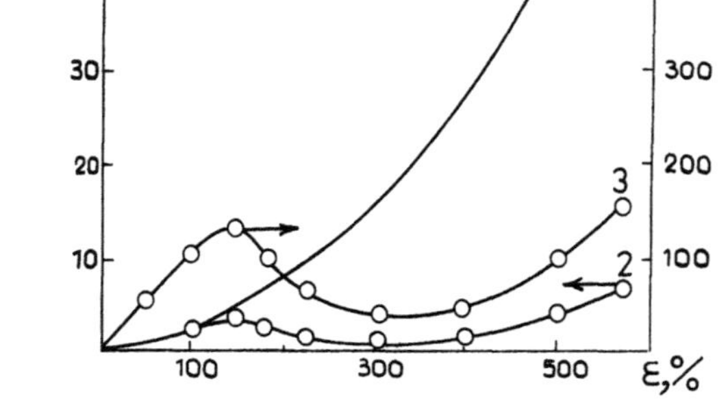

Fig. 17. Work of stretching (*1*), work of contraction (*2*), and recoverable deformation ε' (*3*) as a function of extension ratio ε for crosslinked PDES film [26]

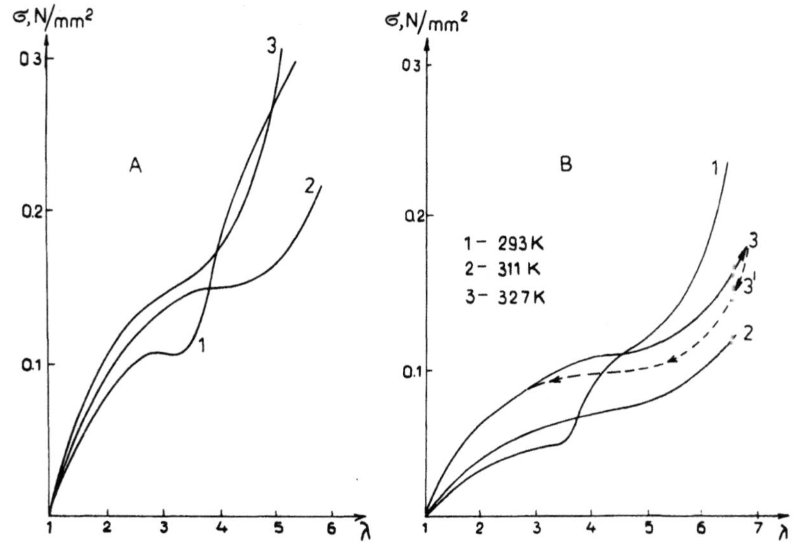

Fig. 18. Stress-strain curves for crosslinked PDES film at different temperatures. Crosslink density: A) $1.31 \cdot 10^{-5}$ mol/cm^3, B) $6.37 \cdot 10^{-6}$ mol/cm^3. Strain rate is 0.0065 s^{-1} [26]

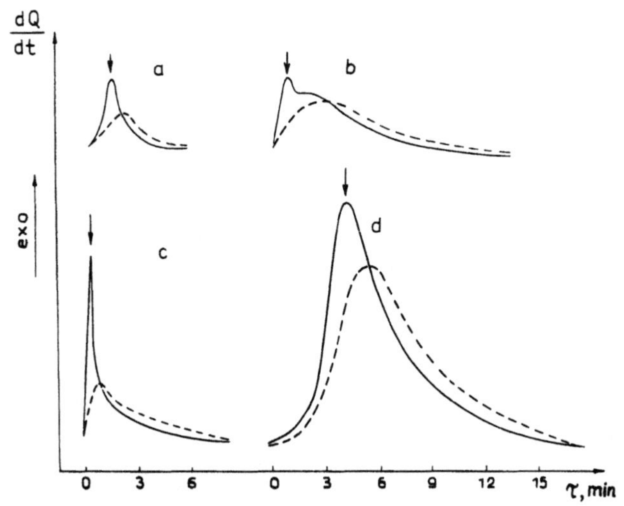

Fig. 19. Kinetic curves of heat evolution upon stepwise extension of crosslinked PDES film. Extension step from λ_1 to λ_2: a) $1.0 \rightarrow 1.80$, b) $2.30 \rightarrow 2.66$, c) $5.45 \rightarrow 6.05$, d) $1.0 \rightarrow 3.65$. Broken lines correspond to experimental curves monitored; solid curves are constructed with allowance for the thermal lag of the calorimeter and represent the real kinetic of heat evolution. Arrows indicate the end of extension [26]

is caused by formation of mesophase as it follows from x-ray data and further thermoelastic data. This increase is accompanied by the development of stress relaxation and the appearance of residual deformation on unloading. The latter is clearly seen in Fig. 17 where works of loading and unloading are compared as well. With increasing temperature the shape of stress-strain curves alters: namely, the second *S*-shape section disappears and an upward turn for stress occurs only at rather high extension ratios. Corresponding curves at temperatures higher than isotropization temperature of the origin noncrosslinked PDES sample are shown in Fig. 18. A significant increase in the stress on the first section of these curves is also observed.

The change of stress-strain characteristics at higher extension ratios is connected with the arising mesophase. The measurement of thermal effect accompanying stretching and contraction of PDES films and comparison of them with the mechanical work enable determination of the change of internal energy as a result of mesophase formation.

According to the first thermodynamic law

$$\Delta U = W + Q \,, \tag{1}$$

where W is the work of stretching, Q is the heat evolved, and ΔU is the internal energy change. Some examples of heat evolution kinetics on stepwise loading are presented in the Fig. 19. Here are experimental curves and real kinetic curves of heat evolution constructed with the allowance for thermal lag of the calorimeter. Note that heat relaxation was observed for all the extension ratios, even on the first mechanically equilibrium section of stress-strain curves. Most heat relaxation takes place at extension ratios corresponding to the onset of stress relaxation. Comparison of the heat and the work of loading shows how mesophase formation proceeds at different λ (see Fig. 20). The heat relaxation at low λ may be connected with formation of a very small quantity of mesophase and with conformational reorganizations. It may also be caused, at least partially, by volume relaxation. All these factors hinder an accurate determination of the energetic intramolecular contribution in the work. Nevertheless, in the first approximation it can be assessed using the data on Q/W presented in Fig. 20. If the ratio of the energetic force f_e to the total force f is assumed to be independent of λ then

$$f_e/f = \Delta U/W = Q/W + 1 \ . \tag{2}$$

In Fig. 20 Q' relates to the heat evolved only in the course of stretching and Q stands for the total heat, that is the sum of Q' and the heat evolved during relaxation.

Consequently, the ratio of energetic force to the total force at initial λ should be between -1.2 and -8. Such value is very high compared with that obtained for other polymers [27]. At higher temperature when the mesophase formation can be fully excluded, f_e/f is equal to about -1.2. As known [27]

$$f_e/f = T\frac{d\ln\langle r^2\rangle}{dT} \ , \tag{3}$$

and therefore, the negative value of f_e/f implies that the temperature coefficient of the mean square end-to-end distance $\langle r^2\rangle$ is also negative and correspondingly, that extended conformation of PDES chain is

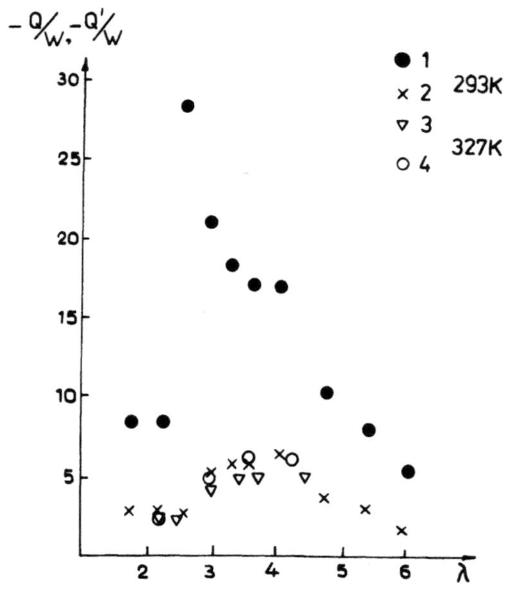

Fig. 20. Ratios of the heat evolved only during stretching (Q') and the total heat evolved during stretching and relaxation (Q) to the work of stretching W upon step-by-step extension of crosslinked PDES films from λ_1 to λ_2 at two temperatures. Crosslink density: (1, 2, 3) $6.37\cdot10^{-6}$ mol/cm^3, (4) $1.31\cdot10^{-5}$ mol/cm^3. Curves 1, 3, 4 correspond to Q/W, and curve 2 to Q'/W [26] (see also Fig. 19)

energetically more favorable in contrast to PDMS macromolecule characterized by positive $f_e/f = 0.28$ [27]. Thus, negative sign of f_e/f, formation mesophase domains and their reorganization and reorientation upon stretching are the main factors determining stress-strain characteristics of PDES network and its relaxation behavior. Consequently, the observed shape of stress-strain curves can be apparently explained in the following way. After the initial equilibrium elongation of crosslinked PDES film the mesophase starts to arise. Chains straightening due to the growth of oriented mesomorphic domains results in leveling off of the stress. Further elongation leads (along with additional mesophase formation) to a further orientation of mesomorphic chains through their plastic drawing. As a result, the second *S*-shape section appears on stress-strain curves. The third section corresponds to stretching oriented chains and at these extension ratios relaxation is caused additionally by reorganization and perfection of mesomorphic domains.

The influence of mesophase formation under stress on relaxation phenomena is displayed in a more striking fashion upon contraction of PDES film. In contrast to uniform deformation upon stretching the contraction of a PDES film proceeds through initial arising of an isotropic part of the film with well-defined boundaries between it and the oriented mesomorphic part, and via further consumption of the mesomorphic part by the growing isotropic part of the sample. Such phenomena appears to be the reverse of necking upon cold drawing of polymers.

Thus, there exists a certain analogy between behavior of PDES network and crystallizable polymer network upon uniaxial stretching. This fact is easily conceivable regarding the mesophase in PDES as a disordered crystal. However, weak intermolecular interaction in the mesophase and subsequently its high plasticity leads to some specific effects useful for the reinforcement of silicon rubbers.

Polyphosphazenes

Finally, we consider another type of mesomorphic inorganic polymers, viz. polyphosphazenes $[-PR_2=N-]_n$. There is a variety of polyphosphazenes with different alkoxy and aryloxy side groups R capable of forming mesophases [2]. Their specific feature is the stability in a rather wide temperature region. For example, in the most investigated poly[bistrifluoroethoxyphosphazene] (PBFP) the mesophase is stable in the temperature interval extended over 150 K (melting point is 367 K, isotropiza-

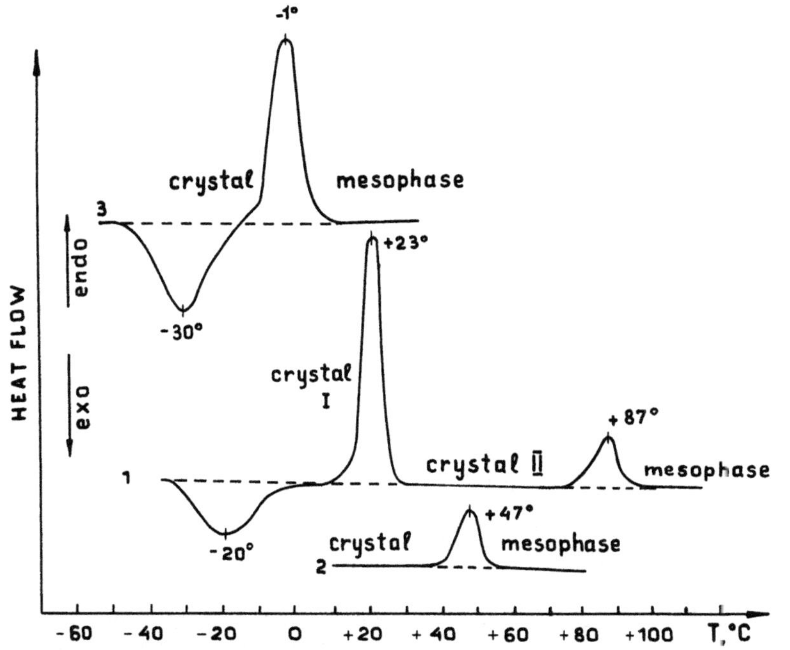

Fig. 21. Main thermal events on DSC traces for polyphosphazenes $[-PR_2=N-]_n$ with different fluorinated alkoxy side groups *R:* 1) $-OCH_2CF_2CF_2H$, 2) $-OCH_2C_3F_6CF_2H$, 3) $OCH_2C_5F_{10}CF_2H$. Heating rate is 8 K/min [31]

tion temperature is near 523 K). Transition from the crystalline modification into the mesophase in this polymer is accompanied by development of high molecular mobilities. According to Alexander et al. [28] the second moment M_2 of ^1H NMR line is equal to $0.4-0.5\,G^2$, and M_2 of ^{34}P NMR line is approximately equal to $0.2\,G^2$. These values correspond theoretically to free rotation of chain units. Our NMR relaxation data support this conclusion [29]. Temperature dependences of the spin-lattice relaxation time T_1 and the spin-lattice relaxation time in local fields T_{1D} point to the fact that the relaxation is determined by a single mode of motion with frequencies higher than 10^9 Hz. According to Magill et al. [30], creep compliance increases sharply in the vicinity of the crystal-mesophase transition by more than one order and, according to our data obtained using a torsion pendulum operating at a frequency of about 1 Hz, the storage modulus decreases from $5\cdot10^8$ to $4\cdot10^7$ Pa.

Polyphosphazenes with enough long fluorinated alkoxy side groups merit separate consideration; they were the subject of our recent investigation. DSC traces for three of such polymers are shown in Fig. 21. X-ray data proved that these polymers (after melting crystalline phases) form thermotropic mesophases which are stable up to destruction temperatures higher than 700 K. Under magnification one can observe specific birefringent textures in nonoriented samples of these polymers. The polymer with side propoxy groups exhibits two crystalline modifications and the thermotropic mesophase forms at heating, as well as in the case of PDES, via subsequent transitions the first crystalline phase into the next and finally, into the mesophase. A peculiar feature of these polymers, in particular polymers with propoxy and pentoxy side groups, is that the transition in mesomorphic state is accompanied by rather small heats of transformation ($3\,Jg^{-1}$ and $1\,Jg^{-1}$, respectively). This fact is apparently connected with the transition into mesophase occurring as a result of small conformational disordering in side groups or small longitudinal shifts of chains with respect to each other. All these transitions manifest themselves upon dynamic mechanical measurements. Our corresponding data obtained using a torsion pendulum operating at about 1 Hz are presented in Fig. 22. Beside loss peaks connected with the above-mentioned transitions, there is the loss peak arising from glass transition in amorphous regions. For all the polyphosphazenes with fluorinated alkoxy side groups glass transition temperature is close to 205 K. Another specific feature of these polymers is the existence of secondary multiple transitions below glass transition temperature. These transitions are displayed clearly in plots of dielectric loss factor against temperature (see Fig. 23). Additional data for two samples of PBFP with different crystallinity are also presented in this figure. Correlation in the intensities of loss peaks of glass transition and secondary transition in PBFP gives evidence that the secondary transition really proceeds in amorphous regions. More detailed study of molecular mobilities in these

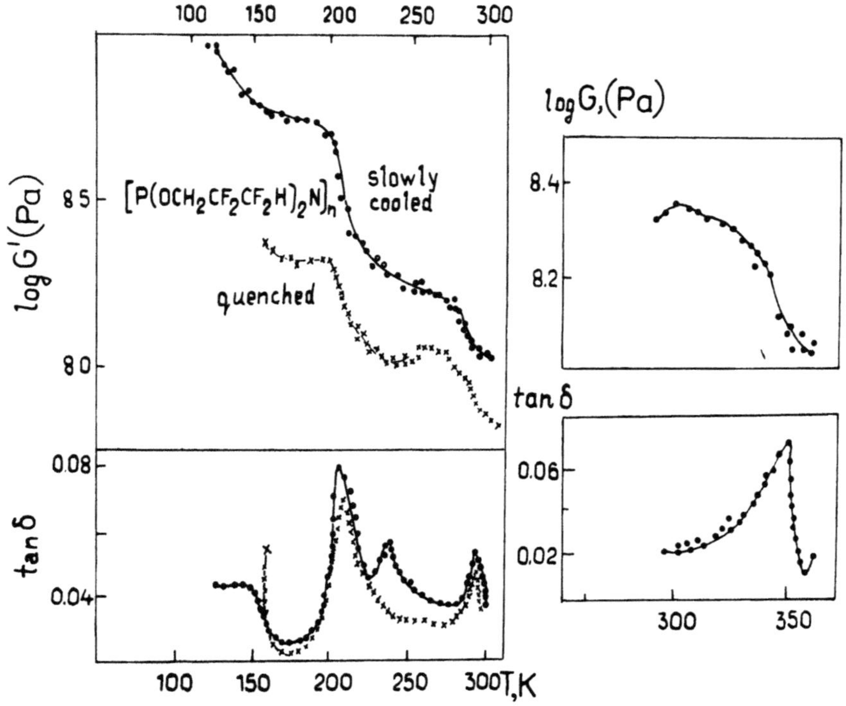

Fig. 22. Plots of the shear storage modulus and the tangent of mechanical losses vs temperature for polyphosphazene (torsion pendulum operating at about 1 Hz)

polymers is in progress. However, it is worthy of noting that enhancing molecular mobilities influence macroscopic mechanical properties of these polymers, as judged by their stress-strain characteristics at different temperatures. As seen in Fig. 24, the shape of all the curves is characteristic of cold drawing semicrystalline polymers. This figure also shows how the transformation from one phase state into another results in jumpwise changes in yield point of these polymers. These changes reflect, of course, different levels of intermolecular interaction in different phases.

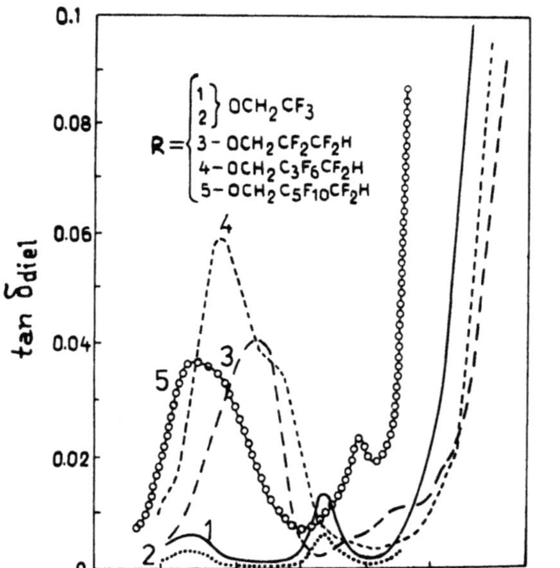

Fig. 23. Dielectric losses recorded at various temperature at frequency 10^3 Hz for polyphosphazenes $[-PR_2=N-]_n$ with different side groups R. Curve 1), corresponds to the sample with crystallinity of 0.4, and curve 2 to the sample with crystallinity of about 0.8

Conclusions

Comparison of various NMR, dynamic mechanical, and dielectric data shows that the subsequent phase transitions in PDES from the low temperature crystalline state into isotropic melt are accompanied by stepwise increase in molecular mobilities. These data reveal very high molecular mobilities in the mesophase of PDES close to molecular mobilities in the amorphous state but, nevertheless, are significantly different from those. The high temperature crystalline polymorphs preceding the mesophase are also characterized by a high level of conformational disorder. All these phases seem to match the definition of condis-crystal.

There exists an analogy between behavior of cross-linked PDES and crystallizable polymer networks upon stretching. This fact is also in line with the treatment of the mesophase in PDES as a disordered crystal.

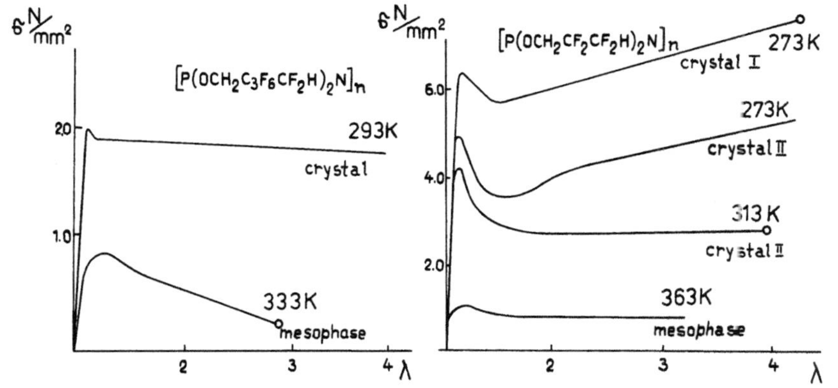

Fig. 24. Stress-strain curves for two polyphosphazenes in different phase states. Strain ratee 0.0065 s^{-1}. Circle is the rupture point of the sample

The transition of the crystalline polyphosphazenes with fluoroalkoxy side groups into mesomorphic state apparently occurs as a result of the introduction of small conformational disordering in side groups. Stress-strain curves of both crystalline and mesomorphic samples of the polyphosphazenes are characteristic of cold drawing of semicrystalline polymers, but the transition into mesomorphic state results in a significant jumpwise decrease in yield point.

Thus, from a structural point of view mesophases in flexible organoelement polymers can be reasonably regarded as disordered crystals characterized by high molecular mobilities and, in respect to macroscopic mechanical properties, as solids with weak intermolecular interaction providing the development of plastic deformation and corresponding relaxation phenomena, even at stress that is not high.

References

1. Godovsky YuK, Papkov VS (1989) Adv Polym Sci 88:129
2. Schneider NS, Desper CR, Beres JH (1978) Blumstein A (ed) In: Liquid crystalline order in polymers. Academic Press, New York Chap. 9, p 299
3. Wunderlich B, Grebowisz J (1984) Adv Polym Sci 60/61:1
4. Smith GW (1975) In: Brown GH (ed) Adv in Liquid Crystals. Academic Press, New York, p 1
5. Vinogradova SV, Tur DR, Minosiants JJ et al. (1982) Acta Polymerica B33:598
6. Korshak VV, Vinogradova SV, Tur DR (1980) Acta Polymerica B31:568
7. Lee CL, Johanson OK, Flanigan OL, Hahn P (1969) ACS Polym Preprints, 10 2:1319
8. Beatty CL, Pochan JM, Froix MF, Hinman DD (1975) Macromolecules 8:547
9. Papkov VS, Godovsky YuK, Svistunov VS, Litvinov VM, Zhadanov AA (1984) J Polym Sci. Polym Chem Ed 22:3617
10. Tsvankin DJa, Papkov VS, Zhukov VP, Godovsky JuK, Svistunov VS, Zhdanov AA (1985) J Polym Sci, Polym Chem Ed 23:1043
11. Froix MF, Beatty CL, Pochan JM, Hinman DD (1975) J Polym Sci, Polym Phys Ed 13:1269
12. Litvinov VM, Laurukhin BD, Papkov VS, Zhdanov AA (1985) Vysokomol Soed A27:1529
13. Burnett LJ, Durrett MG (1980) J Polym Sci, Polym Phys Ed 18:227
14. Jurga S, Jurga K, Pajak Z (1982) J Magn Res 50:175
15. Friedrich J, Rabolt JF (1987) Macromolecules 20:1975
16. Godovsky JuK, Papkov VS (Unpublished results)
17. Litvinov VM, Lavrukhin BD, Papkov VS, Zhdanov AA (1983) Dokl Akad Nauk SSSR 271:900
18. Papkov VS, Svistunov V, Godovsky JuK, Zhdanov AA (1987) J Polym Sci, Polym Phys Ed 25:1859
19. Tsvetkov VN (1964) Birefringence in Flow. In: Ki B (ed) New Methods of Polymer Characterization. Interscience Publisher, New York
20. Papkov VS, Kvachev JuP (1989) Vysokomol Soed, to be published
21. Shirina NG, Zubov JuA, Kostromina SV (1986) Vysokomol Soed B28:379
22. Tsvankin DJa, Zhukov VP, Dedovik AM, Guenin JaV, Levin VYu, Papkov VS (1987) Dokl Akad Nauk SSSR 297:906
23. McCrum NG (1959) J Polym Sci 34:355
24. Bernier GA, Kline PE, Sauer JA (1967) J Macromol Sci B1:335
25. Bulkin AF, Papkov VS, Vysokomol Soed, submitted for publication
26. Papkov VS, Godovsky JuK, Svistunov VS (1989) Vysokomol Soed 31:1547
27. Krigbaum WR, Roe R-J (1965) Rubber Chem Technol 38:1039
28. Alexander MN, Desper CR, Sagalyn PL, Schneider NS (1977) Macromolecules 10:721
29. Litvinov VM, Papkov VS, Tur DR (1986) Vysokomol Soed A28:289
30. Masuko T, Simeone RL, Magill JH, Plazek DJ (1984) Macromolecules 17:2857
31. Papkov VS, Dubovik JJ, Tur DR et al. (1989) Vysokomol Soed, submitted for publication

Received February 13, 1989; accepted July 21, 1989

Authors' address:

Prof. V.S. Papkov
Institute of Synthetical Polymeric Materials
USSR Academy of Sciences
ul. Profsoyznzaya, 70
117393 Moscow

Progr Colloid & Polymer Sci 80:236–244 (1989)

Relaxation currents in insulating polymers and their dependence on the polymer structure

H. Hänsel

Pedagogical University "Karl Liebknecht" Potsdam, Potsdam, GDR

Abstract: The electrical characterization of high-ohmic polymers is of growing interest. In the present paper the advantages and disadvantages of time-domain and frequency-domain measurements are discussed. It will be shown that these methods are also very useful to investigate charge carrier relaxation processes, boundary effects, and ageing effects. The possibility to use these methods to observe microscopic parameters will be discussed.

Key words: Poly(phenylquineoxaline); poly(ethylenterephthalate); poly(methacrylate); ac-measurements; isothermal polarization; thermally stimulated polarization; electrical relaxation

Introduction

An alteration of the external voltage may cause time-consuming electrical relaxation processes in high-ohmic polymers, meaning that in these systems an alteration of the acting electric field corresponds to a retarded reaction of the charge displacement. If we speak of an insulating (high-ohmic) polymer, we mean a polymer in which by a charge-transport experiment an equilibrium between the external field and the charge distribution cannot be attained in the measuring time τ_m. Therefore, the ratio between the dielectric relaxation time

$$\tau_D = \frac{\varepsilon_0 \varepsilon_r}{\sigma} \qquad (1)$$

and the measuring time are an appropriate criterion for the distinction between metallic ($\tau_D \ll \tau_m$), semiconducting ($\tau_D < \tau_m$) or insulating polymers ($\tau_D > \tau_m$) [1].

While metals and usually semiconductors allow experiments under stationary or quasistationary conditions this is in general not possible for insulating polymers. For instance in high-ohmic polymers the current relaxes to equilibrium within times which, depending on temperature, are of minutes, hours, days, or greater. Figure 1 demonstrates this behavior by the time dependence of the current measured in poly(phenylquineoxaline) (PPQ). At $t = 0$ an external voltage was applied and the current relaxes to an equilibrium state. This effect can be covered by ageing, because ageing also give rise to an observed decay of the current (Fig. 2). It is difficult to separate these processes from the relaxation phenomena. The only

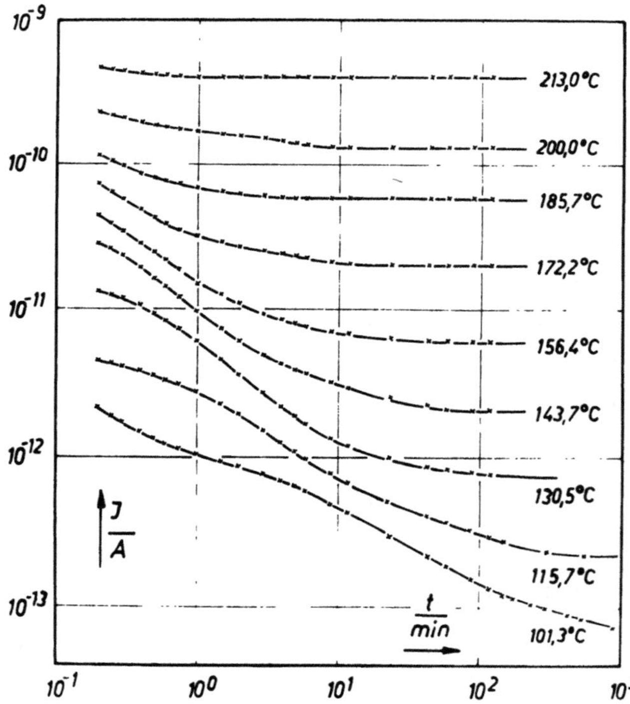

Fig. 1. Current I vs time t after a voltage step at $t = 0$ for PPQ

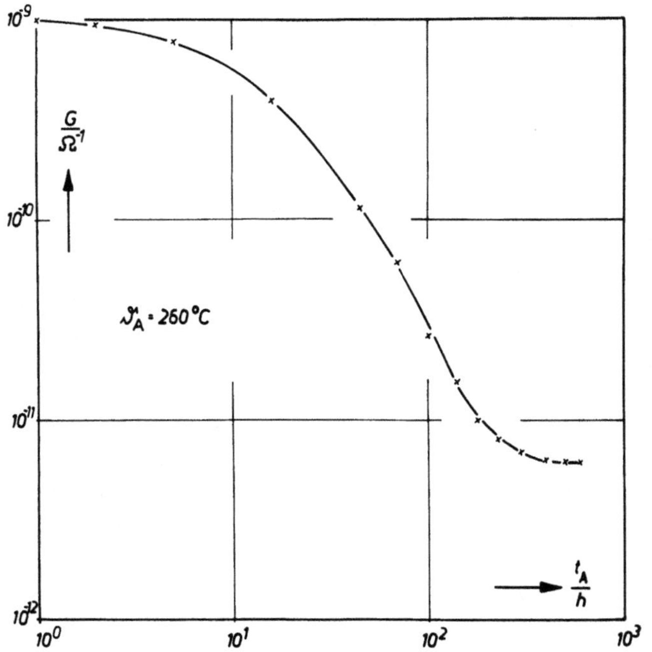

Fig. 2. Electrical conductance G of PPQ in dependence on the annealing time t_a (ϑ_a annealing temperature)

distinction is that the relaxation is reversible, while ageing is not. Reproducible measurements can only be obtained if the sample is annealed below the glass-transition temperature for several hours and the following measuring program takes place in a temperature range below the ageing temperature. A change of the polymer structure during the annealing process is discussed as the cause of this effect, for instance by cross-linking (see Fig. 5). But the entire reason for this behavior is still unknown [2].

The reason for the electrical relaxation processes can be: polarization processes (electronic, ionic, and orientational polarization), movement of equilibrium and excess charge carriers, space charges, and electrode effects.

Debye-relaxation process

Until now, the most discussed relaxation process has been the Debye-relaxation. For this the following relaxation system will be considered:

- only *one* type of non-interacting object (charge carriers or dipoles);
- the transition between two equilibrium positions requires the activation energy W;
- only *one* relaxation time τ, that can depend on temperature; and
- position-independent polarization P.

Charge carrier relaxation processes can only perform these assumptions if internal electrical fields E_i are to be neglected.

Isothermal polarization

Mostly the electrical relaxation processes are classified in single-pulse-, single-step-processes and ac-measurements. Figure 1 gives an example for a single-step process. With $T = \text{const}$ the external electric field E will be changed, but then is kept constant. The current I does not react instantly but adapts gradually to the corresponding new equilibrium state.

Such a process is named isothermal polarization if the step goes from $E = 0$ to $E_P \neq 0$, or isothermal depolarization if it goes from $E_P \neq 0$ to $E = 0$. With that assumption the Debye-relaxation yields for the isothermal polarization and depolarization current density

$$j = j_0 \exp\left(-\frac{t}{\tau}\right) \quad (t > 0) \;, \tag{2}$$

with

$$P_S = \varepsilon_0(\varepsilon_S - \varepsilon_\infty)E_P \tag{2a}$$

$$\varepsilon_r(\omega \to 0) = \varepsilon_S \; ; \quad \varepsilon_r(\omega \to \infty) = \varepsilon_\infty \tag{2b}$$

(E_P = polarization field strength). Hence a plot $\ln j$ vs t yields the relaxation time of this process and j_0, and with this, also the static polarization P_S and the intensity $\Delta\varepsilon = \varepsilon_S - \varepsilon_\infty$.

In the case of a single relaxation process the temperature dependence of the relaxation time is given by the Arrhenius-equation

$$\tau = \tau_0 \exp \frac{W}{kT} \,. \tag{3}$$

Therefore, from measurements carried out at different temperatures the activation energy W of this process can be calculated.

Contrary to this the current in all insulating polymers follows a power law of time dependence (Schweidler law [3])

$$I(t) \sim t^{-n} \tag{4}$$

instead of the exponential relation (2), which characterizes the Debye mechanism. This law is a universal one. It is only weakly influenced by the physical and chemical structure of the polymer. Most-

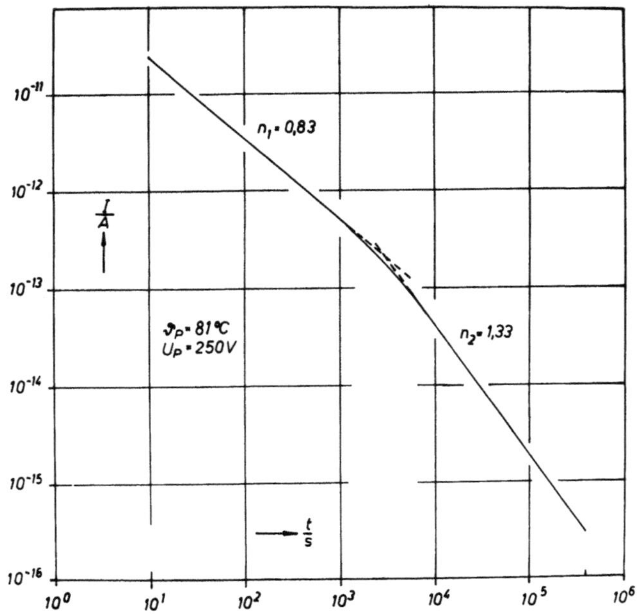

Fig. 3. Logarithm of the isothermal depolarization current I vs logarithm time t for PETP

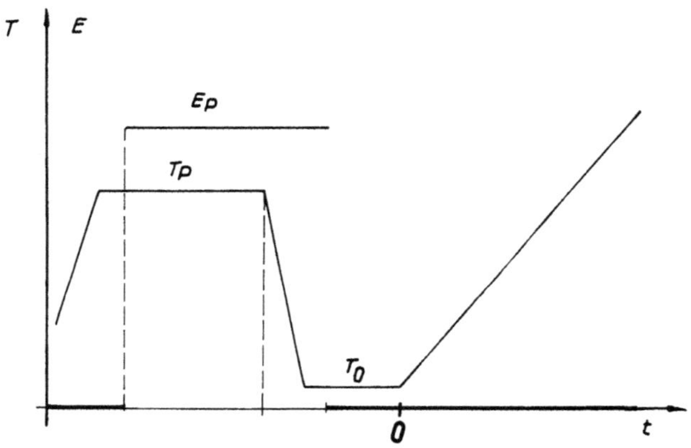

Fig. 4a. $T=f(t)$ and $E=f(t)$ for TSD

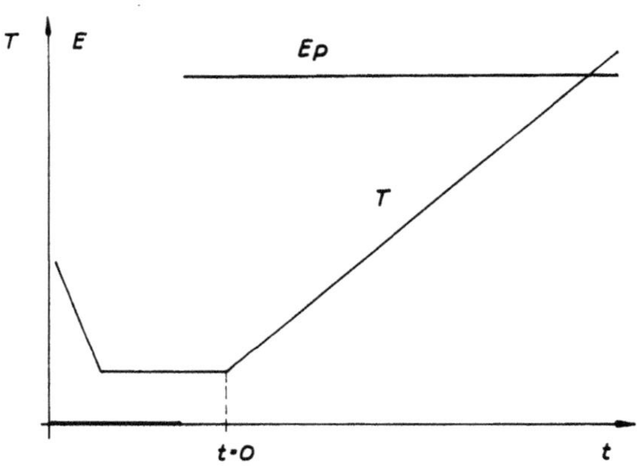

Fig. 4b. $T=f(t)$ and $E=f(t)$ for TSP

ly the plot $\ln I$ vs $\ln t$ exhibits two linear regions with different slopes (Fig. 3).

Thermally stimulated polarization

An even 10 times better sensitivity can be attained by measurements of the thermally stimulated polarization (TSP) or depolarization (TSD) currents instead of the thermal polarization or depolarization [4, 5].

With the TSD method the thermally activated release of the stored polarization is studied. The sample is polarized by an applied polarization field E_P at the temperature T_P (Fig. 4a). This polarization is subsequently frozen in by cooling the sample down to a temperature T_0 sufficiently low to prevent depolarization by thermal energy. The field is then switched off, the sample is short-circuited and warmed-up. For each polarization mechanism an inherent current peak can be detected (see Fig. 5). In the case of the TSP the sample is cooled in the unpolarized state, then an electric field is switched on and during the following heating the current is measured (Fig. 4b). Contrary to the TSD by measuring the TSP the current is superimposed from the dark current, therefore, mostly the TSD is preferred.

In order to evaluate a thermally stimulated process a definite heating regime must be applied. Usually a linear heating rate

$$\frac{dT}{dt} = s = \text{const} \quad \text{or} \quad T = T_0 + st \tag{5}$$

is chosen. Typical heating rates are of several K/min. Only few papers are known that use a other heating regime. For the TSD current yields

$$j(T) = j_0 \exp\left[-\frac{W}{kT} - \frac{s}{\tau_0}\int_{T_0}^{T} \exp\left(-\frac{W}{kT}\right) dT\right] \tag{6}$$

with

$$j_0 = \frac{P_S}{\tau_0} \,. \tag{6a}$$

$j(T)$ is a asymmetrical curve with the maximum condition

$$\tau(T_m) = \tau_0 \exp\left(\frac{W}{kT_m}\right) = \frac{kT_m^2}{sW} \,. \tag{7}$$

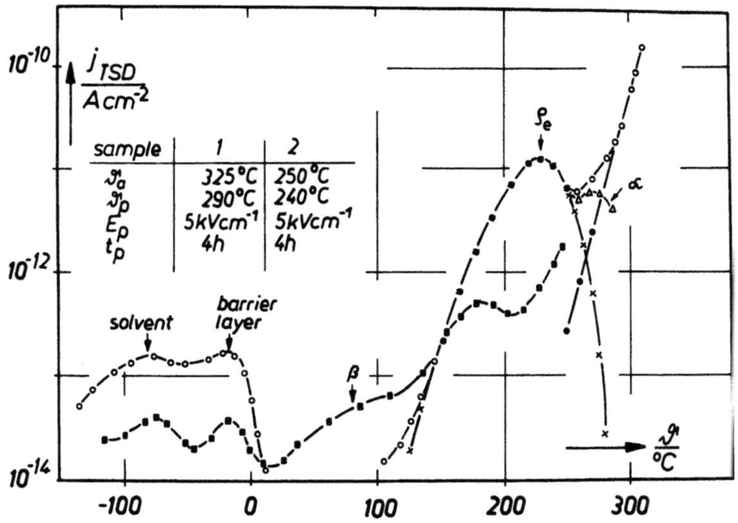

Fig. 5. TSD spectrum of PPQ (t_p polarization time)

The evaluation of interesting parameters from TSD measurements is well known. Therefore we only briefly state that the activation energy can be estimated from:

- the differences between the maximum temperatures T_m by measuring with different heating rates;
- a plot $\log j$ vs T^{-1} of the initial rise of the signal, because then the integral in (6) can be neglected

$$\ln j(T) \approx -\frac{W}{kT} + \ln j_0 \ . \tag{8}$$

Also j_0 can be gotten from this plot.

If W and j_0 are known and a Debye-relaxation is valid, τ_0 and P_S can calculated from (7) and (6a), and with this $\Delta\varepsilon = \varepsilon_S - \varepsilon_\infty$ from (2a). If we have no Debye-relaxation then the distribution function $\tau(T)$ can be estimated from [6, 7] (Bucci-plot)

$$\tau(T) = \frac{\int_T^\infty j(T)\,dT}{s\,j(T)} \ . \tag{9}$$

It can be seen from Eqs. (2a), (6), and (6a) that the TSD current is proportional to the polarization field and the maximum is independent on E_P. This has very often been confirmed experimentally.

In general it can be said that the increase of sensitivity by using thermal-activated polarization (instead of isothermal) causes a decrease in the possibility to theoretically interpret the experimental results.

Comparison of time and frequency domain measurements

The determination of the response to harmonic excitation (ac-measurements) is a powerful alternative approach to the measurement of the time-dependence of the current in response to a step-function field. ac-measurements describe the dielectric behavior of the polymer by the complex permittivity $\hat{\varepsilon}(\omega)$, which is given by the ratio between the complex amplitudes of the current density $\hat{j}$ and the electric field $\hat{E}$

$$\frac{\hat{j}}{\varepsilon_0 \hat{E}\omega} = \hat{\varepsilon}(\omega) = \varepsilon'(\omega) - i\varepsilon''(\omega) \ . \tag{10}$$

Because real part $\varepsilon'(\omega)$ and imaginary (loss) part $\varepsilon''(\omega)$ are interrelated via the Hilbert transform every part contains the complete information [8–10]. Generally, it is convenient to discuss the loss part. Every relaxation region is indicated by a maximum of $\varepsilon''(\omega)$.

The time-dependence of the current in response to a step-function field (*time domain measurement*) and the frequency dependence of the permittivity in response to harmonic excitation (*frequency domain measurement*) are connected by the Fourier integral transformation. Therefore, there is no fundamental difference between the two types of measurements, time domain and ac-measurements, and they complement each other.

Time domain and frequency domain measurements have been applied extensively to the investigation of insulating polymers in order to draw conclusions about the rotation of side groups (β-peaks), the motion of main chains (α-peak) and space charges (ϱ-peak). In general from these investigations it can be said that we have no pure Debye-relaxation processes in insulating polymers because

- for a single-step process the current follows a power law of time-dependence instead of the exponential relation characterizing the Debye-mechanism; and
- the shape of the normalized loss peak is not the shape of the pure Debye peak. It is broader than a pure Debye peak by several orders of magnitude. Therefore, mostly the relaxation time and activation energy were considered as distributed. Some experimental techniques were developed to obtain the distribution of the relaxation time and the activation energy.
- Primarily in the β-peaks the temperature-dependence of the relaxation time obeys the Arrhenius law. Against it for the α-relaxation region the WLF equation

$$\tau(T) = \tau_0' \exp\left(-\frac{C_1(T-T_g)}{C_2+T-T_g}\right) \tag{11}$$

is valid [10].

From these experimental facts we must conclude that the assumption of non-interacting objects is not valid in polymers [10–14].

Whereas powerful measuring techniques exist for ac-measurements, time-domain measuring equipment must be specially built. Isothermal and, if typical heating rates are applied, thermal-activated measurements correspond to loss factor measurements carried out with frequencies $f \approx 10^{-2}\,s^{-1}$ and lower, because for the maximum of ε'', $\omega_{max}\tau = 1$ is valid, and therefore, with (7)

$$\omega_{max} = \frac{W}{skT_m^2} \,. \tag{12}$$

Since ac-measurements at such low frequencies are difficult to carry out and the loss data are increasingly dominated by direct conductivity, time-domain measurements should be preferred. This is possible because between the isothermal depolarization current $j(t)$ and $\dot{\varepsilon}(t)$ the equation

$$j(t)/E_P\varepsilon_0 = \dot{\varepsilon}(t) \quad (t > 0) \tag{13}$$

is valid. Mostly the function $t\dot{\varepsilon}(t)$ is used instead of $\dot{\varepsilon}(t)$, because $t\dot{\varepsilon}(t)$ exhibits a maximum nearly like the function $\varepsilon''(\omega)$ [15].

Furthermore, all charge carriers' relaxation regions and boundary effects are not observable in the frequency domain, because the ac-conductivity is too high for the measuring temperature and therefore, the ac-current covers these peaks. Time-domain measurements must be used to observe these effects. Moreover, time-domain measurements give a good possibility to follow up ageing processes.

Mostly the investigation and interpretation of the several relaxation regions observed is difficult or impossible because they overlap. Also for this reason, it is advantageous to use, for instance, TSD-measurements instead of ac-measurements, because:

- firstly, the TSD signals are narrower than the loss factor signals. Therefore the different relaxation regions can be better separated for TSD-measurements than for ac-measurements
- secondly, experimental methods could be developed to separate them and to calculate the distributed values $\tau_0(T)$ and $W(\tau)$. These methods are: thermal cleaning [16], partial heating [16], and thermal sampling [17, 18].

We will demonstrate these facts on the TSD- and ac-spectrum of PPQ. Because this polymer is amorphous, its TSD-spectrum cannot be influenced by structural changes during the measuring process. Seven relaxation regions may be distinguished in the TSD-spectrum (Figs. 5 and 6) [19]:

- *water* ($\approx -100\,°C$) *and solvent* ($\approx -25\,°C$) *peak*
  The magnitude of these peaks differs from sample to sample. This fact shows that the water and the solvent are not homogeneously distributed in the foil from which the sample was made. The peak due to water vanishes on annealing.
- *β-peak* ($\approx 70\,°C$)
  The magnitude of this peak decreases strongly on annealing. The β-relaxation region may be seen only for $E_P > 10^4\,V\,cm^{-1}$ for annealed samples. Our interpretation of this fact is that the annealing produces crosslinking of the substance.
- *α-peak and ϱ_i-peak* (see Fig. 6)
  We have found that the TSD spectrum includes either the α-relaxation region or a strong increase of the current covering the α-peak depending on annealing above glass-transition temperature T_g. If the polarization is hindered by application of very short polarization times (< 10 min), then this region is separated in the α- and ϱ_i-peaks. The ϱ_i-region is given by ions generated by the degradation of PPQ on annealing.

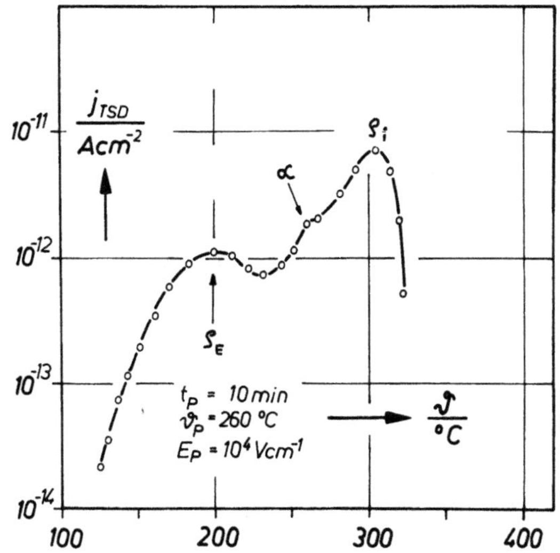

Fig. 6. α- and ϱ_i-peak of PPQ. These two relaxation regions can be separated by using polarization times $t_P < 10$ min

- ϱ_e*-peak*
  The investigation of the temperature and field strength dependencies of this peak shows that electronic charge carriers produce this relaxation region.
- *barrier-layer-peak*
  Another TSD peak at about 0 °C may be seen in annealed samples. We suppose barrier layer effects produce this peak, because
  - the stored charge depends on the direction of the polarizing field;
  - the activation energy of this process depends on the field strength; and
  - $\lg j_{max}$ is not a linear funciton of $\lg E_P$.

The possibility to observe these regions depends strongly on the annealing degree of the samples. Only three of these relaxation regions, namely the solvent-, water- and δ-relaxation regions, were also detected in the tan δ-spectrum [20] (see Fig. 7).

Structural changes and time and frequency domain measurements

Often the aim of these investigations was to explain the macroscopic electrical behavior of the polymer from its chemical and physical structures. It is a well known fact that σ is influenced by the polymer structure. For instance, the plot $\ln \sigma$ vs $1/T$ for PETP is linear with two different slopes (see Fig. 8). The temperature of the intersection of these two straight lines is the glass-transition temperature T_g where the polymer changes its structure. Also, the drift mobility μ of the charge carriers changes with the polymer structure. To evidence this fact, the crystallinity α of the partial crystalline polymer PETP was changed by special annealing processes [22, 23]. The orientation was modified by stretching the foils. The birefringence Δn was used for characterizing the orientation we attained. We found that:

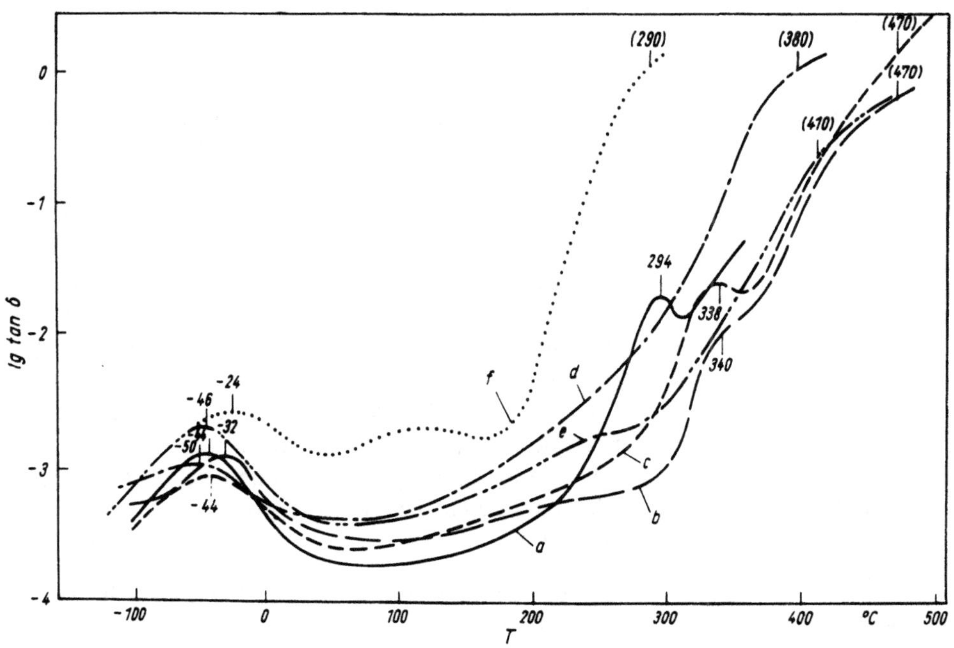

Fig. 7. tan δ-spectrum of PPQ

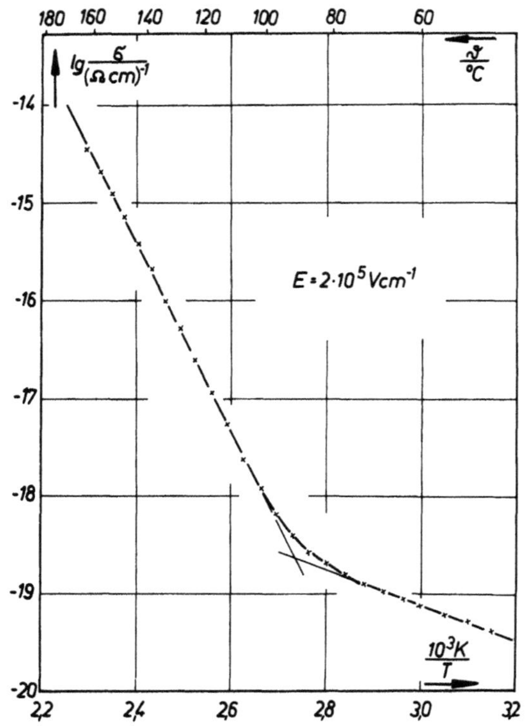

Fig. 8. $\lg \sigma = f(1/T)$ for PETP, the temperature of the intersection of the two straight lines yields T_g

- the mobility increases about three orders of magnitude with increasing of the crystallinity from $\alpha \approx 0$ to $\alpha \approx 0.3$ (Fig. 9).
- the mobility decreases with increasing orientation.

This is shown in Fig. 10 for two crystallinities [23, 24]. Until now it has been impossible to get microscopic

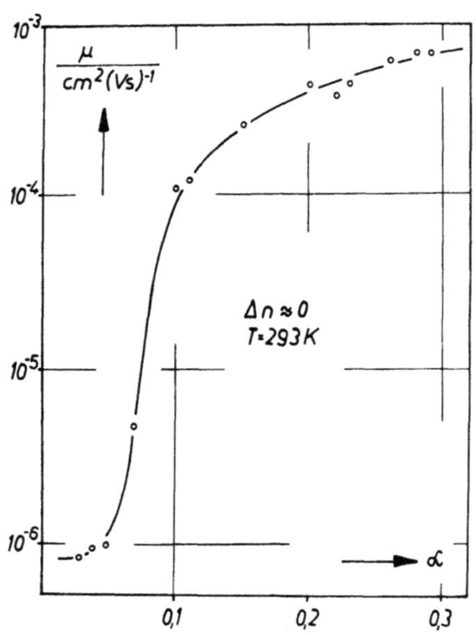

Fig. 9. Drift mobility μ vs the crystallinity α of PETP

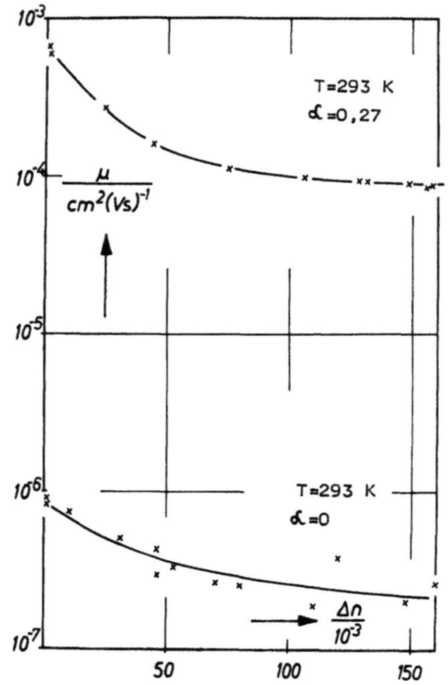

Fig. 10. Drift mobility μ vs birefringence Δn of stretched PETP foils

parameters from the shape of the TSD curves. The situation is a little better because of the interpretation of the ac- or isothermal depolarization measurements, respectively. Although most relaxation processes are not Debye-relaxations the curves of all relaxation regions of ac- and isothermal depolarization measurements can be represented by the empirical relation introduced by Havriliak and Negami [25]

$$\hat{\varepsilon}(\omega) - \varepsilon_\infty = \Delta\varepsilon\,[1 + (i\omega/\omega_0)^b]^{-c} \qquad 0 < b, \quad bc < 1\,, \tag{14a}$$

or

$$\dot{\varepsilon}(t) = \Delta\varepsilon \int_0^\infty [1 + (i\omega/\omega_0)^b]^{-c} e^{i\omega t} d\omega\,, \tag{14b}$$

respectively. b and c characterize the shape of the relaxation curve, ω_0 denote its position, and ε_∞ is $\varepsilon'(\omega)$ for $\omega \gg \omega_0$. The four HN-parameters ω_0, $\Delta\varepsilon$, b, and c contain complete information about the relaxation mechanism. They can be determined by a computer least-squares fit of the HN-function to those measured loss data which are not influenced by neighbored relaxation processes [26–29]. This is possible in the frequency or time domain or in both. The parameters can be interpreted in terms of the structure:

- ω_0 is related to the mean relaxation time of the process;

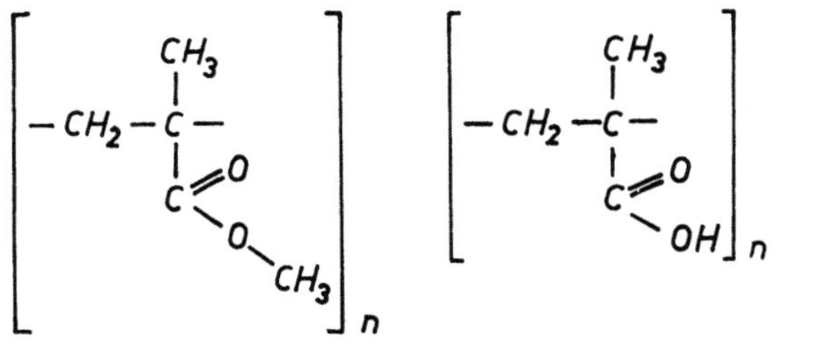

Fig. 11. a) pure PMMA and b) saponificated PMMA

- $\Delta\varepsilon$ is related to effective dipole moment which involves an integral information about intermolecular interaction;
- b and c have been suggested to differentiate between short and large scale interaction in the case of main relaxation [27].

Further information of structural relaxations could be obtained from these parameters [29].

Up to now it has not been possible to get microscopic parameters from the shape of the TSD curves, because in this field an adequate relation to (14) is unknown. Therefore, we have only the possibility to change the chemical and (or) the physical structure of the substance; we observe and try to find correlations with the electric properties, for instance, with the positions of the TSD regions as well as the shape of the TSD curves. Also the dielectric behavior of compressed pellets of hydrated cellulose was studied by TSD measurements in an attempt to determine the binding modes of water molecules [30, 31]. The water content was varied between 1% and 22%; two broad relaxation regions were observed. The region at lower temperature is attributed to the reorientation of sorbed water molecules. The region at higher temperature can be split in two by using thermal sampling technique. The first peak is due to dipolar polarization, the second peak is related to charge carrier relaxations of the cellulose.

The TSD in PETP is influenced by the physical structure. The storage charge for crystallized and stretched foils is greater at temperatures below T_g than for amorphous or unstretched foils, and does not depend much on temperature. The difference in storage charge was observed only in the α-relaxation region [32].

We investigated the TSD-spectrum of pure and saponificated PMMA (Fig. 11) as well. Three relaxation regions can be observed: The β-, α-, and ϱ space charge relaxation (Fig. 12). The β-relaxation region remains unchanged by the saponification, while the peaks of the other relaxation regions shift [33]. Previously it was not possible to connect the shift of the TSD curves and the change of their shape with microscopic parameters. These examples show that much work remains to be done before we are able to interpret the TSD curves at the same theoretical level which we show for the isothermal curves in the frequency and time domains.

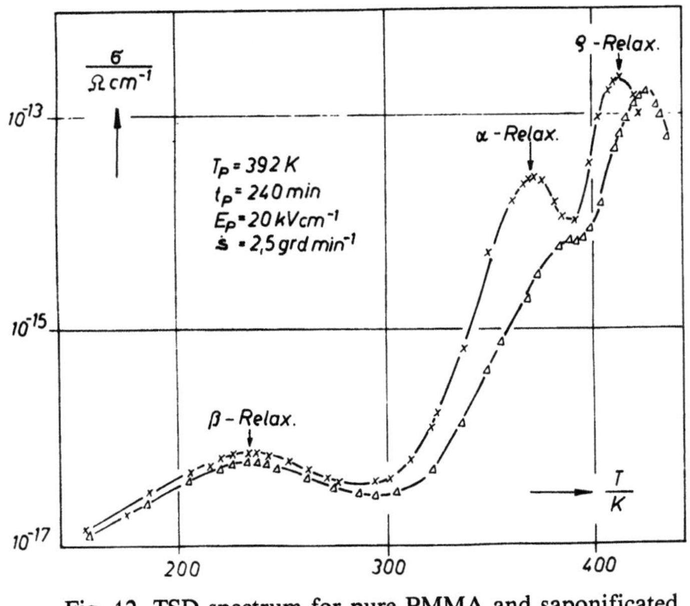

Fig. 12. TSD spectrum for pure PMMA and saponificated PMMA

References

1. Müller P (1981) Phys Stat Sol (a) 67:11
2. Hänsel H (1987) Acta Polymerica 38:378
3. v. Schweidler E (1904) Ann Phys 24:711
4. Bucci C, Fieschi R (1964) Phys Rev Letters 12:16
5. van Turnhout (1972) J Thesis (B), Leiden
6. Bucci C, Fieschi R, Guidi R (1966) Phys Rev 148:816
7. van Turnhout J (1975) Thermally Stimulated Discharge of Polymer Electrets. Elsevier Sci Publ Comp, Amsterdam Oxford New York
8. Müller FH, Schmelzer Ch (1951) Ergebn exakt Naturw 25:359
9. Böttcher CJF, Bordewigk P (1978) Theorie of Electric Polarization. Elsevier Sci Publ Comp, Amsterdam Oxford New York
10. Jonscher AK (1983) Dielectric Relaxation in Solids. Chelsea Dielectric Press, London
11. Jonscher AK (1974) Nature 250:191
12. Jonscher AK (1975) Colloid Polym Sci 253:231
13. Jonscher AK (1977) Nature 267:673
14. Fischer E (1983) Thesis (A), TU Dresden
15. Hamon BV (1952) Proc Instn Electr Engr, Part IV, 99:151, Monograph No 27
16. Creswell RA, Perlman MM (1970) J Appl Phys 41:2365
17. Hino T (1975) J Appl Phys 46:1956

18. Lacabanne C, Chatain D (1975) J Polym Sci. Polym Phys Ed 13:445
19. Stroloke W, Neumann W, Hänsel W (1982) Polym Bull 8:527
20. Schlosser E, Raubach H (1977) Plaste und Kautschuk 24:182
21. Mischok W, Schick Ch, Krämer L (1985) Acta Polymerica 36:47
22. Schick C, Fabry F, Schnell U, Stoll G, Deutschbein L, Mischok W (1988) Acta Polymerica 39:705
23. Hanspach J, Kryszewski M, Hänsel H, Gensch K (1981) Acta Polymerica 32:131
24. Hanspach J, Hänsel H, Kryszewski M, Gaulke R (1985) Acta Polymerica 36:234
25. Havriliak S, Negami S (1966) J Polym Sci (C). Polym Symp 14:99
26. Schlosser E, Schönhals A (1987) Acta Polymerica 38:357
27. Schönhals A, Schlosser E (1989) Colloid & Polymer Sci 267:125
28. Schlosser E, Schönhals A (1989) Colloid & Polymer Sci 267:133
29. Schlosser E, Schönhals A (1989) Colloid & Polymer Sci, in press
30. Pissis P et al. (1985) 5th Intern Symp. Electrets, Heidelberg
31. Pissis P (1985) J Phys D. Appl Phys 18:1897
32. Voß K-L, Neumann W, Hänsel H (1986) Morphology of Polymers. Walter de Gruyter & Co, Berlin New York, 700
33 Gietzel D (1989) Thesis (A), Potsdam

Received February 15, 1989;
accepted July 31, 1989

Authors' address:

Prof. Dr. H. Hänsel
Pädagogische Hochschule
"Karl Liebknecht" Potsdam
Sektion Physik
WB Physik II
Am Neuen Palais
DDR-1571 Potsdam

Discussion

KREMER:
How does electrode polarization caused by ionic charge carriers influence the TSC-spectra?

HÄNSEL:
An electrode polarization caused by ionic charge carriers is seen in the TSD-spectra as a space charge peak (ϱ-peak) above the α-relaxation region. I demonstrated this fact in my lecture with the TSD-spectra of poly(phenylquineoxaline).

HERMS:
What are the effects of atomic and molecular defects on electric relaxation processes, e.g., on charge carrier drift mobility and volume charge relaxations?

HÄNSEL:
Up to now this influence has been practically unknown. Electrical measurements alone usually do not enable the derivation of microscopic models.

HOLZMÜLLER:
Have you considered the influence of coupled dislocation processes overcoming a multitude of low barriers ending in a high barrier?

HÄNSEL:
No, we have not considered this influence. Further, I believe that until now the knowledge of these effects has not been sufficient to do this.

HOLZMÜLLER:
Have you calculated the influence of internal field to discuss electrical conductivity in the sense of Arht, Onsager, Kirkwood, Fröhlich, Gotlib? There will be a time-dependent fluctuation of the internal field.

HÄNSEL:
I discussed the TSD for pure Debye-relaxation processes, that means with the assumption of a position-independent polarization. Charge carrier relaxation processes can only perform this assumption if the internal fields are to be neglected. Because injection phenomena are quite typical in high resistivity materials and, at least in the neighborhood of contacts, deviations of neutrality are to be expected normally in real systems. Therefore, internal fields can be expected. Until now it has not been possible to calculate them. The internal field must be measured, for example, with the equipment from Sessler and Gerhard-Multhaupt. Mostly this is very difficult and also not possible for all high-ohmic materials.

Theoretical problems and dynamic models of the relaxation behavior of macromolecules in anisotropic ordered polymer systems

Y. Y. Gotlib

Institute of Macromolecular Compounds of the Academy of Sciences of the USSR, Leningrad, USSR

Abstract: A multisegmental semirigid chain model with finite extensibility of chain segments is used. The relaxational properties of macromolecules with mesogenic fragments in the main chain in external quadrupole field or immersed in the nematic liquid-cristalline matrix are investigated. The modified theory of self-consistent molecular field (Maier-Saupe approximation) is applied.

A system of linearized equations of motion for the average projection of chain units on the field direction and on the normal direction is developed on the basis of the effective quasi-potential and dissipation function obtained.

The anisotropy of rotational friction of chain elements is taken into account. The "longitudinal" motions are accompanied by surmounting the local potential barriers. In transverse motion no drastic reorientation with respect to the field occurs.

Two types of normal modes and relaxation spectra appear: transverse and longitudinal. Relaxation times and orientation correlation functions for a selected chain element exhibit different dependencies on the initial (zero-field) chain rigidity, final rigidity induced by the external field (or by LC-ordering), on the degree of order, and on the scale of the normal mode.

Key words: Multisegmental semirigid chain model; finite extensibility; relaxational properties; self-consistent molecular field; anisotropy rotational friction; transversal and longitudinal relaxation spectra.

Introduction

The theoretical analysis of molecular mobility of strongly oriented and deformed polymer chains in solutions, melts, and liquid crystal state is an important problem. The solution of this problem can be applied to the explanation of the relaxation properties of polymers in strong shear and stretching hydrodynamic fields, in the strained amorphous regions of semicrystalline polymers, in oriented surface layers (or membranes), and in liquid-crystalline (LC) polymer systems. In the latter case the chains are subjected to the action of molecular internal field ("pseudopotential") which may be overlapped by external electric, magnetic, and mechanical fields. These fields profoundly affect the initial properties of statistically and dynamically labile LC polymer systems.

This paper deals with the theory of relaxation processes corresponding to the linear response of the systems to weak nonstationary perturbation overlapping a strong static external (or "internal") field which actually leads to strong orientation and deformation of the system. In other words, we consider the molecular motion occurring on the background of strong static chain deformation and orientation.

The proposed dynamic model may be directly applicable to the study of the behavior in linear and bilinear fields. This is, first, the "vector" dipole field with the potential

$$U_{\text{dipole}} = \sum_j \mu_j l \cos\theta_j \tag{1}$$

including the case of chain stretching at the ends

$$U_{\text{stretch}} = -fh_x \equiv -fl \sum_j \cos\theta_j = -f \sum_j u_j\ , \tag{2}$$

where θ_j is the orientation angle of the j-th chain element with respect to the field axis and $u_j = l\cos\theta_j =$

$x_{j+1} - x_j$ is the projection of the j-th element on the field axis X.

The quantities x, y, and z are the coordinates of chain elements (in the Cartesian system of coordinates) with the length l.

The quadrupole (tensor) field may be used

$$U_{\text{quadr.}} = -\sum_j \beta_j \cos^2\theta_j = -\sum_j \beta_j (u_j/l)^2 \tag{3}$$

or in another form

$$U_{\text{quadr.}} = -\sum_j \beta_j (3/2)[\cos^2\theta_j - (1/3)] = -\sum_j \beta_j P_2(\cos\theta_j)\ ,$$

where $P_2(\cos\theta_j)$ is the Legendre polynomial of 2nd order.

Another important case is that of hydrodynamic fields with the rate gradient of the flow G, the longitudinal field,

$$v_x = Gx \quad v_y = (-1/2)Gy \quad v_z = -(1/2)Gz \tag{4a}$$

and the shear field

$$v_x = -Gz \quad v_y = v_z = 0\ . \tag{4b}$$

In the studying the equlibrium and relaxation behavior of macromolecules containing mesogenic groups in the main chain in liquid-crystalline state, the modification of self-consistent quadrupole molecular field theory (with Maier-Saupe pseudopotential) is used. In this case (for a homogeneous chain) the parameter $\beta_j \equiv \beta$ [see Eq. (3)] is assumed to be proportional to quantity

$$\beta = \alpha S \tag{5}$$

where S is the mean degree of order (order parameter):

$$S = (3/2)\langle\cos^2\theta - (1/3)\rangle = \langle P_2(\cos\theta)\rangle \approx (1/N)(3/2)\sum_{j=1}^{N} \langle\cos^2\theta_j - (1/3)\rangle\ . \tag{6}$$

The α is the orientational interaction parameter between the orienting particle and internal field in the LC. The quantity S in the LC-system may be determined from the conditions of selfconsistency (cf. [1–3]).

In this paper a modified Hearst-Harris model for Gaussian subchains (GSC) characterized by bending rigidity is used. The parameters of the effective Gaussian potential of this model are chosen in such a manner as to obtain the agreement between the second moments of the GSC model and the second moments of a more rigorous statistical chain model which the GSC model imitates, simultaneously taken into account the action of the external (or internal) orienting field [4].

Instead of second moments one can consider mean quadratic dimensions of the chain and the corresponding parameters of angular correlations between the elements along the chain.

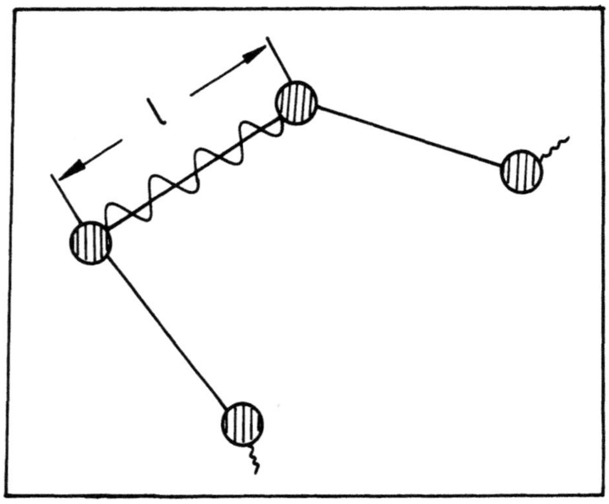

Fig. 1. Dynamic chain model consisting of elements with a fixed length

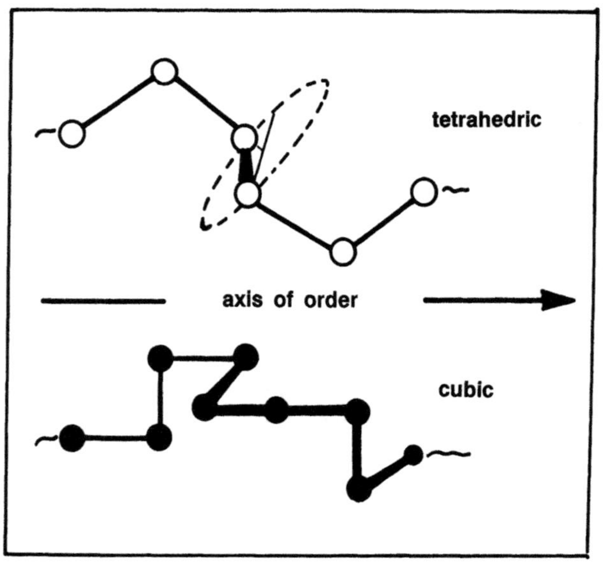

Fig. 2. Lattice chain model

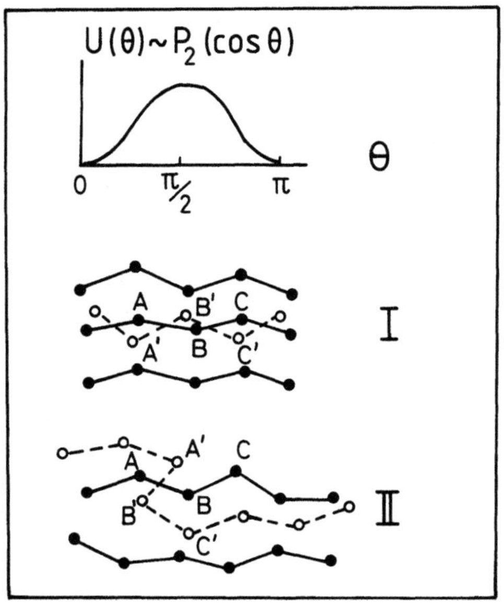

Fig. 3. Quadrupole external (or molecular) field. Rearrangement with or without transitions over the barrier

In the theory, the fact is taken into account that when strong field are superposed on the initial GSC model, the chain elements are stretched (their length increases), whereas in real chains stiff elements exist. These elements are only oriented so that the contour chain length determined in the proper manner remains constant in a field of any arbitrary great strength. In the modified GSC model used in this treatment this fact is taken into account "on the average" (see [5]), through the condition of the constancy (i.e., independence of field strength) of the mean square length of a chain element.

In the case of a quadrupole field, for a homogeneous chain it is convenient to consider the parameters and properties of the model as functions of S (instead of treating them as functions of the field strength).

The effective GSC-potential is equal in general to the sum of the internal chain potential U_{int} and the external (or molecular) field U_{ext}

$$U_{\text{eff}} = U_{\text{int}} + U_{\text{ext}} \; . \tag{7}$$

In the case of quadrupole field which is itself a quadratic function of the coordinates, U_{eff} may be presented in the most general form by expression

$$U_{\text{eff}} = (1/2)K_1 \sum_j u_j^2 + (1/2)K_2 \sum_j (v_j^2 + w_j^2) - E_1 \sum_j u_j u_{j+1} - E_2 \sum_j (v_j v_{j+1} + w_j w_{j+1} \; . \tag{8}$$

The set of second moments of this quasielastic GSC chain model is predetermined by S and by the mean size of the "rigid" chain element l (which is independent of external field or S)

$$l^2 = \langle u_j^2 + v_j^2 + w_j^2 \rangle = (1/N) \sum_{j=1}^{N} \langle u_s^2 + w_s^2 + v_s^2 \rangle \; . \tag{9}$$

The distribution of chain conformations in the symmetry axes becomes anisotropic, and hence two correlation parameters $\eta_\parallel$ and $\eta_\perp$ are defined and given by the equations

$$\eta_\parallel = \langle u_j u_{j+1} \rangle / \langle u_j^2 \rangle \tag{10}$$

$$\eta_\perp = \langle v_j v_{j+1} \rangle / \langle v_j^2 \rangle = \langle w_j w_{j+1} \rangle / \langle w_j^2 \rangle \; . \tag{11}$$

For an isotropic chain ($S = 0$) we have $\eta_\perp = \eta_\parallel = \eta$, where $\eta = \langle \cos(\vec{e}_j \vec{e}_{j+1}) \rangle$ is the mean cosine of the angle between neighboring chain elements. The initial persistent chain length a (or the length of the statistical segment $A = 2a$) is given by expression

$$A = 2a = 1[1+\eta)/(1-\eta)] \; .$$

According to Eqs. (10) and (11) for a chain in an anisotropic oriented system it is natural to introduce two persistent lengths

$$a_\parallel = (1/2) l [(1+\eta_\parallel)/(1-\eta_\parallel)] \; , \quad \text{and} \tag{12}$$

$$a_\perp = (1/2) l [(1+\eta_\perp)/(1-\eta_\perp)] \; . \tag{13}$$

In this case for the mean square of the length of the chain $\langle h^2 \rangle = \langle h_x^2 \rangle + 2\langle h_y^2 \rangle$, taking into account that $\langle u^2 \rangle = (1/3) l^2 (1+2S)$, and $\langle v^2 \rangle = (1/3) l^2 (1-S)$, one obtains the equation

$$\langle h^2 \rangle = (1/3) l^2 \times \left[(1+2S) \frac{1+\eta_\parallel}{1-\eta_\parallel} + 2\,(1-S) \frac{1+\eta_\perp}{1-\eta_\perp} \right] \; . \tag{14}$$

If the values of l^2 and S are fixed and the values of $\eta_\parallel$ and $\eta_\perp$ are also obtained by considering the corresponding statistical model (e.g., from lattice models) then the parameters K_1, K_2, E_1 and E_2 for the "effective" GSC model will be estimated from Eqs. (9), (10), (11). The averaging $\langle\,\rangle$ in these equations is performed with the equilibrium distribution function

$$\Psi(\{u_s, v_s, w_s\}) = \mathbb{C} \exp\left[-(V_{\text{eff}}(S)/kT)\right] \; . \tag{15}$$

For LC systems the dependence of S on the interaction parameter (α/kT), the dependence of the free energy of the system on S, and the types of phase transitions ("the isotropic state – LC") are obtained from the condition of self-consistency (Maier-Saupe approach [1–3, 5]),

$$S = \langle P_2(\cos\theta)\rangle = \int \Psi(l, S|\{u_j, v_j, w_j\}) \times P_2(\cos\theta)\, d\{uvw\} \ . \qquad (16)$$

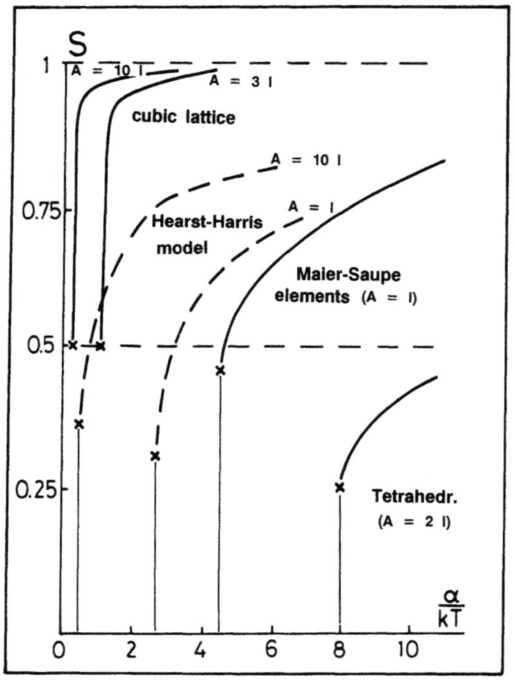

Fig. 4. Degree of order S vs parameter of orientational interaction α/kT for different statistical models and at different initial thermodynamic rigidities

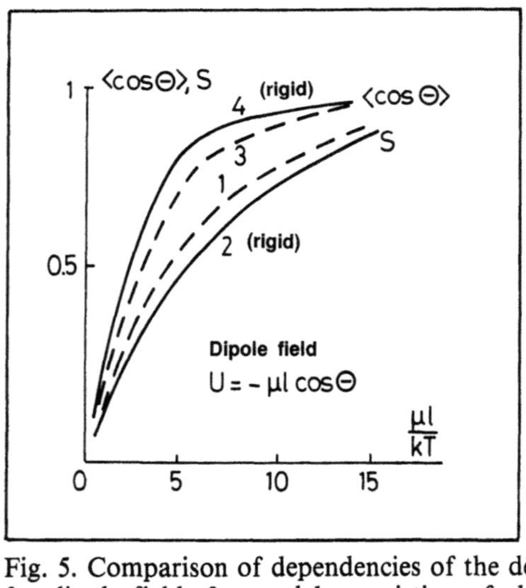

Fig. 5. Comparison of dependencies of the degree of order for dipole fields for models consisting of absolutely rigid elements and elements "fixed on the average"

It is natural that the form of the dependencies $S(\alpha/kT)$ will be determined by the type of the statistical chain model used. Figure 4 shows an example of the dependence $S(\alpha/kT)$ obtained by the above method for several types of lattice chain models if the order axis coincides with one of the main directions of the lattice symmetry axes and for a freely jointed chain model with a fixed "average" element length.

For a freely jointed chain it is possible to compare the values of S or $\langle h^2\rangle/h_{max}^2 = \langle h^2\rangle/L^2$ calculated on the basis of the proposed model with "fixed in average" elements, with those calculated with the aid of a model consisting of absolutely rigid elements. The quantities S and $\langle h^2\rangle/L^2$ are considered as functions of the magnitude of the quadrupole fixed. This comparison indicates the adequacy of the approximation used. Just as good an agreement was obtained for the chain influenced by the vector field $(U = -\mu l \sum_j \cos\theta_j)$. For a freely-jointed chain consisting of rigid elements the behavior in this field is described by the well-known reciprocal Langevin function (Fig. 5).

When the dissipation function $\boldsymbol{R}$ is obtained, the method applied previously in [6] is used. The dissipation function of the chain is summed up additively from those of individual rigid or flexible elements. The coefficients of the dissipation function of an individual element are expressed in turn through the coefficients of rotational and translational diffusion. The values of these coefficients may be obtained from the hydrodynamic models for these elements (a dumbbell with the centers of viscous resistance at the ends, an ellipsoid, a cylinder, etc.).

The proposed dynamic model describes local motions of chain elements which include the reorientation by large angles with respect to the axis of the external field (or the molecular field in LC). Therefore, by analogy with the corresponding approach for small molecules in LC systems it is assumed that for quadrupole fields the coefficient of "longitudinal rotational diffusion" contains the exponential factor $\exp[-U^*(S)/kT]$ depending on the magnitude of the barrier $U^*(S)$ of the orienting (external or internal field).

The coefficient of "transverse" rotational diffusion weakly depends (in the first approximation) or does not depend on the field (or on the degree of order S).

In the simplest case the dissipative function of the polymer chain immersed in anisotropic viscous matrix is given by the equation

$$R \doteq (1/2) \sum_j \{R_{1,\perp} (\dot{y}_j^2+\dot{z}_j^2)+R_{1,\parallel}\dot{x}_j^2 + R_{2,\perp} (\dot{y}_j\dot{y}_{j+1}+\dot{z}_j\dot{z}_{j+1})+R_{2,\parallel}\dot{x}_j\dot{x}_{j+1} \tag{17}$$

where $\dot{x}_j, \dot{y}_j$ and $\dot{z}_j$ are the coordinates of the "junctions" of the model chain. $\dot{x}$ are the coordinates along the selected axis of the quadrupole field (along the director in the LC).

The present paper reports the results for a chain without internal friction. However, in the general case internal friction may be included in the consideration through the modification of the dissipation function (see [4]).

The coefficients of the dissipation function are given by

$$\begin{aligned} (R_{1,2})_\parallel &= (\zeta/2)/[1 \pm (2K_1/D_{\mathrm{rot},\parallel}\zeta)] \\ (R_{1,2})_\perp &= (\zeta/2)/[1 \pm (2K_2/D_{\mathrm{rot},\perp}\zeta)] \\ \zeta = R_{1,\parallel}+R_{2,\parallel} &= R_{1,\perp}+R_{2,\perp} = (kT/D_{\mathrm{transl}}) \ , \end{aligned} \tag{18}$$

where ζ is the mean translational friction coefficient and $(D_{\parallel,\perp})_{\mathrm{transl}}$ and $(D_{\parallel,\perp})_{\mathrm{rot}}$ are the coefficients of translational and rotational diffusion of chain elements, respectively.

For a rigid dumbbell model of a chain element we have

$$(D_{\mathrm{transl}}/D_{\mathrm{rot}}) = (l^2/4) \ . \tag{19}$$

In the general case we have

$$(D_{\mathrm{transl}}/D_{\mathrm{rot}}) \sim l^2 F \ ,$$

where F is the form-factor depending on the shape, geometry, and distribution of viscous resistance along the element.

The ratio $D_{\mathrm{rot},\parallel}/D_{\mathrm{rot},\perp}$ is determined by the factor describing the overcoming of an additional barrier of the local (or external) field

$$D_{\mathrm{rot},\parallel} \sim D_{\mathrm{rot},\perp} \exp\left[-\frac{U^*(S)}{kT}\right] \ . \tag{20}$$

The equations of chain motion at given effective potential U_{eff} [see Eq. (1)] and the dissipative function R (Eq. 17) are obtained from the usual relationships

$$(\partial U_{\mathrm{eff}}/\partial \xi_\alpha)+(\partial R/\partial \dot{\xi}_\alpha) = -f_\alpha \ , \tag{21}$$

where $\xi_\alpha(x_\alpha, y_\alpha, z_\alpha)$, and f_α are additional (electric, magnetic, and mechanical) forces not included in the external (or internal) fields discussed above. It follows from the solution of system (21) that two different sets of equations of motion are developed for dynamic variables x_j, y_j, z_j (or $u_j = x_{j+1}-x_j$; $v_j = y_{j+1}-y_j$; $w_j = z_{j+1}-z_j$). Correspondingly, two types of relaxation spectra ("longitudinal" and "transverse" spectra) and two types of normal modes $Q_\xi(k) \sim \sum_p \xi_p e^{-ipk}$ (where $\xi_p = \{u_p, v_p$ or $w_p\}$ and relaxation times $\tau_\parallel(k)$ and $\tau_\perp(k)$, accordingly, appear in the relaxation properties of macromolecules in the LC or in the oriented by external fields state. Here k is the one-dimensional wave vector (Fig. 6).

Relaxation times $\tau_\parallel(k)$ and $\tau_\perp(k)$ exhibit different dependencies on the thermodynamic rigidity of the chain (i.e., on initial and "induced" final rigidity).

The final thermodynamic "transverse" or longitudinal rigidities, in turn, depend on the field (or the degree of order), on the value of the additional barrier

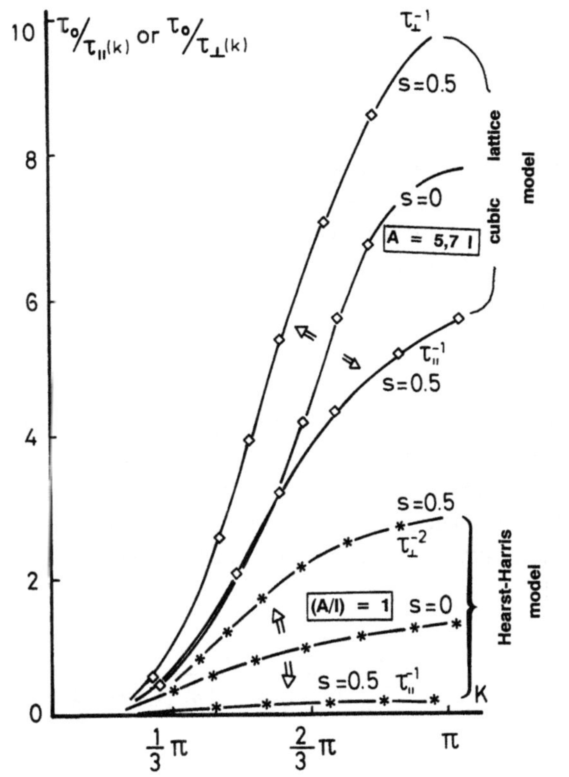

Fig. 6. Dispersion curves $\tau(k)$ for longitudinal and transverse relaxation processes for different statistical chain models

induced by the external field, and on the shape and characteristic scale of the normal mode $Q(k)$ (i.e., on the wave number k). The dependencies $\tau(k)$ for a simplest chain model consisting of "dumbbell" elements may be reduced to

$$\frac{\tau_0}{\tau_\parallel}=\frac{(1+\eta_\parallel^2-2\eta_\parallel \cos k)(1-\cos k)}{(1-\eta_\parallel^2)(1+2S)[1+b_\parallel(1-\cos k)]} \quad (22)$$
$$\frac{\tau_0}{\tau_\perp}=\frac{(1+\eta_\perp^2-2\eta_\perp \cos k)(1-\cos k)}{(1-\eta_\perp^2)/(1-S)[1+b_\perp(1-\cos k)]}$$

where

$$b_\parallel \cong (\eta_\parallel/\eta_\perp) \sim \exp[(3/2)\alpha S] \quad (23)$$
$$b_\perp \sim [(1/2) \div (1/4)] \ .$$

Here α is the interaction parameter. The value $(3/2)\alpha S$ is proportional to the above additional barrier induced by the external or internal field.

For the smallest scale ($k \sim \pi$) and the highest frequency normal mode we have (for high $b_\parallel$)

$$\frac{\tau_0}{\tau_\parallel} \cong \frac{1}{b_\parallel}\frac{1}{1+2S}\frac{1+\eta_\parallel}{1-\eta_\parallel} \quad (24)$$
$$\frac{\tau_0}{\tau_\perp}=\frac{2}{1+2b_\perp}\frac{1}{1-S}\frac{1+\eta_\perp}{1-\eta_\perp} \ .$$

The longitudinal and transverse processes differ in both relaxation times and activation energies.

The value of τ_0 in Eq. (22) is the time scaling factor having the order of value of the time of rotational diffusion in chain elements in the absence of the field [see Eq. (18)]

$$\tau_0 \sim (\zeta l^2/kT) \ .$$

Intensities of relaxation times $J_\parallel(k)$ and $J_\perp(k)$ for normal modes of the same type (with the same k, i.e., relating to the same type of excitation or dipole moment distribution along the chain) also depend on k, the parameters of longitudinal and transverse rigidity (Fig. 7)

$$J_x = J_\parallel(k) = \mathbb{C}\,\frac{1+2S}{3}\,\frac{1-\eta_\parallel^2}{(1-\eta_\parallel)^2+2\eta_\parallel(1-\cos k)}$$

$$J_y = J_z = J_\perp(k)$$
$$= \mathbb{C}\,\frac{1-S}{3}\,\frac{1-\eta_\perp^2}{(1-\eta_\perp)^2+2\eta_\perp(1-\cos k)} \ . \quad (25)$$

For example, if we consider dielectric relaxation in the oriented anisotropic LC polymer if the perturbing electric field is directed along the order axis or the normal direction, then the corresponding dielectric permittivities or, more precisely, the Fourier components for $\varepsilon_\parallel$ and $\varepsilon_\perp$ are $\varepsilon_\parallel(k) \sim J_\parallel(k)$ and $\varepsilon_\perp(k) \sim J_\perp(k)$. The ratios $J_\parallel/J_\perp$ (or $\varepsilon_\parallel/\varepsilon_\perp$) will depend on $\eta_\parallel, \eta_\perp$, and k.

For statistically flexible chains ($\eta_\parallel = \eta_\perp = 0$) we have $(J_\perp/J_\parallel) \sim (1-S)/(1+2S)$. In the case of semiflexible chains the result depend on the choice of the statistical chain model and the value of the field-induced chain stiffening which, in turn, depends on the initial chain rigidity for a given model [3, 5].

In the analysis of relaxation processes of dielectric, NMR, and EPR relaxations and, in particular, the processes occurring in chains with luminescent markers, the orientational relaxation of the selected chain element is manifested. This type of relaxation is also closely correlated with the chain response by the random distribution of values of longitudinal components of dipole moments of units.

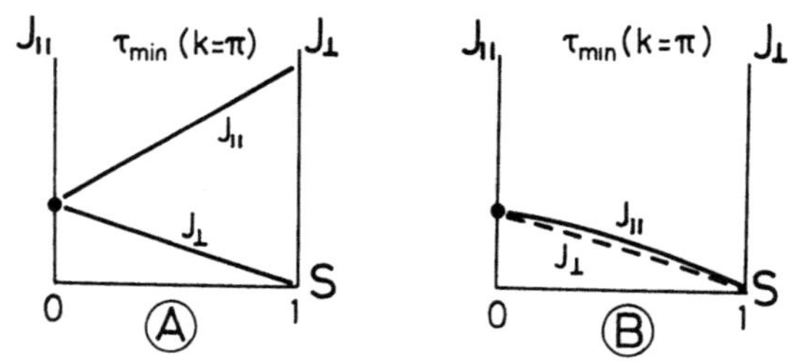

Fig. 7. Intensity of transverse and longitudinal processes degree of order. A) Flexible chain without stiffening ($A/L = 1$); B) Semiflexible chain with induced stiffening (cubic lattice) ($A/L = 2$)

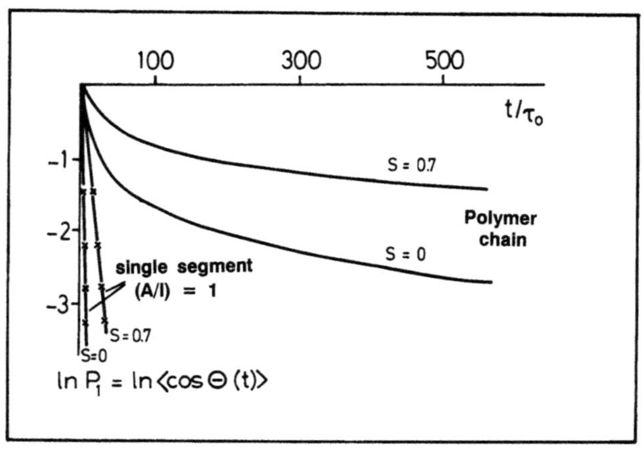

Fig. 8. Correlation function $P_1(t)$ for the rotational mobility of a selected chain element

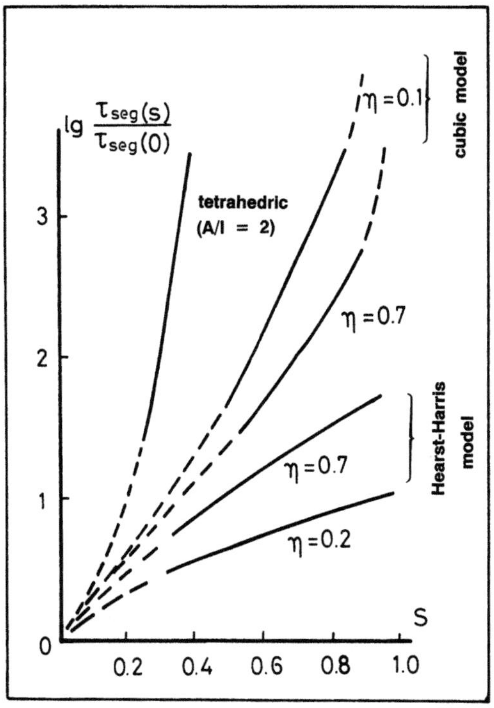

Fig. 9. Characteristic relaxation time of a statistical segment vs degree of order for statistical chain models with induced rigidity

The relationships of changes in the relaxation properties of the selected chain element during ordering are shown in Fig. 8, taking as an example the time evolution of the mean cosine of the rotation angle $\langle \cos\theta(t)\rangle = f(t)$ for a freely jointed chain. In the general case the dependence $\langle \cos\theta(t)\rangle$ is characterized by a broad spectrum of relaxation times with the participation of all normal modes.

If follows from the general theory of relaxation spectra for semistiff chains that the most probable time of the relaxation spectrum for P_1 ($\cos\theta$) is found to be close to the relaxation time of the normal mode of statistical segment (τ_{segm}). The scale of this mode ($\pi/k \sim A/l$) corresponds to the length of the statistical segment (persistent length) of the chain. For statistical models of a chain with a finite initial thermodynamic rigidity, the length of the statistical segment increases with the field (or with the degree of order). In other words, the induced stiffening is an important factor leading to an increase in the most probable relaxation time in $P_1 = \langle \cos\theta(t)\rangle$ (or in τ_{segm}) (Fig. 9).

It should be noted that relaxation times $\tau_{\|}(k)$ usually increase with quadrupole field (or with ordering), both because it is necessary to overcome the field barrier (or the molecular field in LC) and because the chain becomes more rigid as the segment becomes longer. The latter factor is important for normal modes, the size of which is longer than that of the statistical chain segment.

The times $\tau_{\perp}(k)$ decrease with increasing order. The details of the behavior of $\tau_{\|}(k/S)$ and $\tau_{\perp}(k/S)$ depend on the values of the parameters $\eta_{\|}(S)$, $\eta_{\perp}(S)$, S, and $b_{\|}(S)$.

The dynamic model and the corresponding types of mobility considered in the present study refer to much systems in which, under the influence of the external quadrupole field (or with LC ordering), strong orientation or selfordering appear but segmental mobility is retained. Such systems are polymer LC, oriented polymers in the highly elastic state, strained amorphous regions in semi-crystalline polymers at temperatures above their glass transition temperatures.

Finally, the local mobility in strongly oriented systems may be manifested in special types of experiments, e.g., in the study of polarized luminescence (or NMR and EPR) of macromolecules in a strong flow and in other similar situations.

In this study such types of motions are considered during which strong local mutual rearrangements (reorientations with respect to the order axis) take place. Apart from them, in multichain system with a strong interchain orientational interaction, bending or torsional chain motions are possible, that are accompanied by weak mutual orientational displacements or distortions and slight rotations with respect to the axis of the local order.

These motions may also be regarded in the single-chain approximation (e.g., by using the Maier-Saupe molecular field method) by "forbidding" large reorientations in this field. A more rigorous approach involves the evaluation of cooperative multichain motions with slight internal distortions (in the molecular coordinate systems).

On the contrary, the rotations by relative large angles in the laboratory coordinate systems may become possible. However, when the long-range order is retained, these large-angle rotations are very rare. A similar approach has been developed in [7–10], and the complete description of the local dynamics in oriented systems could doubtless include a combination of the above large-angle types of motion and cooperative multichain distortion motions.

Almost all experimental data, e.g., those on dielectric relaxation [11], refer either to low molecular weight LC or to LC polymers with mesogenic groups in side chains [12, 13]. It follows from these data that at least two types of relaxation processes occur ("transverse" and "longitudinal" processes) in the formation of the LC state as compared to a single process of this type in an isotropic polymer at higher temperatures.

In recent years many papers have been published on the study of the dynamics of LC polymers with mesogenic groups in the main chain and flexible spacers between them by NPR methods [14–16]. The results of the present theory are also applicable to these polymers. However, a sufficiently adequate comparison is possible only if the dynamic heterogeneity of real systems is taken into account.

In this papers the existence of longitudinal and transverse (with respect to the order axis) relaxation processes are also shown for the elements of mesogenic groups and for spacer units.

A more detailed comparison with the theory also requires the introduction into the theory of the elements normal to the chain back-bone and the consideration of the dynamics of these elements in oriented chains.

Further development of the theory should take into account additional factors which have not been considered in the present study:

1) The change in the density, in the distribution of the local free volume, and in microviscosity on passing from the isotropic to the oriented state (in LC transitions or under the influence of external orienting fields).
2) The change in molecular packing, local steric and orientational interactions on the molecular level.
3) The change in the types of kinetics units and mechanisms of molecular mobility in ordered state.
4) The explicit evaluation of the cooperative motion of segments of neighboring chains (i.e., taking into account the multichain motions).
5) The evaluation of larger scale interchain steric limitations (entanglements or the tube effect) and taking into account the reptational motion.

The solution of this problem on the molecular level may be facilitated by computer simulation of motions by using the methods of molecular and Brownian dynamics.

References

1. Maier W, Saupe A (1958) Z Naturforsch 13A:569; (1959) ibid 14A:882
2. De Gennes PG (1974) The Physics of Liquid Crystalls. Clarendon Press, Oxford
3. Rusakov VV (1987) Thesis, Perm
4. Gotlib YY, Darinskii AA, Svetlov YE (1986) Fizicheskaya Kinetika Makromolekul (Physical Kinetics of Macromolecules), Ed "Khimiya", Leningrad
5. Gotlib YY (1988) 8th Europ Symp Polym Spectr, Abstract 37–38, Budapest
6. Gotlib YY, Darinskii AA, Klushin LI, Neelov IM (1984) Acta Polymerica 35B:124
7. Maksimov AV, Gotlib YY, Baranov VG (1987) Vysokomol Soedin 26B:822
8. Maksimov AV, Gotlib YY, Baranov VG (1984) Vysokomol Soedin 26A:2512
9. Gotlib YY, Baranov VG, Maksimov AV (1987) Vysokomol Soedin 29A:2620
10. Maksimov AV, Gotlib YY (1988) Vysokomol Soedin 30A:1411
11. Kresse H (1983) "Dielectric Behavior of Liquid Crystals" in Adv Liquid Crystals 6:109–172. Academic Press, New York
12. Parneix JP, Njeumo R, Legrand C, Le Barry P, Dubois JC (1987) Liquid Crystals 2:167
13. Attard GS, Moira-Ramos, Williams G (1987) J Polymer Sci, Polym Phys Ed 25:1099
14. Laupretre F, Nöel C, Jenkins WN, Williams G (1985) Faraday Disc Chem Soc 79:191
15. Müller K, Meier P, Kothe G (1985) Progress in NMR 17:211
16. Müller K, Schleicher A, Kothe G (1987) Mol Cryst Liquid Cryst 153:117

Received February 13, 1989;
accepted July 20, 1989

Author's address:

Prof. Dr. Y. Y. Gotlib
Institute of Macromolecular Compounds
of Academy of Sciences of the USSR
Bolshoi Pr. 31
199004 Leningrad, USSR

Discussion

HOLZMÜLLER:

I am interested in the average dipole moment for oligomers related to one monomeric unit. Starting with one monomer, the second (dimere) will favor the opposite direction, because it is influenced by the internal electrical field of the first, and so on. So I expect a sinusoidal function with an exponential damping factor complicated by the helical structure of the chain. Have you calculated this model?

GOTLIB:

We have calculated the relaxation times for the chain in quadrupole field – or in the nematic liquid crystal – for all wave vectors k. The value for $k = \pi$ can be attributed to the alternating dipole moments along the chain that can be applied to the systems mentioned in your question.

ROTH:

In experimental studies we observe large differences between liquid crystal polymers having the mesogenic groups in the side chain or main chain, respectively. In your lecture you introduced a theoretical treatment in which you do not distinguish between the two types of LC polymers. At the end of your lecture you discussed experimental results both with side chain mesogenic group as well as with the mesogenic group in the main chain. Is it really not necessary to differ in the theoretical treatment between the two types of LC polymers?

GOTLIB:

Our theoretical treatment is made only for the chain with mesogenic groups arranged in the main chain. The statistics and dynamics of our model is therefore rigorously applied only to such type of chains. But any types of results – e.g., the splitting of relaxation spectra, the induced stiffening – may be applied qualitatively to side-chain LC-polymers. But in this case we must introduce special transverse rigidity – only for correlations between side chains. Also the effective dissipative function, for side chain, will have another form. Consequently the results for the side-chain polymer must differ from those of main-chain polymers.

Electron spin resonance and relaxation studies on conducting poly(tetrathiafulvalenes)

H.-K. Roth[1]), H. Gruber[1]), G. Völkel[2]), W. Brunner[2]), and E. Fanghänel[3])

[1]) Leipzig University of Technology, Department of Natural Sciences, Leipzig, GDR
[2]) Karl Marx University Leipzig, Department of Physics, Leipzig, GDR
[3]) Merseburg Technical University "Carl Schorlemmer", Department of Chemistry, Merseburg, GDR

Abstract: Magnetic properties and the relaxation behavior of paramagnetic centers in various poly(tetrathiafulvalenes) with semiconductor properties have been studied by electron spin resonance (ESR) and electron spin echo (ESE) in the temperature range of 2 to 400 K. At room temperature the iodine doped polymers have a DC conductivity between 10^{-9} S cm^{-1} and 10^{-4} S cm^{-1} that increases with temperature.

The existence of two types of paragmagnetic centers is confirmed by ESR line shape simulation and by saturation experiments. The electron spin echo registration of ESR spectra and the measurement of relaxation times allow the separate study of the temperature dependence of the paramagnetic centers with relaxation times longer than 300 ns. The relations between the spin concentration and the conductivity of the samples are discussed by means of a model based on polarons, bipolarons, and trapped radicals.

Key words: Conducting polymers; electron spin resonance; electron spin echo; doping with iodine; poly(tetrathiafulvalene); polarons and bipolarons; radicals

Introduction

Electron spin resonance (ESR) studies and electron spin echo (ESE) investigations have been carried out on various poly(tetrathiafulvalenes) synthesized earlier [1, 2]. The TTF units are linked via a 1,4-phenyl group or via a 1,4,7,8 tetrahydroanthracene bridge. The polymers are doped with iodine by a treatment with the dopant in chloroform. As has been shown in a previous paper [3] the ESR spectrum of the doped polymers of poly(tetrathiafulvalenes) abbreviated in the following to PTTF, appears as a superposition of single lines without hyperfine splitting. Studies on PTTF show that in some doping areas the increase in the number of unpaired spins formed by iodine doping is associated with an increase in the ESR line width and in the conductivity up to a limiting value. These values depend on the molecular structure of the PTTF investigated.

In this paper the studies of the three series of differently doped 1,4-phenylene-bridged poly(tetrathiafulvalenes) A.1, A.2 and A.3 with substituents R = H, R = CH_3 and R = CH_2CH_3, respectively, and of one series of differently doped poly(1,2,5,6-tetrahydroanthra-tetrathiafulvalene) are continued.

Previous studies of the DC conductivity and of the radical concentration led to the following results [3]:

– By doping with iodine the highest conductivity can be achieved at those polymers having a more or less plane structure of "monomer" units (polymers A.1 and B.1).

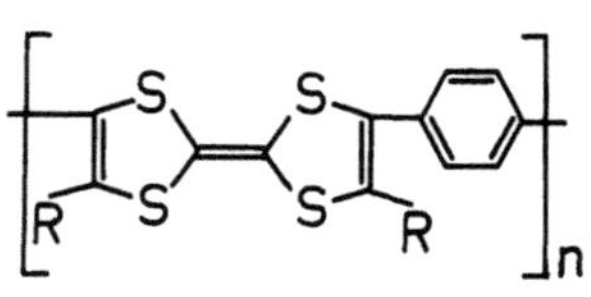

A

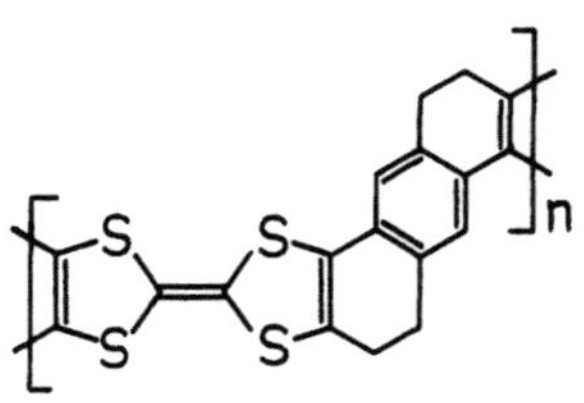

B.1

A.1	R = H	n = 25
A.2	R = CH_3	n = 25
A.3	R = CH_2CH_3	n = 32
B.1		n = 20

- In the case of medium dopant concentration a correlation exists between the increase in spin concentration and the increase in conductivity. The highest radical concentration (10^{20} spins per g), however, could not be observed in polymers with the highest conductivity but in the samples A.2 and A.3, being of a slightly twisted molecular structure due to the steric hinderance between the substituents. In the polymers with plane structure the maximum spin concentration amounts only $4 \cdot 10^{18}$ spin/g.
- In the undoped state the samples A.2 and A.3 show a strongly asymmetric powder spectrum. Doping them with a small amount of iodine effects an increase in conductivity without increasing the spin concentration. But a drastic change in the line shape referring to a change of the radical species is observed.
- All the samples contain a certain amount of free radicals. We know there are different types of radicals, not all of them contributing to conductivity. However, there are no clear relations between radical concentrations, radical properties and conductivity.

The aims of the present paper are:

1. A more detailed characterization of the various paramagnetic centers. Radicals with various magnetic and electric properties have to be distinguished.
2. Relations are to be found between the concentration of the different radicals and the conductivity to be measured for the samples.
3. The connections between molecular structure, radical structure and conductivity have to be investigated. It will have to be tested whether the experimental results at PTTF are explainable by means of the polaron and bipolaron picture as has been done, for example, with doped poly(thiophene) [4] and with poly(p-phenylene) [5].

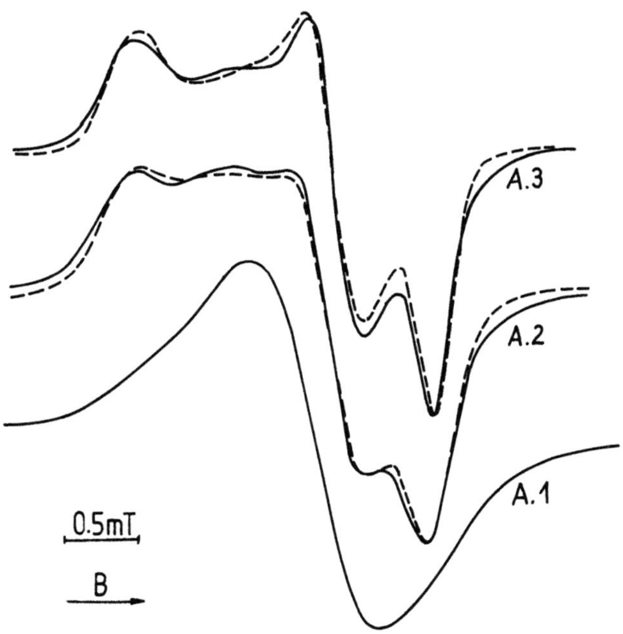

Fig. 1. ESR spectra of undoped samples of the polymers A.1 (R = H), A.2 (R = CH_3), and A.3 (R = CH_2CH_3) recorded at room temperature and by a microware power of 1 mW. The dashed spectra are simulated as a superposition of an asymmetric spectra of radicals. R_i and of a symmetric one of radical R_2. B_i magnetic induction

ESR on undoped PTTF

The ESR spectra of poly(tetrathiafulvalenes) do not show a hyperfine splitting but consists of a superposition of single lines as can be seen in Fig. 1. In the case of the undoped PTTF with CH_3 and C_2H_5 substituents (polymers A.2 and A.3) the predominant ESR signal is asymmetrical, whereas in the unsubstituted PTTF (polymers A.1 and B.1) the predominant signal is almost symmetrical as it is the case with all medium and highly doped PTTF [3].

The spectra of undoped and weakly-doped PTTF are a superposition of a strongly asymmetric spectrum of radicals R_i representing immobile radicals, to be discussed later, and a symmetric spectrum caused by radicals which were called R_2 in our previous paper [3]. We assume the R_2 to be mobile cation radicals or polarons. The latter type of radical has a g-factor between g = 2.0068 (B.1) and g = 2.0078 (A.1) depending on the molecular structure of the PTTF [3]. These values are in agreement with the g-factors of cation radicals in "monomer" TTF [6]. The PTTF A.1 and B.1 have a more or less plane structure. Due to this structure an electronic interaction occurs between the TTF units in the polymer chain which leads to a low oxydation potential of the polymer, so that in the case of an oxygen contaminated planar PTTF predominantly, an electron transfer from the PTTF system to the $^3O_2$ occurs, forming a TTF radical cation R_2 that has the same structure as a radical cation formed by doping with iodine [3].

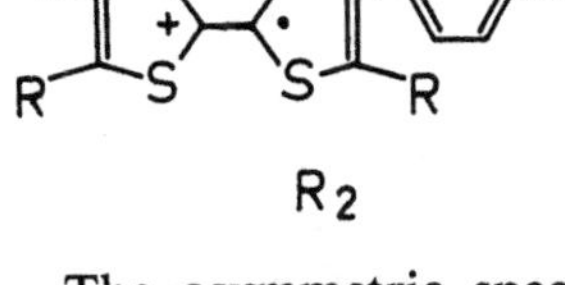

R_2

The asymmetric spectrum predominant in PTTF A.2 and A.3, assigned to radicals R_i, has a typical powder pattern caused by an anisotropic g tensor with a symmetry lower than axial. The mean g-factor is $g = 2.0081$. The value refers to radicals with a large electron spin density of the unpaired electrons near sulphur atoms. g-factors of $g \approx 2.008$ were measured, for example, ón stable radicals of dithiazetyl salts [7] and it was related to a localization of the unpaired spins near sulphur. In the low temperature region the authors also observed asymmetrical spectra in rigid solvents. They refer these spectra of the sulphur and nitrogen containing radicals to the anisotropic hyperfine structure tensor [7]. Such an interpretation of the asymmetric spectrum is nearly impossible in the case of the PTTF because there are no atoms with nuclear spins in the near neighborhood of the sulphur atoms.

The reason for the asymmetric spectrum of the immobile radicals R_i is supposed to be predominantly the g-factor anisotropy. The simulation of the measured spectra of Fig. 1 by superposition of an asymmetric spectrum with $g_x = 2.0147$, $g_z = 2.0067$, and $g_y = 2.0028$, and a symmetric one of R_2 with $g = 2.0071$ yields that the paramagnetic centers of the undoped samples consist (at least to 10%) of the radicals R_i. In the samples A.2 and A.3 the portion of R_i radicals even amounts to 54% and 75%, respectively. As has been discussed in [3], the radicals which effect the asymmetric spectra in the undoped samples are due to the interaction or the reaction of oxygen from the air with TTF macromolecules forming the immobile radicals R_i. The structure of R_i can be realized by homeopolar bond between the oxygen and a sulphur atom of the TTF ring (forming a radical R_1 [3]) or by an ion pair interaction between the cation radical R_2 and the counterion O_2-forming R_2^T, that is, a trapped cation radical. A possible structure of R_1 can be represented by the following structure [3].

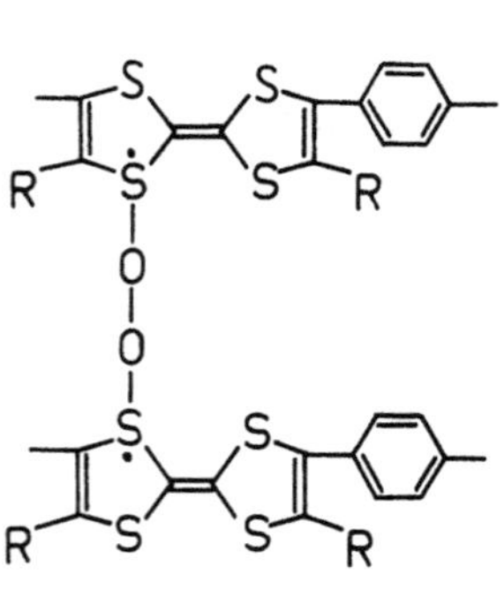

R_1

Studies on various polyacetylenes have shown that oxygen molecules or ions via chemical reactions and via physical interaction with unpaired electrons effect a localization, trapping or the immobilization, respectively, of paramagnetic centers [8–10]. In the case of radicals in polymers with carbon atom backbone like polyacetylene the trapping effects only a line broadening because the spin orbit coupling which may effect the g-factor anisotropy, is (with $\lambda_c = 25\,\mathrm{cm}^{-1}$) very small. Because sulphur has a relatively large spin orbit coupling (with $\lambda_s = 382\,\mathrm{cm}^{-1}$), an asymmetric spectrum is to be expected, in the case of trapped cation radicals R_2^T. Nevertheless, it cannot be to excluded that the observed asymmetric spectrum is partly due to radicals from the type R_1. That is, we assume the immobile radicals R_i can be peroxy radicals R_i and R_2^T. Further investigation will allow an unambiguous interpretation of the asymmetric spectrum.

As has been mentioned the recorded ESR spectra of the undoped PTTF samples are superpositions of an asymmetric and a nearly symmetric one belonging to immobile radicals R_i and mobile radicals R_2, respectively. The superposition of R_i- and R_2-spectra measured is experimentally testable by ESR by saturation of one of the spectra with increasing microwave power.

Figure 2 shows a series of partly saturated spectra. Comparing the three spectra of the sample A.3 – recorded at various microwave powers – with the spectrum of the sample A.1 which is caused nearly exclusively by mobile cation radicals, it is to be seen that the portions of the anisotropic spectrum of the localized radicals in the superposed spectrum decreases by increasing the microwave power.

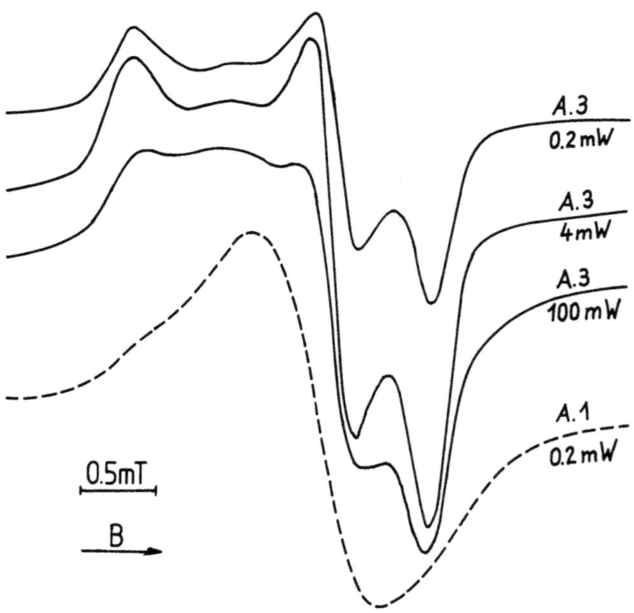

Fig. 2. Influence of the microwave power of the signal shape of ESR spectra of the undoped sample A.3 at room temperature, comparison with the undoped sample A.1. The B_1-field amount about 0.03 mT at 100 mW

The circumstance that at first the asymmetric spectrum of immobile radicals becomes weaker by saturation shows that the relaxation times of the mobile cation radical are shorter than that of the localized radicals. The faster relaxation of the more mobile cation radicals R_2 comes up to our expectations because the mobility of the paramagnetic charge carriers effects additional relaxation mechanisms such as spectral diffusion, spin diffusion, and exchange interactions.

Especially, the different phase memory times of the localized and mobile radicals allow the separation by ESE and relaxation studies of the radicals with longer relaxation time. We took this opportunity in connection with the study of weakly doped PTTF.

ESR, ESE, and conductivity studies on doped PTTF

ESR and conductivity studies

If PTTF is doped by addition of less than 1 iodine atom per monomer unit an increase in the DC conductivity is observed at each dopant concentration in all PTTF.

In the case of the samples A.1 and B.1 the increase is also valid for the radical concentration, though the radical concentration increases much less than the conductivity. The conductivity of B.1 changes by doping from $5 \cdot 10^{-8}$ S cm^{-1} to 10^{-4} S cm^{-1}, whereas the radical concentration increases only from $1 \cdot 10^{18}$ spin/g to $3 \cdot 10^{18}$ spin/g [3]. In the case of the samples A.2 and A.3 the increase of conductivity by doping with 0.2 or 0.4 iodine atoms per monomer unit is not connected with an increase in the radical concentration but it remains almost constant at about $2 \cdot 10^{18}$ spin/g [3]. The addition of iodine, however, effects of drastic change of the line shape as can be seen from Fig. 3, referring to a radical conversion.

This change is due to the fact that the PTTF samples already contain a certain number of paramagnetic species before doping with iodine, but most of them are localized radicals. At very low dopant concentration the iodine atoms effect mainly a conversion of the localized radicals to mobile cation radicals. This results in the increase of conductivity without increasing the total spin concentration. The mobile radicals R_2, as polarons, directly contribute to the conductivity or indirectly contribute by recombination to bipolarons, thereby losing their spin paramagnetism. In the case of low dopant concentration the line shape of the ESR spectra shows that there are still localized paramagnetic centers in the samples. The simulation of ESR spectra measured at room temperature has shown that the addition of 0.2 iodine atoms per monomer unit effects that the spectrum is more than 90% due to mobile R_2 radicals and, to less than 10%, due to localized radicals.

Conductivity measurements as a function of temperature demonstrate that the electrical properties of the PTTF depend on the dopant concentration and on the molecular structure but that there are no significant differences between the polymers A.2 and A.3. Figure 4 contains results from electrical measurements carried out on the polymer A.2 with a constant doping degree of 0.2 iodine atoms per monomer unit. The variation of temperature by 240 K changes the elec-

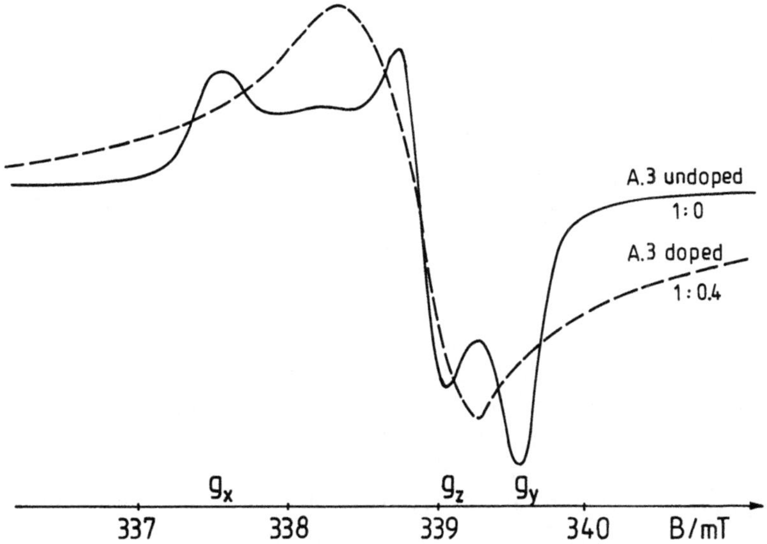

Fig. 3. Comparison of the line shape of spectra of an undoped and a doped sample of the polymer A.3. The spectrum of the undoped sample is about three times higher amplified than that of the doped sample

trical conductivity by five orders of magnitude (from $1.7 \cdot 10^{-10}$ S cm^{-1} at 87 K to $1.2 \cdot 10^{-5}$ S cm^{-1} at 327 K). The polymer A.3 shows a similar behavior.

With the help of these measurements the activation energy E_A of the conductivity was estimated. E_A amounts to 0.03 eV in the temperature region below 120 K and to 0.25 eV in the temperature region above 200 K. This points to two different processes of the electrical conductivity in the investigated temperature range.

The following questions arise from the temperature dependence of the conductivity:

- Is the ESR spectrum strongly dependent on temperature as is the case with the conductivity of the samples?
- What is the behavior of the localized or trapped radicals which contribute to the ESR-spectra with a portion of less than 10%?

Figure 5 shows spectra of the polymer A.3 doped by the addition of 0.2 iodine atoms per monomer unit measured at three different temperatures and normalized to equal amplitudes. Without normalization the signal intensities at 100 K are more than two times stronger than those at room temperature. The signal increase with decreasing temperature is two times stronger than the temperature dependence yielded by the Boltzmann factor.

The superposed ESR spectrum is shifted to the more symmetrical one with lowering temperature. Therefore, the increase of the signal amplitude with lowering the temperature should be governed by the

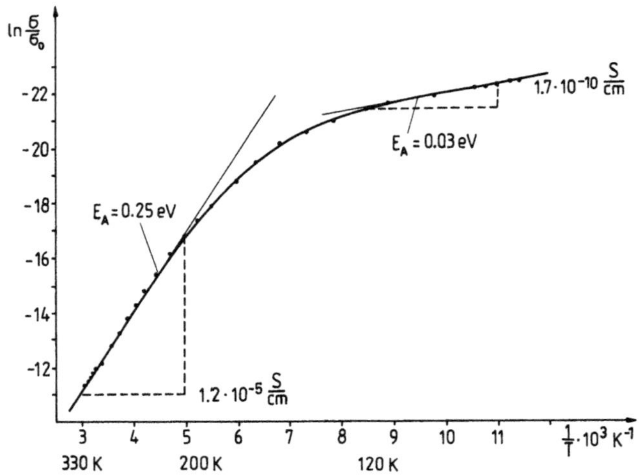

Fig. 4. Electrical conductivity of the sample A.2 doped with 0.2 iodine atoms per monomer unit as a function of temperature. σ_0 is in these diagrams 1 S cm^{-1}, and has no special physical content

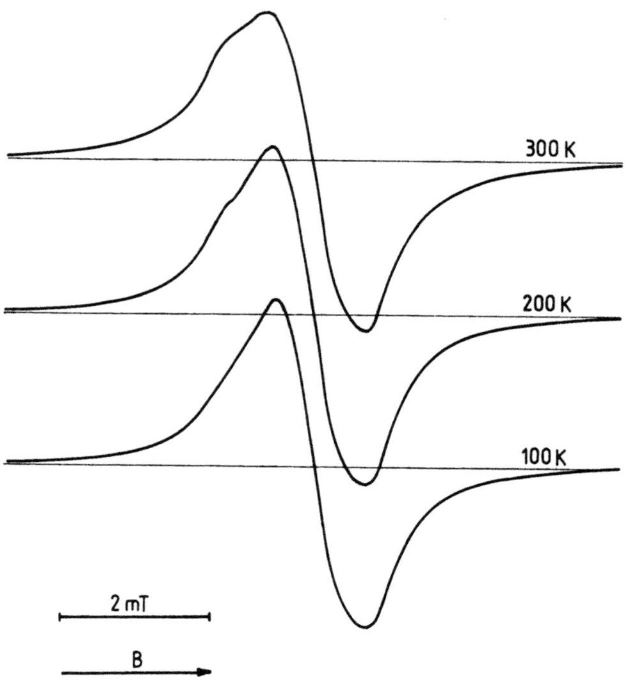

Fig. 5. Line shape comparison of the ESR spectra of the sample A.3 doped with 0.2 iodine atoms per monomer unit and measured at three different temperatures. The spectra are normalized to equal intensity

symmetric part of the spectrum caused by mobile cation radicals R_2 which, as polarons P^{+}, contribute to the electrical conductivity. The expected changes in the intensity of the asymmetric spectrum part – that is the temperature dependence of the concentration of the localized radicals – cannot effect a significant change in the line shape of the superposed ESR spectrum because the contribution of the localized radicals to the total intensity of the spectrum is too small.

ESE studies

The study of the temperature behavior of the localized radicals in weakly-doped PTTF requires the application of a spectroscopic method which allows the separate detection of the immobile radicals. For this purpose we used the electron spin echo (ESE) technique of ESR detection.

The ESE investigations were performed by means of a home-made ESE spectrometer described earlier [11]. It has successfully been applied to study motional and relaxation effects on trapped paramagnetic centers in O_2-contaminated polyacetylenes [9, 10]. With the help of this ESE equipment only those paramagnetic centers are detectable which have a phase memory time $T_M > 250$ ns. From the radicals in PTTF discuss-

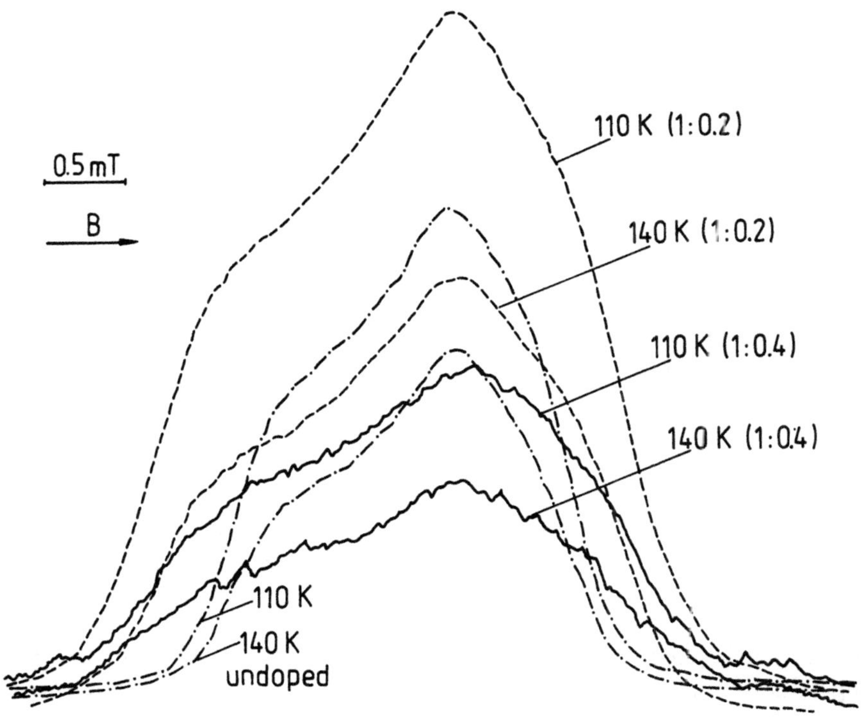

Fig. 6. ESE detected ESR spectra of undoped and weakly doped PTTF; sample A.3 at two different temperatures

ed up to now, only the localized or trapped radicals R_1 and R_2^T were detectable by means of this technique. The phase memory time of the mobile cation radicals R_2 is probably shorter than 100 ns and lies below the time resolution of the equipment used.

The ESE detected ESR spectra of the phenylene-bridged PTTF A.3 doped with 0.2 and 0.4 iodine atoms per monomer unit are presented in Fig. 6. The spectra show the typical powder pattern caused by the anisotropic g tensor of a center with a symmetry lower than axial. The magnetic field distances between the singular points of this powder pattern only approximately agree with those of the EPR spectrum of the immobile radicals trapped by interaction or reaction with oxyen (R_1, R_2^T/O_2^- which, predominantly determines the shape of the spectra A.2 and A.3 in Fig. 1 of the undoped PTTF. There are differences in the g-value anisotropy detectable by ESE of the immobile radicals in the undoped and weakly doped samples of A.3 (as is seen in Fig. 6). The anisotropic spectrum in iodine doped samples is probably due to localized radical R_2^T trapped by an interaction with the counter anion I_x^- (R_2^T/I_x^-).

Also from a chemical point of view it can be assumed that in the iodine doped samples the peroxy species cannot be present because I_x^- is a reducting anion and should destroy R_1 and R_2^T/O_2^-, respectively.

The intensity of the ESE detected powder pattern is strongly temperature dependent (see Fig. 6). The higher the temperature, the lower the concentration of the trapped or immobile radicals.

Because the ESR signal of the immobile radicals is a measure of the magnetic susceptibility which follows

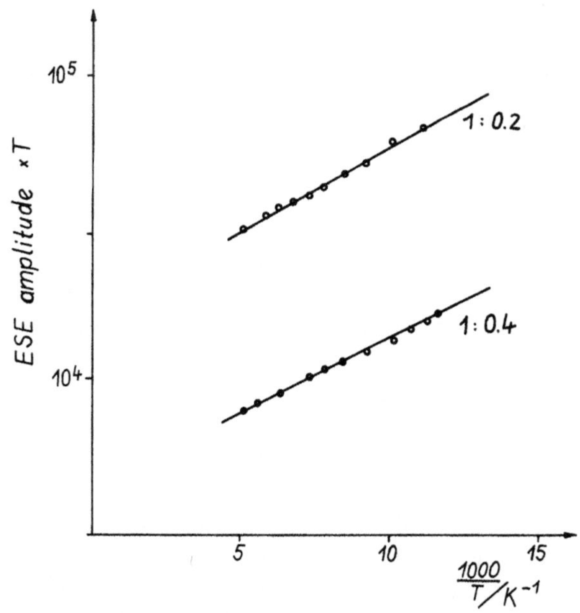

Fig. 7. ESE amplitude times T vs the reciprocal temperature for two weakly doped PTTF; sample A.3

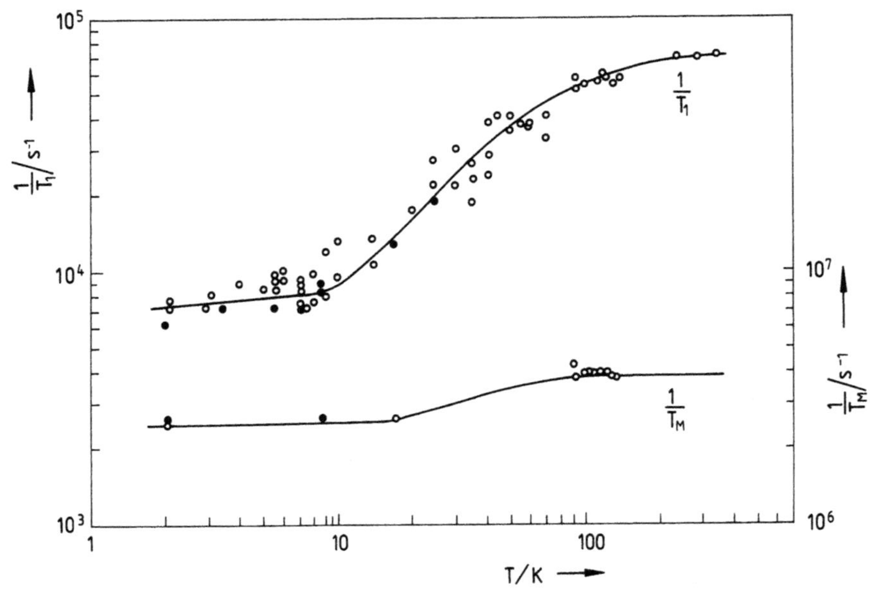

Fig. 8. Electron spin relaxation rates T_1^{-1} and T_M^{-1} as a function of the temperature on two weakly doped PTTF; sample A.3. The measuring points ○ belong to the doping degree 1:0.2 and the points ● to PTTF with 1:0.4

the Curie law the signal intensity was multiplied by the temperature T in Fig. 7. The resulting values, proportional to the relative number of immobile radicals, is plotted in a logarithmic scale vs the reciprocal temperature for two weakly iodine doped samples of A.3.

The number of immobile radicals N_R decreases with rising temperature according to

$$N_R = N_{RO} \exp(-E_A/kT)$$

with an activation energy $E_A = 0.01$ eV, equal for both the samples and independent on the portion of iodine in the sample.

The basic idea of interpretation is that with increasing temperature the R_2^T/I_x^- lose their trapping state and become mobile cation radicals R_2 (or polarons) with such a short relaxation time that they are undetectable by the ESE spectrometer applied, or they form bipolarons, thereby losing their spin paramagnetism. The activation energy E_A for the decrease of localized radicals is of the same order of magnitude as the E_A of increasing electrical conductivity in this temperature region ($E_A = 0.03$ eV). This points out that the radical conversion mentioned really contributes to the increase of conductivity measured.

Furthermore, measurements of the phase memory time T_M and of the spin-lattice relaxation time T_1 were performed in a temperature interval between 2 and 300 K on two doped PTTF samples A.3. The experimental results are presented in Fig. 8. It can be seen that the value of T_M^{-1} slightly increases at higher temperatures which refers to a higher mobility of the localized radicals. This is in good agreement with the measured changes of the line shape of the ESE detected spectra shown in Fig. 9. Concerning the T_1 results, quite an unusually weak temperature dependence can be observed. Usually, the spin-lattice relaxation rate T_1^{-1} in polymers depends linearly on the temperature below 20 K and behaves like T^2 at higher temperatures [12–14]. This behavior can be explained by a direct relaxation process of the paramagnetic radical centers with one lattice phonon at low temperature and by a Raman process at higher temperatures where two lattice phonons are involved. Our results are not explainable by means of the existing theory based on harmonic lattice motion.

The temperature independent part of the spin lattice relaxation rate below 10 K points out that special low-energetic lattice excitations do exist with energies of the order of that of the ESR transitions at 10 GHz. At higher temperatures the harmonic lattice vibration in the ordered micro areas of the samples are stronger excited and their interaction with the unpaired localized spins becomes stronger than that of the special low-energetic excitations. A combined interaction process of one lattice phonon and one localized anharmonic lattice excitation could also explain the linear increase of the spin-lattice relaxation rate T_1^{-1} with temperature T [12–14].

The circumstance that the relaxation times T_1 are independent on the iodine content of the two variously doped samples A.3 points out that the relaxation of the trapped centers is really due to interactions of the unpaired electrons with the polymer excitations and not due to cross-relaxations with the mobile cation radicals or polarons.

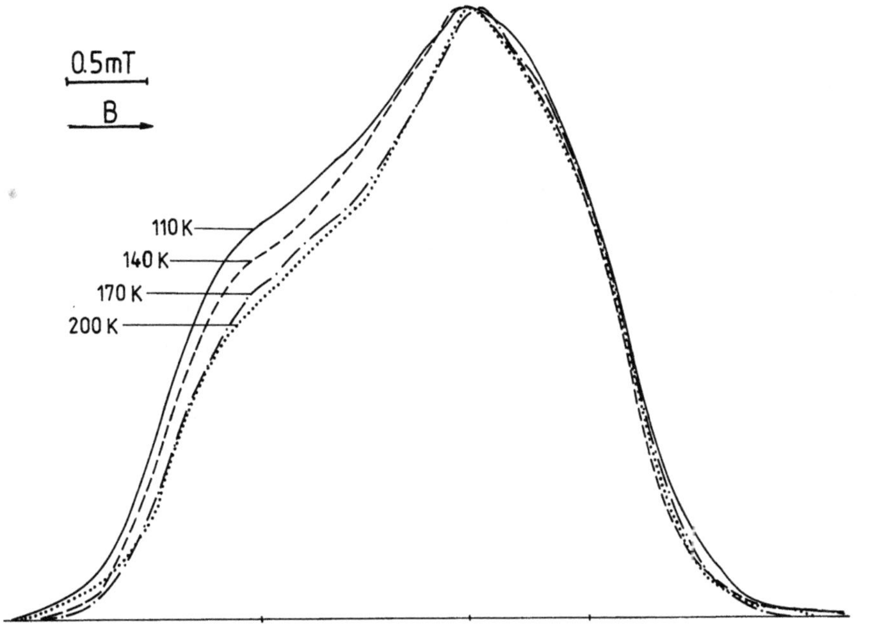

Fig. 9. Line shape comparison of ESE-detected ESR spectra of the sample A.3 doped with 0.2 iodine atoms per monomer unit and measured at four different temperatures. The spectra are normalized to equal intensity

Relationships between conductivity and radical concentration

By investigating the dependence of the radical concentration as well as the conductivity on the dopant concentration we found that in all doped PTTF there exist iodine concentration areas in which the increase of radical concentration is connected with an increase of conductivity. For that reason it is possible to state that species with spin contribute to the conductivity observed. However, not all radicals contribute to the electrical conductivity and not all the charge carriers have a spin.

We can estimate the charge carrier density from the dopant concentration. If we assume the predominant formation of I_3^- as counterions every third iodine atom takes an electron from the PTTF generating a cation R_2; in the samples there is a minimum of about 10^{21} charge carriers per gramm. The radical concentration amounts to 10^{18} to 10^{20} spins per gramm; this means the spin density in the doped PTTF is at least one order of magnitude lower than the charge carrier density.

The localized radicals R_i cannot contribute to the conductivity, but mobile cation radicals R_2 contribute as polarons $P^{\dot{+}}$ to the electrical conductivity. By doping with iodine their number increases up to a limiting value, depending on the molecular structure.

From these ESR measurements it follows that there are, on an average, less mobile radicals $R_2 = P^{\dot{+}}$ in the sample than macromolecules. If a second polaron $P^{\dot{+}}$ is formed on a macromolecule the two polarons can form a diamagnetic bipolaron B^{2+}

$$P^{\dot{+}} + P^{\dot{+}} \rightleftarrows B^{2+} \quad .$$

This process takes place with a higher probability in molecules with plane structure where an electron interaction is favored between the monomer units. This can be assumed because the samples A.1 and B.1 show a radical concentration which is two orders of magnitude smaller than that in the polymers A.2 and A.3 which are of a twisted molecular structure. Nevertheless, the samples A.1 and B.1 have the highest conductivity. These results support the assumption that the bipolarons generated via polarons deliver the main contribution to the conductivity of the doped PTTF.

The formulas on the next page represent possible structures of polarons and bipolarons in the PTTF samples A and B.1:

In the polarons and bipolarons of poly(tetrathiafulvalenes) the positive charges are delocalized, in particular, within the TTF units.

The observed increase in radical concentration with decreasing temperature leads us to the assumption that there is a temperature dependent equilibrium between polarons and bipolarons which tends to shift more to the polaron-side in the low-temperature region.

The possibility of the break-up of bipolarons to form pairs of polarons and separated polarons has been proposed in connection with studies on poly-

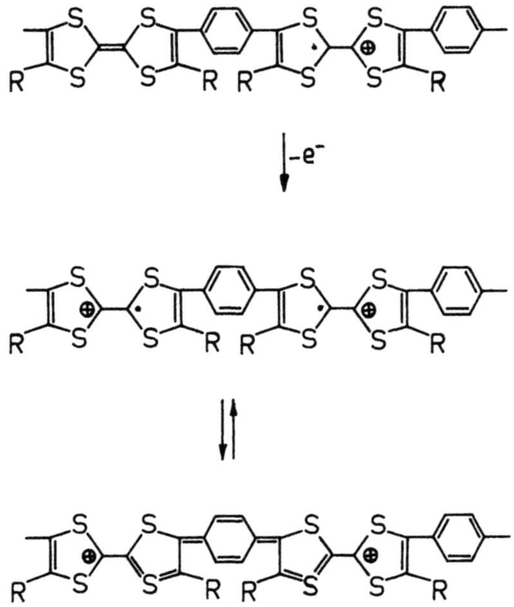

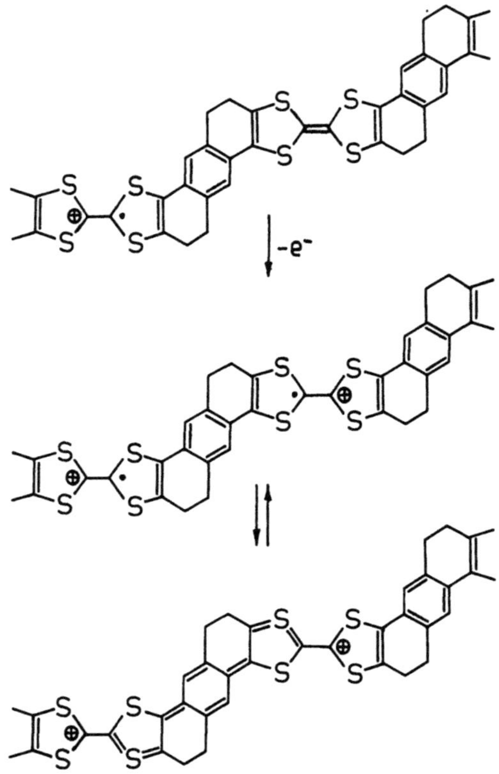

aniline and protonated emeraldine polymeric metals [15, 16]. The origin of the transition from a bipolaron to a polaron lattice may be related to disorder, coulomb repulsion within the charged bipolaron, delocalization energy, dielectric screening, and/or counter ion interactions [15].

An exchange between isolated radicals in an equilibrium with polarons and bipolarons was also suggested in polyphenylene with metallic conductivity [5]. In this conducting polymer the equilibrium was assumed in favor of bipolarons at low temperature near 4 K.

The favored side of the equilibrium at low temperature is presumably dependent on the energy levels of the polaron and bipolaron lattice. These levels differ considerably for the various polymers [17].

In conclusion it can be pointed out that the results of the electron spin resonance and conductivity measurements on various PTTF can be interpreted by application of a model based on polarons and bipolarons. Up to now this model has been predominantly used for polymers with metallic conductivity [4, 5, 15–17] and less frequently for typical semiconducting polymers. However, it has also been applied in sigma-conjugated polymer backbones [18] offering the use for different classes of polymers. It should be mentioned that most of the results received on highly doped polymers, such as polythiophene, seem to be sufficently explained also without the transition of polarons into bipolarons, especially if the polymers show metallic conductivity in microscopic areas. Under this condition it was assumed that the reduction of spin concentration in polythiophene at high doping levels is due to the increasing metallic character of the polarons and not due to the formation of spinless bipolarons [19].

In order to decide which mechanism of conductivity really exists in the PTTF studied, further investigations that include AC measurements in the high frequency region are being undertaken.

In the lecture which formed the basis for this paper, we also included results on laser modified poly(bis-alkylthio-acetylenes) with metallic conductivity ($5 \cdot 10^2$ S cm^{-1}) [20]. This topic will be treated in an additional paper.

Acknowledgement

The authors thank Mr. Mario Schrödner and Mr. Jörg Patzsch for their support with ESR measurements and spectra simulation. They also wish to thank Professor Juliusz Sworakowski and Mr. Maciej Orczyk from the Technical University of Wroctaw for useful discussions and the opportunity to carry out some conductivity measurements in their laboratory.

References

1. Le van Hinh, Schukat G, Fanghänel E (1979) J Prakt Chem 321:299, 321:475
2. Schukat G, Richter AM, Fanghänel E (1987) Sulfur reports 7:155
3. Roth HK, Gruber H, Fanghänel E, Trinh vu Quang (1988) Progr Colloid & Polymer Sci 78:75

4. Maraes F, Davidov D, Kobayashi M, Chung TC, Chen J, Heeger AJ (1985) Synth Met 10:169.
Heeger AJ (1985) Phil Trans Sic Lond A 314:17
5. Kispert LD, Joseph J, Tang J, Bowman MK, Van Brakel GH, Norris J (1987) Synth Met 17:617
6. Fitzky HG, Hocker J (1986) Synth Met 13:355
7. Mayer R, Bleisch S, Domschke G, Tkac A, Staiko A, Bartl (1979) Org Magn Resonance 12:532
8. Sixl H, Hübsch H, Rühle W, Roth S (1985) Mol Cryst Liq Cryst 117:479
9. Völkel G, Dzuba SA, Bartl A, Brunner W, Fröhner J (1984) Phys stat sol A 85:257
10. Völkel G, Brunner W (1986) Phys stat sol A 94:673
11. Brunner W, Gentzsch F (1971) Exper Tech Phys 19:39
12. Fatkullin NF (1980) Vys Mol Soed 228:816
13. Kurzin SP, Tarasov BG, Fatkullin NF, Aseeva RM (1982) Vys Mol Soed A 24:117
14. Kurzin SP, Solovev BV, Haibullin PJ, Tarasov BG, Kobryanski VM (1986) Vys Mol Soed B 28:539
15. Epstein AJ, MacDiarmid AG (1988) Mol Cryst Liq Cryst 160:165
16. Epstein AJ, MacDiarmid AG (1988) J Mol Electr 4:161
17. Stafström S, Bredes JL (1988) Mol Cryst Liq Cryst 160:405
18. Jeyader S, Philluot SR, Rice MJ (1988) Mol Cryst Liq Cryst 160:433
19. Schärli M, Kiess H, Harbecke G, Berlinger W, Blazey KM, Müller KA (1988) Synth Met 22:317
20. Gruber H, Richter AM, Fanghänel E, Roth HK, Friedrich H, Maras S (1986) GDR Patent HO1B/2949758, (1988) Europapatent 88105658/4, (1988) US-Patent F-3119, DN-14151

Author's address:

Prof. Dr. H.-K. Roth
Department of Natural Sciences
Leipzig University of Technology, PSF 66
7030 Leipzig, GDR

Received April 14, 1989;
accepted July 28, 1989

Discussion

REINEKER:

Is the mechanism of electron spin relaxation known?

Comment: Radical cation salts experiments of Dormann et al. in Bayreuth indicate that an essential path for electron spin relaxation is mediated by paramagnetic impurities which were associated with crystal imperfections.

ROTH:

Our knowledge about the mechanism on the electron spin relaxation is limited to that mentioned in the chapter devoted to the ESE studies. The weak temperature dependence of T_1, especially the temperature independent low temperature part, points out that the mechanism of the spin-lattice relaxation for the trapped centers in PTTF is different from that in usual polymeric solids where the phonon dynamics determines the relaxation behavior via one- and two-phonon processes. The low-temperature part points to the dominant influence of special low-energetic lattice excitations which could be related with the special polaron dynamics or with strongly anharmonic local motions in the neighborhood of the trapped centers. Paramagnetic impurities with very short T_1 which may influence the T_1 of the investigated immobile centers via cross-relaxation are not of importance in the polymer system studied which is nearly amorphous.

HÄNSEL:

You measured the HALL mobility of the charge carriers. The given value is in a range in which these measurements are easy to manage. Have you measured the temperature dependence of the mobility and from this depenence calculated the charge carrier dependence of temperature and compared it with the measured spin density?

ROTH:

The HALL mobility has been measured only for the laser modified poly(bis-alkylthio-acetylenes) and, up to now, only at room temperature; however, it is planned to carry out such investigations also as a function of temperature because of its high information content in the direction mentioned by Prof. Hänsel. In the case of the lower conductive poly(tetrathiafulvalenes) it was impossible to obtain reproducible data from HALL experiments. We can estimate the charge carrier density from the dopant concentration. If we assume the predominant formation of I_3^- as conterions, every third iodine atom takes an electron from the PTTF generating a cation; in the samples there is a minimum of about 10^{21} charge carriers per gram. The radical concentration amounts to 10^{18} to 10^{20} spins per gram. That means the spin density in the doped PTTF is at least one order of magnitude lower than the charge carrier density. This corresponds with our model based on polarons and bipolarons.

CHUDACEK:

I think that some temperature and dopant concentration dependencies are in connection with molecular mobilities and/or structure of your polymeric systems. Have you some comment about it?

ROTH:

We have stated a destined dependence of the concentration and the properties of radicals on the molecular structure of the PTTF by comparing the polymers A.1 and B.1 (which have an almost plane molecular structure) with those of A.2 and A.3 which have a slightly twisted structure. They differ not only in their behavior at room temperature, but also in the temperature dependence of electric and magnetic properties. Because the molecular structure determines the polymer morphology and with it the molecular mobility in various temperature regions, we can indirectly conclude that there are some dependencies of the density and properties of the studied radicals on the molecular mobility. A clear answer about the molecular mobility dependence of the magnetic and electric properties of the PTTF requires systematic investigations of the molecular mobility of these speciality polymers, which we have not yet carried out.

What I mentioned in my lecture was the fact that the *electron spin relaxation* does not so sensitively depend on the molecular mobility, as has been shown from mechanical and NMR relaxation.

Paraelasticity of organic colloid systems

W. Pechhold, H. P. Grossmann, K.-H. Hanus and E. Jürgens*)

Abteilung Angewandte Physik, Universität Ulm, Ulm, FRG

Abstract: The structure and the thermodynamic and electric properties of dense, organic colloid systems should be studied in conjunction with their mechanical properties to understand the role of interfaces and to have a further check on the structural models assumed. Such study can be carried out experimentally and by model theories. In this paper the dynamic shear compliance of two anisotropic colloids is reported (for a clay organic complex and an oriented stack of DPPC-lamellae) and interpreted by dislocation-mediated shear across the interfaces. An additional relaxation process in DPPC is attributed to a switching of the tilted unit cells between equivalent states. Moreover, the shear compliance in the "rotor phase" of the *n*-alkanes and the plateau-compliance of polymer melts can be described by the combined shear modes of cubic blocks, the size of which is determined by the molecule length (in the paraffins) or by the meander size (meander theory for polymer melts). It is worth mentioning that a single formula can explain, to a first approximation, all of the four examples reported.

Key words: Paraelasticity; clay organic complexes; lecithin-bilayer; *n*-alkanes; waxy state

1. Introduction

In dense colloid systems the interfaces between the particles (e.g., domains, grains, lamellae, etc.) determine the mechanical properties. This is accomplished mainly in two ways: i) The misfit between the particles allows the particles to deform or to interchange between equivalent states without increasing the energy; and ii) the interfaces easily permit the creation and annihilation of dislocation loops, mediating additional shear-deformation or even viscous flow between the particles.

In the following the paraelastic behavior of two anisotropic colloids (a clay-organic mineral and an oriented stack of DPPC-lamellae) and two isotropic colloids (paraffin wax and polymer melt) will be discussed.

Before presenting these examples the meaning of paraelasticity shall be elucidated by the well-known Snoek-Effect [1] (Fig. 1):

Chemical impurities (e.g., carbon atoms) statistically distributed on the edges of a *bcc*-lattice (e.g., α-

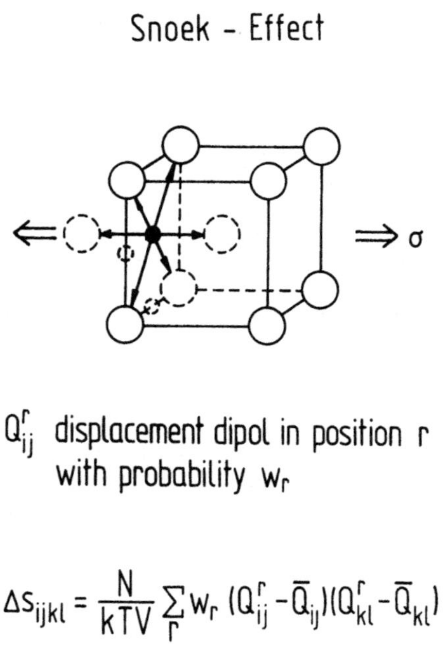

Fig. 1. Snoek-Effect and its paraelastic relaxation strengths. The black dot represents an interstitial impurity, acting as a force dipole (tensor) on the surrounding lattice atoms. The small circles indicate the other two possible positions for the interstitial

*) Now with Bayer AG, Leverkusen

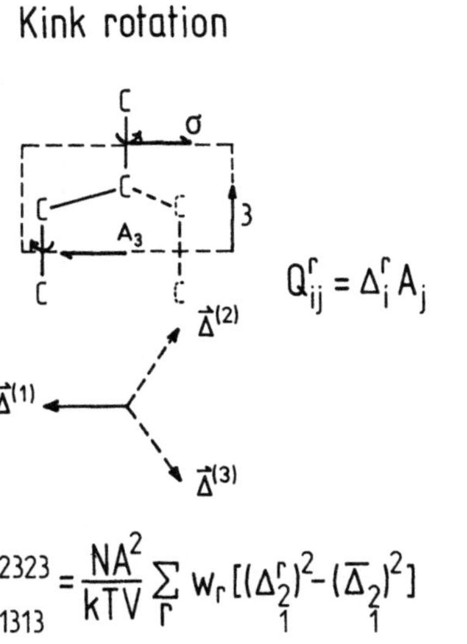

Fig. 2. Kink rotation and its paraelastic relaxation strengths. A_j representative area of displacement. Δ^r_i displacement (i-direction) in position r. $\bar{\Delta}_i = \Sigma w_r \Delta^r_i$, average displacement in i-direction

iron) can be described as displacement dipoles Q^r_{ij} (or force dipoles P^r_{ij}) indicating the local deformation (or stress field) of the lattice by the interstitial in position r. If an external stress is applied, the normal elastic deformation occurs instantly, followed by a relaxation process (with relaxation time τ) during redistribution of the impurities. They prefer the extended edges and so give rise to an additional "paraelastic" deformation. Its contribution to the elastic coefficients, the paraelastic relaxation strength Δs_{ijkl}, is derived from the theory of elasticity [2], yielding the expression in Fig. 1 for N interstitials in the volume V.

This formula can also be applied to the para- or anelastic effect due to the rotation of N/V kink defects in trans-polyethylene under the influence of a shear stress σ. The displacement dipoles read $Q^r_{ij} = \Delta^r_i A_j$, but must be symmetrized (yielding the factor 2 on the lefthand side of the expression in Fig. 2).

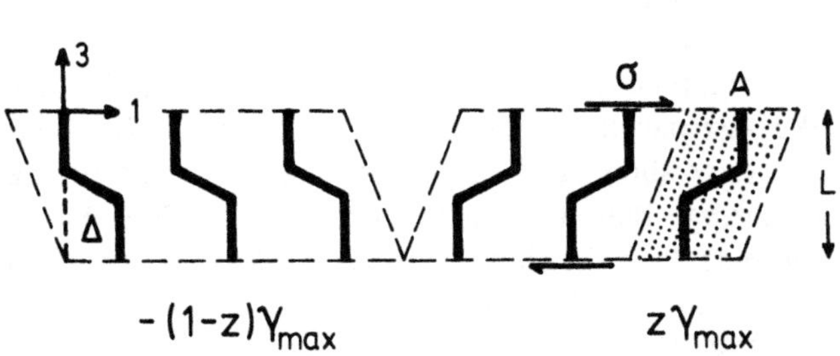

Fig. 3. Two-state model of kink-block orientation: probability z for positive shear $\gamma_{max} = \Delta/L$, volume of representative element $v = A \cdot L$

2. Paraelastic relaxation strength

To derive a formula for the relaxation strength of domains (kink-blocks, tilted chain-blocks), switching between two equivalent states (Fig. 3), one may simply consider the free energy of shear deformation,

$$\frac{\Delta g_{def}}{kT} = z \ln z + (1-z) \ln (1-z) + \frac{v\sigma}{kT} [\gamma - (2z-1)\gamma_{max}] \quad (1)$$

asking for the equilibrium z (and hence γ) via minimization. The first two terms comprise the ideal mixing entropy of both states, the third term represents the necessary condition for the total shear γ, with $v\sigma/kT$ as the Lagrangian multiplier. The latter relation follows from the definition $\partial(\Delta g/v)\partial\gamma = \sigma$, v being the volume of the statistical element (e.g., a kink in a small kink-block). Minimization yields

$$\gamma = \gamma_{max} \cdot \tanh (v\sigma\gamma_{max}/kT) \approx v\sigma\gamma^2_{max}/kT(\sigma \to 0) \ , \quad (2)$$

and the relaxation strength becomes (with $v = A \cdot L$, $\gamma_{max} = \Delta/L$)

$$2\Delta s_{1313} \equiv \partial\gamma/\partial\sigma \approx \frac{v}{kT} \gamma^2_{max} = \frac{A \cdot L}{kT} \left(\frac{\Delta}{L}\right)^2 = \frac{N}{V} \frac{A^2\Delta^2}{kT} \ , \quad (3)$$

where $1/v = N/V$ is the number of elements per unit volume.

It is obvious that (3) coincides with the formula in Fig. 2 (assuming $w_r = \frac{1}{2}$).

3. Clay organic complexes

Lagaly and Weiss [6] have prepared n-alkylammonium-n-alkanol-beidellite- or vermiculite-complexes by cation exchange in Ca^{2+}-beidellite or vermiculite with n-alkylammonium, and subsequent drying and swelling in n-alkanol. They have described the organic bilayer-structures formed, and carefully measured their basal spacings from the melting point of the n-alkanol to higher temperatures. In Fig. 4 the preparation of the model systems and the results obtained from the x-ray-diffraction are summarized as taken from [3, 6]. With rising temperature the basal spacing

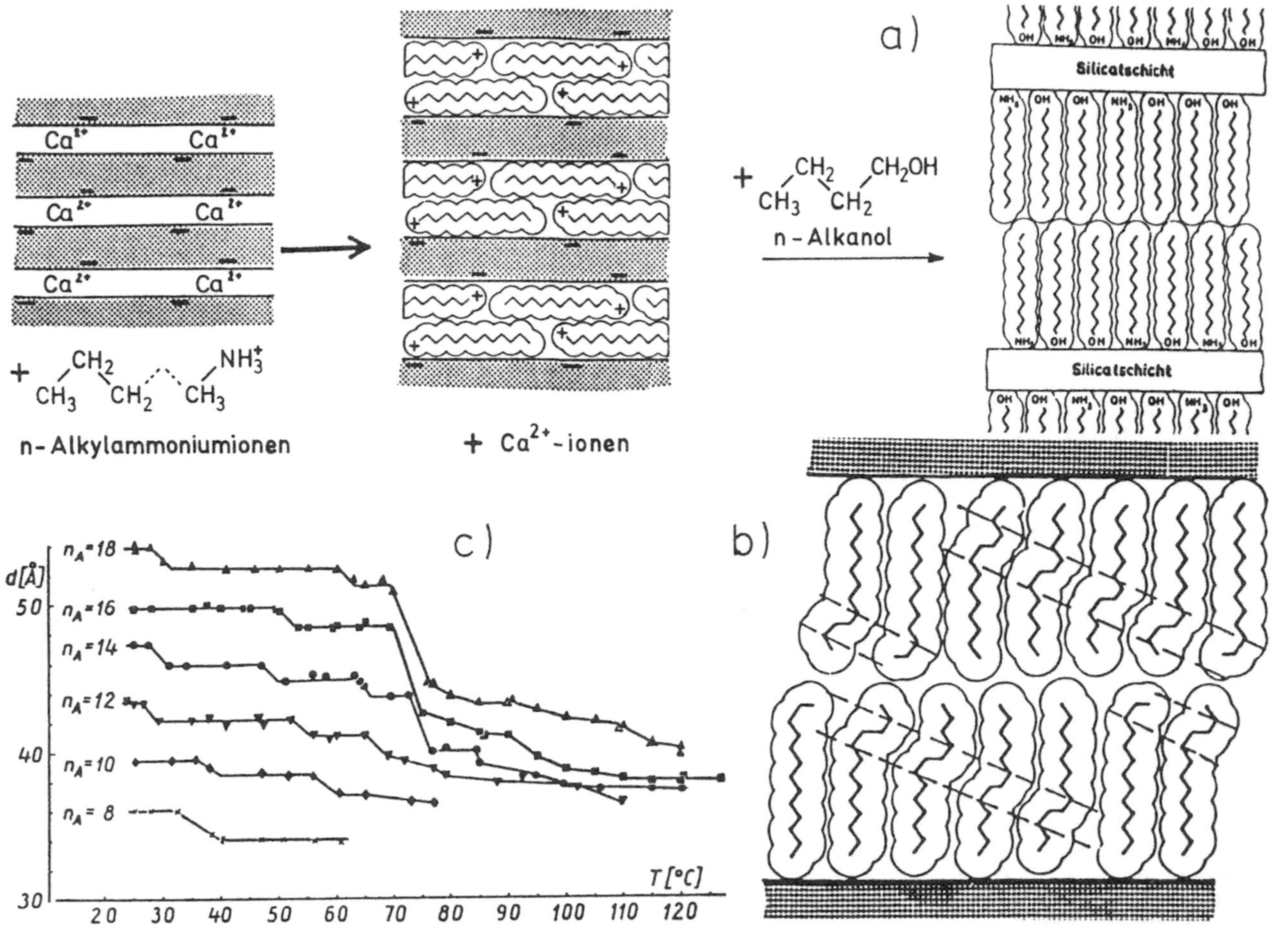

Fig. 4. Formation and structure of bimolecular films of long-chain alkanols and alkylammonium ions between inorganic layers: a) all-trans conformation, b) kink-blocks, c) temperature dependence of the basal spacings d for dodecylammonium/n_A-alkanol-vermiculites. (After [3, 6])

decreases in distinct steps of 1.2 ± 0.1 Å up to a transition temperature from the β- into the α-phase (indicated by a larger step of $4-6$ Å [4–6]).

Lagaly and Weiss [6] explained these steps as the consequence of cooperatively formed kink-blocks

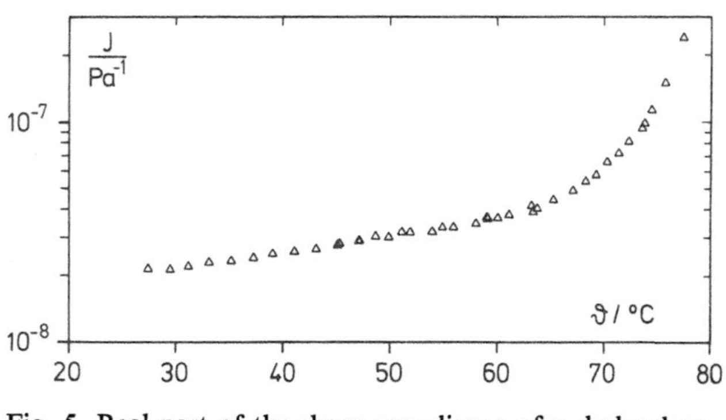

Fig. 5. Real part of the shear compliance of n-dodecylammonium-n-dodecanol-vermiculite at 12 kHz

[3–6], and Pechhold et al. [7] confirmed this view by a statistical model theory applying cluster-entropy-hypothesis (CEH). It should be kept in mind that there is a thermodynamic driving force towards kink-block formation in these clay organic complexes: the available cross-section per all-trans chain is $23-25$ Å^2 (due to the pockets in the pseudohexagonal silicate layers) compared with that of the best intermolecular fit (18.2 Å^2 in PE).

Mechanically these clay organic complexes behave like wax parallel to the basal planes: at low frequencies they show a visco-plastic behavior, but in the kHz-range they are elastic as reported in Fig. 5: the real part J' of the shear compliance (by a factor of 10 larger than the loss-compliance J'') is smoothly depending on temperature with increasing slope only near to the β/α-transition. A first check of the kink-block model discussed above requires a comparison of the (nearly, i.e., within the present experimental error) missing J-steps (at 28° and 55 °C) with the prediction from model theory. Applying formula (3) with an additional factor of $\frac{1}{2}$ (assuming half of the unit cells to contribute in one shear direction) one may use the following data

$\Delta^2 = (\delta_y^2 + \delta_z^2) \approx 4.73$ Å^2 from ideal lateral kink displacements δ_y, δ_z [8]

$T \approx 320\,K$,

$A = 2 \cdot 24.5$ Å^2 (two pairs of molecules/unit cell) ,

$L \approx 43$ Å ,

and gets $\Delta J_1 = 2\Delta s_{1313} \approx 6.1 \cdot 10^{-10}\,\mathrm{m^2/N}$ for the single unit cell as the statistical element. More precise measurements of the shear compliance simultaneous with the sample thickness of perfectly swollen complexes will test this relaxation strength prediction. The view that the whole kink-blocks might act as statistical elements would lead to $\Delta J_b \approx \Delta J_1 \cdot L/2c \approx 5 \cdot 10^{-9}\,\mathrm{m^2/N}$ and can probably be excluded by the experimental result (Fig. 5).

The second question concerns the absolute value of the shear compliance and its smooth temperature dependence (Fig. 5). Since a shear compliance of $2 \ldots 5 \cdot 10^{-8}\,\mathrm{m^2/N}$ cannot be explained by energy elasticity, it is supposed to be due to a paraelastic shear in the central interface of the bilayers. The model proposed here (Fig. 6) assumes creation and annihilation of dislocation loops across these interfaces. The loop density (or size) in equilibrium can be estimated by free energy consideration (applying CEH), thereby taking into account dislocation self-energy and interaction as well as the relation between the total shear and the dislocation density of both signs. This model theory, which also explains the temperature dependence of the shear compliance, will be published elsewhere. It is worth noting that the elastic modulus included in the dislocation energy is due to chain bending. This makes it clear why such interface paraelasticity can only occur in lamellae with less dense packing than in the ideal chain crystal. To have a rough estimation of the shear compliance here one may assume average slip areas $A \approx (L/2)^2$ to exist in three states and produce local shears of d/L, O, $-d/L$, respectively. Then one can apply formula (3) (with an additional factor of $\frac{2}{3}$ which follows from a similar derivation). Using the data

$\Delta \equiv d = \sqrt{24.5}$ Å for the average chain distance,

$A = (L/2)^2$, $L = 43$ Å and $T \approx 320\,K$,

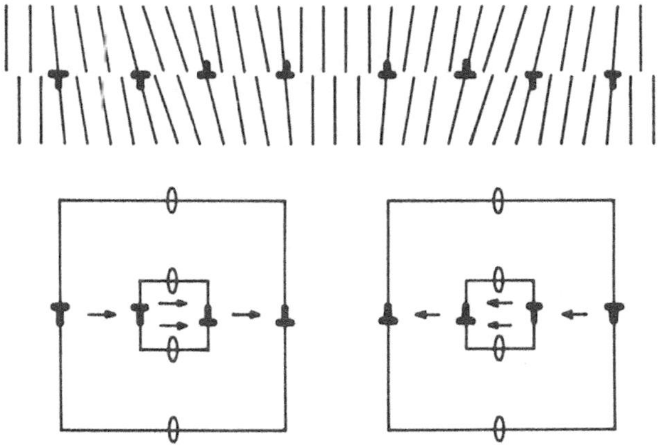

Fig. 6. Dislocation-loop model to explain interface paraelasticity of less dense organic lamellar systems. The symbols ⊥, ⊤ indicate the positions of step-dislocations which form closed loops by changing into the screw types

one gets $J = 2\Delta s_{1313} \approx 4 \cdot 10^{-8}\,\mathrm{m^2/N}$, a value not far from the experimental result (but with a wrong dependence on temperature). The increasing slope of the compliance, starting at 65 °C, indicates the onset of larger shear fluctuations which finally lead to tilted chain domains and a larger step in the basal spacing.

4. Oriented stacks of lipid lamellae

Dipalmitoyl-phosphatidylcholine (DPPC) was incorporated with water at 90° to 120 °C (i.e., in the LC-phase) in the vapor pressure of K-acetate/water solutions, pressed to about 0.2-mm thickness between

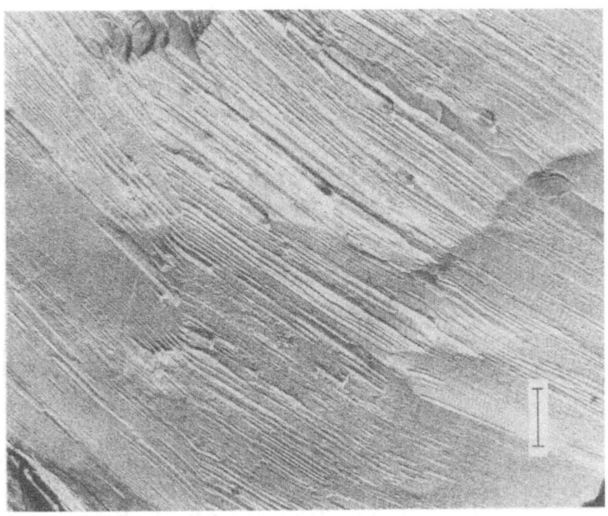

Fig. 7. Freeze-fracture electronmicrograph of oriented DPPC. The bar indicates 0.2 μm

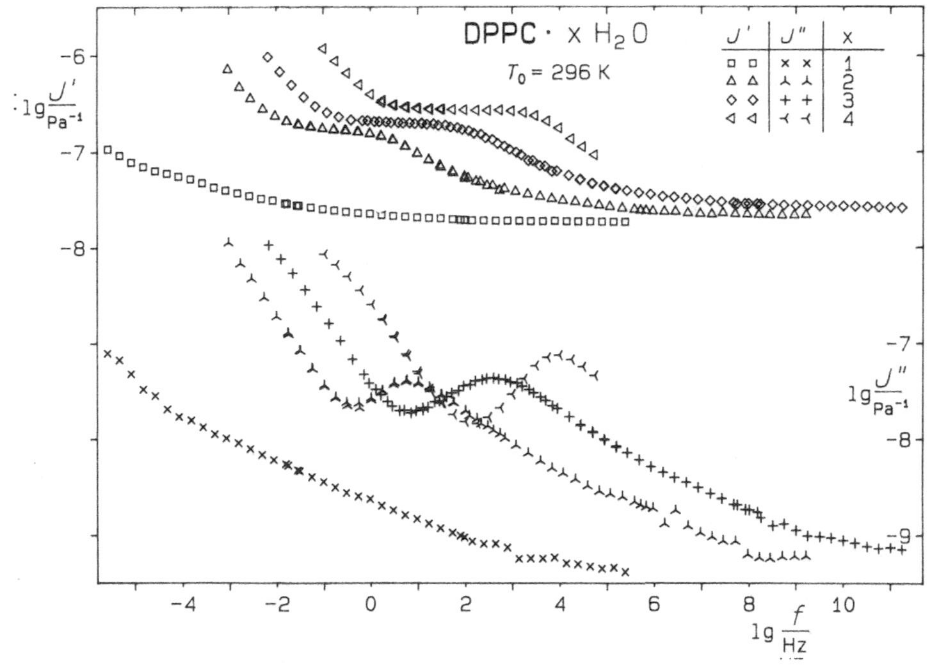

Fig. 8. Shear compliance mastercurves of oriented DPPC in the $L_{\beta'}$-phase reduced to 296 K at different water contents

silanated glass slides, and after 4 h annealing, slowly cooled to room temperature [9] [1]). Figure 7 shows a freeze-fractured and *Pt/C*-shadowed electron-micrograph, almost perpendicular to the multilayers [9]. Such samples of different water content were glued in pairs into the double sandwich holder of the mechanical spectrometer and measured in the frequency range 10^{-4} to 200 Hz at different temperatures (30 to $-30\,°C$), i.e., in the crystalline $L_{\beta'}$ – phase). The water content was kept about constant by appropriate vapor pressure in the spectrometer cell. Quantitatively, the water content was determined on excess samples before and after the measurement by coulombmetric titration using the Karl-Fischer-method (Mitsubishi moisture meter, model CA-02).

The frequency-curves measured on each sample at different temperatures could be combined to give a mastercurve ($T_0 = 296$ K). Four of them are shown in Fig. 8 and will now be discussed. i) The high frequency plateau of J' is about $2 \cdot 10^{-8}\,m^2/N$, a value similar to that of the clay organic complexes. Therefore its interpretation probably must be the same, i.e., a dislocation-loop mechanism within the hydrophobic interface of the lecithin-bilayer. ii) At very low frequencies the onset of flow relaxation and viscous flow can be seen well. It seems to be characteristic for the waxy state as well as for the polymer melt, even though the retarding mechanisms will be different. iii) In the intermediate frequency range a strongly developed relaxation process takes place, which probably is due to a switching of the tilted unit cells between energetically equivalent states (in differently tilted domains).

Figure 9 gives a schematic drawing (according to the structure analysis in [10]) of two such domains tilted by an angle Θ. Applying formula (3) with an additional factor of $\frac{1}{2}$ (assuming half of the unit cells to switch in one shear direction) one gets for the relaxation strength

$$\Delta J \equiv 2 \Delta s_{1313} \approx \gamma_{max}^2 \cdot A \cdot L / 2kT \, . \qquad (3a)$$

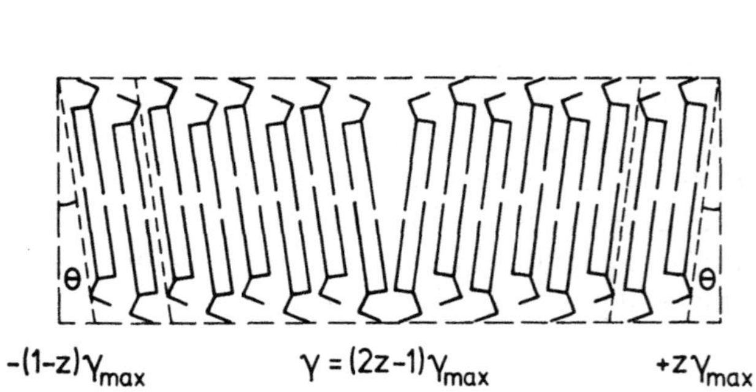

Fig. 9. Schematic drawing of the DPPC-bilayer with oppositely tilted domains

[1]) The phase diagram of DPPC-H_2O-mixtures shows [9]: i) a crystalline lamellar phase $L_{\beta'}$ below 70° to 30 °C, depending on water content; ii) a smectic A phase L_α (with hydrocarbon chains molten) below 80° to 150 °C, and iii) an optically isotropic cubic phase Q_α at higher temperatures.

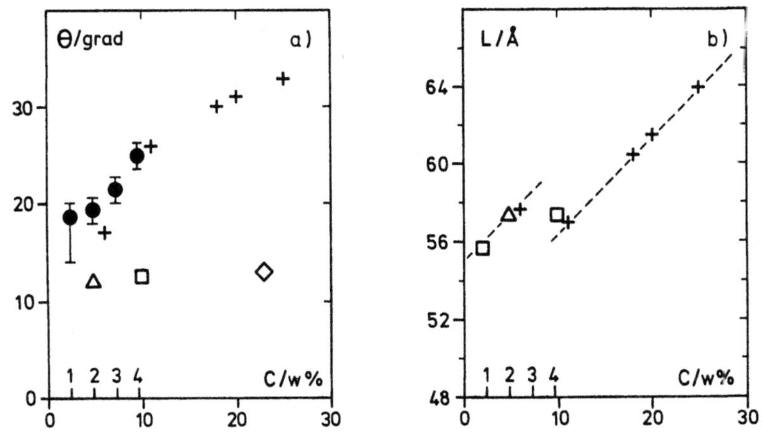

Fig. 10. a) Tilt angle Θ_h of the hydrocarbon chains in DPPC, Δ [10], □ [11], ◇ [12]. Overall tilt angle Θ from *x*-ray and specific volume + [13] and from paraelastic relaxation strength ● [this work], all data refer to the crystalline $L_{\beta'}$-phase; b) lamellar repeat L in DPPC in dependence on water content (from the same sources)

In Fig. 10a are plotted the tilt angles $\Theta = \arctan \gamma_{\max}$ calculated from the experimental relaxation strengths, using the data

$$L = 57\ \text{Å}\ , \quad A = 4\cdot 39\ \text{Å}^2 \quad \text{(4 pairs of molecules/unit cell)}\ ,$$
$$T = 296\ \text{K}\ .$$

This tilt-angle Θ includes the tilting of hydrocarbon chains Θ_h as well as the tilting of the polar heads Θ_p. In the phase $L_{\beta'}$, $\Theta_h \approx 12 \cdots 13°$ obviously does not depend on water content [10–12]. Since the lamellar repeat L (Fig. 10b) stays about constant between the di- and tetrahydrate it must be assumed that Θ_p as well as Θ increases. The latter result is also shown by the tilt angles Θ deduced from the mechanical relaxation strengths via (3a).

5. The *n*-alkanes in the waxy state

The *n*-alkanes exhibit up to four ($A-D$) modifications below ultimate melting. A detailed description of the molecular packing (lateral and end-to-end) and motion in these phases has been given for $C_{33}H_{68}$ [14], mainly from *x*-ray, Raman, dielectric, NMR, and INS analyses.

Dynamic mechanical measurements at 12 kHz [15] (cp. Fig. 11) show that only the quasihexagonal modification (the so-called "rotor phase" below the melting point) exhibits strong paraelasticity (and waxy flow at low frequencies). Its interpretation in [15] – assuming switching of highly concentrated kink-blocks (without applying CEH at those days) – was disproved by Strobl's finding [14] of only 0.4–0.7 kinks per chain in the *D*-phase. No further explanation for the paraelastic plateau in the kHz-range was put forward since that time.

Taking Strobl's analysis of the *D*-phase, the following model can now be suggested for its paraelasticity (Fig. 12): assuming a cubic block of chains as the statistical element for shear deformation of the lamellae, there will be three contributions of anisotropic relaxation strength.

i) Two contributions of interlamellar shear

$$2\Delta s^{(1)}_{1313} = 2\Delta s^{(1)}_{2323} = (d/2L)^2 \cdot L^3/kT \qquad \text{(3b)}$$

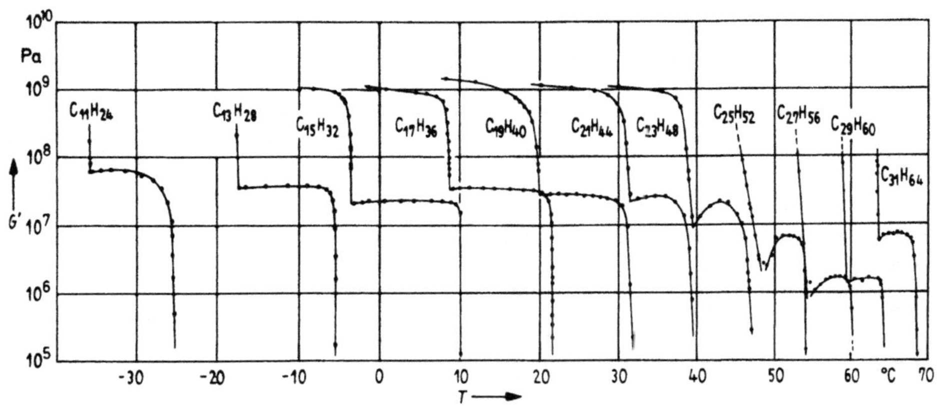

Fig. 11. The real part G' of the shear modulus at 12 kHz of isotropic *n*-alkanes near to the melting point. (G'' is some 5 to 10 times smaller than G'). (After [15])

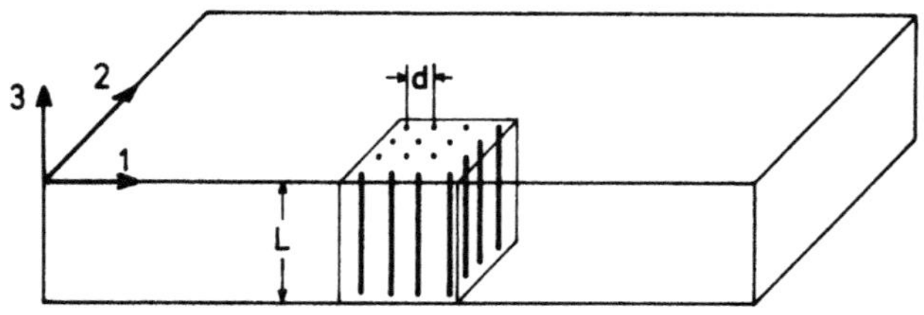

Fig. 12. A cubic block of chains as the statistical element for paraelastic shear in the "rotor phase" of n-alkanes

due to the dislocation loop mechanism, with two states of local shear ($\pm d/2L$) assumed, and

$$2\Delta s^{(2)}_{1313} = 2\Delta s^{(2)}_{2323} = (\tan\Theta)^2 A\cdot L/2kT \qquad (3\,c)$$

due to the switching of tilted chains, assuming half of the unit cells to interchange in one shear direction.

ii) A paraelastic shear of the cube in direction of the n-alkane chains, consisting of multiples of $c/2 = 1{,}27$ Å glide steps, which every layer of molecules may undergo relative to its neighbor:

$$2\Delta s_{3131} = 2\Delta s_{3232} = \tfrac{1}{3}m(m+1)(c/2d_\perp)^2 d_\perp L^2/kT. \qquad (3\,d)$$

This formula follows from a derivation similar to that of (3).

The maximum number m of steps to each side is restricted by one-half of the overall thickness D_t of the void layers between the lamellae at one hand, and by the increase in average thickness of voids Δd_{av} at the other limit, $D_t/2$ and Δd_{av} have been determined for $C_{33}H_{68}$ with SAXS by Strobl [14] to be $\sim 10/2$ Å and 1.5 Å, respectively; therefore $m = 2$ is most probable.

To calculate the isotropic shear compliance one favorably applies Reuss-averaging, which in this case reads

$$\Delta J_{iso} = \tfrac{1}{5}(\Delta s_{1313} + \Delta s_{2323} + \Delta s_{3131} + \Delta s_{3232}) \; . \qquad (4)$$

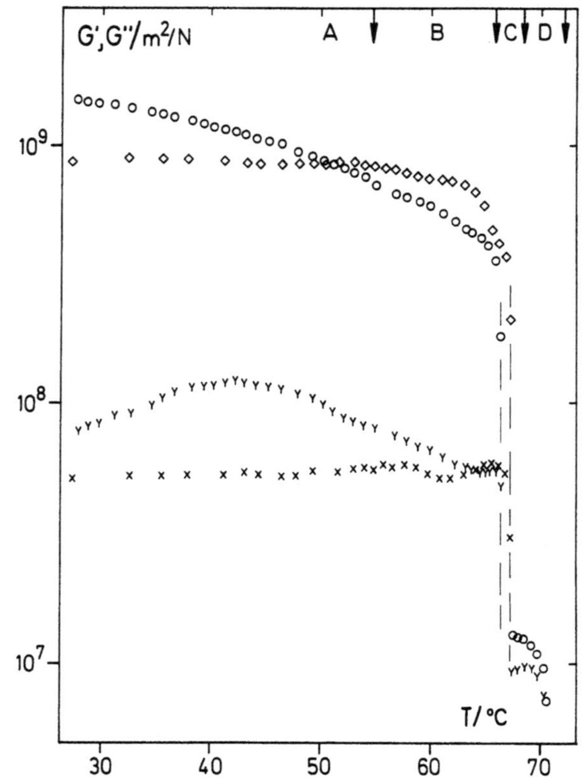

Fig. 13. Dynamic shear modulus of $C_{33}H_{68}$ measured at 1.2 kHz. ◇, x pressed powder, crystallized from petrol ether dilute solution; ○, Y after heating into the waxy state

Using the data [14]: $A = 2\cdot 20.8$ Å^2, $d = 4.9$ Å (chain distance), $d_\perp = d\sqrt{3}/2 = 4.25$ Å (distance between lattice planes), $L/$Å $= 1.27\,n$ (length of trans-chain, n number of C-atoms), $c = 2.54$ Å (identity period), $\Theta = 19.5°$ (tilt angle), $m = 2$, the formulae (3 b – d, 4) yield the isotropic shear compliances/moduli in dependence of the chain length which are listed in Table 1. The respective temperatures were taken from Fig. 11.

The agreement of theoretical prediction (Table 1) and experiment (Fig. 11) is quite good, including

Table 1. Paraelastic shear compliance/modulus relaxation strength in the "rotor phase" calculated from model theory in dependence on chain length (cp. text)

n-alkane	T_D/K	$2\Delta s^{(1)}_{1313}$	$2\Delta s^{(2)}_{1313}$	$2\Delta s_{3131}$	$\Delta J\cdot 10^7/\frac{m^2}{N}$	$\approx G'\cdot 10^{-7}/\frac{N}{m^2}$
$C_{11}H_{24}$	240	0.253	0.110	0.447	0.162	6.17
$C_{15}H_{32}$	275	0.301	0.131	0.726	0.232	4.32
$C_{19}H_{40}$	300	0.350	0.152	1.067	0.314	3.19
$C_{23}H_{48}$	315	0.403	0.175	1.49	0.414	2.42
$C_{29}H_{60}$	335	0.478	0.208	2.23	0.583	1.72
$C_{33}H_{68}$	343	0.531	0.231	2.82	0.716	1.40

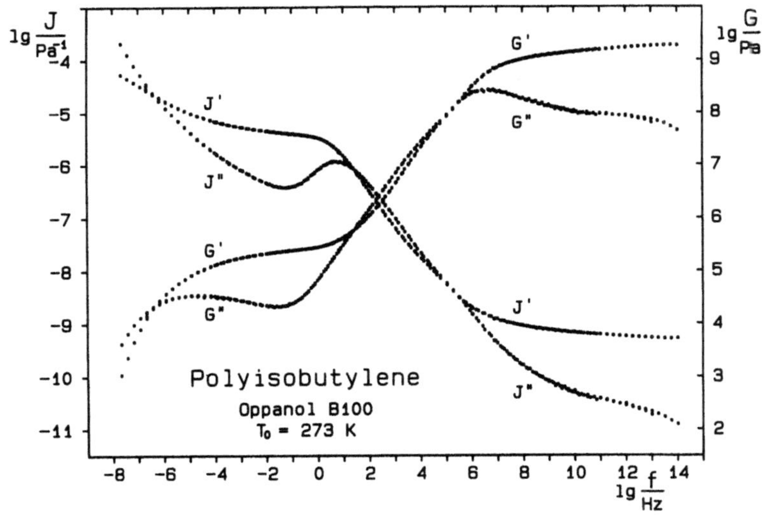

Fig. 14. Shear compliance and shear modulus mastercurves as deduced from measurements over six decades ($10^{-4}-200$ Hz) at six temperatures [18]

$C_{33}H_{68}$ which has been measured separately at 1.2 kHz (Fig. 13), therefore G'' is much higher than at 12 kHz. A large deviation is observed in the chain length range $C_{25} \dots C_{31}$ (Fig. 11) which one might speculate may be due to $m = 4$ $c/2$-glide steps in each chain direction, combined with a macroscopic orientation (the n-alkanes arranging parallel to the surface of the thin sample).

6. Polymer melts and networks in the rubbery state (plateau compliance)

The most remarkable property of polymer melts or networks is rubber elasticity, which can only be explained by assuming paraelastic segmental movement yielding entropic restoring forces. This can be studied by analyzing dynamic shear compliances vs frequency. For melts and amorphous networks the frequency/temperature superposition is valid, as a rule, yielding mastercurves (over more than 20 decades in frequency (Fig. 14)) and the proper shift factors on both axes.

In the coil model the plateau-compliance (also well defined in higher molecular weight polymer melts between the glass-relaxation and the flow region) is attributed to the number ν of network strands per unit volume between entanglements or crosslinks and reads in the simple case

$$J_{eN} = (R_0^2/R^2)\nu kT \; . \tag{5}$$

In the meander model the plateau compliance can be explained by the relaxation-strengths of intra-meander shear deformation modes (Fig. 15) applying Reuss-averaging. In the statistical derivation of the bulk anisotropic compliances, layers of molecules are allowed to be displaced by one chain distance d against each other. This theoretical model has been worked through in detail in preceding papers [16–18] and yields, for the plateau compliance J_{eN}^0 of an uncrosslinked polymer melt

$$J_{eN}^0 = \frac{1}{5}\left\{\begin{matrix} s_{1212} + s_{2121} \\ + s_{3131} + s_{3232}\end{matrix}\right\} = \frac{d(r+x)^2}{9kT} \; . \tag{6}$$

Taking J_{eN}^0 from the decomposition of master curves of various polymer melts, Eq. (6) represents probably the best method to determine the cube side length $r+x$ or the average number of quasiparallel stems $(r+x)/d$ in a layer of molecules [19].

Conclusion

The study and interpretation of paraelastic properties of organic colloid systems is a powerful tool for understanding the role of interfaces and to further check the structural models assumed. The examples presented here shall be looked at as a first attempt towards this aim. The similarity in the approach to the "rotor phase" of the n-alkanes (though still crystalline) with that to the polymer melts and networks gives additional support to the meander model of amorphous polymers. (The rotation of the meander cubes across their space diagonals does not contribute to the plateau compliance.) The tilt-switch relaxation in the DPPC lamellae could be the important mechanism driving the gate-proteins in biological

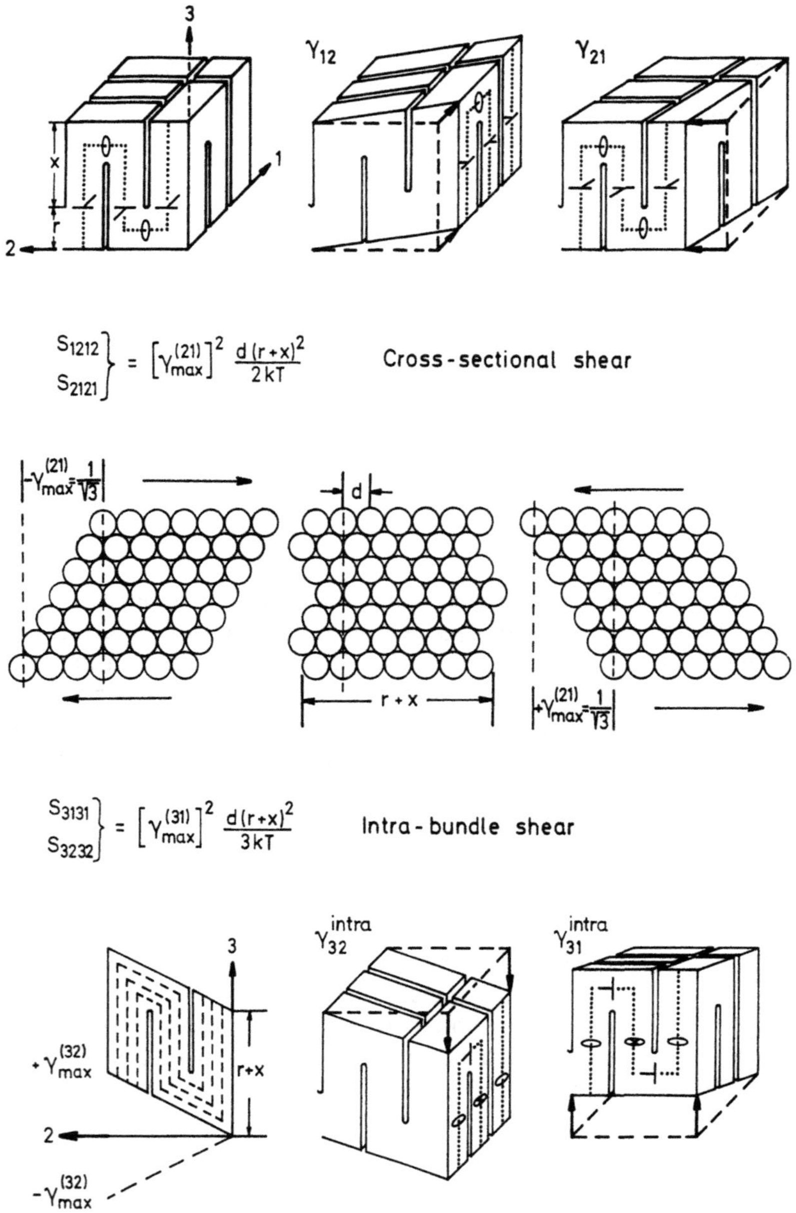

Fig. 15. The two main intra-meander shear modes demonstrated on single meander cubes with dislocations indicated on the surfaces [18]

membranes. Further examples of paraelasticity to be dealt with in the future will include the deformation and swelling behavior of biopolymers and biological structures, e.g., the cornea and sclera of the eye.

Acknowledgement

The authors wish to thank the Deutsche Forschungsgemeinschaft and the Fonds der Chemie for financial support.

References

1. Zener C (1948) Elasticity and anelasticity of metals. University Press, Chicago
2. Kröner E (1958) Kontinuumstheorie der Versetzungen und Eigenspannungen. Springer, Berlin Göttingen Heidelberg
3. Lagaly G, Fitz S, Weiss A (1975) Prog Colloid Polym Sci 57:54–60
4. Lagaly G, Fitz S, Weiss A (1975) Clays and Clay Minerals 23:45–54
5. Lagaly G (1976) Angew Chemie 88:628–639
6. Lagaly G, Weiss A (1971) Kolloid Z u Z Polym 248:979–985
7. Pechhold W, Liska E, Grossmann HP, Hägele PC (1976) Pure & Appl Chem 46:127–134
8. Pechhold W (1968) Kolloid Z u Z Polym 228:1–38
9. Jürgens E (1981) Thesis Universität Ulm
10. Hauser H, Pascher I, Pearson RH, Sundell S (1981) Biochim Biophys Acta 650:21–51

11. Stamatoff JB, Graddick WF, Powers L, Moncton DE (1979) Biophys J 25:253 – 262
12. Hentschel M, Hosemann R, Helfrich W (1980) Z Naturforschung 35 a:643 – 644
13. Tardieu A, Luzzati V, Reman FC (1973) J Mol Biol 75:711 – 733
14. Ewen B, Strobl GR, Richter D (1980) Farad Disc Chem Soc 69:19 – 31
15. Pechhold W, Dollhopf W, Engel A (1966) Acustica 17:61 – 72
16. Pechhold WR (1980) Colloid Polym Sci 258:269 – 288
17. Pechhold W (1984) Makromol Chem, Suppl 6:163 – 194
18. Pechhold W, Böhm M, v Soden W (1987) Progr Colloid Polym Sci 75:23 – 44
19. Pechhold W, Sautter E, v Soden W, Stoll B, Großmann HP (1979) Makromol Chem, Suppl 3:247 – 275

Received May 31, 1989;
accepted August 30, 1989

Authors' address:

Dr. W. Pechhold
Abteilung Angewandte Physik
Universität Ulm
Oberer Eselsberg
7900 Ulm, FRG

Author Index

Subject Index

GPSR Compliance

The European Union's (EU) General Product Safety Regulation (GPSR) is a set of rules that requires consumer products to be safe and our obligations to ensure this.

If you have any concerns about our products, you can contact us on ProductSafety@springernature.com

In case Publisher is established outside the EU, the EU authorized representative is:

Springer Nature Customer Service Center GmbH
Europaplatz 3
69115 Heidelberg, Germany

Batch number: 10578403

Printed by Printforce, the Netherlands